AF606602

Global Climatology and Ecodynamics

Anthropogenic Changes to Planet Earth

Arthur P. Cracknell, Vladimir F. Krapivin,
Costas A. Varotsos

Global Climatology and Ecodynamics

Anthropogenic Changes to Planet Earth

Professor Arthur P. Cracknell
Department of Applied Physics and Electronic Engineering
University of Dundee
Dundee
UK

Professor Vladimir F. Krapivin
Institute of Radioengineering and Electronics
Russian Academy of Sciences
Moscow
Russia

Professor Costas A. Varotsos
University of Athens
Faculty of Physics
Department of Applied Physics
Laboratory of Upper Air
Athens
Greece

The photograph reproduced on the back cover of Kirill Kondratyev, to whom this book is dedicated, is reproduced with the kind permission of his widow, Svetlana Kondratiev

SPRINGER–PRAXIS BOOKS IN ENVIRONMENTAL SCIENCES
SUBJECT *ADVISORY EDITOR*: John Mason B.Sc., M.Sc., Ph.D.

ISBN 978-3-540-78208-7 Springer Berlin Heidelberg New York

Springer is part of Springer-Science + Business Media (springer.com)

Library of Congress Control Number: 2008926500

Printed in Germany

Cover design: Jim Wilkie
Project management: Originator Publishing Services, Gt Yarmouth, Norfolk, UK

Printed on acid-free paper

Contents

Preface

Uncertainties of information on the problems of global climatology are a principal barrier for adequate understanding of the anthropogenic effects on global ecodynamics. The purpose of the book is to summarize existing information and assess the level of these uncertainties as well as to stimulate readers to think in the longer term about climate change and the ecological damage that is being done to the planet Earth in the hope that it may remain fit for human habitation and a satisfying life style for future generations, not just the next generation or two.

This is a scholarly book which is concerned with climate change and the various aspects of ecology, all in relation to the sustainability of life, especially human life, on Earth. It is widely accepted that global warming, due to anthropogenic greenhouse gas emissions, represents a threat to the sustainability of human life on Earth. However, there are many other threats that are, potentially, just as serious; these include atmospheric pollution, ozone depletion, water pollution, the degradation of agricultural land, deforestation, the depletion of the world's mineral resources, and population growth. This books aims to redress the balance and discuss the scientific aspects of many of these other threats.

The book takes its inspiration from the life and work of the late Academican K.Ya. Kondratyev who pioneered research in a number of fields related to these problems, namely satellite meteorology, atmospheric physics, climatology, and global change. His work at all stages of his life was seminal and the work described in this book is in areas that were initiated or stimulated by him. The book is not just a eulogy of a great man, but is a study of numerous fields of work that owe their origins to, or have been stimulated by, him.

Kondratyev's work has been poorly recognized outside Russia/the former Soviet Union, primarily for two reasons. First, a lot of the earlier work was done in days when there was great secrecy surrounding a lot of Soviet science. Second, over the last nearly 20 years he has stood out against the conventional wisdom adopted by many climatologists and politicians, as embodied in the IPCC (Intergovernmental Panel on

Climate Change). The IPCC concentrated the resources of hundreds of climatologists on the question of anthropogenically produced greenhouse gases and their consequences in terms of global warming. Undoubtedly the achievement of the IPCC is that now most sensible people accept that human activities do lead to global warming and that it is occurring at an increasing rate. But the downside is that other threats to the existence of life and our standard of living have been virtually ignored. Kondratyev stood out against that and argued that the various forms of pollution, degradation, and consumption of the world's mineral resources and population growth are all part of global change and pose a very serious combined threat to the future of (human) life on Earth.

Arthur P. Cracknell, Costas A. Varotsos, Vladimir F. Krapivin

Figures

Tables

Abbreviations and acronyms

ABL	Atmospheric Boundary Layer
AIRS	Atmospheric InfraRed Sounder
AIS	Airborne Imaging System
AO	Arctic Oscillation
APAR	Absorbed Photosynthetically Active Radiation
ARM	Atmospheric Radiation Measurement
AUCF	Avalanche Unstable Crack Formation
AVHRR	Advanced Very High Resolution Radiometer
AVIRIS	Airborne Visible/InfraRed Imaging Spectrometer
BAMRS	Basin Administration of the Marine Rescue Service
BESEX	BEring Sea EXperiment
CAENEX	Complex Atmospheric ENergetics EXperiment
CASI	Compact Airborne Spectrographic Imager
CCD	Charge Coupled Device
CCSP	Climate Change Science Program
CCSR	Center for Climate System Research
CCSS	Carbon–Climate–Society System
CDM	Clean Development Mechanism
CERES	Clouds and the Earth's Radiant Energy System
CFC	ChloroFluoroCarbon
CKD	Correlated *k*-Distribution
COPs	Conferences of the Parties
COSPAR	Committee on Space Research
DDGCF	Density Dependent Growth Control Factor
DS	Dynamical System
EA	East Atlantic
EAJ	EA Jet
EAWR	East Atlantic/West Russia

ECHAM	European Centre Hamburg Model
EIE	Estimation of the Impact on the Environment
ENSO	El Niño Southern Oscillation
EOF	Empirical Orthogonal Function
EOS	Earth Observing System
ERB	Earth Radiation Budget
ERBE	Earth Radiation Budget Experiment
EROEI	Energy Return On Energy Invested
ES	Explosive Substance
EUROTRAC	EUREKA Project on the TRansport And Chemical Transformation of Environmentally Relevant Trace Constituents in the Troposphere over Europe
FC	Federal Center
FGGY	First GARP Global Year
FIRE-ARMS	Fine InfraRed Explorer of Atmospheric Radiation MeasurementS
FOV	Field Of View
FR	Full Resolution
GAAREX	Global Atmospheric Aerosol Radiation EXperiment
GARP	Global Atmospheric Research Program
GATE	GARP Atlantic Tropical Experiment
GCM	General Circulation Model
GCOS	Global Climate Observing System
GCP	Global Carbon Project
GEO	Group on Earth Observations
GIMS	Geo-Information Monitoring System
GIS	Geographic Information System
GISS	Goddard Institute for Space Studies
GLO-PEM	GLObal Production Efficiency Model
GMNSS	Global Model of the Nature–Society System
GOOS	Global Ocean Observing System
GPP	Gross Primary Production
GPS	Global Positioning System
GSM	Global Simulation Model
GTOS	Global Terrestrial Observing System
GWL	*GrossWetterLagen*
GWP	Global Warming Potential
GWT	*GrossWetterTypen*
HELCOM	Helsinki Commission
HIRIS	HIgh Resolution Imaging Spectrometer
HITRAN	High-resolution Transmission Model
HWHM	Half Width at Half Maximum
IAF	International Astronautical Federation
IAMAP	International Association of Meteorology and Atmospheric Physics

IAO SB RAS	Institute for Atmospheric Optics Siberia Russian Academy of Science
IASI	Infrared Atmospheric Sounder Interferometer
ICGGM	International Center on Global Geoinformation Monitoring
ICRCCM	Intercomparison of Radiation Codes in Climate Models
ICSU	International Council of Scientific Unions
IE	Industrial Enterprise
IEA	International Energy Agency
IGBP	International Geosphere–Biosphere Program
IGM	Inverted Gaussian Model
IGOS	Integrated Global Observing Strategy
IHDP	International Human Dimensions Program
ILS	Instrumental Line Shape
IMO	International Maritime Organization
INDOEX	INDian Ocean EXperiment
INTAS	INTernational ASsociation for the Promotion of Co-operation with Scientists from the New Independent States or NIS of the former Soviet Union
IPCC	Intergovernmental Panel on Climate Change
IRC	International Radiation Commission
IUGG	International Union of Geodesy and Geophysics
LAI	Leaf Area Index
LB	Living Biomass
LTE	Local Thermodynamic Equilibrium
LWP	Liquid Water Path
MB	Management Body
MES	MicroEcoSystem
METOP	METeorological Operational Polar
MFRSR	Multi-Filter Rotating Shadowband Radiometer
MGBN	Modulated Generalized Binary Noise
MGOC	Model of the Global Oxygen Cycle
MLS	Microwave Limb Sounder
MODIS	MODerate Resolution Imaging Spectroradiometer
MPC	Maximum Permissible Concentration
MPD	Maximum Permissible Discharge
MSSA	Multichannel Singular Spectrum Analysis
MTCI	MERIS Terrestrial Chlorophyll Index
NAO	North Atlantic Oscillation
NASA	National Aeronautics and Space Administration
NASDA	National Space Development Agency (Japan)
NCAR	National Center for Atmospheric Research
NCEP	National Center for Environment Protection
NDVI	Normalized Difference Vegetation Index
NEGP	North European Gas Pipeline

NIERSC	Nansen International Environmental and Remote Sensing Center
NIR	Near-InfraRed
NIS	New Independent States
NN	Neural Network
NOAA	National Ocean and Atmospheric Administration
NPP	Net Primary Production
NSS	Nature–Society System
OLR	Outgoing Longwave Radiation
PAR	Photosynthetically Active Radiation
PDI	Potential Destructiveness Index
PE	Polar/Eurasian
POP	Principal Oscillation Pattern
PS	Polluting Substance
REP	Red Edge Position
RFBR	Russian Foundation for Basic Research
RR	Reduced Resolution
RSS	Rotating Shadowband Spectroradiometer
RVSI	Red-edge Vegetation Stress Index
S	Scandinavia
S-theorem	Similarity theorem
SAP	Surface Atmospheric Pressure
SAT	Surface Air Temperature
SB RAS	Siberian Branch of Russian Academy of Sciences
ScaRaB	Scanner for Radiation Budget
SCIAMACHY	Scanning Imaging Absorption Spectrometer for Atmospheric Chartography
SeaWiFS	Sea-viewing Wide Field-of-view Sensor
SGP	Southern Great Plains
SGR	Specific Growth Rate
SIS	Shuttle Imaging Spectrometer
SOM	Suspended Organic Matter
SPECTRA	Surface Process and Ecosystem Changes Through Response Analysis
SRES	Special Report on Emissions Scenarios
SST	Sea Surface Temperature
STD	STandard Deviation
SVD	Singular Value Decomposition
VEI	Volcanic Explosivity Index
WCRP	World Climate Research Program
WEO	World Energy Outlook
WMO	World Meteorological Organization

Contributors

BARTSEV, SERGEY I.
Institute of Biophysics, Krasnoyarsk, Siberian Branch of Russian Academy of Sciences, Tomsk, Russia
bartsev@online.ru

BERKOVITS, A.V.
Scientific Research Center for Ecological Safety, Russian Academy of Sciences, St. Petersburg, Russia
berkovits@mail.ru

BINENKO, VICTOR I.
Scientific Research Center for Ecological Safety, Russian Academy of Sciences, St. Petersburg, Russia
vibinenko@mail.ru

BOEHMER-CHRISTIANSEN, SONJA
Department of Geography, University of Hull, Hull, U.K.
Sonja.B-C@hull.ac.uk

CRACKNELL, ARTHUR P.
University of Dundee, Universiti Teknologi Malaysia
apcracknell774787@yahoo.co.uk

CURRAN, P.J.
Bournemouth University, Fern Barrow, Talbot Campus, Poole, Dorset, U.K.
p.curran@soton.ac.uk

DASH, JADUNANDAN
School of Geography, University of Southampton, Southampton, U.K.
j.dash@soton.ac.uk

DEGERMENDZHI, ANDREY G.
Institute of Biophysics, Siberian Branch of Russian Academy of Sciences, Krasnoyarsk, Russia
ibp@ibp.ru

EROKHIN, DMITRY V.
Institute of Biophysics, Siberian Branch of Russian Academy of Sciences, Krasnoyarsk, Russia
ibp@ibp.ru

FIRSOV, KONSTANTIN M.
Volgograd State University, Volgograd, Russia
fkm@iao.ru

FOODY, GILES M.
School of Geography, University of Southampton, Southampton, U.K.
g.m.foody@soton.ac.uk

GOLOVKO, VLADIMIR A.
Scientific Research Center "Planet", Moscow, Russia
golovko@mail.mipt.ru

GUBANOV, VLADIMIR G.
Institute of Biophysics, Siberian Branch of Russian Academy of Sciences, Krasnoyarsk, Russia
ibp@ibp.ru

IPPOLITOV, IVAN I.
Institute of Monitoring of Climatic and Ecological Systems, Siberian Branch of Russian Academy of Sciences, Tomsk, Russia
science@iao.ru

IVLEV, LEV S.
Laboratory of Aerosol Physics at Fok's Institute of Physics, St. Petersburg State University, St. Petersburg, Russia
lev@aero.phys.spbu.ru

KELLEY, JOHN J.
Institute of Marine Science, School of Fisheries and Ocean Sciences, University of Alaska Fairbanks, Fairbanks, AK, U.S.A.
ffjjk@uaf.edu

KOMAROV, VALERYS.
V.E. Zuev Institute of Atmospheric Optics, Siberian Branch of Russian Academy of Sciences, Tomsk, Russia
popov@iao.ru

KRAPIVIN, VLADIMIR F.
Institute of Radioengineering and Electronics, Russian Academy of Sciences, Moscow, Russia
vfk@ms.ire.rssi.ru

LOMAKINA, NATALY YA
V.E. Zuev Institute of Atmospheric Optics, Siberian Branch of Russian Academy of Sciences, Tomsk, Russia
science@iao.ru

OUGOLNITSKY, GENNADIY A.
South Federal University, Rostov-on-Don, Russia
ougoln@math.rsu.ru

PKHALAGOV, YURY A.
Institute of Atmospheric Optics, Siberian Branch of the Russian Academy of Sciences, Tomsk, Russia
science@iao.ru

POKROVSKY, OLEG M.
Main Geophysical Observatory, St. Petersburg, Russia
pokrov@main.mgo.rssi.ru

RODIMOVA, OLGA B.
Institute of Atmospheric Optics, Siberian Branch of Russian Academy of Sciences, Tomsk, Russia
rod@iao.ru or *ztb@iao.ru*

ROMANENKO, FEODOR A.
Geographical Department of M.V. Lomonosov Moscow State University, Russia
shil_o@mail.ru

SAVINYKH, VICTOR P.
Moscow State University of Geodesy and Cartography, Moscow, Russia
svp@miigaik.ru

SHEVIRNOGOV, ANATOLY P.
Institute of Biophysics, Siberian Branch of Russian Academy of Sciences, Krasnoyarsk, Russia
ecoinf@ibp.ru

SHILOVTSEVA, OLGA A.
Geographical Department of M.V. Lomonosov, Moscow State University, Russia
shil_o@mail.ru

TVOROGOV, STANISLAV D.
Institute of Atmospheric Optics, Siberian Branch of Russian Academy of Sciences, Tomsk, Russia
rod@iao.ru

USOV, ANATOLIY B.
South Federal University, Rostov-on-Don, Russia
usov@math.rsu.ru

UZHEGOV, VICTOR N.
Institute of Atmospheric Optics, Siberian Branch of the Russian Academy of Sciences, Tomsk, Russia
science@iao.ru

VAROTSOS, COSTAS A.
Department of Applied Physics, Athens University, Greece
covar@phys.uoa.gr

ZAKHAROV, VYACHESLAV I.
Global Ecology & Remote Sensing Laboratory, Physics Department, Ural State University, Yekaterinburg, Russia
vz@uraltc.ru

ZHURAVLEVA, TATYANA B.
Institute of Atmospheric Optics, Siberian Branch of Russian Academy of Sciences, Tomsk, Russia
ztb@iao.ru

Authors

ARTHUR P. CRACKNELL graduated in physics from Cambridge University in 1961 and then obtained his D.Phil. at Oxford University on "Some band structure calculations for metals" in 1964. He worked as a lecturer in physics at Singapore University (now the National University of Singapore) from 1964 to 1967 and at Essex University from 1967 to 1970, before moving to Dundee University in 1970, where he became a professor in 1978. He retired from Dundee University in 2002 and now holds the title of emeritus professor there. He has been the editor in chief of the *International Journal of Remote Sensing* for over 25 years. He and his colleagues and research students have published around 280 research papers and he is the author or co-author of about 30 books, both on theoretical solid state physics, and remote sensing and the physics of the environment.

VLADIMIR F. KRAPIVIN was educated at the Moscow State University as a mathematician. He received his Ph.D. in geophysics from the Moscow Institute of Oceanology in 1973. He became Professor of Radiophysics in 1987 and Head of the Applied Mathematics Department at the Moscow Institute of Radioengineering and Electronics in 1972. He was appointed Grand Professor in 2003 at the World University for Development of Science, Education, and Society. He is a full member of the Russian Academy of Natural Sciences and Balkan Academy of Sciences, New Culture, and Sustainable Development. He has specialized in investigating global environmental change by the application of modeling technology and has published 20 books in the fields of ecoinformatics, game theory, and global modeling.

COSTAS A. VAROTSOS received his B.Sc. in Physics in 1980 from Athens University and Ph.D. in Atmospheric Physics in 1984 from Thessaloniki University. He was appointed Assistant Professor in 1989 (subsequently Associate Professor in 1999 and Professor in 2008) at the Laboratory of Meteorology of the Physics Department of Athens University, where he also set up the Laboratory of the Middle and Upper

Atmosphere, with special interests in studies of atmospheric ozone. He is an associate editor of the *International Journal of Remote Sensing* and an advisor to the *Environmental Science & Pollution Research* journal. He has published more than 300 papers and 20 books in the fields of atmospheric physics, atmospheric chemistry, and global change.

1

The seminal nature of the work of Kirill Kondratyev

Arthur P. Cracknell, Vladimir F. Krapivin, and Costas A. Varotsos

1.1 INTRODUCTION

The motivation for the writing of this book came from the influence and inspiration of the late Kirill Kondratyev (see Figure 1.1). Kondratyev, a full member of the Russian Academy of Sciences (Academician) from 1982, was a prominent scientist in the field of atmospheric and environmental sciences. Throughout his whole career he was involved in research on atmospheric radiation problems relevant to the physical basis of the Earth's climate. He contributed greatly to the development of remote-sensing techniques for environmental studies as well as to global change research. He was the author or co-author of over one thousand scientific papers and a hundred monographs.

Kondratyev was born on June 14, 1920 in Rybinsk, which is about 300 km northeast of Moscow. He obtained his primary and secondary schooling in Leningrad (now, once more, St. Petersburg). In 1938 he entered Leningrad State University to study physics, mathematics, and chemistry. However, in 1941, he had to interrupt his studies to join the Russian army, where he experienced the blockade of Leningrad and the ensuing starvation; he was wounded three times on the front line before being released from active duty in 1944. After returning to the University in Leningrad, Kondratyev graduated in atmospheric physics in 1946. Between 1946 and 1978 he occupied successively the posts of lecturer, associate professor, professor, and head of the Department of Atmospheric Physics, University Vice-Rector for science and research, and finally Rector. From 1958 to 1981 he was Head of the Department of Radiation Studies at the Main Geophysical Observatory (in Leningrad), and from 1982 to 1992 he was the Head of the Remote Sensing Laboratory at the Institute for Lake Research. A full member of the Russian Academy of Sciences, he was a Counsellor of the Russian Academy of Sciences in the Research Centre for Ecological Safety in St. Petersburg from 1992, and he helped to create the Nansen International

Figure 1.1. Kirill Yakovlevich Kondratyev.

Environmental and Remote Sensing Centre (NIERSC) in St. Petersburg. Kondratyev died on May 1, 2006.

A very valuable source of information about Kondratyev and his work will be found in an interview published in an issue of the *WMO Bulletin* (WMO, 1998). We have made very extensive use of that text in preparing this chapter as well as information from his wife Svetlana Ivanovna Kondratyeva.

1.2 EARLY RADIATION STUDIES

The very beginning of Kondratyev's scientific work was connected with the preparation of his diploma work (Master's Thesis) before graduating from the University. At that time (1945–1946) a problem of crucial practical significance for the country was the development of agriculture. An important task for agriculture in the southern part of Russia was the protection of grape vines against the damaging impact of early morning frosts during spring time. It was empirically discovered that a reliable enough protection was guaranteed by the formation of an artificial smoke layer above the vineyards. To study the problem further it had been decided to organize an expedition to the region of Rostov-on-Don (in the Northern Caucasus) and to undertake relevant simulation modeling. This problem was offered to Kondratyev, who also participated in the expedition. In the context of this problem, Kondratyev undertook a study involving

- consideration of the atmospheric greenhouse effect formation in the presence of a surface layer polluted by smoke; and
- development of a theory of the surface layer thermal regime taking account of both radiative transfer and turbulent mixing (this was, in fact, the first attempt of this kind).

The solution of this problem had, for a substantial period, become the motivation to develop a better (i.e., a more reliable) technique for longwave radiation flux and flux divergence calculations, to model greenhouse effect formation under conditions of a multi-component atmosphere (water vapor, carbon dioxide, ozone, and aerosols), to assess the interactions between radiation and dynamics in the formation of vertical temperature profiles (Borisenkov and Kondratyev, 1988; Kondratyev, 1991, 1992; Kondratyev and Johannessen, 1993; Kondratyev and Varotsos, 2000; Kondratyev *et al.*, 1977a, 2005a). As well as theoretical studies, this work involved relevant experimental studies in the laboratory and under field conditions.

As far as the protection of vineyards against frosts is concerned, the principal conclusion was something of a paradox. It was shown that the smoke layer's impact was determined not by the effect of warming due to the smoke-enhanced greenhouse effect but by the attenuation of solar radiation during the early morning hours when it was important to protect partly frozen vegetation from rapid heating by solar radiation (Kondratyev and Fedchenko, 1982; Kondratyev *et al.*, 1983b, 2002b).

The studies mentioned above have become very important, however, because of a number of other reasons and especially in the context of climate change research. One curious scientific result (published in 1947–1948) was a simple theory of the conventional greenhouse which showed that warming inside the greenhouse is due to the absence of sensible heat exchange between the soil surface and the atmosphere (turbulent mixing cut off by the glass of the greenhouse), but not those processes which are responsible for the formation of what is called the atmospheric greenhouse effect (in this respect such a terminology is misleading). This conclusion was supported by an experiment made much earlier by the famous American physicist Robert Williams Wood (Wood, 1934), who replaced the conventional glass in a greenhouse by quartz glass and did not find any difference.

Kondratyev published the results connected with the development of a new parametrization of longwave radiation transfer in his first monograph (Kondratyev, 1950). Later on a revised and enlarged edition of this monograph was published by Pergamon Press (Kondratyev, 1965a, b).

Many of Kondratyev's studies have been devoted to the development of the theory of radiative infrared (thermal) transfer in the atmosphere which has important applications not only in atmospheric physics, but also in practical problems. In this connection, he undertook a detailed study of various factors determining infrared radiative transfer, the atmospheric greenhouse effect, and approximate techniques for calculating the quantitative characteristics of the field of atmospheric thermal and shortwave radiation. Substantial research was also carried out in numerical modeling in order to obtain data characterizing the radiation field for various atmospheric conditions (and for different planets) (Kondratyev, 1990b; Kondratyev *et al.*, 1987). Numerical modeling of atmospheric absorption spectra was also performed. A number of studies were devoted to remote sensing of the environment from space. Relevant results have been discussed in a series of monographs by Kondratyev (1969, 1970).

Specific attention was given to numerical modeling of the greenhouse effect for various planetary atmospheres (Earth, Mars, Venus, Jupiter, Saturn, and Titan). In collaboration with Dr. N.I. Moscalenko (of the Institute of Applied Optics, Kazan) the absorption spectra of many greenhouse gases under conditions of varying temperature and pressure (to simulate Mars and Venus) were measured, and these data were used for radiative transfer calculations as well as climatic assessments for the Earth and other planets on the basis of radiative–convective model calculations. The results were discussed in the monographs by Kondratyev (1983a, b). Earlier on, the subject of the comparative meteorology of planets was also discussed in books by Kondratyev (Kondratyev, 1972a–d, 1976a–c, 1980).

After becoming assistant professor in 1946 Kondratyev was heavily involved in teaching, delivering a number of lecture courses including Meteorology and Atmospheric Physics, Dynamical Meteorology, Synoptic Meteorology, Geophysics, and Hydrodynamics. One result of this was participation in the preparation of a basic textbook (Kondratyev, 1950).

The second half of the 1950s was devoted to preparations for the interpretation of satellite observation data (a curious fact is that in 1958 Kondratyev published in the Finnish journal *Arkhimedes* the very first scientific paper on satellite observations of the upper atmosphere). A productive collaboration with the leaders of Soviet space research, such as Academician M.V. Keldysh (not being identified by name, he was described at that time as the "Chief Scientist"), Academician S.P. Korolev ("Chief Designer"), Academician V.P. Glushko (all these scientists have passed away), Academician W.P. Mishin, and others via the U.S.S.R. Academy of Sciences Council on Space Research, opened the way to active participation in that part of space research which is relevant to investigations of the Earth and other planets.

A broad program of theoretical and experimental studies was developed at the Department of Atmospheric Physics stimulated by the launch of meteorological and Earth resource satellites in the U.S.A. High-altitude balloon and aircraft observations were an important part of this program.

1.3 BALLOON AND AIRCRAFT OBSERVATIONS IN THE CONTEXT OF CLIMATE STUDIES

While still at the University, Kondratyev became involved with the Main Geophysical Observatory in Leningrad where he held various posts, including senior research scientist and Head of the Department of Radiation Studies. This provided greater opportunities for research than he had at the University. Following the launch of Sputnik-1 in 1957 it became clear that an important development was about to occur in the form of the newly emerging field of satellite meteorology, or remote sensing of the atmosphere. Kondratyev was heavily involved in experiments with balloons and aircraft-flown instrumentation. This work was aimed at investigating both infrared and shortwave radiative transfer in the atmosphere and studying the absorption spectra of various active components, such as water vapor, carbon dioxide, ozone, and various other minor atmospheric constituents.

A number of years were spent in designing and manufacturing balloon instrumentation (solar spectrometers, pyrheliometers, pyranometers, aerosol impactors, and filters, etc.), and this made it possible to conduct during the 1960s a series of 22 high-altitude large-balloon (about 800 kg weight) flights launched from a test site in the middle Volga River region. The principal purpose was to obtain data on vertical profiles (up to 30 km–33 km) of the spectral transparency of the atmosphere, total direct solar radiation, and downward and upward shortwave radiation fluxes with simultaneous information on aerosol properties, such as number concentration, size distribution, and chemical composition (i.e., complex refraction index). The interpretation of this (still unique) set of observational data resulted in many journal publications and has been discussed in the WMO monograph (Kondratyev, 1972b).

An unexpected result of processing balloon data was the discovery of anomalous absorption of solar radiation in the stratosphere which was later interpreted as resulting from the nuclear explosions (tests) in the atmosphere conducted during the late 1950s and early 1960s. The tests produced substantial amounts of NO_2 which strongly absorbs solar radiation, leading to the conclusion of the reality of a "little nuclear winter" during the first half of the 1960s. This was discussed in detail in two monographs by Kondratyev (1988a, b). Balloon observational data were also helpful in substantiating a new hypothesis concerning the impact of solar activity (i.e., of the sunspot cycle) on climate (relevant concluding results have been published in monographs by Kondratyev, 1999a–c) and in obtaining the first directly measured value of the solar constant.

In connection with preparation for the launch of Soviet meteorological satellites, a few flying laboratories consisting of IL-18 four-engine turbojets were designated for testing satellite onboard instrumentation (TV cameras, scanning radiometers, and Earth radiation budget or ERB sensors). One of the aircraft was given to the Main Geophysical Observatory (Leningrad) and then offered to the Department of Radiation Studies (where Kondratyev was the Head of the Department). This aircraft was equipped as a multi-purpose flying laboratory with three aims in view:

(1) to test prototypes of satellite instrumentation;
(2) to test and apply remote-sensing instrumentation (mainly for the microwave wavelength region); and
(3) to investigate radiation processes in the free atmosphere which are responsible for climate formation (especially from the viewpoint of aerosol and cloud impact on climate).

In connection with the last subject two 5-year programs were devised and accomplished during the first and second halves of the 1970s:

- CAENEX, the Complex Atmospheric Energetics Experiment; and
- GAAREX, the Global Atmospheric Aerosol Radiation Experiment;

and they resulted in a substantial number of publications, including several monographs (Kondratyev, 1972d, 1983b, 1986).

The CAENEX and GAAREX field campaigns of combined aircraft and surface observations during the 1970s covered various parts of the U.S.S.R. under different climatic conditions (the Central Asian desert, the southern Russian steppe, the Arctic, the urban environment, etc.). A substantial contribution resulted from the participation in the GATE (GARP Atlantic Tropical Experiment) program (Kondratyev, 1973). Of special interest were two field campaigns conducted within the program of FGGY (First GARP Global Year) during the two special observing periods to study:

(1) desert aerosol and its impact on climate (involving an expedition to the Kara-Kum desert); and
(2) the interaction of extended cloudiness and radiation in the Arctic.

A field experiment over the industrial region of Zaporozhye (Ukraine) should also be mentioned; this resulted in a unique set of data on the properties of "dirty" (polluted) clouds, including cloud chemistry, microphysics, and radiation characteristics. Important results included the conclusions:

(1) on the average, solar radiation absorption by aerosols in the clear atmosphere is close to absorption by water vapor;
(2) the heat balance of the summer atmospheric boundary layer in the steppe region is dominated by longwave radiative flux divergence (but not sensible heat exchange);
(3) cloud cover is always characterized by significant solar radiation absorption, which becomes very strong in the case of "industrial dirty clouds";
(4) Saharan dust transport (during dust storms) to the Atlantic Ocean radically changes the radiative regime of the free troposphere; and
(5) the development of extended cloudiness in the Arctic in spring and the process of interaction between cloudiness and radiation were monitored on the basis of observations and numerical modeling; a mesoscale model developed for this purpose was discussed in two papers by Kondratyev *et al.* (1992a) and Pozdnyakov *et al.* (2002).

The principal part of remote-sensing efforts on the basis of aircraft observations was connected with the development of microwave passive and active remote-sensing techniques to retrieve properties of the atmosphere (total water vapor and liquid water content), ice cover (concentration, age), natural waters (sea state, surface temperature), and soil (moisture). Relevant results were discussed in many papers and a number of monographs, including a recent book (Kondratyev, 1998).

1.4 SATELLITE REMOTE SENSING

A project on the so-called "small optical satellites" was undertaken, which resulted in the launch of a satellite in 1965 with onboard instrumentation to measure the optical characteristics of the surface–atmosphere system and Earth radiation budget components. The results were published as a separate volume of collected papers and books (Kondratyev, 1981, 1985; Kondratyev and Galindo, 1997; Kondratyev and Nikolsky, 1970; Kondratyev *et al.*, 1973, 1979, 1983a, 1986; Marchuk *et al.*, 1986).

A significant part of this work was connected with the development of satellite meteorology and environmental observations. In particular, the principles of the interpretation of meteorological satellite data were developed. The available observational data were used to study the basic factors of the Earth's radiation balance and to determine the net radiation of the Earth as a planet. Kondratyev was the first scientist to propose and substantiate a statistical approach to the analysis of satellite measurement results on the Earth radiation budget. The results of these studies have been summarized in various monographs (Kondratyev, 1956, 1965b, 1969, 1972a–d, 1983a, b; Kondratyev and Timofeyev, 1970). All these books have been published in English as NASA Technical Translations.

The publication by Kondratyev (1965a) contained for the first time a discussion of the meteorological importance of the Earth's pictures obtained by Soviet cosmonauts. One significant difference between Earth observation (remote sensing) from space in the former U.S.S.R. and in the West was the greater involvement of manned Earth-orbiting space missions in the former U.S.S.R. The original results of later studies in the field of remote sensing from manned spacecraft have been discussed in the monograph by Kondratyev (1972b).

There was a substantial effort connected with working out the program for the development of instrumentation and conducting observations from manned spacecraft and orbital stations. Since relevant results have been discussed in a recent monograph (Kondratyev, 1998), it is enough to mention briefly the principal achievements in the following directions:

(1) visual observations by cosmonauts;
(2) interpretation of twilight and day-time horizon spectra as well as occultation data in terms of vertical profiles of stratospheric aerosol and minor gas components (water vapor, ozone); and
(3) interpretation of the surface atmosphere reflectance spectra at nadir to develop atmospheric correction techniques to recognize various types of terrain.

The problems mentioned have been considered on the basis of observation data obtained with the help of two instruments: a hand-held spectrograph for the visible wavelength region and a complex of solar spectrometers for the visible and near-infrared. Close collaboration with the cosmonauts (including Beregovoy, Grechko, Khrunov, Nikolaev, Savinykh, Serebrov, Sevastyanov, Shatalov, Volkov, and Volynov) was a very important part of the study. A number of interesting results were obtained through visual observations by cosmonauts of various phenomena in the atmosphere (especially near the edge of the planet) and on the Earth's surface. Visible wavelength radiation was used on a number of manned spacecraft (starting from Soyuz-5), and orbiting spacecraft made it possible, for the first time, to determine spectra of the Earth and the atmosphere at the horizon. Techniques were developed to retrieve vertical aerosol profiles in cases of brightness measurements for twilight horizon (primary scattering dominating) and day-time horizon (prevailing multiple scattering). An occultation technique to retrieve aerosol properties has also been applied in the case of measuring attenuation of solar radiation by the atmosphere during sunsets and sunrises relative to a spacecraft (Kondratyev, 1965b). An occultation technique for a case of absorption was also developed to retrieve vertical water vapor and ozone profiles in the stratosphere using visible and near-infrared data. Visible wavelength spectra were used to develop and verify a technique to recognize various terrains, taking account of atmospheric correction. For solving the latter problem a number of approaches to numerical modeling of radiative transfer in the atmosphere with multiple scattering were substantiated (these results were discussed in detail in the monograph on atmospheric correction by Gorshkov *et al.*, 1994). A complex field program was accomplished over the Kara-Kum desert with the use of simultaneous data from three manned spacecraft, aircraft, and the surface. The most reliable approach to making atmospheric corrections was to use observation data close to the sea–land surface border when there are two surfaces with quite different albedos (in this case the desert and the Caspian Sea).

Visible wavelength data were also used to support an approach to the optimal selection of wavelength (channels) for the purpose of terrain identification. This direction of studies has been intensively pursued later on to support the optimization of combined conventional and satellite observations (relevant results have been described in the monograph by Kondratyev, 1998).

1.5 LIMNOLOGICAL STUDIES

Kondratyev spent ten years (1982–1991) in the Institute for Lake Research of the U.S.S.R. Academy of Sciences. These were devoted mainly to the development of the three new directions of research:

(1) using remote sensing to study limnological environmental dynamics (principally the development of eutrophication processes);
(2) the use of lakes as test sites to verify remote-sensing techniques; and
(3) the consideration of lakes as natural simulation models to study similar processes in seas and oceans.

As far as the latter problem is concerned, it has been discussed in detail by

Kondratyev (1987, 1990a, 1998). The development of limnological remote sensing was directed at the determination of various parameters (water surface state and temperature, snow and ice cover properties, etc.), but the most important efforts were devoted to studying remote-sensing techniques for the retrieval of the basic properties of natural waters (phytoplankton, suspended matter, and dissolved organic matter concentrations). Relevant field experiments were conducted on Ladoga Lake, Onega Lake (to the northeast of St. Petersburg), and Sevan Lake (in Armenia). Passive (surface brightness measurements) and active (fluorescence induced by laser light) remote-sensing techniques for the observation of the water surface were developed on the basis of numerical simulation modeling (in which Monte Carlo sensitivity calculations played a special role) and field observations. An intercomparison was made successfully between the limnological environments of the American Great Lakes and the Russian Great Lakes (Baikal, Ladoga, and Onega). Relevant results were published in cooperation with Canadian colleagues in the monograph on the optical properties and remote sensing of natural waters (Kondratyev *et al.*, 1992a). Two international expeditions to the Rybinsk reservoir on the Volga River marked an international contribution to this direction of research.

1.6 GLOBAL CHANGE STUDIES

Kondratyev's work on clouds and atmospheric greenhouse gases necessarily led him into the study of climate change (Kondratyev and Binenko, 1981, 1984; Kondratyev and Zhvalev, 1981; Kondratyev *et al.*, 2001). However, an important strand to his work in later years was his concern that people had become obsessed with global warming and climate change and that not enough attention was being paid to various other changes that are being brought about by human activities and which threaten various ecological systems and the viability of the future standards of living, and indeed the very continued existence, of human life (Grigoriev and Kondratyev, 2001a–c; Kondratyev *et al.*, 2002c, 2006). We shall consider this matter in more detail in Chapter 2.

At the beginning of the 1970s, when the Club of Rome was developing its program of studying global change, Kondratyev organized regular seminars to discuss relevant problems. Steadily, independent research efforts were being pursued with the purpose of determining key issues of global change and the requirements for observations. A cornerstone aspect was the development by Gorshkov (1990) of the concept of biotic regulation of the environment. These efforts resulted in two books (Kondratyev, 1989, 1990a) as well as a recent monograph (Kondratyev and Cracknell, 1998) An important aim of these writings was an analysis of conceptual issues of such international programs as the World Climate Research Program (WCRP), the International Geosphere–Biosphere Program (IGBP), as well as the outcome of the Second U.N. Conference on Environment and Development (UNCED) in the context of the concept of the biotic regulation of the environment (Kondratyev, 1982). An important step was the completion of the monographs by Kondratyev (Kondratyev, 1998; Kondratyev *et al.*, 1997, 2003b, c, 2005b; Krapivin and Kondratyev, 2002) of which the principal aim was an analysis of the interaction between societal and environmental dynamics. Special emphasis was placed on the analysis of the role and place of global climate change studies in the context of global change; this was necessary in the light of certain overemphasis in UNCED documents and Intergovernmental Panel on Climate Change (IPCC) reports on climate change and greenhouse gases reduction. Another conceptual aspect is connected with the problem of optimizing global environmental observing systems of combined conventional and satellite observations (Kondratyev and Cracknell, 1998; Kondratyev and Galindo, 2001; Kondratyev and Krapivin, 2004; Kondratyev and Moskalenko, 1984; Kondratyev *et al.*, 1996, 2002a; Marchuk and Kondratyev, 1992).

1.7 INTERNATIONAL COLLABORATION

A very important part of research efforts is international scientific collaboration. During the period before the Iron Curtain came down, when communication between Soviet and Western scientists was not easy, Kondratyev contributed to the exchange of scientific ideas both by inviting key Western scientists to St. Petersburg (Leningrad as it was then known) and also by his own participation in the activities of various international bodes, such as the World Meteorological Organization (WMO), the International Astronautical Federation (IAF), and the International Radiation Commission (IRC). When travels abroad were very limited for citizens of the U.S.S.R., Kondratyev developed a program of regular visits to the University of Leningrad of outstanding scientists from various countries including H.-J. Bolle (Germany), M. Bossolasco (Italy), Prof. R.M. Goody (U.S.A.), Dr. J.N. Howard (U.S.A.), Prof. J. Lenoble (France), Prof. J. London (U.S.A.), Prof. J. Van Mieghem (Belgium), Prof. F. Möller (Germany), Prof. V.E. Suomi (U.S.A.), Prof. V. Väisä (Finland), Prof. G. Yamamoto (Japan), and many others. A substantial contribution, at the later stage, involved participation in international conferences such as IAF and COSPAR congresses, IAMAP and IUGG Assemblies, etc.

A very productive stage was the participation in various activities of the World Meteorological Organization (WMO, 1998), especially its Advisory Committee (in the 1960s) which was also responsible for the development of the Global Atmospheric Research Program (GARP) as a precursor of WCRP (World Climate Research Program) (GARP-climate). The memorable events were receiving the WMO gold medal and the delivery of a lecture for the WMO Congress with the subsequent publication of the WMO monograph (Kondratyev, 1972b). A similar honor was obtained later from the International Astronautical Federation; in the 1960s Kondratyev initiated the organization of the IAF Committee on Application Satellites which functioned successfully for more than a decade. A significant part of his international efforts were connected with various activities of the IAMAP International Radiation Commission, where Kondratyev served as a member for a long time and during the period 1964–1968 he was President. A notable event in the International Radiation Commission's history was the International Radiation

Symposium in 1964 in Leningrad; similar events were the COSPAR Symposium in 1970 in Leningrad, and the IAF Congress in 1974 in Baku, Azerbaidjan.

Undoubtedly outstanding stages in the history of environmental studies were bilateral Soviet-American agreements on environmental cooperation and space research. The environmental agreement signed in 1972 survived successfully for more than 20 years of the Cold War era. Kondratyev's participation was connected with the Working Group on climate studies and included such efforts as the CAENEX and GAAREX programs, as well as a number of joint Soviet-American expeditions in the U.S.S.R. and U.S.A. (see above).

A very important event of the cooperation in space research was the preparation and accomplishment of the Bering Sea Experiment (BESEX) to develop remote-sensing techniques for the retrieval of atmospheric parameters, ice cover properties, and sea state characteristics. Two ships (an American ice-breaker and a Soviet meteorological research vessel) and three aircraft (two Soviet ones, an IL-18 and an AN-32, and an American Convair-990) participated in the expeditions. The results of BESEX were published by Kondratyev *et al.* (1977b) and Rycroft (1977).

Later on, Kondratyev served as co-chairman (with Dr. S. Tilford from NASA as the other co-chairman) of the Soviet-American Working Group on Remote Sensing during the 5-year period 1988–1993. A rather broad cooperative programme included studies of Kamchatka volcanoes, remote sensing of Siberian forests, preparations to install American TOMS ozone instrumentation onboard the Russian meteorological satellite Meteor-3M and the accomplishment of this task in 1992, preparations of an international Earth resource module for the space station Mir (the module was launched in 1996).

In connection with his scientific and international activities Kondratyev was awarded the U.S.S.R. State Prize, the World Meteorological Organization Prize and Gold Medal, and the Symons Medal of the Royal Meteorological Society (U.K.). He was an Honorary Foreign Member of the German Natural Science Academy "Leopoldina", the American Academy of Arts and Sciences, the American Meteorological Society, and the Royal Meteorological Society, a full member of the International Academy of Astronautics, Academia Scientiarum et Artia Europaea, and an honorary doctor of the Universities of Lille (France), Budapest (Hungary), and Athens (Greece).

1.8 THE RESEARCH CENTER OF ECOLOGICAL SAFETY AND THE NIERSC

The St. Petersburg Scientific and Research Center of Ecological Safety was established within the Russian Academy of Sciences in order to carry out interdisciplinary research aimed at trying to understand the large body of information concerning the environment when it is exposed to technological and human activities. Ecological safety is an interdisciplinary area of knowledge. The Center's activities include theoretical work and field experiments on numerous issues pertaining to ecological safety. Kondratyev joined the Center in 1992 and worked on various environmental problems from global to local scales. As a member of the Russian Academy of Sciences aged over 65 years he had the title of "Counsellor of the Academy". As such, he had a full salary and was free to do whatever work he chose to do with a small group of assistants.

The Nansen International Environment and Remote Sensing Center (NIERSC) was created as a joint venture between Russian and other scientists to deal with environmental problems, rather than basic fundamental science. It was established through collaboration between the St. Petersburg Research Center for Ecological Safety in Russia and the Nansen Environmental and Remote Sensing Center in Bergen, Norway. Later on, other institutions, such as the Environmental Research Institute of Michigan, in the U.S.A., and the Max-Planck Institute for Meteorology, in Germany, also joined. In December 1993, a special agreement was signed between NIERSC and the Joint Research Center of the Commission of the European Communities represented by the Space Applications Institute in Ispra, Italy. NIERSC is an independent non-profit-making institution and its aim is to study and monitor regional and global pollution and environmental and ecological problems. Its function is to serve as an international focal point establishing collaboration between Russian scientists and the rest of the scientific world. Its programs are funded by multinational agencies, research foundations, government councils, and other organizations. The ultimate goal is the integration of efforts for establishing a remote-monitoring service in the St. Petersburg region.

The scientific activities of NIERSC focus on environmental and pollution monitoring and the modeling of the atmosphere, land, and inland and oceanic (including ice-covered) waters. The geographical region concerned extends from the northwestern Russian region, including the Kara and Barents Seas, to the land and water system of the western Siberian coast, the St. Petersburg region, and the Baltic Sea. Apart from regional ecological studies, NIERSC also conducts research in the field of global change, including the human dimension.

1.9 CONCLUSION

The main point that we would like to make in this chapter is that the topic of this book on *Global Climatology and Ecodynamics: Anthropogenic Changes to Planet Earth* owes many of its origins to the lifelong work of the great Soviet and Russian scientist Kirill Kondratyev. His work provided the initial stimulus for much of the work that is described in the various chapters of this book. He was responsible for the development of various important national and international research programs in meteorology and atmospheric physics. In the field of satellite meteorology, he made remarkable efforts in connection with environmental observations and the interpretation of data. He was the first scientist to propose and substantiate a statistical approach to the analysis of satellite measurements of the Earth's radiation budget. In the field of climate change, he was a fervent advocate of the principle of "multidimensional global change" (Kondratyev *et al.*, 2003a, 2004), which aims at an analysis of the interaction between societal and environmental dynamics (see Chapter 2).

The memory of Kondratyev as a man of high and noble soul, thinker, and encyclopaedist through the long span of his life (86 years) will be always with us. The chapters in this book were written by scientists who maintain an invisible but important connection with him. He was a very communicative person; he loved life and always helped young scientists from Russia and other countries to explore the frontiers of science. He always was a man of principle, searching after and defending the truth.

1.10 REFERENCES AND LIST OF SELECTED PUBLICATIONS BY K.YA. KONDRATYEV

Borisenkov E.P. and Kondratyev K.Ya. (1988). *Carbon Cycle and Climate*. Hydrometeoizdat, Leningrad, 320 pp. [in Russian].

Gorshkov V.G. (1990). *Energetics of the Biosphere and Environmental Stability*. ARISTI, Moscow, 237 pp.

Gorshkov V.G., Kondratyev K.Ya., and Losev K.S. (1994). *The Natural Biological Regulation of the Environment*. Springer-Verlag, Berlin, 340 pp.

Grigoryev Al.A. and Kondratyev K.Ya. (2001a). *Ecological Disasters*. St. Petersburg Scientific Center of RAS, St. Petersburg, 206 pp. [in Russian].

Grigoryev Al.A. and Kondratyev K.Ya. (2001b). *Ecological Catastrophes*. St. Petersburg Scientific Center of RAS, St. Petersburg, 661 pp. [in Russian]

Grigoryev Al.A. and Kondratyev K.Ya. (2001c). *Natural and Anthropogenic Ecological Disasters*. St. Petersburg Scientific Center of RAS, St. Petersburg, 688 pp. [in Russian].

Kondratyev K.Ya (1950). *Long-wave Radiation Transfer in the Atmosphere*. Gostechizgat, Leningrad, 288 pp. [in Russian].

Kondratyev K.Ya. (1956). *Radiant Sun Energy*. Hydrometeoizdat, Leningrad, 600 pp. [in Russian].

Kondratyev K.Ya. (1965a). *Actinometry*. Hydrometeoizdat, Leningrad, 691 pp. [in Russian].

Kondratyev K.Ya. (1965b). *Radiative Heat Exchange in the Atmosphere*. Pergamon Press, New York, 350 pp.

Kondratyev K.Ya (1969). *Radiation in the Atmosphere*. Academic Press, New York, 912 pp.

Kondratyev K.Ya. (1970). *The Constants of Gas-Phase Reactions Speed*. Science, Moscow, 350 pp. [in Russian].

Kondratyev K.Ya. (ed) (1972a). *Explorations of the Environment from Manned Spacecraft*. Hydrometeoizdat, Leningrad, 297 pp. [in Russian].

Kondratyev K.Ya. (1972b). *Radiation in the Atmosphere*. WMO Monograph No. 309, Geneva, 214 pp.

Kondratyev K.Ya. (ed.) (1972c). *Studies of Natural Environment from Manned Orbital Stations*. Hydrometeoizdat, Leningrad, 400 pp. [in Russian].

Kondratyev K.Ya. (1972d). *The Complex Energetics Experiment (CAENEX)*. Obninsk Information Center, Obninsk, 79 pp. [in Russian].

Kondratyev K.Ya. (1973). *The Complete Atmospheric Energetics Experiment*, GARP Publ. Series No. 12. WMO, Geneva, 38 pp.

Kondratyev K.Ya. (1976a). *New Results in Climate Theory*. Hydrometeoizdat, Leningrad, 64 pp. [in Russian].

Kondratyev K.Ya. (1976b). *The Complete Radiation Experiment*. Hydrometeoizdat, Leningrad, 239 pp. [in Russian].

Kondratyev K.Ya. (1976c). *The Present Climate Changes and Their Determining Factors*. ARISTI, Moscow, 203 pp. [in Russian].

Kondratyev K.Ya. (1980). *Radiative Factors of Present Global Climate Changes*. Hydrometeoizdat, Leningrad, 280 pp. [in Russian].

Kondratyev K.Ya. (1981). *Stratosphere and Climate*. ARISTI, Moscow, 223 pp. [in Russian].

Kondratyev K. Ya. (1982). *The World Climate Research Programme: The State and Perspectives, and the Role of Spaceborne Observational Means*. ARISTI, Moscow, 274 pp. [in Russian].

Kondratyev K.Ya. (1983a). *Satellite Climatology*. Hydrometeoizdat, Leningrad, 264 pp. [in Russian].

Kondratyev K.Ya. (1983b). *The Earth's Radiation Budget, Aerosol, and Clouds*. ARISTI, Moscow, 315 pp. [in Russian].

Kondratyev K.Ya. (1985). *Volcanoes and Climate*. ARISTI, Moscow, 204 pp. [in Russian].

Kondratyev K.Ya. (1986). *Natural and Anthropogenic Changes of Climate*. ARISTI, Moscow, 349 pp. [in Russian].

Kondratyev K.Ya. (1987). *Global Climate*. ARISTI, Moscow, 313 pp. [in Russian].

Kondratyev K.Ya. (1988a). *Climate Shocks: Natural and Anthropogenic*. Wiley/Praxis, Chichester, U.K., 296 pp.

Kondratyev K.Ya. (1988b). *Comparative Meteorology of the Planets*. ARISTI, Moscow, 138 pp. [in Russian].

Kondratyev K.Ya. (1989). *Global Ozone Dynamics*. ARISTI, Moscow, 212 pp. [in Russian].

Kondratyev K.Ya. (1990a). *Key Problems of Global Ecology*. ARISTI, Moscow, 454 pp. [in Russian].

Kondratyev K.Ya. (1990b). *Planet Mars*. Hydrometeoizdat, Leningrad, 368 pp. [in Russian].

Kondratyev K.Ya. (ed.) (1991). *Aerosols and Climate*. Hydrometeoizdat, Leningrad, 542 pp. [in Russian].

Kondratyev K.Ya. (1992). *Global Climate*. Science, St. Petersburg, 359 pp. [in Russian].

Kondratyev, K.Ya. (1998). *Multidimensional Global Change*. Wiley/Praxis. Chichester, U.K., 771 pp.

Kondratyev, K.Ya. (1999a). *Atmospheric Ozone Variability*. Springer/Praxis, Chichester, U.K., 592 pp.

Kondratyev, K.Ya. (1999b). *Climate Effects of Aerosols and Clouds*. Springer/Praxis, Chichester, U.K., 272 pp.

Kondratyev K.Ya. (1999c). *Ecodynamics and Geopolicy, Vol. 1: Global Problems*. St. Petersburg Scientific Center of RAS. St. Petersburg, 1,036 pp. [in Russian].

Kondratyev K.Ya. and Binenko V.I. (1981). *Polar Aerosols, Extended Cloudiness, and Radiation*. Hydrometeoizdat, Leningrad, 150 pp. [in Russian].

Kondratyev K.Ya. and Binenko V.I. (1984). *Effect of Clouds on Radiation and Climate*. Hydrometeoizdat, Leningrad, 240 pp. [in Russian].

Kondratyev K.Ya. and Cracknell A.P. (1998). *Observing Global Climate Change*. Taylor & Francis, London, 562 pp.

Kondratyev K.Ya. and Fedchenko P.P. (1982). *Spectral Reflection Ability and Vegetation Recognition*. Hydrometeoizdat, Leningrad, 216 pp. [in Russian].

Kondratyev K.Ya., and Galindo I. (1997). *Volcanic Activity and Climate*. A. Deepak, Hampton, VA, 382 pp.

Kondratyev K.Ya. and Galindo I. (2001). *Global Change Situations: Today and Tomorrow*. Universidad de Colima, Colima, Mexico, 164 pp.

Kondratyev K.Ya. and Johannessen O. (1993). *The Arctic and Climate*. PROPO, St. Petersburg, 140 pp. [in Russian].

Kondratyev K.Ya. and Krapivin V.F. (2004). *Global Carbon Cycle Modeling*. Science, Moscow, 335 pp. [in Russian].

Kondratyev K.Ya. and Moskalenko N.I. (1984). *Greenhouse Effect of the Atmosphere and Climate*. ARISTI, Moscow, 262 pp. [in Russian].

Kondratyev, K.Ya. and Nikolsky, G.A. (1970). Solar radiation and solar activity. *Quarterly Journal of the Royal Meteorological Society*, **96**, 509–522.

Kondratyev K.Ya. and Nikolsky G.A. (2005). Influence of solar activity on the Earth's structural components, 1: Meteorological conditions. *Research of the Earth from Space*, **3**, 22–31 [in Russian].

Kondratyev K.Ya. and Nikolsky G.A. (2006a). Further about impact of solar activity on geospheres. *Il Nuovo Cimento C.*, NCC9200, **29C**(6), 695–708.

Kondratyev K.Ya. and Nikolsky G.A. (2006b). Impact of solar activity on structure component of the Earth, I: Meteorological conditions. *Il Nuovo Cimento, Geophysics and Space Physics*, **29C**(2), 253–268.

Kondratyev K.Ya. and Timofeyev Yu.M. (1970). *Thermal Sounding of the Earth from Space*. Hydrometeoizdat, Leningrad, 421 pp. [in Russian].

Kondratyev K.Ya. and Varotsos C.A. (2000). *Atmospheric Ozone Variability: Implications for Climate Change, Human Health, and Ecosystems*. Springer/Praxis, Chichester, U.K., 758 pp.

Kondratyev K.Ya. and Zhvalev V.F. (eds.) (1981). *First Global GARP Experiment, Vol. 2: Polar Aerosols, Extended Cloudiness, and Radiation*. Hydrometeoizdat, Leningrad, 150 pp. [in Russian].

Kondratyev K.Ya., Vasilyev O.B., Ivlev L.S., Nikolsky G.A., and Smokty O.I. (1973). *The Effect of Aerosol on Radiation Transfer: Possible Climatic Consequences*. Leningrad State University, Leningrad, 266 pp. [in Russian].

Kondratyev K.Ya., Marchuk G.I., Buznikov A.A., Minin I.N., Mikhailov G.A., Nazarliev M.A., Orlov V.M., and Smokty O.I. (1977a). *The Radiation Field of the Spherical Atmosphere*. Leningrad State University, Leningrad, 214 pp. [in Russian].

Kondratyev K.Ya., Nordberg W., Rabinovich Yu.I., and Melentyev V.V. (1977b). The USSR/USA Bering Sea Experiment BESEX. *Proceedings of the 18th Plenary Meeting of COSPAR, May 31–June 7, 1975, Varna, Bulgaria*. Pergamon Press, London, pp. 456–461.

Kondratyev K.Ya., Grigoryev A.A., Rabinovich Yu.I., and Shulgina E.M. (1979). *Meteorological Sensing of the Land Surface from Space*. Hydrometeoizdat, Leningrad, 274 pp. [in Russian].

Kondratyev K.Ya., Grigoryev Al.A., Pokrovsky O.M., and Shalina E.V. (1983a). *Satellite Remote Sensing of Atmospheric Aerosol*. Hydrometeoizdat, Leningrad, 216 pp. [in Russian].

Kondratyev K.Ya., Moskalenko N.I., and Pozdnyakov D.V. (1983b). *Atmospheric Aerosol*. Hydrometeoizdat, Leningrad, 224 pp. [in Russian].

Kondratyev K.Ya., Kozoderov V.V., and Fedchenko P.P. (1986). *Aero-space Investigations of Soils and Vegetation*. Hydrometeoizdat, Leningrad, 232 pp. [in Russian].

Kondratyev K.Ya., Krupenio N.N., and Selivanov A.S. (1987). *Planet Venus*. Hydrometeoizdat, Leningrad, 279 pp. [in Russian].

Kondratyev K.Ya. , Bondarenko V.G., and Khvorostyanov V.I. (1992a). A three-dimensional numerical model of cloud formation and aerosol transport in an orographically inhomogeneous atmospheric boundary layer. *Boundary-Layer Meteorology*, **61**(3), 265–285.

Kondratyev K.Ya., Melentyev V.V., and Nazarkin V.A. (1992b). *Remote Sensing of Water Areas and Water Heads (Microwave Methods)*. Hydrometeoizdat, St. Petersburg, 248 pp. [in Russian].

Kondratyev K.Ya., Johannessen O.M., and Melentyev V.V. (1996). *High Latitude Climate and Remote Sensing*. Wiley/Praxis, Chichester, U.K., 200 pp.

Kondratyev, K.Ya., Moreno Pena F., and Galindo I. (1997). *Sustainable Development and Population Dynamics*. Universidad de Colima, Mexico, 128 pp.

Kondratyev K.Ya., Demirchian K.S., Baliunas S., Adamenko V.N., Bohmer-Christiansen S., Idso Sh.B., Postmentier E.S., and Soon W. (2001). *Global Climate Changes: Conceptual Aspects*. St. Petersburg Scientific Center of RAS, St. Petersburg, 125 pp. [in Russian].

Kondratyev K.Ya., Krapivin V.F., and Phillips G.W. (2002a). *Global Environnmental Change: Modelling and Monitoring*. Springer, Berlin, 319 pp.

Kondratyev K.Ya., Krapivin V.F., and Phillips G.V. (2002b). *Problems of High-latitude Environmental Pollution*. St. Petersburg State University, St. Petersburg, 280 pp. [in Russian].

Kondratyev K.Ya., Grigoryev Al.A., and Varotsos, C.A. (2002c). *Environmental Disasters: Anthropogenic and Natural*. Springer/Praxis. Chichester, U.K., 484 pp.

Kondratyev K.Ya., Krapivin V.F., and Savinykh V.P. (2003a). *Perspectives of Civilization Development: Multidimensional Analysis*. Logos, Moscow, 546 pp. [in Russian].

Kondratyev K.Ya., Losev K.S., Ananicheva M.D., and Chesnokova I.V. (2003b). *Natural Science Fundamentals of Life Stability*. ARISTI, Moscow, 240 pp. [in Russian].

Kondratyev K.Ya., Losev K.S., Ananicheva M.D., and Chesnokova I.V. (2003c). *Stability of Life on Earth: Principal Subject of Scientific Research in the 21st Century*. Springer/Praxis, Chichester, U.K., 152 pp.

Kondratyev K.Ya., Krapivin V.F., Savinykh V.P., and Varotsos C.A. (2004). *Global Ecodynamics: A Multidimensional Analysis*. Springer/Praxis, Chichester, U.K., 658 pp.

Kondratyev K.Ya., Ivlev L.S., Krapivin V.F., and Varotsos C.A. (2005a). *Atmospheric Aerosol Properties: Formation, Processes and Impacts*. Springer/Praxis, Chichester, U.K., 572 pp.

Kondratyev K.Ya., Krapivin V.F., Lakasa H., and Savinikh V.P. (2005b). *Globalization and Sustainable Development: Ecological Aspects*. Science, St. Petersburg, 240 pp. [in Russian].

Kondratyev K.Ya., Krapivin V.F., and Varotsos C.A. (2006). *Natural Disasters as Components of Ecodynamics*. Springer/Praxis, Chichester, U.K., 625 pp.

Krapivin V.F. and Kondratyev K.Ya. (2002). *Global Changes of the Environment*. St. Petersburg University, St. Petersburg, 724 pp. [in Russian].

Marchuk G.I. and Kondratyev K.Ya. (1992). *Priorities in Global Ecology*. Science, Moscow, 264 pp. [in Russian].

Marchuk G.I., Kondratyev K.Ya., Kozoderov V.V., and Khvorostyanov V.I. (1986). *Clouds and Climate*. Hydrometeoizdat, Leningrad, 512 pp. [in Russian].

Pozdnyakov D.V., Kondratyev K.Ya., and Petterson L.H. (2002). Earth observation and remote sensing. *Boundary-Layer Meteorology*, **105**, 384–409.

Rycroft M.J. (ed.) (1977). *Proceedings of the 18th Plenary Meeting of COSPAR, May 31–June 7, 1975, Varna, Bulgaria*. Pergamon Press, London, 1,097 pp.

WMO (1998). The Bulletin interviews Professor K.Ya. Kondratyev. *WMO Bulletin*, **47**(1), January.

Wood R.W. (1934). *Physical Optics*, third edition. MacMillan, New York, 259 pp.

2

Kirill Kondratyev and the IPCC: His opposition to the Kyoto Protocol

Sonja A. Boehmer-Christiansen and Arthur P. Cracknell

2.1 INTRODUCTION

In Chapter 1 we outlined the scientific work of Kirill Kondratyev in the fields of atmospheric physics, meteorology, and the pioneering of remote-sensing methods in these sciences, work which occupied a period of nearly 50 years from the mid-1940s. It involved considerable international cooperation and led to widespread international recognition (see Section 1.7). However, in the last 15–20 years when he was no longer involved in front-line fundamental scientific research, he turned his attention to ecology, climate change, and global change. His research work at the St. Petersburg Scientific and Research Center of Ecological Safety and at the Nansen International Environment and Remote Sensing Center was then concerned with environmental problems in general, and especially those that might be arising from human activities. In this chapter we examine Kondratyev's relationship with the Intergovernmental Panel on Climate Change (the IPCC) and the Kyoto Protocol. The operation of the IPCC was established in 1988 by the WMO (the World Meteorological Organization) and UNEP (the United Nations Environment Program), with the help of ICSU and many other research bodies. By the late 1980s Kondratyev had achieved widespread international recognition, including by the WMO, for his scientific work. Therefore, one might have supposed that he would play a leading role in the IPCC, but that was not so. There is, of course, a large amount of literature published by and about the IPCC. We shall not attempt to give a general discussion of the IPCC and its work, but will confine ourselves to examining Kondratyev's relationship with and his views of the IPCC, of which he was an intelligent and informed critic. We shall consider the reasons for this in Section 2.3. The Kyoto Protocol is an agreement made under the United Nations Framework Convention on Climate Change (UNFCCC); this is an international environmental treaty which was produced at the United Nations Conference on Environment and Development (UNCED) in Rio de Janeiro in 1992. The stated objective is "to achieve stabilization of greenhouse gas concentrations in the atmosphere at a low enough level to prevent dangerous anthropogenic interference with the climate system." However, the treaty itself sets no mandatory limits on greenhouse gas emissions for individual nations; limits, enforcement conditions, and penalties are provided for in updates, of which the principal update is the Kyoto Protocol. We shall discuss this in Section 2.4. Kondratyev, as we shall see, was highly critical of much of the work of the IPCC, of what is generally pronounced to be its scientific consensus, and therefore of the Kyoto Protocol. It may be of interest, however, to stress at the outset that he was not a critic of the global emission reduction effort from an "anti-environmentalist" perspective but from the deeper "green" or Gaia side. For him the postulated enhanced global warming due to increasing greenhouse gas emissions as a result of human activities remained an unproven hypothesis and was in any case not the most serious to human life on Earth (Kondratyev *et al.*, 2004).

2.2 KONDRATYEV'S LIFE FROM CIRCA 1990 TO 2006 AND HIS INVOLVEMENT WITH CLIMATE SKEPTICS

2.2.1 The last 15–20 years of Kondratyev's life

In the last decades of his life Kondratayev turned his attention to ecology, a scientific field, as well as "climate change" and "global change" which are essentially research agenda to which a large range of environmental policy prescriptions have become attached, ranging from anti-industrial policies to technological innovation, from the decarbonization of energy supplies to major efforts at afforestation, waste reduction, recycling, reduced consumption, and even population control. On many of these topics he developed productive cooperation with colleagues from various countries. His prolonged visits to Germany (Max-Planck Institute for Meteorology of the University of Hamburg), Greece (Athens University), and Mexico (University of Colima) resulted in the completion of a number of monographs (Kondratyev and Galindo, 1997, 2001; Kondratyev and Grassl, 1993; Kondratyev and Johannessen, 1993; Kondratyev and Varotsos, 2000; Kondratyev *et al.*, 2002a, b) An especially intensive international collaboration was developed with the University of Athens via INTAS (the International Association for the Promotion of Co-operation with Scientists from the New Independent States or NIS of the former Soviet Union) and EUROTRAC (the EUREKA Project on the Transport and Chemical Transformation of Environmentally Relevant Trace Constituents in the Troposphere over Europe) programs on problems of stratospheric and tropospheric ozone, including UV-B variability and relevant biological impacts.

A 6-month stay at the Center for Climate System Research (CCSR) of the University of Tokyo provided an opportunity for productive cooperation with some Japanese scientists; in collaboration with Prof. A. Sumi and Prof. T. Nakajima a detailed survey on global climate change problems was completed. He also (with Dr. T. Tanaka) produced a paper on perspectives of remote sensing in Japan in connection with the development of the ADEOS-II remote-sensing satellite which is useful for studying water quality problems, stratospheric and tropospheric ozone dynamics, atmospheric aerosols, and sea ice. A similar, but more general, survey on priorities in global change and development of remote sensing in Japan was prepared with Prof. Sumi, Prof. Nakajima, and Dr.Tanaka.

A significant part of Kondratyev's international work consisted of participation in editorial boards of a number of scientific journals, including *Geofisica Pura e Applicata*, *Zeitschrift für Meteorologie*, *Climatic Change*, *Energy and Environment*, the *International Journal of Climatology*, *Boundary-Layer Meteorology*, and *Idöjárás*. As Editor-in-Chief of the Russian journal *Studying the Earth from Space* (this journal was published in English in the U.S. under the title *Earth Observation and Remote Sensing*), he invited a number of scientists from other countries to participate on the editorial board and stimulated the publication of papers by foreign scientists. The journal *Energy and Environment*, in which some of Prof Kondratyev's later papers were published, is an interdisciplinary journal which began publication in 1990. Since 1995 the editor has been Dr. S.A. Boehmer-Christiansen, and by which date its IPCC critical perspective had been established. Kondratyev joined the Editorial Board because it lacked an atmospheric scientist who would offer advice on the controversial science of global warming with its huge implications for the energy industries and policy.

When Dr. Boehmer-Christiansen first met him in 2001, Kirill Kondratyev was in his early seventies: upright, clean shaven, and in excellent physical condition, still very handsome. They were involved in a common cause, the reasoned opposition to the Kyoto Protocol and Russia's signature of it, which brought them into closer contact and which will be discussed in Section 2.4.

One cannot write about Kirill Konratyev in his later years without mentioning his wife, Svetlana, for he could not have been as active and prolific as he was without her constant help, care, and attention. Having visitors would not stop him from working, for Svetlana, his much younger second wife and a former scientist with fluent English, did most of the entertaining and caring. He was indeed fortunate to have a helpmate in his research and writing, while also remaining embedded in the wider life of St. Petersburg and his Institute. In private he had become a lone scholar who had turned his very active mind away from fundamental research on atmospheric physics towards encouraging younger researchers and synthesizing the available knowledge of ecology, humanity, and of understanding the Earth, in order to manage it sustainably. His publications on this subject are numerous, both in English and Russian. He was however generally reticent about discussing his political views, at least with foreigners.

Visiting the Kondratyev's flat in St. Petersburg (Korpusnaya Street) and also their dacha in the forest northwest of St. Petersburg, one could not fail to notice that both places were meant for reading and writing, filled with books, journals (very many in English and from academic and U.N. sources all over the world) and of course there was a computer. The flat in St Petersburg was part of a large apartment block close to Leningrad University built especially for academics during the Stalin era and facing the Baltic Sea. The flat was tiny for a former rector of Leningrad

University (Vice-Chancellors and Presidents in Western countries expect to live in mansions), a hero of the last war (his medals did help with obtaining transport), and renowned academician; it comprised three rooms plus a small bathroom, an enclosed balcony, small kitchen, and a spacious hall mainly used for exhibiting his many books. The Kondratyevs nevertheless enjoyed considerably more space than the average citizen, as Sonja Boehmer-Christiansen learnt from several visits to artist friends. The dacha at Kammarov was within walking distance from the railway line to Finland (the Kondratyevs did not own a car); it too was small compared with the new mansions and large houses of party officials and the nouveau riche. It was one of a number of small terrace houses especially built for academicians by Stalin soon after the war. Kondratyev worked both in his flat and when at his dacha. He did so largely alone with few breaks during the day, and certainly without any vast number of students doing the work for him, as one American colleague had alleged when trying to explain his prolific output. For about 15 years after his official retirement he had worked largely from home and from his small office at the Institute for Ecological Security, with some secretarial help, and that of Svetlana.

Dr. Sonja Boehmer-Christiansen met Kirill Kondratayev four times before his death in May 2006. In the summer of 2001 she visited his Institute in St. Petersburg to make a presentation on the politics of climate change. One year later she accompanied him to a meeting he had organized in Rostov-on-Don ("Round Table: Global Environmental Dynamics Now and in 21st Century", Chairman Thor Heyerdahl, May 2001, Rostov-on-Don, Russia), and in 2003 they both attended the Third World Climate Change Conference in Moscow (September 29–October 3). She last met Kirill Kondratyev in the spring of 2005 on a research trip to Moscow to explore, rather unsuccessfully because of bad timing, Russia's climate policy. In Rostov-on-Don there was ample evidence of the decline of infrastructure and of industrial activity and also the opportunity to learn from younger Russian environmental scientists and economists who all, at that time at least, bemoaned the decline of Russian research and their growing dependence on funding from abroad, or even going abroad to find work. Obtaining grants for any research at all had become the overriding issue, and at that time environmental research money came mainly from collaboration with the EU or North America. In Moscow in September 2003 at the World Climate Change Conference in Moscow, Dr Boehmer-Christiansen and Prof. Kondratyev faced a press conference together after Prof. Kondratyev had addressed a large crowd of scientists in front of President Putin. He warned against taking precipitate action against fossil fuels because of the lack of evidence for man-made climate change, pointed to serious uncertainties, and encouraged the assembled scientists to read his books. Mankind would have to work much harder to understand ecological damage and then regulate itself according to ecological principles and targets defined by research. The biosphere needed protection rather than emissions reduction! The evolution of the Russian attitude to the Kyoto Protocol will be discussed in Section 2.4. He was obviously a grand old man among Russian scientists, highly respected including by a considerable number of people from the West, some of whom had made their peace with Working Group 1 of the Intergovernmental Panel on Climate Change (IPCC) either because they believed its scientific consensus, or because public opposition would have endangered their funding and cordial relations with national governments.

Judging by his books and articles, rather than from discussion, Kondratyev appeared to be a serious critic of the prevailing environmental policies advocated at this stage by Western governments at the U.N., including the ideas that global warming was actually taking place (some areas of Siberia appeared not to be warming, see Chapter 12), that it would be dangerous to humanity, and that it could be attributed mainly to the emission of greenhouse gases into the atmosphere. The policies advocated by the U.N. and hence major governments in the West, were not, he argued, scientific enough or directed to the main issues, which included (for him) ecological damage, pollution, depletion of resources, overpopulation, etc. He was a man with a deep belief in ecological principles and the power of science to shape human behavior in a top-down fashion. He hoped that the U.N., advised by scientists from many countries, would and could decide in the interest of all humanity. Advocating this with much passion and learning, as well as intellectual consistency, meant that it seemed to some that Kondratyev paid too little attention to the realities of politics and economics, and especially to the deep divisions of humanity.

2.2.2 The journal *Energy and Environment*

A particular objective of *Energy and Environment* is to cover the social, economic, and political dimensions of issues relating environment to energy at the local, national, and international level. Papers are published in it that cover energy-related aspects of wider environmental questions, such as the use of fuel wood and the impacts of de-forestation. A major aim of *Energy and Environment* is to act as a forum for constructive and professional debate between scientists and technologists, social scientists, and economists from academia, government, and the energy industries on energy and environment issues in both a national and international context. Particular attention is given to ways of resolving conflict in the energy and environment field. This journal has by now something of a reputation for attracting critical views of the policies adopted by assorted governments, especially of policies that are justified with reference primarily to the alleged scientific consensus of the IPCC.

Two major papers by Kondratyev were published in *Energy and Environment*, one on "Key issues of global change at the end of the second millennium" (Kondratyev, 1997) and the other on "Key aspects of global climate change" (Kondratyev, 2004). They might have been published in larger, better known journals had these accepted his skeptical views of what was presented increasingly by the IPCC as a consensus. A third paper on "Uncertainties of the global climate change observations and numerical modeling" is an extension of "Key aspects of global climate change" and is, as yet, unpublished.

2.3 KONDRATYEV AND THE IPCC

As we noted at the beginning of this chapter, the Intergovernmental Panel on Climate Change was set up in 1988 by the WMO, the World Meteorological Oganization, and

UNEP, the United Nations Environment Progam. Its original purpose was to prepare a report for the Second World Climate Conference (October 20–November 7, 1990, held in Geneva, Switzerland). The report was to characterize the current understanding of the observed regularities of climate change and possibly to forecast the climate, its impact on the environment and human activity, as well as the economic actions needed to prevent undesirable climate changes. To prepare the report for the conference three working groups were formed. According to the IPCC's website (*http://www.ipcc.ch*), the present roles of these working groups are:

- "IPCC Working Group I (WG1) assesses the physical scientific aspects of the climate system and climate change.
- IPCC Working Group II assesses the vulnerability of socio-economic and natural systems to climate change, the negative and positive consequences of climate change, and options for adapting to it.
- IPCC WG3 assesses options for mitigating climate change through limiting or preventing greenhouse gas emissions and enhancing activities that remove them from the atmosphere."

There is now also a Task Force: the *Task Force on National Greenhouse Gas Inventories* is responsible for the IPCC National Greenhouse Gas Inventories Program. The initial task of IPCC Working Group 1 was to look at the various climate models and their computer outputs, and in 1990 this Working Group produced its first report (Houghton *et al.*, 1990). There were conclusions relating to temperature, precipitation, soil moisture, snow, and ice. After the Geneva conference the IPCC continued its work, and now the main activity of the IPCC is to provide at regular intervals an assessment of the state of knowledge on climate change.

In a succession of reports over the period since 1990 the IPCC has come more and more firmly to the view that human activities are contributing significantly to global warming, and the Fourth Assessment Report (IPCC, 2007) says that "most of the observed increase in global average temperatures since the mid-20th century is *very likely* due to the observed increase in anthropogenic greenhouse gas concentrations." In December 2007 the IPCC shared the Nobel Peace Prize with the former American Vice-President Al Gore. A whole double issue of the journal *Energy and Environment* (Volume 18, Nos. 7/8, December 2007) was devoted to "The IPCC: Structure, Process and Politics", and the first article in that issue gives a particularly good account of the IPCC and also summarizes both sides of the arguments of those who support the IPCC and those who are against it (Zillman, 2007).

Kondratyev was out of sympathy with much of the work of the IPCC, especially its heavy reliance on computer climate models for predicting future climates, while neglecting several factors in the modeling. It is worth noting that Russian scientists at that time did not have access to large powerful computers (only the Russian military had that), and so they could not participate in the computer modeling experiments.

Fairly soon Kondratyev was largely marginalized by many of the leading figures in the IPCC. Why did this occur? He was no longer involved directly and personally in front-line experimental or theoretical research. But this cannot have been the main reason; there were several people in very senior positions in the IPCC who were also not themselves conducting front-line research in person. We believe that there are three reasons there seemed to be such a wide gulf between Kondratyev and the IPCC.

First, he claimed that there were various processes, etc. that were not included in the computer climate models on which the IPCC relies so much. It is only recently that his views on this point are attracting more attention. More scientists outside the modeling "community" are beginning to realize that while the models can deal quite successfully with gradual change they are not able to predict abrupt changes that suddenly occur. For instance, changes in feedback due to the changes in albedo arising from the collapse of an Antarctic ice sheet can be accommodated after the event, but such a collapse would not be predicted by the models. We shall return to this question in Section 18.4.2.

Second, Kondratayev strongly believed that global warming induced by excessive carbon dioxide production by the burning of fossil fuels is only one, and possibly even only a minor one, of a large number of serious threats facing humanity. These threats may come from overpopulation, pollution of the atmosphere, pollution of water sources, contamination and degradation of the land, damage to the biosphere and the extinction of many species, the depletion of fossil fuel sources, the depletion of non-fuel mineral resources, the destruction of stratospheric ozone, etc., etc. (Kondratyev *et al.* 2004). It can seriously be argued that the success of the IPCC in making people generally aware of the threat of global warming induced by the burning of fossil fuels has led to these many other threats being largely ignored at various levels of policy-making and human behavior. Kondratyev labored to bring to people's attention the whole question of global change in general and its threats to human life. Kondratyev's warnings in this area still go largely unheeded.

The third point of dispute with the IPCC was that Kondratyev was skeptical about the interpretation of the experimental evidence that was adduced for global warming. Was it selected to confirm a hypothesis already assumed as true for political reasons? He was rightly cautious, as many other people have been. A discussion of the arguments against the scientific basis of the IPCC's conclusions is given by Singer (2008).

Kondratyev developed and expounded his views on global change and the threats to our way of life in various monographs and these are included in the list of selected publications at the end of Chapter 1. To understand his views on the IPCC, it is best to consider those of his writings that were more specifically concerned with the IPCC and its dependence on computer models and its use of climate-related observations. These writing are mostly to be found in some articles of his that were published in the journal *Energy and Environment*, which we have already mentioned in the previous section.

We must consider Kondratyev's position with respect to global warming, the IPCC, and the Kyoto Protocol from his own writing. Following the Second World Climate Conference in Geneva in 1990 the United Nations Conference on Environment and Development (UNCED), informally known as the Earth Summit, was held in Rio de Janeiro in 1992. The stated objective was "to achieve stabilization of greenhouse gas concentrations in the atmosphere at a low enough level to prevent dangerous anthropogenic interference with the climate system." This is vague, it does not say how stabilization is to be achieved, nor does it define what is meant by the "low enough level" to which it refers. This conference produced the United Nations Framework Convention on Climate Change (UNFCCC), which came into force in 1994 but made virtually no demands on any country. As far as emission reduction policies were concerned, only three countries did reduce their emissions: the U.K. by a massive switching from coal to gas, Germany through re-unification and the collapse of the East German energy demand, and Russia as a result of its de-industrialization following the collapse of communism.

The UNFCCC was criticized, quite forcefully, by Kondratyev (1997) for concentrating so much on greenhouse gas emissions. In his paper on "Key issues in global change" (Kondratyev, 1997) he wrote,

> "The most discussed problem is global warming—it is more appropriate to call it global climate change—and ... specifically the growth of greenhouse gas emissions into the atmosphere. World carbon emissions from fossil fuel burning are still growing although some countries have undertaken certain measures to reduce emissions."

He then went on to study in detail the carbon dioxide emissions of various countries. He thought that people were devoting far too much of their attention to the increase of carbon dioxide in the atmosphere and a predicted catastrophic scenario of global warming. But we know that the biosphere assimilates a great deal of carbon dioxide emitted in the atmosphere and helps to guarantee future ecological safety. "If we destroy the biosphere which functions as a sink for carbon, we create an ecological catastrophe ...," he said in the interview with the WMO (1998). However, he stressed that carbon dioxide emissions and global warming are not the only problem, or even the most serious problem facing the future of mankind.

> "Undoubtedly, one of the most worrying features of the present time is the continuing growth of the global population. Two specific features of this growth have been the concentration in developing countries and the growth of urban populations ..." (WMO, 1998).
>
> "An important question in this context is the adequacy of the UNFCCC recommendation to reduce greenhouse gas emissions. On the one hand, it is obvious that, generally speaking, the reduction of greenhouse gas emissions is a very useful measure. But, on the other hand, it is equally clear that such a measure is not a panacea against global change dangers" (Kondratyev, 1997).
>
> "The problem of global change cannot be solved without using a system's approach comprising all processes involved. Studying carbon dioxide or ozone in isolation will serve little purpose. Such studies should be made in the context of the overall problem" (WMO, 1998).
>
> "As far as global change science is concerned, it is important to recognize that present-day numerical climate modelling (even in the case of 3-D coupled global models) remains far from being able to reliably simulate real climate change and,

> consequently, to identify the contributions of various climate-forming factors, including the enhanced greenhouse effect. Though it is well known that climate change results from interaction between all components of the climate system, the relative influence of various factors cannot be defined precisely and 'new' influences are still being added to the climate equation" (Kondratyev, 1997).

To be specific, aerosols are highly variable, both spatially and temporally, and it is very difficult to build their effect reliably into the models; Kondratyev had himself done a lot of work earlier on atmospheric aerosols and this work was persistently ignored. He continued,

> "As far as climate change is concerned, the key task must be to study climate in all its complexity without an overemphasis on certain individual factors such as the greenhouse effect. But it is also necessary to identify the place and the role of climate change within the more general framework of global change."

He argued that it had been shown by Gorshkov (1995) that the basic processes which regulate environmental dynamics are founded on the principle of the biotic regulation of the environment. If we accept such a concept then the priority order given in Table 2.1 was suggested as a basis for further discussion. This preliminary scheme of priorities demonstrates a subordinate role for climate change within a much more general framework of concern about global change; we shall return to this question in Chapter 18.

2.4 KONDRATYEV AND THE KYOTO PROTOCOL, INCLUDING RUSSIA'S SIGNING OF THE PROTOCOL

From 1750 till now the CO_2 concentration in the atmosphere has increased by a little over one-third, reaching the highest level for the last 420,000 years (and, probably, for the last 20 million years), which is illustrated by the data of ice cores (IPCC, 2001). About two-thirds of the growth of CO_2 concentration in recent years is explained by emissions to the atmosphere from fossil fuel burning and the remaining one-third is due to deforestation and cement manufacture. It is of interest that by the end of 1999, CO_2 emissions in the U.S.A. exceeded the 1990 level by 12%, and by 2008 their further increase should raise this value by 10% more (Victor *et al.*, 1998). Meanwhile, according to the Kyoto Protocol, emissions should be reduced by 7% by the year 2008 with respect to the 1990 level which requires their total reduction by about 25% which is of course utterly unfeasible. According to the IPCC (2001), the probable levels of CO_2 concentration by the end of the century will range from 540 ppm to 970 ppm (pre-industrial and present values are, respectively, 280 ppm and 385 ppm).

As mentioned at the beginning of this chapter, the Kyoto Protocol was the first attempt to implement the stabilization of greenhouse gas emissions referred to in the UNFCCC. Once a sufficient number of countries had ratified the UNFCCC and it

Table 2.1. Priorities (from Kondratyev, 1997).

#1. Biotic regulation of environmental dynamics #1.1 Biosphere dynamics: biogeochemical cycles; use of biospheric resources (thresholds); monitoring of both terrestrial and marine biosphere dynamics (relevant indicators). #1.2 Consumption of biospheric resources: use of renewable and non-renewable resources; developed and developing countries; countries with a transitional economy. #1.3 Life standards: "Golden Billion" and the rest of the world; sustainability: ecological, socio-economic, political, and ethical. #1.4 Carrying capacities on natural, regional, and global levels for various components of nature: land, forests, seas, inland waters, etc. #1.5 Sustainable development and population dynamics.
#2. Water Drinking water deficit.
#3. Energy Fossil fuels vs. renewable sources (wind, tides, etc.); future of nuclear and hydrogen energy; coupled energy production development and environmental dynamics.
#4. Food Impacts of environmental dynamics.
#5. Environment Key issues for life support: #5.1 Climate change (internal variability, external impacts: anthropogenic contribution, greenhouse gases and aerosols, volcanic eruptions, solar activity). #5.2 Stratospheric ozone depletion. #5.3 Tropospheric ozone increase (UV-B enhancement and subsequent impacts on humankind and ecosystems). #5.4 Polluted urban atmospheres and health. #5.5 Environmental economics (ecological taxes, etc.). #5.6 Environmental ethics.

therefore came into force in 1994, there have been annual Conferences of the Parties (COPs). In December 1997 in Kyoto (Japan) the third Conference of the Representatives, COP-3, of the countries that had signed the UNFCCC (over 160) met and engaged in lengthy and hot debates on the need to recommend a 5% CO_2 emissions reduction by 2008–2012 for industrially developed countries (relative to the 1990 level). It was at this conference that the Kyoto Protocol was adopted. However, before it could become legally binding it had to be ratified by a required number of countries, and there was a considerable time lapse before that occurred.

The text of the Kyoto Protocol can conveniently be found in the book by Grubb *et al.* (1999). Updating information is always available, for instance, from the Wikipedia website (*http://en.wikipedia.org/wiki/Kyoto_Protocol*). The following summary is adapted from the article in Wikipedia:

- The Kyoto Protocol is underwritten by governments and is governed by international law enacted under the aegis of the United Nations.
- Governments are separated into two general categories: developed countries, referred to as Annex I countries (which have accepted greenhouse gas emission reduction obligations and must submit an annual greenhouse gas inventory); and developing countries, referred to as Non-Annex I countries (who have no greenhouse gas emission reduction obligations but may participate in the Clean Development Mechanism).
- Any Annex I country that fails to meet its Kyoto obligation will be penalized by having to submit emission allowances in a second commitment period for every ton of greenhouse gas emissions they exceed their cap in the first commitment period (i.e., 2008–2012).
- By 2008–2012, Annex I countries have to reduce their greenhouse gas emissions by a collective average of 5% below their 1990 levels (for many countries, such as the European Union member states, this corresponds to some 15% below their expected greenhouse gas emissions in 2008). While the average emissions reduction is 5%, national limitations range from an 8% average reduction across the European Union to a 10% emissions increase for Iceland; but since the European Union's member states each have individual obligations, much larger increases (up to 27%) are allowed for some of the less developed European Union countries. Reduction limitations expire in 2013.
- Kyoto includes "flexible mechanisms" which allow Annex I economies to meet their greenhouse gas emission limitation by purchasing greenhouse gas emission reductions from elsewhere. These can be bought either from financial exchanges, from projects which reduce emissions in non-Annex I economies under the Clean Development Mechanism, from other Annex 1 countries under Joint Implementation (see below), or from Annex I countries with excess allowances. Only Clean Development Mechanism Executive Board-accredited Certified Emission Reductions can be bought and sold in this manner. Under the aegis of the United Nations, the Bonn-based Clean Development Mechanism Executive Board was established to assess and approve projects (CDM Projects) in Non-Annex I economies prior to awarding Certified Emission Reductions. (A similar scheme called the Joint Implementation scheme applies in transitional economies mainly covering the former Soviet Union and Eastern Europe.)

Given that the objective of the UNFCCC is "to achieve stabilisation of greenhouse gas concentrations in the atmosphere ..." (see above) the Kyoto Protocol is a step in that direction. But the controversy did not end with the conference in Kyoto. Opposition to the Kyoto Protocol has come from various directions. There is the position of the developing countries. Naturally, the position of the developing countries gives primary consideration to socio-economic development, including the overcoming of poverty and its consequences. They argued, not unreasonably, that it is the industrialized countries which have caused most of the human-induced global warming so far, and that their own development or progress towards industrialization should not be held back because of a problem that they have not themselves created. Developing countries are not prepared to accept greenhouse gas emissions reduction; their point of view was respected and they were not required by the Kyoto Protocol to accept reductions in their emissions.

Opposition to the Kyoto Protocol has come from some people who see it as an attempt to reduce the growth of the world's industrial economies. The former Australian Prime Minister, John Howard, refused to ratify the Kyoto Protocol on the grounds that it would curtail development and cost Australian jobs; his successor, Kevin Rudd, ratified the Kyoto Protocol in December 2007. U.S. President G.W. Bush rejected the Kyoto Protocol because: (1) ostensibly this document lacks scientific substantiation; (2) its adoption would cause serious economic damage to the U.S.A. (whose energy supply is based mainly on the use of hydrocarbon fuels) without providing any marked positive impact on the environment. Of these two reasons, it is fairly clear that the second one, which is naked self-interest on the part of the U.S.A, was the dominant reason.

To come into force the Kyoto Protocol needed to be ratified by countries responsible for at least 55% of global carbon dioxide emissions. Since the U.S.A. had refused to ratify the Protocol this minimum could only be achieved if Russia decided to ratify it. A problem arises from the choice of 1990 as the baseline for calculating reductions of carbon dioxide emissions. In 1990 the former Soviet Union had done little to raise its energy efficiency; shortly after that came the collapse of communism and the downturn in the economy and a consequent reduction in energy consumption and greenhouse gas emissions. On the other hand, Japan, as a net importer of oil and other raw materials, had become very energy-efficient by 1990. Such factors were ignored and the subsequent inactivity of the former Soviet Union, following the collapse of communism, meant that it could then look forward to generating an income by trading its surplus emissions allowance. This did not prevent many Russians from seeing the Kyoto Protocol as an attempt to hold back regeneration of their economy. At the Moscow World Conference on Climate Change (September 29–October 3, 2003) the Kyoto Protocol was attacked on two fronts that were rather similar to President Bush's points. President Putin's economic adviser, Andrei Illarionov, said that ratification would stall Russia's economic growth, it would "doom Russia to poverty, weakness, and backwardness." The Kyoto Protocol calls for countries to reduce their level of greenhouse gas emissions by certain amounts which are specified individually for the various countries. If a country exceeds the emissions level, it could be forced to cut back industrial production. This would be likely to conflict with President Putin's goal of doubling Russia's gross domestic product by 2010. The economic concerns were supported at the Moscow Conference by several top Russian climate scientists, including Kondratyev. His long paper, on "Key aspects of global climate change", was submitted just prior to the World Climate Change Conference in Moscow in 2003 and was published in the following year (Kondratyev, 2004). This paper defines his almost entirely scientific objections to climate models and the Kyoto Protocol and also demonstrates the aim of his work during the last years of his life. President Putin told the conference that his Cabinet had not yet decided whether or not Russia would ratify the Protocol. It appears (Walker and King, 2008) that, at the suggestion of the U.K., a deal was proposed in which the European Union would support Russia in its quest to join the World Trade Organisation (WTO) in return for Russia ratifying the Kyoto Protocol. In November 2004 President Putin ratified the Kyoto Protocol. Finally, it came into force in February 2005, following its ratification by Russia. Although it was adopted nearly seven years before that, the Kyoto Protocol had until then remained a statement of intent, rather than a legally binding document. Once Russia had signed the Protocol, it then became a legally binding document on the signatories. Countries which failed to meet the target cuts in carbon dioxide emissions would face penalties and have to cut back on their production. Thus, eventually the U.S. failure to ratify the Kyoto Protocol has not prevented its adoption, with the requirements to reduce greenhouse gas emissions.

Kondratyev's second article published in *Energy and Environment* (Kondratyev, 2004) was, as we have already mentioned, prepared in anticipation of the Moscow World Climate Change Conference (September 29–October 3, 2003). This is a lengthy article, and it is not possible to recount here all the detail it contains. He was concerned with the question of whether the Kyoto Protocol should be considered as a scientifically justified document:

> "Confusion reigns and is caused, in particular, by the lack of sufficiently clear and agreed terminology. Ignoring the very complicated notion of climate itself (which needs a separate discussion), one should remember, for instance, that in the UNFCCC climate change was defined as being anthropogenically induced. One of the main unsolved problems is the absence of convincing quantitative estimates of the contribution of anthropogenic factors to the formation of global climate, though there can be no doubt that anthropogenic forcings of climate do exist."

Some international documents containing analyses of the present ideas of climate refer to the prevalent idea of a consensus with respect to scientific conclusions as enshrined in these documents. This wrongly assumes that the development of science is determined not over time by different views and relevant debates and discussions, but by a general agreement and even voting. Apart from the question of definitions, the issue of uncertain conceptual estimates concerning various aspects of climate problems remains of importance. In particular, this refers to the main conclusion in the summary of IPCC (2001) which claims that " ... An increasing body of observations gives a collective picture of a warming world and most of the observed warming over the last fifty years is likely to have been due to human activities."

The Earth's climate system has indeed changed markedly since the Industrial Revolution, with some changes being of anthropogenic origin. The consequences of climate change do present a serious challenge to the policy-makers responsible for the environmental ("ecological" in Russian) policy, and this alone makes the acquisition of objective information on climate change, of its impact and possible response, most urgent.

The IPCC had, by the time of the Moscow World Climate Change Conference in 2003, prepared three detailed reports (in 1990, 1996, and 2001), as well as several special reports and technical papers. Griggs and Noguer (2001) made a brief review of the first volume of the IPCC Third Assessment Report (IPCC, 2001) prepared by WG1 for the period June 1998–January 2001 with the participation of 122 leading authors and 515 experts, each with their materials. Four hundred and twenty experts reviewed the first volume and 23 experts edited it. Moreover, several hundred reviewers and representatives of many governments made additional remarks. With the participation of delegates from 99 countries and 50 scientists recommended by the leading authors, the final discussion of the Third Assessment Report was held in Shanghai on January 17–20, 2001. The "Summary for decision-makers" was approved after a detailed discussion by 59 specialists.

Kondratyev (2004) continued with a discussion of the political challenge and ten questions raised by Prof. A.N. Illarionov, Economic Adviser to President Putin, at the Moscow Conference. He then argued that "the main cause of contradictions in studies of the present climate and its changes is the inadequacy of the available observational databases." He cited in particular surface air temperature, ground surface temperature, the extent of snow and ice cover, sea level and the heat content of the upper layer of the oceans, precipitation, He also alluded to abrupt changes in the climate and the fact that the models do not predict such events. We shall discuss this further in Section 2.5. The final section of the paper (Kondratyev, 2004) deals with the results of numerical climate modeling and their reliability; hopefully, their reliability has improved since that paper was written.

Carbon dioxide is, of course, not the only greenhouse gas. The other major greenhouse "gas" is water vapor and the whole question of anthropogenic effects on the hydrological cycle, atmospheric water vapor, and cloud patterns is very difficult to study. There are also many other greenhouse gases: CH_4, various oxides of nitrogen (collectively referred to as NO_x), H_2S, SO_2, SF_6, DMS (dimethyl sulphide, $(CH_3)_2S$), CFCs (chlorofuorocarbons), etc., some of which occur naturally and some of which are of anthropogenic origin. Climate models are usually run on the basis of taking these gases into account by considering their carbon dioxide equivalent, in terms of global warming, and adding it to the actual predicted concentration of carbon dioxide itself. The Kyoto Protocol, however, appears only to concern itself with carbon dioxide emissions and makes no reference to any attempt to restrict the emissions of these other gases.

Comparisons are sometimes made between the Kyoto Protocol and the Montreal Protocol. The Montreal Protocol came about as a result of the scientific evidence for human-induced depletion of the ozone layer, and especially the famous "ozone hole" which appears in the Antarctic each spring. This was rapidly accepted to be a result of the escape of CFCs (chlorofluorocarbons) into the atmosphere. The world's leaders came together and in the Montreal Protocol agreed to phase out the production of CFCs and to replace them by other "ozone-friendly" substances. The reasons for the relative success of the Montreal Protocol are neatly summarized in box 21.2 of the Stern (2007) Review. Twenty-four countries signed the original Protocol in 1987, and by October 2006 74 countries had ratified the Protocol and this included the major developing countries. Emissions of CFCs have largely been brought under control, but of course the ozone layer will not recover immediately; it is expected to take up to 100 years to do so.

There were several factors which contributed to the success of the Montreal Protocol. First, there was a high degree of scientific consensus and evidence that there was a problem that required urgent political action, and public opinion galvanized politicians. The Protocol used expert advice to establish targets and timetables to phase out the use of ozone-depleting chemicals, based on recommendations of expert panels including government and industry representatives. Second, developing countries participated partly because of the convincing nature of the science, but also because of the financial support provided to help them to make the transition to phase out harmful substances (albeit at a slower pace than that for developed countries). Third, the Montreal Protocol recognized the importance of stimulating and developing new technologies so that industry could manufacture alternatives to harmful ozone-depleting chemicals, and providing access to these technologies to developing countries. Finally, groups of like-minded countries came together to provide fora to examine the complex issues involved in and to consider the consequences of taking action.

The Kyoto Protocol has been different for several reasons. First, there was much more hesitation by governments to accept the need for action to curb carbon dioxide emissions. This was partly because of doubts about the science and these doubts were stimulated by vested interests. It was also because of fears about the restrictions that the Kyoto Protocol would cause on economic activity and industrial development, both in industrialized countries and in developing countries. Second, it has become more and more apparent that the restrictions on carbon dioxide emissions proposed in the Kyoto Protocol were far too small to deal with the problem of human-induced global warming. Third, there are some countries where some warming would actually be welcome for economic or social reasons.

The Kyoto Protocol commits its signatories to a 5.2% reduction in carbon dioxide emissions, relative to 1990, by 2012. However, it is becoming more and more clear that such a small reduction is far too small to reduce global warming to what might be regarded as an "acceptable" level. From the data of approximate numerical modeling, even the complete achievement of Kyoto Protocol recommendations would provide a decrease of the mean global mean annual surface air temperature not exceeding several hundredths of degree. Perhaps the most extreme evaluation is that of George Monbiot (2006) who proposed that a *reduction* of 90% (note *of* 90%, not *to* 90%, which would be a reduction of 10%) in carbon dioxide emissions by 2030 is necessary, and he examines how in one country, the U.K. as an example, this target might be able to be achieved. It should be pointed out that Monbiot is not suggesting that for the U.K. alone to reduce its emissions by this amount will achieve very much in global terms; what he is doing is illustrating (for the example of one country) the likely problems that very many countries would face in meeting such a target. The Kyoto Protocol can only be regarded as a first and very tentative step towards making the necessary reductions in carbon dioxide emissions to enable us to avoid dramatic climate change. Kondratyev's view was that it was such a tiny first step as to be dangerously misleading in the sense that people might think that the problem had been solved once these targets were met. More realistic targets need to be established. Moreover, governments and peoples have got to learn to work together to tackle this serious problem.

2.5 CONCLUSION

It is extremely difficult to understand the scientific laws governing the present climate system and even more more so to assess potential climate changes in future. This is confirmed by the lack of reliable estimates of the contribution of anthropogenic factors to the formation of the present climate and, even more so, to any understanding of why the anthropogenically induced enhancement of the atmospheric greenhouse effect (due to the growth of greenhouse gas concentrations in the atmosphere) should cause certain changes of global climate. In this connection a primitive understanding of global warming as a general increase of temperature increasing with latitude is rather dangerous. An analysis of the observed data obtained in high latitudes of the northern hemisphere (Adamenko and Kondratayev, 1990) has shown that such claims do not correspond to reality.

In order to assess the reality of climate predictions, it is critically important to test the adequacy of models from the perspective of their ability to reproduce the present observed changes and paleo-dynamics of climate (from proxy data). As for the use of present-day observed data, the situation is rather paradoxical: the experience of testing the adequacy is confined to the use of average temperatures while it would be necessary to use different information and moments of a higher order. Goody (2001) drew attention to the prospects of using space-based observations of the spectral distribution of outgoing longwave radiation. Unfortunately, the issue of an adequately planned climate observation system has not yet been recognized (Kondratyev, 1998; Kondratyev and Cracknell, 1999; Kondratyev and Galindo, 1997). The present confused paradoxical situation is characterized by a huge amount of poorly systematized satellite observations combined with the degradation of conventional (*in situ*) observations as mentioned above.

It is very difficult to test the adequacy of global climate models by comparing the results of numerical modeling with observational data. Most often, this problem is solved by comparing a long data series of the global annual average surface air temperature. The main conclusion, despite the substantial (sometimes radical) differences in the consideration of climate-forming processes, is practically always the same: on the whole, results of calculations agree with observation data. Another characteristic feature of such testing is the invariable conclusion in support of the considerable (or even dominating) climate-forming contribution of anthropogenic factors, above all of the greenhouse effect. Yet the necessary quantitative substantiation remains lacking. Such an approach to verification of the models cannot be taken seriously because

(1) the present climate models are still very imperfect from the viewpoint of an interactive account of biospheric processes, aerosol–cloud–radiation interaction, and many other factors;
(2) the only long-term (100–150 years) series of surface air temperature observations is far from being adequate, from the viewpoint of calculations of the global annual average surface air temperature values.

Beven (2002) discusses the conceptual aspects of the numerical modeling of the environment connected with analysis of the possibilities of simulation modeling from the viewpoint of realistic simulation of natural processes. At present, computer modeling is widely developed and is actively used as an instrument of theoretical studies of the environment as well as to solve various practical problems and to substantiate recommendations for decision-makers. Of special interest are predictions of potential impacts of global climate changes and of the functioning of groundwater use systems, as well as long-term geomorphological predictions and assessments of the impacts of underground repositories of radioactive emissions. In all these cases it is assumed that the problems being studied can be solved despite the non-linearity and the open nature of the natural systems considered as well as various assumptions that serve as a basis for numerical modeling.

Of course, such an assumption is rather naive, since from the methodical ("philosophical") and scientific points of view, it proceeds from the presumption that the considered systems have been sufficiently studied. Clearly, many natural systems are so complicated that the existing ideas of them are far from being adequate. It always happens that real natural systems are much more complicated than their analogs which are described by numerical models. One of the most vivid examples in numerical climate modeling is connected with the use of a sub-grid parameterization of many climate-forming processes (on the land surface, in the atmosphere, etc.). This entails not only sometimes-far-from-real representations of the processes being considered, but also the necessity to introduce a great number of insufficiently reliably determined empirical parameters.

Recent developments associated with the global research programs GCOS (Global Climate Observing System), GOOS (Global Ocean Observing System), GTOS (Global Terrestrial Observing System), and IGOS (Integrated Global Observing Strategy) are useful, but they still do not contain adequate grounds for an optimal global observing system, as discussed in detail in the monographs of Kondratyev (1998) and Kondratyev and Cracknell (1999) and quite recently by Goody (2001, 2002) and Goody *et al.* (1998, 2002). The main cause of such a situation is the imperfection of climate models which should serve as the conceptual basis in planning the observations that are to be specified as the models are being improved. In this connection, it should be emphasized that it is not illusory statements about sufficient adequacy of the global climate models that are needed, but an analysis of their differences when compared with observations. This would reveal the "weak points" of the models. It is clear that a totality of climate parameters should be considered (and not only surface air temperature), with emphasis on the models' capability to simulate climate changes including, at least, moments of the second order.

Preparations of a strategic plan of the Climate Change Science Program planned for 10 years were started in the U.S.A. in July 2002 and completed in 2003. The program has five main goals (CCSP, 2003):

1. To get a deeper knowledge of the past and present climates and the environment, including natural variability as well as to improve an understanding of the causes of observed climatic variability.
2. To obtain more reliable quantitative estimates of the factors determining the Earth's climate changes and changes of related systems.
3. To reduce the levels of uncertainties of the prognostic assessments of future changes of climate and related systems.
4. To better understand the sensitivity and adjustability of natural and regulated ecosystems as well as anthropogenic systems to climate and to global changes in general.
5. To analyze possibilities to use and recognize the limits of understanding how to control risks in the context of climate changes.

The CCSP indicates concrete ways of how to reach these goals. In this connection, it was pointed out that the priorities of perspective developments should include a decrease of the levels of uncertainties in such problems as the properties of aerosol and its climatic implications; climatic feedbacks and sensitivity (mainly, for polar regions); and the carbon cycle. Among the key priorities in the CCSP will also be developments concerning climate-observing systems (it was very important to organize an *ad hoc* Group on Earth observations or GEO) and further development of numerical climate modeling (mainly, for a more adequate consideration of the physics and chemistry of climate).

2.6 REFERENCES

Adamenko V.N. and Kondratyev K.Ya. (1990). Global climate changes and their empirical diagnostics. In: Yu.A. Izrael, G.V. Kalabin, and V.V. Nikonov (eds.), *Anthropogenic Impact on the Nature of the North and Its Ecological Implications*. Apatity: Kola Scientific Center, Russian Academy of Sciences, pp. 17–34 [in Russian].

Beven K. (2002) Towards a coherent philosophy for modeling the environment. *Proc. Roy. Soc. London, A.*, **458**(2026), 2465–2484.

CCSP (2003). *Vision for the Program and Highlights of the Science Strategic Plan* (a report). Climate Change Science Program and the Subcommittee on Global Change Research, Washington, D.C., July, 34 pp.

Goody R. (2001). Climate benchmarks: Data to test climate models. *Studies of the Earth from Space*, **6**, 87–93 [in Russian].

Goody R. (2002). Observing and thinking about the atmosphere. *Annu. Rev. Energy Environ.*, **27**, 1–20.

Goody R., Anderson J., and North G. (1998) Testing climate models: An approach. *Bull. Amer. Meteorol. Soc.*, **79**, 2541–2549.

Goody R., Anderson J., Karl T., Miller R.B., North G., Simpson J., Stephens G., Washington W. (2002). Why monitor the climate? *Bull. Amer. Meteorol. Soc.*, **83**, 873–878.

Gorshkov, V.G. (1995). *Physical and Biological Bases of Life Stability: Man, Biota, Environment*. Springer-Verlag, Berlin.

Griggs D.J. and Noguer M. (2001). Climate change 2001: The scientific basis (contribution of Working Group I to the Third Assessment Report of the Intergovernmental Panel on Climate Change). *Weather*, 2002, **57**, 267–269.

Grubb M., Vrolijk C., and Brack D. (1999). *The Kyoto Protocol: A Guide and Assessment*. Royal Institute of International Affairs/Earthscan, London.

Houghton J.T., Jenkins G.J., and Ephraums J.J. (1990). *Climate Change: The IPCC Scientific Assessment*. Cambridge University Press, Cambridge, U.K., 365 pp.

IPCC (2001). *Third Assessment Report, Vol. 1: Climate Change 2001. The Scientific Basis*. Cambridge University Press, Cambridge, U.K., 881 pp.

IPCC (2007). Summary for policymakers. In: *Climate Change 2007: The Physical Science Basis. Contribution of Working Group I to the Fourth Assessment Report of the Intergovernmental Panel on Climate Change*. Cambridge University Press, Cambridge, U.K., 18 pp.

Kondratyev K.Ya. (1997). Key issues in global change. *Energy and Environment*, **8**, 5–9.

Kondratyev K.Ya. (1998). *Multidimensional Global Change*. Wiley/Praxis, Chichester, U.K., 761 pp.

Kondratyev K.Ya. (2004). Key aspects of global climate change. *Energy and Environment*, **15**, 469–503.

Kondratyev K.Ya. and Cracknell A.P. (1999). *Observing Global Climate Change*. London: Taylor & Francis, 562 pp.

Kondratyev K.Ya. and Galindo I. (1997). *Volcanic Activity and Climate*. A. Deepak, Hampton, VA, 382 pp.

Kondratyev K.Ya. and Galindo I. (2001). *Global Change Situations: Today and Tomorrow*. Universidad de Colima, Colima, Mexico, 164 pp.

Kondratyev K.Ya. and Grassl, H. (1993). *Global Climate Change in the Context of Global Ecodynamics*. PROPO, St. Petersburg [in Russian].

Kondratyev K.Ya. and Johannessen O. (1993). *The Arctic and Climate*. PROPO, St. Petersburg, 140 pp. [in Russian].

Kondratyev K.Ya. and Varotsos C. A. (2000). *Atmospheric Ozone Variability: Implications for Climate Change, Human Health, and Ecosystems*. Springer/Praxis, Chichester, U.K., 614 pp.

Kondratyev K.Ya., Krapivin V.F., and Phillips G.W. (2002a). *Global Environmental Change: Modelling and Monitoring*. Springer-Verlag, Heidelberg, Germany, 316 pp.

Kondratyev K.Ya., Krapivin V.F., and Phillips G.V. (2002b). *Problems of High-latitude Environmental Pollution*. St. Petersburg State University, St. Petersburg, 280 pp. [n Russian].

Kondratyev K.Ya., Krapivin V.F., and Savinykh V.P. (2003a). *Prospects for Civilization Development: Multi-dimensional Analysis*. Logos, Moscow, 575 pp. [in Russian].

Kondratyev K.Ya., Losev K.S., Ananicheva M.D., and Chesnokova I.V. (2003b) *Natural-Scientific Basis for Life Stability*. VINITI, Moscow, 240 pp. [n Russian].

Kondratyev K.Ya., Krapivin V.F., and Varotsos C.A. (2003c) *Global Carbon Cycle and Climate Change*. Springer/Praxis, Chichester, U.K., 370 pp.

Kondratyev K.Ya., Losev K.S., Ananicheva M.D., and Chesnokova I.V. (2004) *Stability of Life on Earth*. Springer/Praxis, Chichester, U.K., 165 pp.

Monbiot G. (2006). *Heat: How to Stop the Planet Burning*. Allen Lane, London, 304 pp

Singer S.F. (ed.) (2008). *Nature, Not Human Activity, Rules the Climate: A Critique of the UN-IPCC Report of May 2007*. The Heartland Institute, Chicago.

Stern N. (2007). *The Economics of Climate Change: The Stern Review*. Cambridge University Press, Cambridge, U.K., 692 pp.

Victor B.G., Raustiala K., and Skolnikoff E.B. (eds.) (1998) *The Implementation and Effectiveness of International Environmental Commitments: Theory and Practice*. MIT Press, Cambridge, MA, 737 pp.

Walker G. and King D. (2008). *The Hot Topic: How to Tackle Global Warming and Still Keep the Lights On*. Bloomsbury, London.

WMO (1998). The Bulletin interviews Profesor K.Ya. Kondratyev. *WMO Bulletin*, **47**(1), January.

Zillman, J.W. (2007). Some observations on the IPCC Assessment Process 1988–2007. *Energy and Environment*, **18**, 869–891.

3

The Earth radiation budget, 20 years later (1985–2005)

Vladimir A. Golovko

3.1 INTRODUCTION

The Earth radiation budget at the top of the atmosphere is a key parameter which measures the energy exchange between the Earth's climate system and space. It must be taken into account when constructing any climate model, whether that model is being used to describe the present climate, or whether it is being used to predict future changes in the climate, and whether those changes are natural or are due to the influence of human activities. We have already discussed Kirill Kondratyev's initial seminal work on this subject in Chapter 1. In the present chapter we consider subsequent Soviet/Russian contributions to the study of the Earth radiation budget. In particular, this chapter describes a Soviet/Russian project ScaRaB (Scanner for Radiation Budget) which formed a bridge across the gap between two United States NASA (National Aeronautics and Space Administration) programs ERBE (Earth Radiation Budget Experiment) and CERES (Clouds and Earth Radiation Energy System).

Cloud radiative forcing (Charlock and Ramanathan, 1985; Coakley and Baldwin, 1984; Ramanathan *et al.*, 1989) is a very important parameter of the role of different types of cloud in the energy balance of the climate system. The high spatial variability and the constantly changing state of the atmosphere give rise to highly variable and constantly changing cloud cover, and therefore to a highly variable value of cloud radiative forcing. Needless to say, the cloud radiative forcing parameter must be correctly simulated by a general circulation model (GCM) of the present climate if that model is to be judged valid. This is a necessary, although probably not sufficient, condition for obtaining a correct estimate of the sensitivity of the climate system (Cess *et al.*, 1990). Thus, in a model of the present climate one needs to determine the temporal average of cloud radiative forcing at a spatial resolution appropriate to the grid spacing of the model one is using. If one is to look for evidence of changes in cloud radiative forcing and in the Earth radiation budget due to human activities, then one needs to have a long time series of measurements available. Simulating cloud radiative forcing is not straightforward since the cloud radiative forcing parameter integrates the results of many different processes (cloud generation, cloud microphysics, cloud geometry, etc.) that lead to varied cloud occurrence and radiative effects. In any event, both at the instantaneous and at the monthly mean timescale, broadband radiative fluxes at the top of the atmosphere result from an integration of many surface and atmospheric processes. The necessary evaluation of the validity of the representation of cloud processes in GCMs must therefore involve the verification of a large number of parameters, including the diurnal variation of cloud occurrence, which may profoundly influence the radiative effect in the shortwave region and other effects incorporated in the cloud radiative forcing parameter. For these reasons, and also to detect and study variations of the clear-sky greenhouse effect (Raval and Ramanathan, 1989), the broadband longwave and shortwave radiances must be measured at a spatial scale of a few tens of kilometers.

Measured variations of the Earth radiation budget may indicate changes in the climate system, but they also require additional observations to define the precise nature of, and the reasons for, the change. On the one hand, various global radiation fluxes at the top of the atmosphere may correspond to the same average temperature at the Earth's surface; on the other hand, different surface and atmospheric temperature distributions may yield the same top-of-the-atmosphere radiation fluxes. This is because of the complex action of cloudiness (Fouquart *et al.*, 1990; Stephens and Greenwald, 1991) and of atmospheric composition and structure (Bony and Duvel, 1994; Bony *et al.*, 1995; Duvel *et al.*, 1997) on the global greenhouse effect. The accuracy of these Earth radiation budget measurements is also probably not sufficient to monitor the eventual imbalance in the global fluxes resulting from slow warming or cooling of the global climate system (Stowe, 1988; Wielicki *et al.*, 1996). One may note, for example, that the estimated ocean warming since 1950 corresponds to a global mean flux (at the surface—not top of the atmosphere) of the order 0.3 $\mathrm{W\,m^{-2}}$ (Levitus *et al.*, 2000). However, other large-scale parameters, such as the meridional distribution of zonal mean radiative fluxes, may certainly be monitored by Earth radiation budget experiments, giving information on modulations or trends in the meridional energy transfer by the atmosphere and the ocean. It is thus important to have a continuous set of Earth radiation budget measurements, not only to do such monitoring, but also to give more sampling of typical climate variations including strong El Niño or La Niña events or other large perturbations due, for example, to atypical monsoons or to volcanic eruptions. This point is especially important to test the sensitivity of GCMs in regard to a large spectrum of climate conditions.

The top-of-the-atmosphere Earth radiation budget may be estimated by using measurements from space of broadband reflected solar or shortwave (0.2 μm–4 μm) radiation and of outgoing infrared or longwave (4 μm–100 μm) radiation (Hartmann *et al.*, 1986; House *et al.*, 1986; Jacobowitz *et al.*, 1984; Kandel, 1990; Marchuk *et al.*, 1988; Raschke *et al.*, 1973; Stephens *et al.*, 1981). Such space measurements of regional radiation fluxes are required to document the origin and variability of the distribution of radiative energy sources and sinks over the Earth's surface.

3.2 THE ScaRaB PROJECT AND INSTRUMENT

The Scanner for Radiation Budget (ScaRaB) project was initiated in 1986 in the framework of what was then the French–Soviet Cooperation for Space Research. Germany joined the project in 1988. ScaRaB became the joint responsibility of Russia, France, and Germany early in 1992. The original aim was to provide measurements of the broadband shortwave and longwave fluxes with spatial resolution adequate for the estimation of cloud radiative forcing, in particular, but not solely in order to ensure continuity of coverage after the NASA/ERBE scanner operation ended (Barkstrom *et al.*, 1989) and well before the start of NASA/CERES scanner observations (Wielicki *et al.*, 1996). In fact, the ERBE scanner onboard ERBS operated successfully for over 5 years, until February 28, 1990; the first of the CERES scanners was launched onboard the NASA–NASDA TRMM satellite in November 1997. In addition to having the cooperating partners, Russia, France and Germany, the ScaRaB program was also assisted by the International ScaRaB Scientific Working Group (Kandel *et al.*, 1998).

The first ScaRaB flight model was integrated on the Russian operational weather satellite Meteor-3/7 and launched on January 24, 1994 from the Plesetsk spaceport in northwest Russia. Earth observations began on February 24, 1994 and continued (with some interruptions) until March 6, 1995 (Kandel *et al.*, 1998). The second flight model of the ScaRaB instrument was integrated on the Russian satellite Resurs 01-4 and launched on July 10, 1998 from the Baikonour (Kazakhstan) spaceport. Unfortunately, because of transmitter failures on the Resurs 01-4 platforms, the data collected from this second flight model are relatively sparse, even in the period of measurement from October 1998 to April 1999. However, the data collected are of excellent quality and are available to the broader scientific community for scientific use.

The ScaRaB instrument was a cross-track scanning radiometer with four channels. A detailed description of the instrument is given by Monge *et al.* (1991) and by Kandel *et al.* (1998). There were two broadband channels, the shortwave (0.2 μm–4 μm) channel and the total radiance (0.2 μm–100 μm) channel, from which the longwave (4 μm–100 μm) is deduced. During nighttime, longwave radiance is directly given by the total radiance channel. During daylight, however, longwave radiance is given by the difference between total radiance and shortwave radiance measurements. Since the ScaRaB total radiance and shortwave channels had very similar spectral response in the shortwave spectral domain, no additional spectral correction was necessary to determine longwave radiance from such a difference. However, as for ERBE or CERES, an excellent cross-calibration between the shortwave and the total radiance channel is required. An interesting and original characteristic of the ScaRaB instrument is the inclusion of two additional narrowband channels in the visible (0.55 μm–0.65 μm) and in the thermal infrared window (10.5 μm–12.5 μm) in order to test the cloud/clear-sky detection (Briand *et al.*, 1997). Various applications of these auxiliary narrowband channels have been described (Stubenrauch *et al.*, 1993; Li and Trishchenko, 1999; Duvel and Raberanto, 2000; Duvel *et al.*, 2000). These narrowband channels were especially useful for precise assessment of the error resulting when narrowband data are used to estimate the Earth radiation budget (Duvel *et al.*, 2000). Also, the thermal infrared channel may be used to compute or verify the cross-calibration between the shortwave and total radiance channels (Duvel and Raberanto, 2000) or to estimate the angular correction for the determination of the longwave flux from longwave and thermal infrared radiance measurements (Stubenrauch *et al.*, 1993).

Onboard calibration of the ScaRaB radiometers was nominally performed using a calibration module containing high-quality blackbodies for the longwave part of the spectrum and lamps for the shortwave part. For the first ScaRaB flight model, scanner characterization and calibration of the onboard blackbody simulators were carried out in a vacuum chamber at the Institut d'Astrophysique Spatiale in Orsay, France. The calibration procedure described by Kandel *et al.* (1998) checks the linearity of the radiometer response and determines the emissivity of onboard calibration blackbodies and the temperature dependence of the detector gains. The shortwave sources (lamps) were calibrated using the solar Ground Calibration Unit operated (for ScaRaB–Resurs) at Odeillo in southwest France. This calibration procedure, described by Mueller *et al.* (1993, 1996, 1997), gives a verification of the spectral response of radiometers by comparing the detector gains obtained using known infrared and solar sources. The accuracy of ground calibration is estimated to be 0.4% for onboard blackbodies and 1.5% for onboard lamp sources. For the ScaRaB–Resurs instrument, additional calibrations were performed (Dinguirard *et al.*, 1998, Duvel and Raberanto, 2000). In-flight operation modes include Earth measurement and calibration modes. Each rotation of the scanning mirror includes an Earth scan of 102°, a space look, and an observation of onboard blackbodies and lamps (which are turned on for certain cycles). Note that for the ScaRaB-2 onboard Sun-synchronous Resurs, true space looks are obtained also for the shortwave channel and not only for the total radiance and thermal infrared channels as was the case for ScaRaB–Meteor (ScaRaB-1). There were two principal inflight calibration modes. The first one, activated every 12 h, improved the calibration of the radiometers by looking at blackbodies, lamps, and space during a longer period of time and by looking at lamps that are not observed during Earth observation mode. The second calibration mode was activated once a month and measured the shortwave gain on less frequently used reference lamps.

ScaRaB–Meteor (ScaRaB-1) had a polar orbit at 1,200 km with an inclination of 82.5° and thus a precession of the orbit with a period of around 7 months. Resurs 01-4 (ScaRaB-2) was a Sun-synchronous satellite in polar orbit (inclination 98.8°), with perigee at 815 km and apogee at 818 km. The local time of equatorial crossing was around 22:15 LST. With the sun-synchronous orbit of Resurs, most of the variable bias in the products, due to changes in the local time of observation, is eliminated. With ScaRaB-2 at the lower altitude of 815 km, the swath of each scan was smaller, giving gaps from one orbit to the next near the equator; the nadir projection on the ground of the instantaneous field of view is a 41 km square. The distance between two adjacent pixels is 29 km. The data processing of ScaRaB is an ERBE-like processing (Viollier *et al.*, 1995) that is only summarized here. The first step after determination of filtered shortwave and longwave radiances is to determine the scene type (i.e., cloud cover estimate) using a maximum likelihood technique. Then a spectral correction is applied to deduce shortwave radiance from filtered shortwave radiance. This spectral correction is necessary in the shortwave because of the imperfect flatness of the spectral response. The next step is to apply scene type-dependent angular correction models to deduce the shortwave and longwave fluxes of the pixel as a function of the measured radiances. These fluxes are averages over a geographical area of $2.5^\circ \times 2.5^\circ$ latitude and longitude. Diurnal models are then applied in order to compute regional monthly mean values of mean and clear-sky fluxes. The ScaRaB instrument was operating nominally during the entire mission and available satellite housekeeping data indicate that it was still functioning after 2000. Unfortunately, no scientific data were transmitted.

The availability of the ScaRaB–Resurs data at the same time as CERES Tropical Rainfall Measurement Mission (TRMM) data gave an opportunity to cross-calibrate the two instruments. This is extremely important if one seeks to study long-term trends. The CERES instrument was turned on for a few orbits during periods favorable for comparison with ScaRaB-Resurs. The scanning azimuth of the CERES instrument was rotated so as to obtain parallel scans for the two instruments. This is necessary to compare shortwave radiances precisely, which are very sensitive to the viewing and solar zenith angles and to the azimuth between the Sun and the satellite. This cross-calibration exercise, described by Haeffelin *et al.* (2001), shows that radiances in the shortwave domain are in agreement within $(1.5 \pm 1)\%$ (at the 95% confidence level) with ScaRaB radiances being larger. Radiances in the longwave domain are in agreement within $(0.7 \pm 0.1)\%$ during daytime and $(0.5 \pm 0.1)\%$ during nighttime, with CERES radiances being larger. In the shortwave domain, this good agreement demonstrates the consistency of the very different calibration and spectral correction procedures between the two instruments. In the longwave domain, the good agreement for both daytime and nighttime radiance measurements confirms first the good absolute calibration of the longwave part of both CERES and ScaRaB radiometers. In addition, consistency between the longwave daytime and nighttime comparisons demonstrates that the ScaRaB and CERES procedures to obtain longwave radiance from total and shortwave radiance measurements during daylight perform consistently. The comparisons also support the absolute character of the calibration of these two instruments for both the longwave and shortwave spectral domains.

3.3 EARTH RADIATION BUDGET OBSERVATIONS FOR CLIMATE RESEARCH

The long time series of highly accurate radiation data provides a basis for scientific understanding of the mechanisms and factors that determine long-term climate variations and trends. Outgoing radiation is one of the major climatological factors, which determines to a great extent the dynamics of natural phenomena in the underlying surface–atmosphere system of the Earth (Golovko *et al.*, 2000). Space monitoring of outgoing radiation characteristics is considered an important element of global observations for the practical purpose of outlining current conditions of the climatic system and serves as a main information product for diagnostics and predictability of climate change. Any detection of climatic signals regarding energetic processes and the relevant identification of their relationships with anomalous natural phenomena is based on temporal datasets of space observations of Earth radiation budget components.

Considering only the tropical zone (20°N, 20°S) offers us the opportunity to compare interannual variations of ERBE, ScaRaB, and the CERES–TRMM dataset (Duvel *et al.*, 2001; Golovko *et al.*, 2003a–c). This comparison is shown in Figure 3.1, together with the continuous evolution of National Oceanic and Atmospheric Administration (NOAA) outgoing longwave radiation (OLR) (Liebmann and Smith, 1996).

For the period between 1994 and 1999, NOAA outgoing longwave radiation (OLR) is underestimated by about 8.5 W m^{-2} compared with ScaRaB and CERES instruments (the point of September 1994 is suppressed from the figure because there was a technical problem with the ScaRaB instrument at that time). Apart from this constant underestimate of 8.5 W m^{-2}, there is a very good agreement between the NOAA OLR time series, the ScaRaB measurements for the two flight models, and the CERES–TRMM measurements. In particular, the relatively large gap (of order 5 W m^{-2}) between the monthly tropical mean measured by CERES–TRMM and ScaRaB Resurs is fully consistent with the variation of the NOAA OLR product.

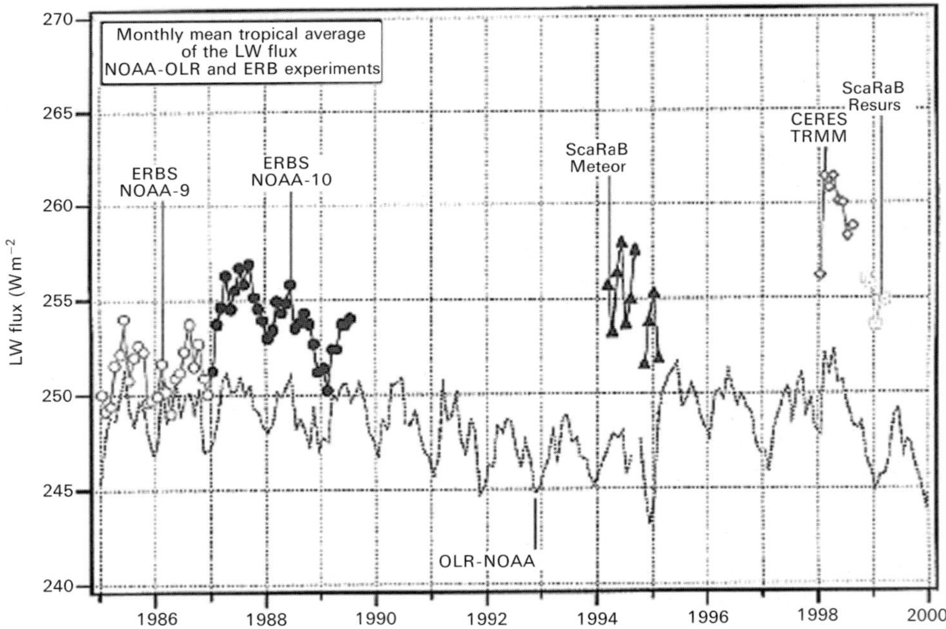

Figure 3.1. Monthly mean tropical (20°N, 20°S) average of outgoing longwave fluxes for ERBE, CERES, and ScaRaB experiments, and for the NOAA OLR time series.

This difference is thus certainly a real geophysical signal of the transition from El Niño to La Niña between 1998 and 1999. However, compared with ERBE measurements, the NOAA OLR is underestimated by about $4\,\mathrm{W\,m^{-2}}$ (instead of $8.5\,\mathrm{W\,m^{-2}}$ for ScaRaB and CERES–TRMM). A similar comparison, done using the ERBE non-scanner wide-field-of-view dataset extending from 1985 through 1998, shows good consistency between the continuous ERBS dataset and corresponding ERBE, ScaRaB–Meteor, and CERES–TRMM scanner data products (Wielicki *et al.*, 1999). The fact that, after a period of relative consistency with ERBE products between 1985 and 1989, the difference between NOAA OLR and broadband products showed an increase in following years (Figure 3.1) may be related in part to changes in equatorial crossing time already discussed by Waliser and Zhou (1997). This is however more likely due to absolute calibration and narrowband to broadband problems and illustrates the difficulty resulting from the use of a multi-platform radiometer to estimate precisely (better than $5\,\mathrm{W\,m^{-2}}$) long-term variations of outgoing longwave fluxes. Note also that the difference between the NOAA-9 and NOAA-10 period of ERBE appears to have no equivalent in the NOAA OLR time series, confirming a likely calibration and processing problem in the determination of longwave radiance.

Complete (after several reanalysis procedures) monthly mean tropical (20°N, 20°S) average datasets of the outgoing longwave and shortwave fluxes are shown in Figure 3.2. Reconstructed outgoing longwave and shortwave flux time series, based on updated different data sources (Figure 3.2), are represented in Figure 3.3.

3.3.1 Trends

It was detected (Duvel *et al.*, 2001) that during the past two decades global outgoing longwave radiation increased, with considerable energy emitted from the tropics (Chen *et al.*, 2002; Wielicki *et al.*, 2002) and certain regions of the northern hemisphere. Time series of global outgoing radiation fields have been reconstructed for the last 20 years using satellite observations from various space systems including the Russian systems Meteor and Resurs (Golovko, 2003). Reconstructed monthly mean global values of outgoing longwave radiation for the past two decades and main natural phenomena for this period are shown in Figure 3.4 (top).

Regional (2.5° × 2.5°) outgoing longwave radiation trends are shown in Figure 3.4 (bottom). The absolute maximum and minimum (most positive and negative regional trends) were located in the Tropics. So the maximum positive trend ($11.8\,\mathrm{W\,m^{-2}}$/decade) was observed over east Indonesia (1.25°S, 161.25°E) and the maximum negative trend ($-2.5\,\mathrm{W\,m^{-2}}$/decade) over northwest Indonesia (11.25°N, 96.25°E). Also shown in Figure 3.4 are several extensive regions with large positive trends near the Middle East, China, Mongolia, the U.S.A., and Brazil. For example, the positive trend over southeast Iran ($11.2\,\mathrm{W\,m^{-2}}$/decade) is only slightly less than the absolute maximum. The significance of the detected trends has been proved based on statistical tests.

Based on these reconstructed data, statistical models have been elaborated of the spatio-temporal variability of Earth radiation budget components; the results

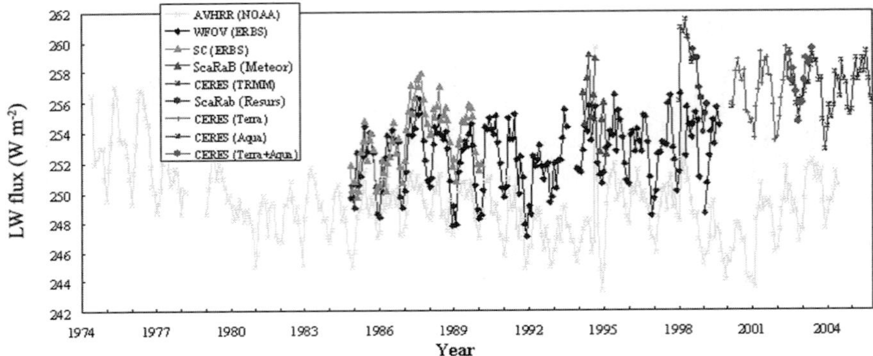

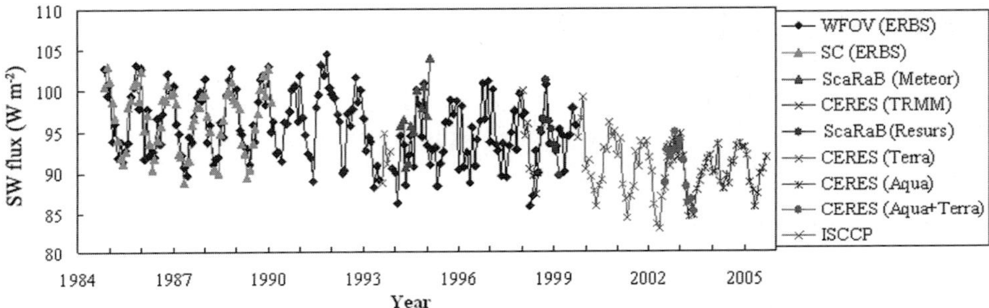

Figure 3.2. Monthly mean tropical (20°N, 20°S) average of outgoing longwave fluxes (top) and reflected shortwave fluxes (bottom) based on updated different data sources

obtained could reflect many characteristic features of anomalous natural phenomena in the dynamically unstable climatic system (Golovko, 2004; Golovko and Kondranin, 2005).

3.3.2 Mathematical modeling for spatio-temporal variability of outgoing radiation fields

Spatio-temporal statistical methods have been applied extensively in climatological sciences. The climate system is composed of many processes that exhibit complicated variability over a vast range of spatio-temporal scales. Datasets of measurements collected on this system are typically very large, and their analysis requires dimension reduction in space and/or time. Descriptive statistical techniques aid in the summary and interpretation of these data. The focus here is on a subset of two of the most useful methodologies: Multichannel Singular Spectrum Analysis (MSSA) and Principal Oscillation Pattern (POP) Analysis. Details of these statistical investigations may be found in the paper by Golovko (2004).

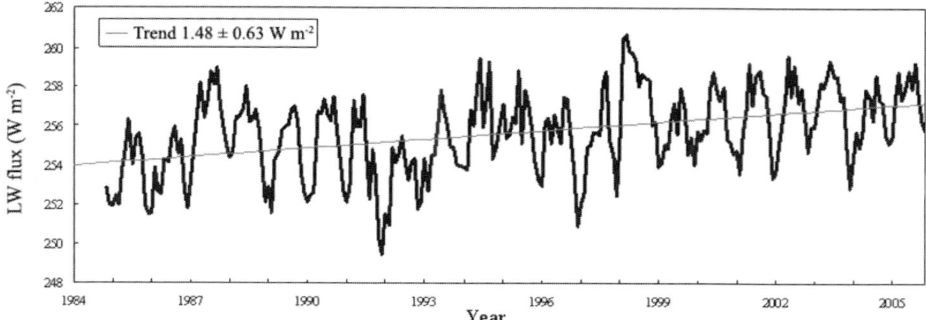

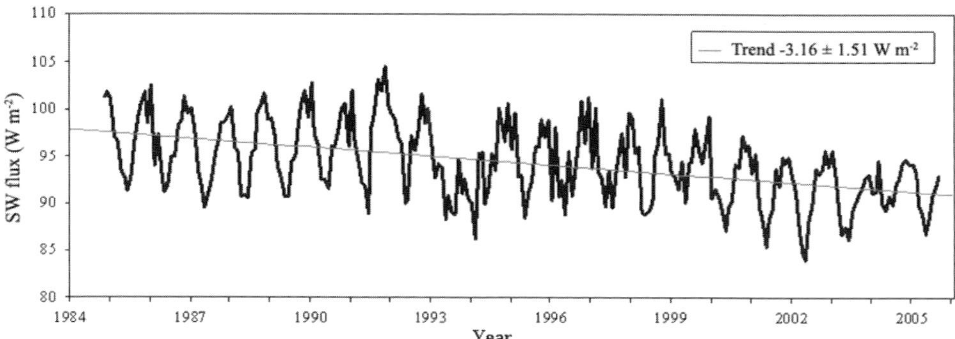

Figure 3.3. Reconstructed monthly mean tropical (20°N, 20°S) average of outgoing longwave fluxes (top) and reflected shortwave fluxes (bottom).

3.3.3 Problem of climate signal detection

The motivation for exploratory methods of data analysis in climate comes from the need to separate climate "signals" from background climate variability or "noise". This decomposition of the data is done with the hope of identifying the physical processes responsible for the generation of the signal. A fundamental characteristic of the statistical methods for signal detection is their ability to represent spatially distributed data in a compressed way such that the physical processes behind the data, or their effects, can be best visualized by the researcher. Signal detection in climate is useful to achieve four main goals in climate research:

(1) to recognize the patterns of natural climate variability and distinguish them from presumed anthropogenic or other external effects;
(2) to use the physical mechanisms inferred from the detected signals to construct numerical climate models;
(3) to validate numerical climate models by comparing the fundamental characteristics of the modeled data with those of the observed data; and
(4) to use the signals themselves to forecast the behavior of the system in the future.

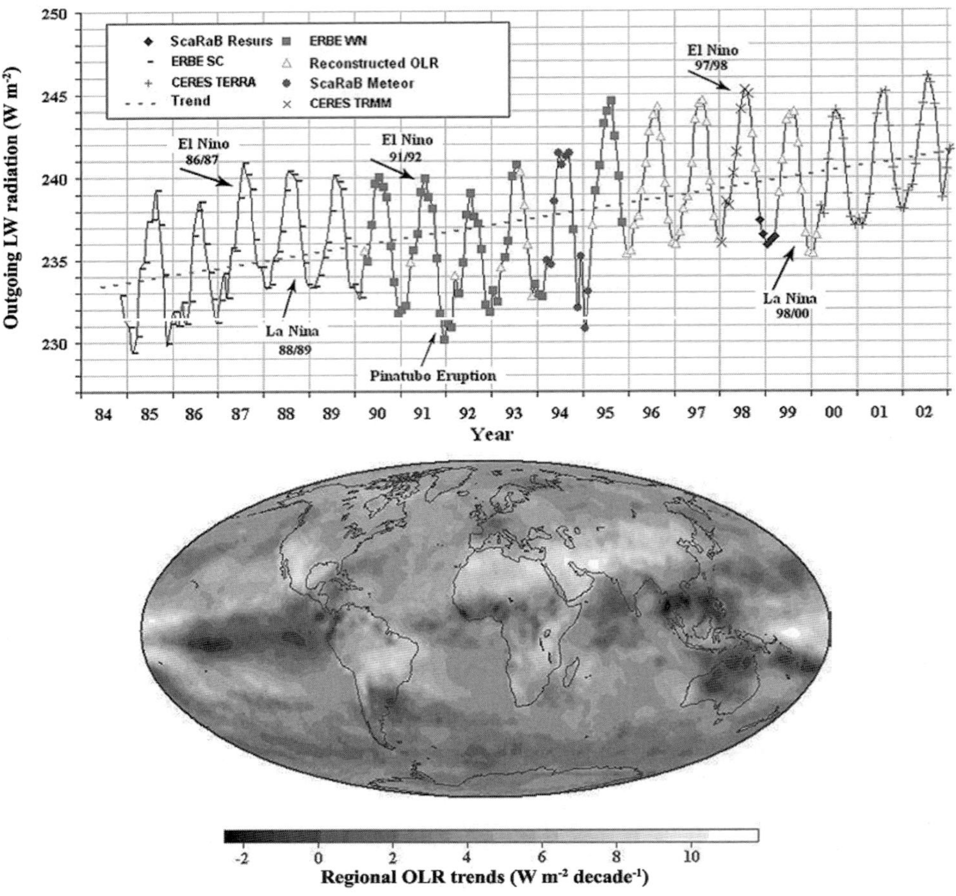

Figure 3.4. Reconstructed monthly mean global values of outgoing longwave radiation for the past two decades, and main natural phenomena for this period (top). Regional trends of outgoing longwave fluxes for the same period (bottom).

For all these reasons, the detection and description of climate signals represents a problem of increasing interest in the scientific community.

3.3.4 Methods of signal detection

The statistical techniques described here belong to a category of analysis called "Exploratory Analysis'. The aim of these techniques is to summarize the dominant characteristics of a field, such as the dominant space and/or time patterns, and discriminate between the signal of interest and the unrelated processes or noise (Venegas, 2001). These methods attempt to exploit the information available in spatially distributed datasets and involve eigenvalue decompositions. The most traditional technique is empirical orthogonal function (EOF) analysis. There are two different approaches for performing EOF decomposition: the covariance matrix approach and the singular value decomposition approach.

3.4 MULTICHANNEL SINGULAR SPECTRUM ANALYSIS (MSSA)

There are two good reasons for choosing the singular value decomposition approach rather than the covariance matrix approach in order to perform EOF analysis. First, it provides a one-step method for computing all the components of the eigenvalue problem, without having to compute and store large covariance matrices. Second, the resulting decomposition is computationally more stable and robust.

In this approach, a singular value decomposition is performed directly on the rectangular data matrix $\mathbf{F}$ that consists of M rows (spatial points) and N columns (temporal samples). The singular value decomposition of a matrix is based on the concept that any rectangular $M \times N$ matrix $\mathbf{F}$ can be written as the product of three matrices: an $M \times M$ matrix $\mathbf{U}$, an $M \times N$ diagonal matrix $\mathbf{S}$ with positive or zero elements, and the transpose ($\mathbf{V}^T$) of the $N \times N$ matrix $\mathbf{V}$:

$$\mathbf{F} = \mathbf{USV}^T. \tag{3.1}$$

Matrix $\mathbf{S}$ is a rectangular $M \times N$ matrix with zero elements outside the diagonal and positive or zero elements on the diagonal. The scalars on the diagonal, s_k, are called the singular values and are typically placed in decreasing order of magnitude. The singular values s_k are related to the eigenvalues λ_k such that $\lambda_k = s_k^2$. There is a maximum of $K < \min(N, M)$ non-zero singular values, which defines the maximum number of EOF modes we can determine, so that the effective dimension of matrix $\mathbf{S}$ is $K \times K$.

The columns of the quadratic $M \times M$ matrix $\mathbf{U}$ are orthogonal and are called the left singular vectors of $\mathbf{F}$. They are identical to the eigenvectors obtained from the covariance matrix $\mathbf{R}_{\mathbf{FF}}$ ($\mathbf{R}_{\mathbf{FF}} = \mathbf{RR}^T$) (i.e., they are the EOF patterns associated with each singular value). There are only K useful left singular vectors associated with the K non-zero singular values, hence the effective dimension of matrix $\mathbf{U}$ is $M \times K$.

The rows of the quadratic $N \times N$ matrix $\mathbf{V}^T$ are also orthogonal and are called the right singular vectors of $\mathbf{F}$. They are proportional to the principal components $\mathbf{Z}$, and the constants of proportionality are the singular values s_k, such that $Z^k(t) : \mathbf{Z} = \mathbf{SV}^T$.

An alternative to the visualization of re-normalized EOFs is the presentation of the patterns as correlation maps. A correlation map for mode k is a map of correlation coefficients r_m^k, between the principal component $Z^k(t)$ and the values of field $F_m(t)$ at each location $m = 1 \cdots M$: $r_m^k = [Z^k(t), F_m(t)]$, where brackets [ ] indicate temporal correlation, k indicates the mode, and m indicates the location. The contours of such a map show the distribution of the centers of action of the mode scaled as correlation coefficients, which is more meaningful than the dimensionless EOFs. The distribution of centers of action in the correlation map is basically the same as that in the EOF spatial pattern.

3.5 MUTUAL EVOLUTION OF THE OUTGOING LONGWAVE AND SHORTWAVE RADIATION ANOMALIES FOR THE LAST TWO DECADES

It was detected that during the 1990s global outgoing longwave radiation increased considerably. The reflected solar radiation decreased by a larger amount. Mutual evolution of the outgoing longwave and shortwave radiation anomalies for this period is shown in Figure 3.5. More special details are given by Golovko (2004).

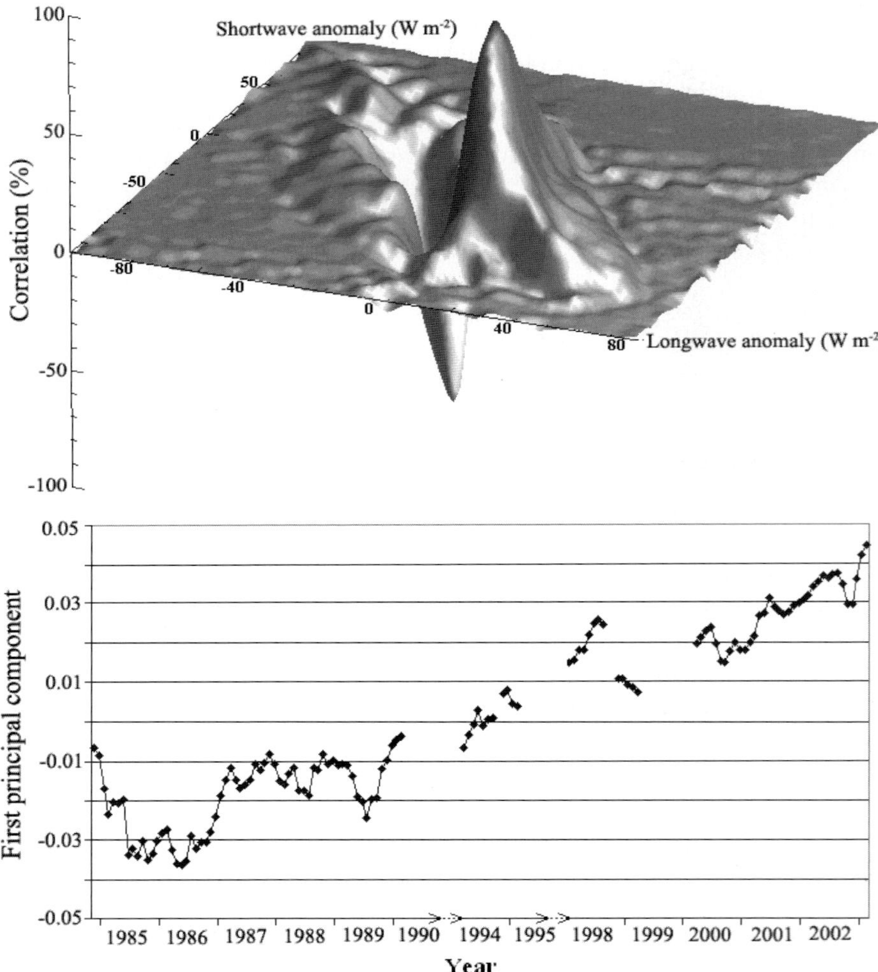

Figure 3.5. Correlation map for the first singular vector explaining much of the variance of the two-dimensional histogram for outgoing longwave and shortwave radiation anomalies (top). First principal component (time dependence), corresponding to the first singular vector, explaining most variance in the data (bottom).

As can be seen from the first principal component (time dependence) in Figure 3.5 (bottom) there is a time period (1985–1986) for the beginning of a very important redistribution of global outgoing longwave and shortwave radiation.

Typical spatio-temporal structures of the outgoing longwave radiation anomalies describing the dynamics of specific processes in the radiation field of the Earth have been identified while utilizing the techniques of Multichannel Singular Spectrum Analysis (MSSA). A correlation map for the first singular vector (spatial structure) and first principal component (time dependence) for global outgoing longwave radiation anomalies is shown in Figure 3.6.

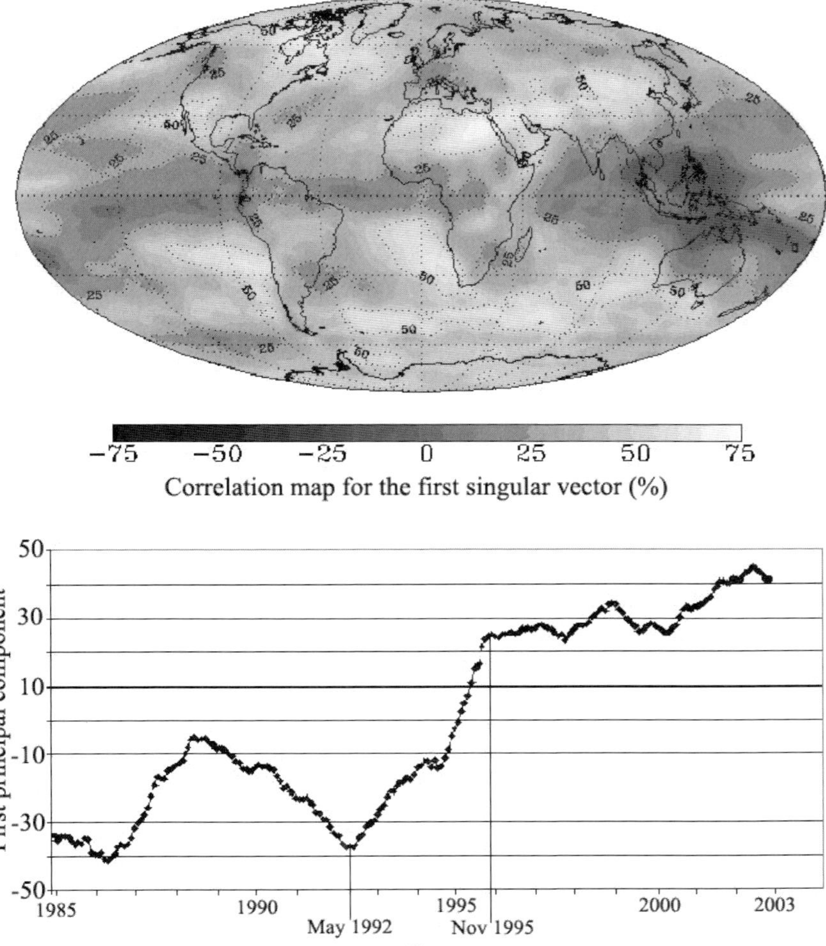

Figure 3.6. Correlation map for the first singular vector (spatial structure) and first principal component (time dependence) explaining much of the (13.4%) variance of global outgoing longwave radiation anomalies.

3.6 PRINCIPAL OSCILLATION PATTERN (POP) ANALYSIS

Principal Oscillation Pattern (POP) analysis is a multivariate technique to infer empirically the characteristics of spacetime variations of a possibly complex system (von Storch *et al.*, 1995). The basic idea is to identify a linear system with a few free parameters that are fitted to the data. Then, the spacetime characteristics of this simple system are regarded as being the same as those of the full system. POP analysis is nowadays a routinely used tool to diagnose the spacetime variability of the climate system. It should be noted that the POP formalism (conventional, cyclostationary, and complex POP analysis) may be applied to linear systems whose system matrices are estimated from data or whose system matrices are derived from theoretical dynamical considerations.

The normal modes of a linear discretized real system,

$$\mathbf{F}(t+1) = \mathbf{AF}(t), \tag{3.2}$$

are the eigenvectors $\mathbf{p}$ of the matrix $\mathbf{A}$. In general, $\mathbf{A}$ is not symmetric and some or all of its eigenvalues λ and eigenvectors $\mathbf{p}$ are complex. However, since $\mathbf{A}$ is a real matrix, the conjugate complex quantities λ^* and $\mathbf{p}^*$ satisfy also the eigenequation $\mathbf{Ap}^* = \lambda^*\mathbf{p}^*$. In most cases, all the eigenvalues are different and the eigenvectors form a linear basis. So, state $\mathbf{F}$ at any time t may be uniquely expressed in terms of the eigenvectors:

$$\mathbf{F} = \sum_j z_j \mathbf{p}_j, \tag{3.3}$$

The coefficients of the pairs of conjugate complex eigenvectors are conjugate complex too. Inserting (3.3) into (3.2) we find that the coupled system (3.2) becomes uncoupled, yielding n single equations, where n is the dimension of the process $\mathbf{F}$,

$$z(t+1)\mathbf{p} = \lambda z(t)\mathbf{p} \tag{3.4}$$

so that, if $z(0) = 1$,

$$z(t)\mathbf{p} = \lambda^t \mathbf{p}. \tag{3.5}$$

The contribution $\mathbf{P}(t)$ of the complex conjugate pair $\mathbf{p}, \mathbf{p}^*$ to the process $\mathbf{F}(t)$ is given by

$$\mathbf{P}(t) = z(t)\mathbf{p} + [z(t)\mathbf{p}]^*. \tag{3.6}$$

Writing $\mathbf{p} = \mathbf{p}^r + i\mathbf{p}^i$ and $2z(t) = z^r(t) - iz^i(t)$, this reads

$$\mathbf{P}(t) = z^r(t)\mathbf{p}^r + z^i(t)\mathbf{p}^i = \rho^t(\cos(\eta t)\mathbf{p}^r - \sin(\eta t)\mathbf{p}^i), \tag{3.7}$$

with $\lambda = \rho\exp(-i\eta)$ and $z(0) = 1$. The geometric and physical meaning of (3.7) is the trajectory spirals in the space spanned by $\mathbf{p}^r$ and $\mathbf{p}^i$ with the period $T = 2\pi/\eta$ and the e-folding time $\tau = -1/\ln(\rho)$ in consecutive order

$$\Lambda \rightarrow \mathbf{p}^r \rightarrow -\mathbf{p}^i \rightarrow -\mathbf{p}^r \rightarrow \mathbf{p}^i \rightarrow \mathbf{p}^r \rightarrow \Lambda. \tag{3.8}$$

The pattern coefficients z_j are given as the dot product of $\mathbf{F}$ with the adjoint patterns

$\mathbf{p}_j^A$, which are the normalized eigenvectors of $\mathbf{A}^T$:

$$(\mathbf{p}_j^A)^T\mathbf{F} = \sum_k z_k(\mathbf{p}_j^A)^T\mathbf{p}_k = z_j. \tag{3.9}$$

All the information used so far consists of the existence of a linear equation (3.2) with some matrix **A**. No assumption was made as to from where this matrix originates. In dynamical theory, the origins of (3.2) are linearized and discretized differential equations. In the case of the POP analysis, the relationship

$$\mathbf{F}(t+1) = \mathbf{AF}(t) + \text{noise} \tag{3.10}$$

is hypothesized. Multiplication of (3.10) from the right-hand side by the transposed $\mathbf{F}^T(t)$ and taking expectations E leads to

$$\mathbf{A} = E[\mathbf{F}(t+1)\mathbf{F}^T(t)][E[\mathbf{F}(t)\mathbf{F}^T(t)]]^{-1}. \tag{3.11}$$

The eigenvectors of (3.11), or the normal modes of (3.10), are called principal oscillation patterns. The coefficients z are called POP coefficients. Their time evolution is given by (3.4), superimposed by noise:

$$z(t+1) = \lambda z(t) + \text{noise}. \tag{3.12}$$

The stationarity of (3.12) requires $\rho = |\lambda| < 1$. In practical situations, when only a finite time series $\mathbf{F}(t)$ is available, **A** is estimated by first deriving the sample lag-1 covariance matrix $\mathbf{X}_1 = E[\mathbf{F}(t+1)\mathbf{F}^T(t)]$ and the sample covariance matrix $\mathbf{X}_0 = E[\mathbf{F}(t)\mathbf{F}^T(t)]$, and then forming $\mathbf{A} = \mathbf{X}_1\mathbf{X}_0^1$. The eigenvalues of this matrix always satisfy $\rho < 1$.

The first principal oscillation pattern represented by real and imaginary parts, as well as the schematic diagram of the time evolution for global outgoing longwave radiation anomalies are shown in Figure 3.7.

3.7 POP AS A PREDICTIVE TOOL

The POP technique is naturally suited for making predictions because of the forecast equation (3.12) for POP coefficients, namely,

$$\mathbf{z}(t+1) = \rho \exp\left(-i\left(\frac{2\pi}{T}\right)\right)\mathbf{z}(t), \tag{3.13}$$

with the period $T = \dfrac{2\eta}{\eta} = \dfrac{2\pi}{\tan^{-1}(\operatorname{Im}\lambda/\operatorname{Re}\lambda)}$ and $\rho = |\lambda|$. Equation (3.13) describes the damped persistence of a trajectory in the complex plane. Thus, in the framework of POP prediction, it is only necessary to identify the location in the complex state space of the system at a given time to predict future locations. For a limited time this prediction might be useful, but at longer lead times the built-in linearity of the POP analysis as well as the unpredictable noise will result in a deterioration of the forecast skill.

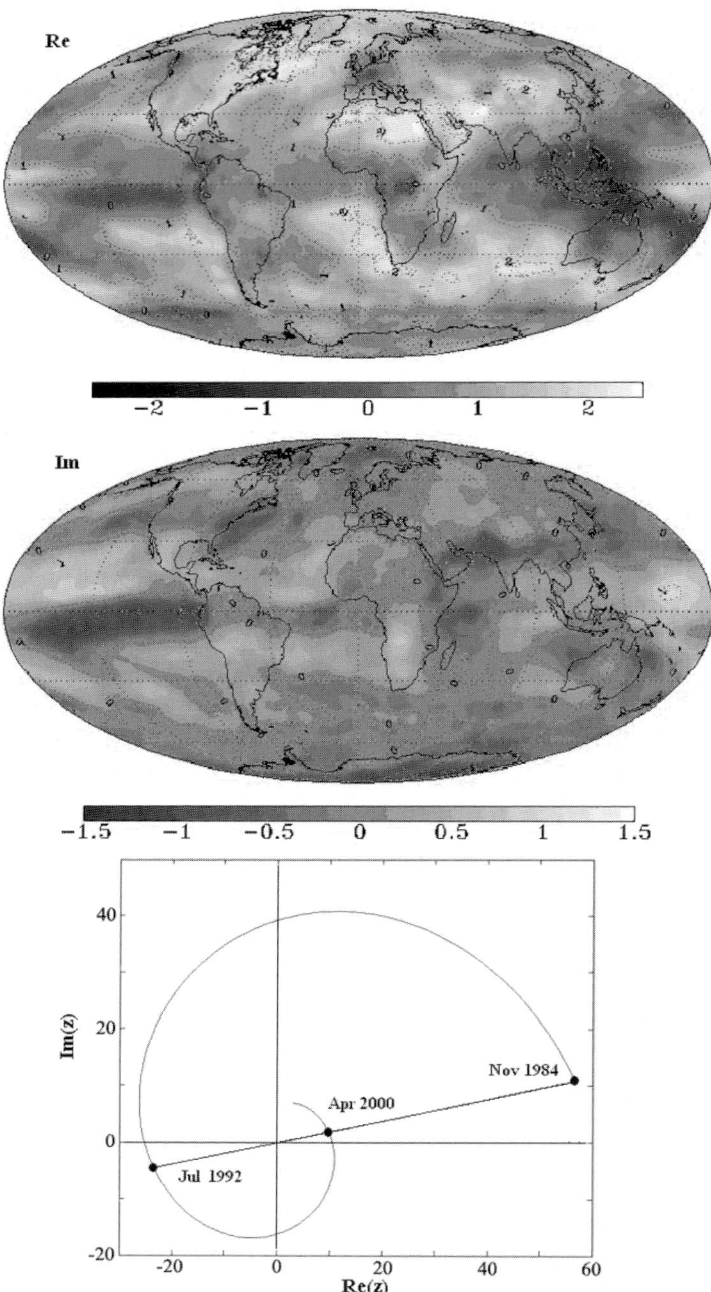

Figure 3.7. First principal oscillation pattern represented by real (top) and imaginary (middle) parts, as well as schematic diagram of the time evolution (bottom) for global outgoing longwave radiation anomalies.

Equation (3.13) always forecasts a decay of the amplitude (because of $\rho < 1$). However, for a stationary time series the probability of a decay at any given time equals the probability of an intensification, namely 50%. Therefore, we respecify ρ in (3.13) as $\rho = 1$ so that the forecast becomes amplitudewise a persistence forecast. Thus, we may expect a non-trivial forecast only for a regularly changing phase. However, a prediction of phase is valuable even if the amplitude is not well predicted.

To measure the quality of POP forecasts, two measures of skill are used: the correlation skill score $S(\tau)$ and the rms skill score $R(\tau)$:

$$S(\tau) = \frac{\langle z_1^*(t) z(t) \rangle}{\sqrt{\langle |z_1(t)|^2 \rangle \langle |z(t)|^2 \rangle}}, \tag{3.14}$$

$$R(\tau) = \sqrt{\langle |\tilde{z}_1(t) - z(t)|^2 \rangle}, \tag{3.15}$$

where $\tilde{z}_1(t)$ denotes the (complex) forecast issued for τ time steps in advance; $z(t)$ is the (complex) verifying quantity; and the angle brackets indicate ensemble averages.

The correlation skill score $S(\tau)$, being insensitive to amplitude errors, is an indicator of phase errors only. With respect to amplitude the POP forecast is a persistence forecast. Therefore, $S(\tau)$ is an adequate skill score of the POP method; $R(\tau)$ is sensitive to both phase and amplitude errors. It may be anticipated, therefore, that the POP forecast appears less successful if measured in terms of root-mean-square error.

As a research result of the ScaRaB project, several mathematical models have been elaborated for spatio-temporal variability of the Earth's outgoing radiation (Golovko, 2004). Results of the forecast for 12 months of OLR integral values for the Tropics by the two methods MSSA and POP are shown on Figure 3.8. A forecast field (for 5 months) by the POP predictive technique is represented in Figure 3.9.

These models describing dynamics of specific processes in the radiation field of the Earth could diagnose the basic characteristics of radiation change and realize the medium-range prediction of outgoing radiation (up to 6 months).

3.8 THE EARTH RADIATION BUDGET AND GLOBAL WARMING

The Earth's climate system has considerable thermal inertia. The effect of the thermal inertia is to delay the Earth's response to climate forcings (i.e., changes of the planet's energy balance that tend to alter global temperature). The primary symptom of the Earth's thermal inertia, in the presence of increasing climate forcing, is an imbalance between the energy absorbed and emitted by the planet. This imbalance provides an invaluable measure of the net climate forcing acting on the Earth. Improved ocean temperature measurements in the past decade permit an indirect but precise quantification of the Earth's energy imbalance. The lag in the climate response to a forcing is a sensitive function of equilibrium climate sensitivity, varying approximately as the square of the sensitivity, and it depends on the rate of heat exchange between the ocean's surface mixed layer and the deeper ocean. The lag could be as short as a

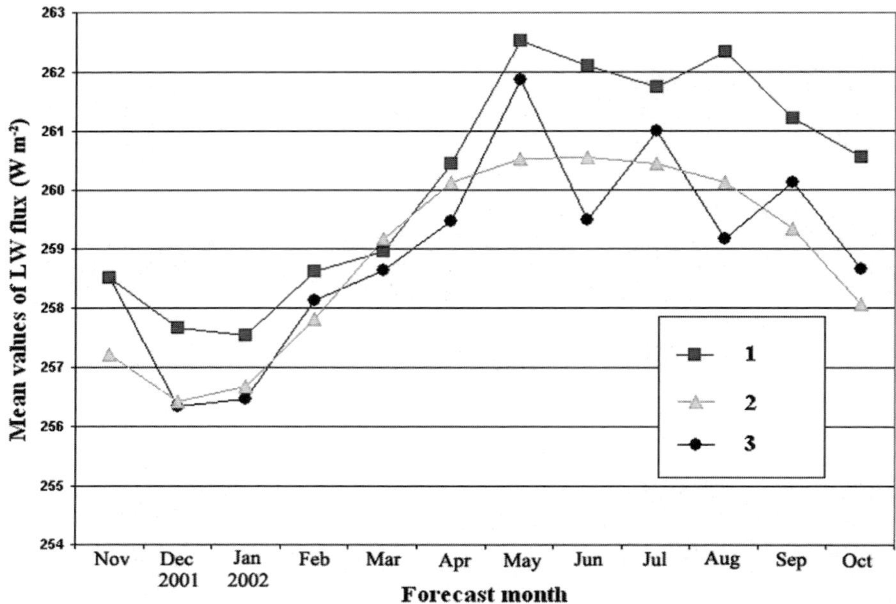

Figure 3.8. Quality of the forecasts for 12 months of OLR integral values for the Tropics by three methods (MSSA-1, POP-2, observed-3).

decade, if climate sensitivity is as small as 0.25°C per W m^{-2} of forcing, but it is a century or longer if climate sensitivity is 1°C per W m^{-2} or larger. Evidence from the Earth's history and climate models suggests that climate sensitivity is (0.75 ± 0.25)°C per W m^{-2}, implying that 25 to 50 years are needed for Earth's surface temperature to reach 60% of its equilibrium response.

The Earth's energy balance was investigated via computations with the current global climate model of the NASA Goddard Institute for Space Studies (GISS) (Hansen *et al.*, 2005). The model and its simulated climatology have been documented, as has its response to a wide variety of climate forcing mechanisms. According to this modeling the net change of effective forcing between 1880 and 2003 is +1.8 W m^{-2}, with a formal uncertainty of $\pm$0.85 W m^{-2}.The planetary energy imbalance in the model did not exceed a few tenths of 1 W m^{-2} before the 1960s. Since then, except for a few years following each large volcanic eruption (Krakatau, Pinatubo), the simulated planetary energy imbalance has grown steadily. According to the model, Earth is now absorbing (0.85 ± 0.15) W m^{-2} more solar energy than it radiates to space as heat. We infer from the consistency of observed and modeled planetary energy gains that the forcing is still driving climate change (i.e., the forcing not yet responded to averaged ~0.75 W m^{-2} in the past decade and was ~(0.85 ± 0.15) W m^{-2} in 2003). This imbalance is consistent with the total forcing of ~1.8 W m^{-2} relative to that in 1880 and climate sensitivity of ~2/3°C per W m^{-2}. The observed 1880 to 2003 global warming was 0.6°C to 0.7°C, which is the full

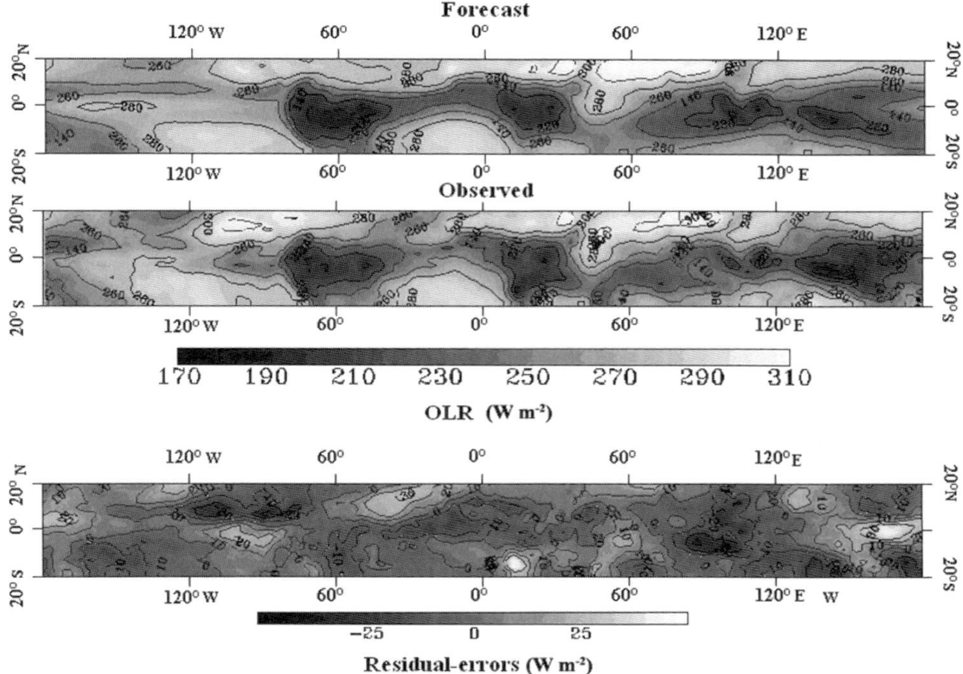

Figure 3.9. Forecast field (for 5 months) by POP predictive technique and observed OLR values for the Tropics as well as residual errors.

response to nearly 1 W m^{-2} of forcing. Of the 1.8 W m^{-2} forcing, 0.85 W m^{-2} remains (i.e., additional global warming of $0.85 \times 0.67 \sim 0.6^{\circ}$C is "in the pipeline" and will occur in the future even if atmospheric composition and other climate forcings remain fixed at today's values; Hansen *et al.*, 2005).

The present 0.85 W m^{-2} planetary energy imbalance, its consistency with estimated growth of climate forcings over the past century, and its consistency with the temporal development of global warming is based on a realistic climate sensitivity. If climate sensitivity, climate forcings, and ocean mixing are taken as arbitrary parameters, one may find other combinations that yield warming comparable with that of the past century. However, climate sensitivity is constrained by empirical data; model-simulated depth of penetration of ocean-warming anomalies is consistent with observations (Figure 3.10a), thus supporting the modeled rate of ocean mixing; and despite ignorance about aerosol changes, there is sufficient knowledge to constrain estimates of climate forcings.

Theory and modeling predict that hurricane intensity should increase with increasing global mean temperatures. Although the frequency of tropical cyclones is an important scientific issue, it is not by itself an optimal measure of tropical cyclone threat. Total power dissipation and consequently actual monetary loss in wind storms rise roughly as the cube of the wind speed, which is integrated over the

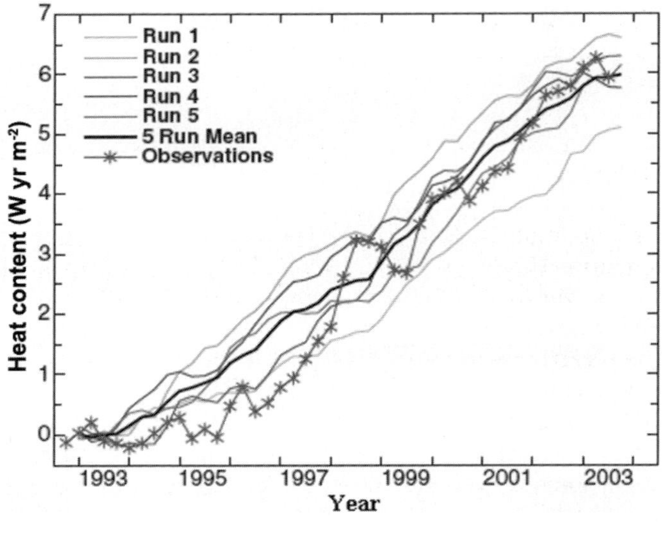

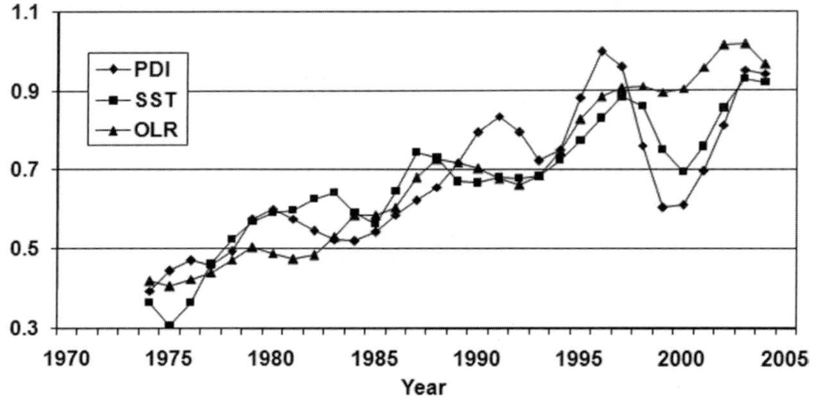

Figure 3.10. Ocean heat content change between 1993 and 2003 in the top 750 m of the World Ocean. Five model runs are shown for the GISS coupled dynamical ocean–atmosphere model (a, top). Normalized time series data of OLR, SST, and PDI for the western North Pacific in the period 1974–2004 (b, bottom).

surface area affected by a storm and over its lifetime. In a novel approach to the estimation of hurricane activity the potential destructiveness index (PDI) is used (Emanuel, 2005). The PDI is based on the total dissipation of power, integrated over the lifetime of the cyclone. The PDI has increased markedly since the mid-1970s (Figure 3.10b). This trend is due to both longer storm lifetimes and greater storm intensities. The record of net hurricane power dissipation is highly correlated with tropical sea surface temperature and outgoing longwave radiation, reflecting well-documented climate signals, including multidecadal oscillations and global warming

(Golovko, 2006). Preliminary results suggest that future warming may lead to an upward trend in tropical cyclone destructive potential and a substantial increase in hurricane-related losses.

3.9 CONCLUSIONS

The Scanner for Radiation Budget (ScaRaB) was a cooperative project of Russia, France, and Germany. The two flight models of ScaRaB instruments were flown on the Russian satellites Meteor-3/7 and Resurs-01/4. ScaRaB partly filled a gap in polar narrow-field-of-view coverage that otherwise would have extended for nearly 10 years in the 1990s. The objective of this program was to make space measurements of the Earth radiation budget to complement those provided by the Earth Radiation Budget Experiment (ERBE) and the Clouds and Earth Radiant Energy System (CERES) missions.

Broadband radiance measurements made by the ScaRaB instruments exhibit strong consistency with earlier Earth radiation budget measurements, while at the same time revealing significant regional and zonal anomalies specific to the 1998/1999 El Niño–La Niña transition. There is excellent agreement with the CERES/TRMM instrument, as demonstrated by direct comparison of simultaneous collocated codirectional broad-band radiance measurements. The highly accurate ScaRaB broadband radiances are therefore valuable for case studies of the natural hazards related to 1998 El Niño events or as a complement to field experiments such as the Indian Ocean Experiment (INDOEX) in a period when no other Earth radiation budget scanner was in operation (Golovko and Kozoderov, 2000; Golovko *et al.*, 2003b; Kozoderov and Golovko, 1999). This dataset makes it possible to quantify the regional radiative anomalies in the winter of 1999, particularly in regard to the La Niña phase of the Southern Oscillation.

Time series of global outgoing radiation fields have been reconstructed for the last 20 years using satellite observations from various space systems including some Russian systems. Observed data indicate that the energy components of the Earth radiation budget have varied substantially in the past two decades. It was detected that during the 1990s global outgoing longwave radiation increased (with considerable energy emitted from the Tropics and from certain regions of the northern hemisphere). Reflected solar radiation decreased by a larger amount. The significance of the detected trends has been proved based on statistical tests. The period 1985–1986 has been established as the beginning of a very important redistribution of global characteristics for outgoing longwave and shortwave radiation. Current investigations are critical for improving our scientific understanding of whether these changes represent a long-term anomalous trend for the climate system or are part of a natural fluctuation of climate.

Based on time series satellite observations of global outgoing longwave radiation fields, statistical models have been elaborated for spatio-temporal variability of the main Earth radiation budget component; the results obtained could reflect many characteristic features of anomalous natural phenomena in the dynamically unstable climatic system. Typical temporal and spatial structures of outgoing longwave radiation anomalies describing the dynamics of specific processes in the radiation field of the Earth have been identified while utilizing techniques of Multichannel Singular Spectrum Analysis (MSSA) and Principal Oscillation Patterns (POP) analysis. An opportunity has been used to make a medium-range prediction for the spatial distribution of outgoing longwave radiation in different latitudinal zones while taking these structures into account.

One climate model, driven mainly by increasing forcings, calculates that the Earth is now absorbing $(0.85 \pm 0.15)\,\mathrm{W\,m^{-2}}$ more energy from the Sun than it is emitting to space. This imbalance is confirmed by precise measurements of increasing ocean heat content over the past 10 years. Implications include the expectation of additional global warming of about 0.6°C without further change of atmospheric composition, confirmation of the climate system's lag in responding to forcings, implying the need for anticipatory actions to avoid any specified level of climate change, and the likelihood of acceleration of ice sheet disintegration and sea level rise. Tropical cyclone intensity and duration are increasing worldwide, in concert with increasing tropical ocean temperature. Changing tropical cyclone activity may eventually affect the ocean's thermohaline circulation, moderating tropical warming but accelerating high-latitude warming.

3.10 REFERENCES

Barkstrom B.R., Harrison E.F., Smith G.L. Green R., Kibler J., and Cess, R.D. (1989). Earth Radiation Budget Experiment (ERBE) archival and April 1985 results. *Bull. Amer. Meteor. Soc.*, **70**, 1254–1262.

Bony S. and Duvel J.-Ph. (1994). Influence of the vertical structure of the atmosphere on the seasonal variation of precipitable water and greenhouse effect. *J. Geophys. Res.*, **99**, 12963–12980.

Bony S., Duvel J.-Ph., and Le Treut H. (1995). Observed dependence of the water vapor and clear-sky greenhouse effect on sea surface temperature: Comparison with climate warming experiments. *Climate Dyn.*, **11**, 307–320.

Briand V., Stubenrauch C.J., Rossow W.B., Walker A., and Holz R. (1997). Scene identification for ScaRaB data: The ISCCP approach. In: J.D. Haigh (ed.), *Satellite Remote Sensing of Clouds and the Atmosphere*. SPIE, Bellingham, WA, pp. 242–252.

Cess R.D., Potter G.L., Blanchet J.P., Boer G.J., Del Genio A.D., Déqué M., Dymnikov V., Galin V., Gates W.L., Ghan S.J. *et al.* (1990). Intercomparison and interpretation of climate feedback processes in 19 atmospheric general circulation models. *J. Geophys. Res.*, **95**, 16601–16615.

Charlock T.P. and Ramanathan V. (1985). The albedo field and cloud radiative forcing produced by a general circulation model with internally generated cloud optics. *J. Atmos. Sci.*, **42**, 1408–1429.

Chen J., Carlson B.E., and Del Genio A.D. (2002). Evidence for strengthening of the tropical general circulation in the 1990s. *Science*, **295**, 838–840.

Coakley J.A. and Baldwin D.G. (1984). Toward the objective analysis of clouds from satellite imagery data. *J. Climate Appl. Meteor.*, **23**, 1065–1099.

Dinguirard M., Mueller J., Sirou F., and Tremas T. (1998). Comparison of ScaRaB ground calibration in the short wave and long wave domains. *Metrologia*, **35**, 597–601.

Duvel J.-Ph. and Raberanto P. (2000). A geophysical crosscalibration approach for broadband channels: Application to the ScaRaB experiment. *J. Atmos. Oceanic Technol.*, **17**, 1609–1617.

Duvel J.-Ph., Bony S., and Le Treut H. (1997). Clear-sky greenhouse effect sensitivity to sea surface temperature changes: An evaluation of AMIP simulations. *Climate Dyn.*, **13**, 259--273.

Duvel J.-Ph., Bouffiès-Cloché S., and Viollier M. (2000). Determination of shortwave earth reflectances from visible radiance measurements: Error estimate using ScaRaB data. *J. Appl. Meteor.*, **39**, 957–970.

Duvel J.-Ph., Viollier M., Raberanto P., Kandel R., Haeffelin M.,. Pakhomov L.A., Golovko V.A., Mueller J., and Stuhlmann R. (2001). The ScaRab-Resurs Earth Radiation Budget Dataset and first results. *Bulletin of the American Meteorological Society*, **82**(7), 1397–1408.

Emanuel K.A. (2005). Increasing destructiveness of tropical cyclones over the past 30 years. *Nature,* **436**, 686–688.

Fouquart Y., Buriez J.C., Herman M., and Kandel R.S. (1990). The influence of clouds on radiation: A climate-modeling perspective. *Rev. Geophys.*, **28**, 145–166.

Golovko V.A. (2003) Global redistribution of Earth radiation budget components. *Russ. J. Remote Sens.*, **6**, 3–13 [in Russian].

Golovko V.A. (2004). Diagnostic and prediction of spatial variation dynamics for outgoing longwave radiation of the Earth. *Russ. J. Remote Sens.*, **5**, 3–14 [in Russian].

Golovko V.A. (2006). Mathematical modeling of hurricane activity based on radiance observations from space. *Russ. J. Remote Sens.*, **5**, 12–37 [in Russian].

Golovko V.A. and Kondranin T.V. (2005). Anomalous global redistribution of the Earth radiation budget components. *Proceedings of the 31st International Symposium on Remote Sensing of Environment (31st ISRSE 2005), June 20–24, 2005, St. Petersburg*. St. Petersburg University, St. Petersburg, pp. 1–4.

Golovko V.A., and Kozoderov V.V. (2000). Earth radiation budget: New applications to study natural hazards from space. *Russ. J. Remote Sens.*, **1**, 29–41 [in Russian].

Golovko V.A., Kozoderov V.V., and Ovchinnikov S.K. (2000). Earth radiation budget: New applications of ScaRaB data for natural hazards investigations. *IRS 2000: Current Problems in Atmospheric Radiation: Proceedings of the International Radiation Symposium, July 24–29, 2000, Sankt-Petersburg*. St. Petersburg University, St. Petersburg, pp. 13–516.

Golovko V.A., Pakhomov L.A., and Uspensky A.B. (2003a). Earth radiation budget monitoring from METEOR-3 and RESURS-01 Satellites. *Meteorology and Hydrology*, **12**, 56–73 [in Russian].

Golovko V.A., Kozoderov V.V., and Kondranin T. V. (2003b). Mathematical modeling of anomalous natural phenomena using space data about Earth radiation budget components. *Proceedings of the World Climate Change Conference (WCCC-2003), September 29–October 3, 2003, Moscow*. Science, Moscow, pp. 491–492.

Golovko V.A., Pakhomov L.A., and Uspensky A.B. (2003c). The research results of the Russian–French scientific project for global monitoring of the Earth radiation budget from Russian satellites. *Proceedings of the World Climate Change Conference (WCCC-2003), September 29–October 3, 2003, Moscow*. Science, Moscow, pp. 401–402.

Haeffelin M., Wielicki B., Duvel J.-Ph., Priestley K., and Viollier M. (2001). Intercalibration of CERES and ScaRaB Earth radiation budget datasets using temporally and spatially collocated radiance measurements. *Geophys. Res. Lett.*, **28**, 167–170.

Hansen J., Nazarenko L., Ruedy R., Sato M., Willis J., Del Genio A., Koch D., Lacis A., Lo K., Menon S., Novakov T., Perlwitz J., Russell G., Schmidt G. A., and Tausnev N. (2005). Earth's energy imbalance: Confirmation and implications. *Science*, **308**, 1431–1435.

Hartmann D.L., Ramanathan V., Berroir A., and Hunt G.E. (1986). Earth radiation budget data and climate research. *Rev. Geophys.*, **24**, 439–468.

House F.B., Gruber A., Hunt G.E., and Mecherikunnel A.T. (1986). History of satellite missions and measurements of the Earth Radiation Budget. *Rev. Geophys.*, **24**, 357–377.

Jacobowitz H., Soule H.V., Kyle H.L., and House F.B. (1984). The Earth Radiation Budget (ERB) experiment: An overview. *J. Geophys. Res.*, **89**, 5021–5038.

Kandel R.S. (1990). Satellite observations of the Earth radiation budget and clouds. *Space Sci. Rev.*, **52**, 1–32.

Kandel R., Viollier M., Raberanto P., Duvel J.-Ph., Pakhomov L.A., Golovko V.A., Trishchenko A.P., Mueller J., Raschke E., and Stuhlmann, R. (1998). The ScaRaB Earth Radiation Budget Dataset. *Bulletin of the American Meteorological Society*, **79**(5), 765–783.

Kozoderov V.V. and Golovko V.A. (1999). Interpretation and analysis of the earth radiation budget components from space. *Proceedings of Third International Scientific Conference on the Global Energy and Water Cycle, June 16–19, 1999, Beijing*. China Meteorological Office, Beijing, pp. 173–174.

Levitus S., Antonov J.I., Boyer T.P., and Stephens C. (2000). Warming of the world ocean. *Science*, **287**, 2225–2229.

Li Z. and Trishchenko A. (1999). A study toward an improved understanding of the relationship between visible and shortwave albedo measurements. *J. Atmos. Oceanic Technol.*, **16**, 347–360.

Liebmann B. and Smith C.A. (1996). Description of a complete (interpolated) outgoing longwave radiation dataset. *Bull. Amer. Meteor. Soc.*, **77**, 1275–1277.

Marchuk G.I., Kondratyev K.Ya., and Kozoderov V.V. (1988). *The Earth Radiation Budget: Key Aspects*. Science, Moscow, 224 pp.

Monge J.L., Kandel R.S., Pakhomov L.A., and Adasko V.I. (1991). ScaRaB earth radiation budget scanning radiometer. *Meteorologia*, **28**, 261–284.

Mueller J., Stuhlmann R., Raschke E., Monge J.L., Kandel R., Burkert P., and Pakhomov L.A. (1993). Solar ground calibration of ScaRaB preliminary results. In: D.K. Lynch (ed.), *Passive Infrared Remote Sensing of Clouds and the Atmosphere*. SPIE, Bellingham, WA, pp. 129–139.

Mueller J., Stuhlmann R., Becker R., Raschke E., Monge J.L., and Burkert P. (1996). Ground-based calibration facility for the Scanner for Radiation Budget instrument in the solar spectral domain. *Meteorologia*, **32**, 657–660.

Mueller J., Stuhlmann R., Becker R., Raschke E., Rinck H., Burkert P., Monge J.-L., Sirou F., Kandel R., Tremas T., and Pakhomov L.A. (1997). Ground characterization of the Scanner for Radiation Budget (ScaRaB) Flight Model 1. *J. Atmos. Oceanic Technol.*, **14**, 802–813.

Ramanathan V., Cess R.D., Harrison E.F., Minnis P., Barkstrom B.R., Ahmad E., and Hartmann D. (1989). Cloud-radiative forcing and climate: Results from the Earth Radiation Budget Experiment. *Science*, **243**, 57–63.

Raschke E., Vonder Haar T.H., Bandeen W.R., and Pasternak M. (1973). The annual radiation balance of the earth–atmosphere system during 1969–1970 from Nimbus-3 measurements. *J. Atmos. Sci.*, **30**, 341–364.

Raval A. and Ramanathan V. (1989). Observational determination of the greenhouse effect. *Nature*, **342**, 758–761.

Stephens G.L. and Greenwald T.J. (1991). The Earth's radiation budget and its relation to atmospheric hydrology, 1: Observations of the clear sky greenhouse effect. *J. Geophys. Res.*, **96**, 15311–15324.

Stephens G.L., Campbell G.G., and Vonder Haar T.H. (1981). Earth radiation budgets measurements from satellites and their interpretation for climate modeling and studies. *J. Geophys. Res.*, **86**, 9739–9760.

Stowe L.L. (ed.) (1988). *Report of the Earth Radiation Budget Requirements Review 1987*, NOAA Tech. Rep. NESDIS-41. National Oceanic and Atmospheric Administration, Washington, D.C., 103 pp.

Stubenrauch C.J., Duvel J.-Ph., and Kandel R.S. (1993). Determination of longwave anisotropic emission factors from combined broad- and narrow-band radiance measurements. *J. Appl. Meteor.*, **32**, 848–856.

Venegas, S.A. (2001). *Statistical Methods for Signal Detection in Climate*. University of Copenhagen, Copenhagen, Denmark, 96 pp.

Viollier M., Kandel R., and Raberanto P. (1995). Inversion and space-time averaging algorithms for ScaRaB (Scanner for Earth Radiation Budget): Comparison with ERBE. *Ann. Geophys.*, **13**, 959–968.

von Storch H., Burger G., Schnur R., and von Storch J. (1995). Principal oscillation patterns: A review. *J. Climate*, **8**, 377–400.

Waliser D.E. and Zhou W. (1997). Removing satellite equatorial crossing time biases from the OLR and HRC datasets. *J. Climate*, **10**, 2125–2146.

Wielicki B.A., Barkstrom B.R., Harrison E.F., Lee III R.B., Smith G.L., and Cooper J.E. (1996). Clouds and the Earth's Radiant Energy System (CERES): An Earth Observing System experiment. *Bull. Amer. Meteor. Soc.*, **77**, 853–868.

Wielicki B.A., Wong T., Young D.F., Barkstrom B.R., Lee R.B., and Haeffelin M. (1999). Differences between ERBE and CERES tropical mean fluxes: ENSO, climate or calibration? (Preprints). *Proceedings of the Tenth Conference on Atmospheric Radiation, June 28–July 2, 1999, Madison, WI*. American Meteorological Society, Washington, D.C., pp. 48–51.

Wielicki B.A., Wong T., Allan R.P., Slingo A., Kiehl J.T., Soden B.J., Gordon C.T., Miller A.J., Yang S.-K., Randall D.A., Robertson F., Susskind J., and Jacobowitz, H. (2002). Evidence for large decadal variability in the tropical mean radiative energy budget. *Science*, **295**, 841–842.

4

Aerosol and atmospheric electricity

Yury A. Pkhalagov, Victor N. Uzhegov, and Ivan I. Ippolitov

4.1 INTRODUCTION

It is well known that the range of scientific interests of Kirill Kondratyev was very wide. It is sufficient to say that in the period 2001–2006 he published about 20 reviews in the journal *Atmospheric and Ocean Optics* covering different problems of atmospheric optics. In particular, a significant part of the material in these reviews is devoted to the problems of aerosol generation, the spatio-temporal variability of aerosols in the atmosphere under the influence of different factors, and the role of aerosols as a climate-forcing component of the atmosphere.

Among the factors affecting weather and climate, Kondratyev considered the smoke from forest fires which occur frequently in different regions of the Earth and emit large quantities of greenhouse gases and aerosols into the atmosphere (Kondratyev and Grigoryev, 2004; Kondratyev and Isidorov, 2001). It is generally assumed that about 30% of tropospheric ozone, carbon monoxide, and carbon dioxide contained in the atmosphere comes from forest fires. Aerosol emissions related with forest fires can significantly affect the radiation budget of the Earth due to direct extinction of incoming solar radiation, due to the change of the albedo of the atmosphere and due to the change of the optical properties of clouds in their interaction with aerosol particles (indirect aerosol effect). In this context, field experiments on the study of aerosol extinction of optical radiation in the visible and infrared wavelength ranges under conditions of the presence of smoke in the atmosphere are doubtless of interest. As smoke aerosols take part in the cloud formation process, the efficiency of which essentially depends on the presence or absence of charge on water vapor particles (Ivlev and Khvorostovskii, 2000), it seems important to study experimentally the peculiarities of the interaction of the atmospheric electric field with aerosols in general, and with smoke aerosols in particular.

In this chapter we discuss the results of such investigations carried out in the region of Tomsk in 1997–2004 following on from the pioneer work of Kondratyev and recognizing the need to improve the treatment of aerosols in climate modeling.

4.2 THE RELATION OF AEROSOL EXTINCTION OF OPTICAL RADIATION WITH THE ELECTRIC FIELD UNDER HAZE CONDITIONS

It is known that the electric field in the atmosphere exists continuously and is characterized by a wide spectrum of natural variations caused by different factors. To date, there are several physical hypotheses explaining the phenomenon of the existence of the electric field in the atmosphere, a detailed discussion of which is given, for example, by Imyanitov and Shifrin (1962). The mean value of the electric field in the near-ground air layer is about $+130\,\mathrm{V\,m^{-1}}$. In precipitation, thunderstorms, snowstorms, and dust storms the electric field varies over a very wide range (up to $10{,}000\,\mathrm{V\,m^{-1}}$), and the sign of the field often changes. The important factor causing the variability of the magnitude of the atmospheric electric field, E, under good weather conditions (as in the case of atmospheric haze) is the variation of the concentration of ionized air molecules (light ions) due to their absorption by atmospheric aerosol. Ionization of air molecules leading to the appearance of electrical conductivity of the atmosphere occurs mainly due to natural gamma radiation and radiation from space. Light ions settling on neutral aerosol particles are effectively transformed into heavy ions, which do not take part significantly in charge transfer. The decrease of the concentration of light ions by their attachment to aerosol particles leads to a decrease in the electric conductivity of the air and, hence, to an increase of the electric field.

The existence of a correlation between dust content in the air and the strength of the electric field under good weather conditions was considered for the first time by Imyanitov and Shifrin (1962). It is based on an ionization–recombination equation for light particles:

$$\frac{dn_+}{dt} = I - an_+n_- - bn_+N_- - cn_+N_0, \tag{4.1}$$

where n_+ and n_- are the concentrations of light positive and negative ions, respectively; I is the intensity of the ionization; N_- is the concentration of negatively charged heavy ions; N_0 is the concentration of neutral particles; and a, b, and c are the respective recombination coefficients.

In the stationary state and in immobile air $\dfrac{dn_+}{dt} = 0$, thus

$$n_+ = \frac{I}{an_- + bN_- + cN_0}. \tag{4.2}$$

It follows from Equation (4.2) that under good weather conditions the quantity of light ions decreases with the increase of the total concentration of heavy ions and neutral particles, and the electric field should increase.

Considering the optical characteristics of the atmosphere, taking into account the relationship (4.2), the correlation between the aerosol extinction coefficient $\beta(\lambda)$ and the electric field of the atmosphere E under good weather conditions can be written as

$$E = k\beta(\lambda), \tag{4.3}$$

where k is the coefficient of proportionality. Equation (4.3) relating the electric and optical characteristics of the atmosphere is also known as the electro-optical relation. Note that it was obtained assuming that variations of the parameter $\beta(\lambda)$ are caused only by variations in the concentration of particles. But the value of $\beta(\lambda)$ really depends not only on the number of particles but also on their size and the material of which they are composed, which can vary significantly depending on circumstances, including meteorological conditions. So, it is interesting to study how the electrical and optical parameters of the atmosphere are related to each other under conditions of real atmospheric haze, and what is the role of particles of different size in the mechanism of the attachment of light ions to aerosol particles.

For this purpose, a 20-day cycle of round-the-clock synchronous measurements of atmospheric spectral transmission near the ground $T(\lambda)$, in the wavelength range $\lambda = 0.44\,\mu\text{m}$–$12.2\,\mu\text{m}$, along an 830 m long near-ground path, of the electric field E, of the relative humidity of the air RH, and of the air temperature θ, was carried out in August–September 1997 in the region of Tomsk (Pkhalagov *et al.*, 1999). All measurements were performed six times per day. The electric field was measured by means of a string dynamic sensor (Struminskii, 1981). Spectral measurements of $T(\lambda)$ were carried out with a two-channel filter photometer (Pkhalagov *et al.*, 1992). The total extinction coefficients $\varepsilon(\lambda)$ were obtained from the measured values of $T(\lambda)$, and they were then used to calculate the aerosol extinction coefficients, $\beta(\lambda)$, in the entire wavelength range using a multiple linear regression (Pkhalagov and Uzhegov, 1988). We note that the value of the parameter $\beta(\lambda)$ in this wavelength range makes it possible to use optical parameters to estimate the efficiency of particles of different sizes as sinks for light ions.

4.3 RESULTS OF MEASUREMENTS

The aerosol extinction coefficient in the visible wavelength range $\beta(0.55)$ varied during the period of measurements from $0.07\,\text{km}^{-1}$ to $0.35\,\text{km}^{-1}$ (that corresponds to a change of visibility range from 55 km to 11 km), the electric field varied from $8\,\text{V}\,\text{m}^{-1}$ to $132\,\text{V}\,\text{m}^{-1}$, the air temperature varied from 0.1°C to 19°C, and the relative humidity of the air varied from 33% to 95%. Such wide ranges of variation of the parameters is evidence of the statistical significance of the array.

The mean values of these parameters, their r.m.s. deviations, and correlation coefficients in the period of measurements are shown in Table 4.1. Two items in Table 4.1 attract our attention. The first is the significant positive correlation of the electric field with the aerosol extinction coefficient in the visible wavelength range ($\rho_{\beta(0.55),E} = 0.50$); this corresponds, in general, to the electro-optical relation (4.3). The second is the unexpected high correlation of the electric field with the relative

Table 4.1. Mean values, r.m.s. deviations, and correlation coefficients of parameters $\beta(0.55)$, RH, θ, and E in the total data array. The level of significant correlation is ~0.20.

Measured parameter	*Mean value*	*r.m.s deviation*	*Correlation coefficients*			
			$\beta(0.55)$ (km^{-1})	RH (%)	θ (°C)	E ($V m^{-1}$)
$\beta(0.55)$ (km^{-1})	0.141	0.053	1.00	—	—	—
RH (%)	69.97	18.43	0.41	1.00	—	—
θ (°C)	9.55	3.42	−0.016	−0.710	1.00	—
E ($V m^{-1}$)	76.13	34.01	0.50	0.727	−0.38	1.00

humidity of the air ($\rho_{RH,E} = 0.73$). As a preliminary hypothesis to explain the correlation of the variations of E and RH in the near-ground layer of the atmosphere, one can consider a mechanism where the concentration of particles of nanometer dimension strongly increases with increase of relative humidity of air. These particles are not seen in scattering (because of the small size), but they do serve as an additional sink for light ions and that leads to an increase of the electric field.

To examine this hypothesis we analyse the mutual dynamics of variations of the coefficients $\beta(\lambda)$ at different wavelengths and the field E in haze of different density. To perform this analysis, the total data array was divided into three subarrays according to the degree of atmospheric turbidity. The extinction coefficient at the wavelength of 0.55 μm was taken as a criterion of the degree of turbidity. The first subarray consisted of the spectra obtained at $\beta(0.55) = 0.11\ km^{-1}$, the second was at $\beta(0.55) = 0.11\ km^{-1}$–$0.2\ km^{-1}$, and the third was at $\beta(0.55) = 0.2\ km^{-1}$. The mean values and r.m.s. deviations of the parameters $\beta(0.55)$, RH, θ, and E in each subarray are presented in Table 4.2. It is seen that an increase in the turbidity of air, which occurs with an increase in the relative humidity of the air, leads to a significant increase of the mean atmospheric electric field. The rate of growth of $\beta(0.55)$ and E dramatically increases at a relative humidity greater than 70%. The mean spectral

Table 4.2. Mean values of parameters $\beta(0.55)$, RH, θ, and E corresponding to the mean spectra of the aerosol extinction coefficients in Figure 4.1 (N is the number of realizations).

Number of curve in Figure 4.1	$\overline{\beta(0.55)}$ (km^{-1})	$\overline{RH}$ (%)	$\overline{\theta}$ (°C)	$\overline{E}$ ($V m^{-1}$)	N
1	0.103	63.6	8.36	55.0	18
2	0.153	69.1	9.87	76.8	64
3	0.302	91.3	4.41	117.3	9

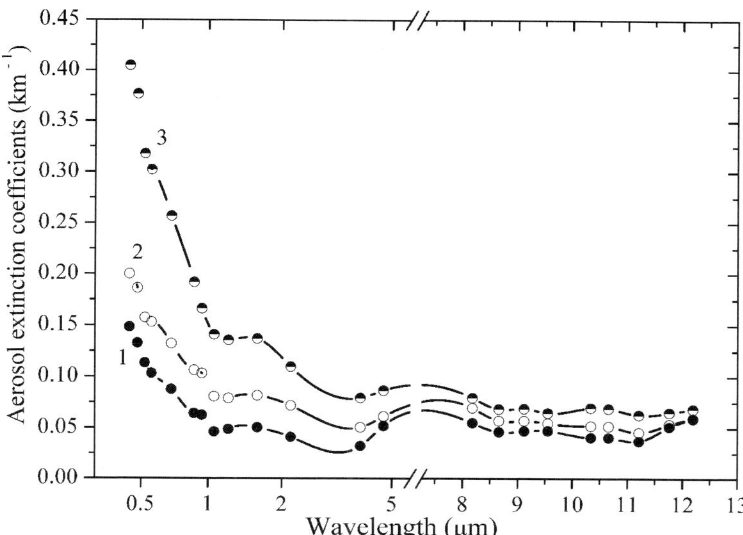

Figure 4.1. The spectral structure of aerosol extinction coefficients $\beta(\lambda)$ in atmospheric haze of different density. The values of the parameters $\beta(\lambda)$, RH, θ, and E for Curves 1, 2, and 3 are presented in Table 4.2.

dependencies of the aerosol extinction coefficients corresponding to these subarrays are shown in Figure 4.1. It is seen that the greatest significant variations of the coefficients $\beta(\lambda)$ occur in the wavelength range 0.44 μm to 2.0 μm, where particles of submicron dimension make the main contribution to aerosol extinction, and they are essentially less in the range 8 μm–12 μm, where coarse aerosol (radius of particles $r > 1$ μm) makes the main contribution. Comparison of the amplitudes of variations of the coefficients $\beta(\lambda)$ at different wavelengths and the electric field in haze of different density confirms the assumption about the prevalent role of submicron particles in the process of attachment of light ions to atmospheric aerosol particles.

For a comparative estimate of the role of aerosol particles of different size in their interaction with the electric field, the spectral dependence of the correlation coefficient between $\beta(\lambda)$ in the range of λ from 0.44 μm to 12 μm and the electric field E for the total data array constructed from the data of Table 4.3 is shown in Figure 4.2. As can be seen, the coefficient $\beta(\lambda)$ has significant correlation with the electric field in the wavelength range $\lambda = 0.44$ μm–3.92 μm with a maximum in the shortwave range. The correlation between $\beta(\lambda)$ and E is broken in the range $\lambda > 4$ μm. These data are evidence of the fact that under good weather conditions the main sink for light ions determining the magnitude of the electric field in the atmosphere occurs on submicron and microdispersed ($r < 0.1$ μm) aerosol particles. The change of concentration of coarse aerosol in near-ground haze weakly affects the electric field that is related with low concentration of coarse particles in the atmosphere.

On the whole, the experimental data obtained show that, even under good weather conditions, the correlation coefficient between variations in aerosol

Table 4.3. Mean values of coefficients $\beta(\lambda)$, their rms deviations $\sigma_{\beta(\lambda)}$, and correlation coefficients between $\beta(\lambda)$ and the parameters RH, θ, and E in autumn haze in western Siberia.

λ (μm)	$\overline{\beta(\lambda)}$ (km^{-1})	$\sigma_\beta(\lambda)$ (km^{-1})	$\rho_{\beta(\lambda),\mathrm{RH}}$	$\rho_{\beta(\lambda),\theta}$	$\rho_{\beta(\lambda),E}$
0.44	0.2087	0.0770	0.461	−0.118	0.511
0.52	0.1639	0.0612	0.423	−0.051	0.502
0.56	0.1406	0.0526	0.413	−0.016	0.498
0.69	0.1353	0.0506	0.406	−0.003	0.495
0.87	0.1066	0.0403	0.327	0.125	0.471
1.06	0.0801	0.0320	0.222	0.203	0.423
1.22	0.0789	0.0315	0.149	0.205	0.342
1.60	0.0816	0.0325	0.160	0.196	0.337
2.17	0.0700	0.0284	0.165	0.233	0.381
3.97	0.0503	0.0221	0.079	0.232	0.308
4.69	0.0623	0.0258	−0.158	0.315	0.112
8.18	0.0682	0.0276	−0.277	0.496	0.056
9.12	0.0569	0.0242	−0.259	0.405	0.056
10.34	0.0522	0.0228	−0.198	0.417	0.096
11.21	0.0459	0.0206	−0.197	0.337	0.121
11.76	0.0549	0.0234	−0.378	0.422	−0.071

extinction in the visible wavelength range and the electric field is comparatively small ($\rho_{\beta(\lambda),E} \sim 0.5$). This means that in the general case there is no unambiguous relation between these parameters in real atmospheric haze. However, undoubtedly, there is an effect of atmospheric aerosol on the electric field. This effect is more pronounced under conditions of dense haze at high relative humidity of the air.

4.4 CORRELATION BETWEEN AEROSOL EXTINCTION OF RADIATION AND THE ATMOSPHERIC ELECTRIC FIELD UNDER SMOKE CONDITIONS

To study the peculiarities of the interaction of aerosol and atmospheric electricity under conditions of smoke in the atmosphere related with forest fires, simultaneous

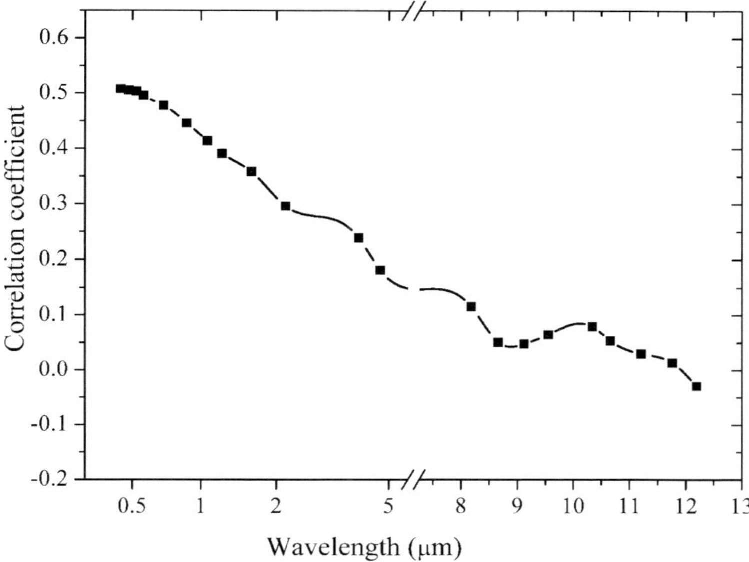

Figure 4.2. Mean spectral structure of the correlation coefficient of aerosol extinction $\beta(\lambda)$ with the electric field E in atmospheric haze.

round-the-clock measurements of the spectral aerosol extinction coefficient $\beta(\lambda)$ in the range $\lambda = 0.44\,\mu\text{m}$–$12\,\mu\text{m}$ and the electric field E were carried out in May 2004 in the region of Tomsk. Measurements of $\beta(\lambda)$ were carried out every 2 hours by means of a filter photometer (Pkhalagov *et al.*, 1992). The electric field was measured with the stationary electrostatic fluxmeter Pole-2 installed on a metallic grid at the end of the measurement path. All the optical and electrical measurements were accompanied by standard meteorological observations, and measurements of the concentration of carbon monoxide, mass concentration of aerosol containing soot, gamma-ray background, and ozone concentration were carried out at the permanently operating stations of the Institute of Atmospheric Optics, Siberian Branch of the Russian Academy of Sciences. Thus, a set of 140 values of all the aforementioned parameters was obtained.

4.5 DISCUSSION OF RESULTS

Typical for May, very variable weather was observed in the region during the period of measurements. The air temperature varied from 1°C to 35°C, the relative humidity from 20% to 97%, and the visibility range from 3 km to 30 km (Pkhalagov *et al.*, 2006).

To illustrate the temperature–humidity regime in the region during the period of measurements, Figure 4.3 shows the temporal behavior of parameters θ and RH. It

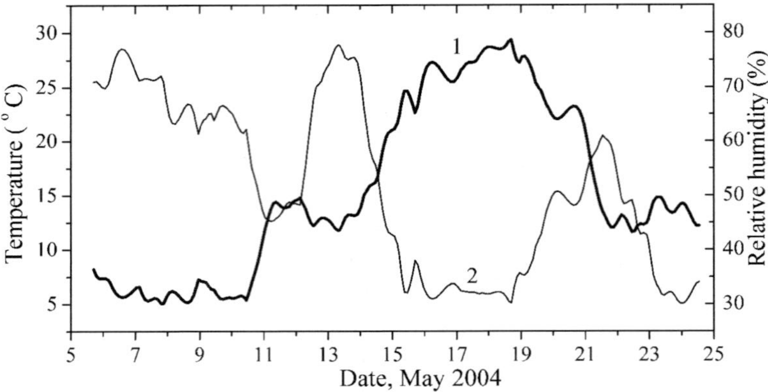

Figure 4.3. Smoothed inter-day dynamics of temperature (Curve 1) and relative humidity of air (Curve 2) during measurements (May 2004).

can be seen in Figure 4.3 that the maximum air temperatures ($\theta > 25°C$) and minimum relative humidity (RH = 35%–50%) were observed in the region in the period from May 15 to 21. The highest content of smoke in the atmosphere was observed in the region in this hot and dry period. It can be seen in Figure 4.4, where the temporal dynamics of the aerosol extinction coefficients in three wavelength intervals ($\lambda = 0.45\,\mu m$, $1.6\,\mu m$, and $3.9\,\mu m$) is shown, that the maximum is also observed in the period from May 15 to 21 in the temporal behavior of the coefficient $\beta(\lambda)$ at all wavelengths. It is the most pronounced in the shortwave range ($\lambda = 0.45\,\mu m$). Synchronic change of temporal variations of $\beta(\lambda)$ at these three wavelength attracts our attention. The quantitative measure of the revealed synchronic variations is the

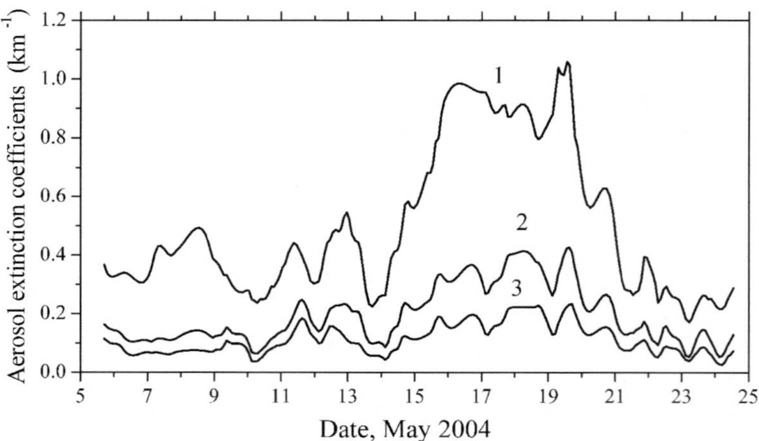

Figure 4.4. Temporal variability of aerosol extinction coefficients in the wavelength range $\lambda = 0.45\,\mu m$ (Curve 1), $1.60\,\mu m$ (Curve 2), and $3.9\,\mu m$ (Curve 3) in the period May 5–25, 2004.

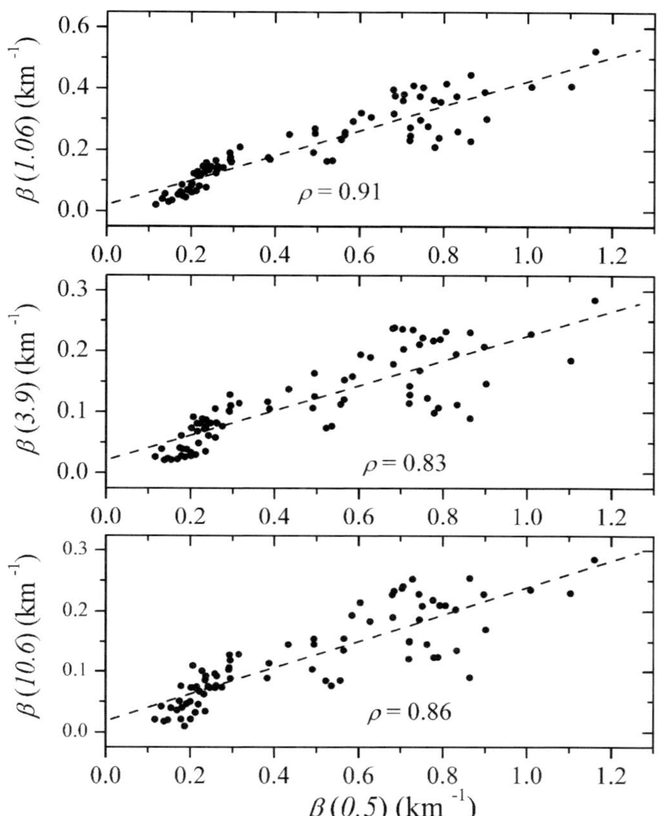

Figure 4.5. Statistical correlation of variations of the aerosol extinction coefficients in the visible and infrared wavelength range. ρ is the correlation coefficient.

autocorrelation coefficients of the parameter $\beta(0.45)$ with the parameters $\beta(1.06)$, $\beta(3.91)$, and $\beta(10.6)$ shown in Figure 4.5 and taking the values 0.91, 0.83, and 0.86, respectively. Such high autocorrelation coefficients are evidence of the presence of the general factor in aerosol extinction of visible and infrared radiation under conditions of forest fire smoke. Most likely, the general factor here is the fact that smoke particles are formed from the gaseous phase, while larger particles coming to the atmosphere from the underlying surface at the location of the fire are emitted into the atmosphere simultaneously during a fire.

Figure 4.6 shows the temporal behavior of the carbon monoxide concentration and the mass concentration of aerosol containing soot, which also have a well-pronounced maximum in the period from May 15 to 21; this is unambiguous evidence of smoke as the origin of the maximum in aerosol extinction (Figure 4.4).

To study the correlation of variations in aerosol extinction and the electric field E under smoke/haze conditions, the smoothed temporal dynamics of the coefficient

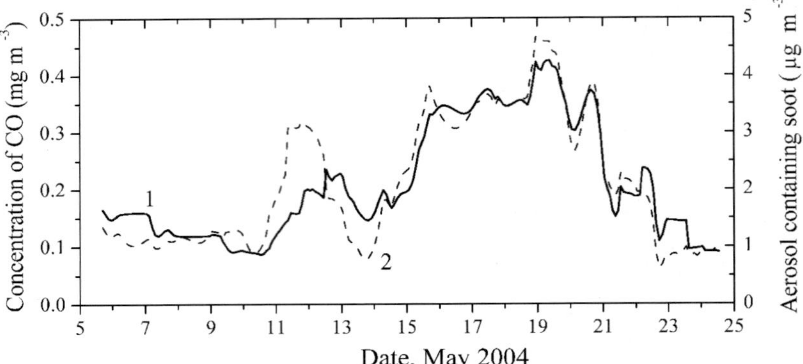

Figure 4.6. Temporal behavior of the concentration of carbon monoxide (CO) (Curve 1) and aerosol containing soot (Curve 2).

$\beta(0.5)$ and the parameter E during measurements is shown in Figure 4.7. It can be seen in Figure 4.7 that significant decrease of the electric field is observed in the period of the strongest smoke content in the atmosphere (i.e., during smoke/haze the parameters $\beta(0.50)$ and E vary oppositely). This obviously contradicts the electro-optical relation (4.3), according to which the field strength should increase with increase in atmospheric turbidity under good weather conditions.

To explain this effect, calculations of the correlation coefficients ρ between the parameter E and the coefficients $\beta(\lambda)$ were carried out in the visible and infrared wavelength ranges. The results of these calculations are shown in Table 4.4.

It can be seen that the correlation coefficients between E and $\beta(\lambda)$ for the site of the fire are negative and are quite close over the entire wavelength range. The

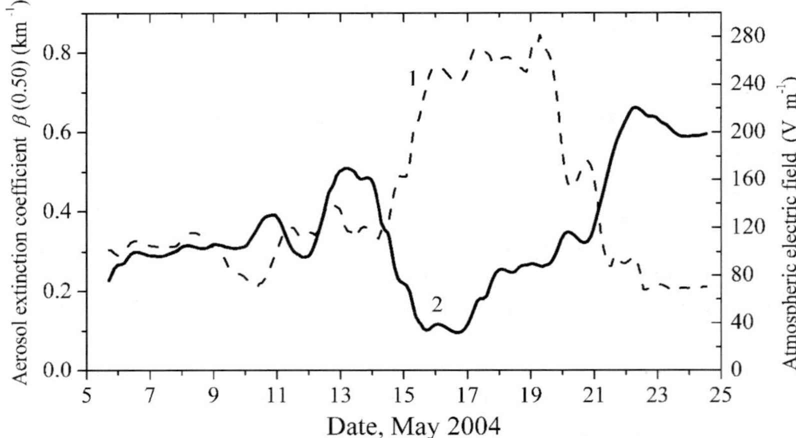

Figure 4.7. Smoothed temporal behavior of the aerosol extinction coefficient $\beta(0.50)$ (Curve 1) and the atmospheric electric field E (Curve 2) during the period of measurements.

Table 4.4. Mean values of coefficients $\beta(\lambda)$, their rms deviations $\sigma_\beta(\lambda)$, and spectral correlation coefficients between $\beta(\lambda)$ and the electric field E under conditions of smoke in the atmosphere.

λ (μm)	$\overline{\beta(\lambda)}$ (km^{-1})	$\sigma_\beta(\lambda)$ (km^{-1})	$\rho_{\beta(\lambda),E}$	
			Smoke (Pkhalagov *et al.*, 2006)	*Haze* (Krechetov and Shamanskii, 2005)
0.45	0.529	0.31	−0.67	0.51
0.50	0.467	0.276	−0.67	0.51
0.55	0.416	0.246	−0.67	0.50
0.63	0.344	0.205	−0.66	0.49
0.69	0.315	0.189	−0.67	0.49
0.87	0.246	0.146	−0.64	0.47
1.06	0.208	0.122	−0.62	0.42
1.6	0.158	0.094	−0.62	0.34
2.17	0.134	0.078	−0.57	0.38
3.91	0.116	0.068	−0.55	0.31
8.18	0.121	0.073	−0.6	0.06
10.6	0.121	0.071	−0.59	0.14

correlation coefficients between E and $\beta(\lambda)$ presented in the last column, obtained by Krechetov and Shamanskii (2005) for measurements in haze, are quite different. They are positive and decrease with wavelength; this is evidence of the fact that the main sink of light ions under good weather conditions occurs on small aerosol particles, and this mechanism under these conditions regulates the variability of the atmospheric electric field. Therefore, it is quite clear that electro-optical relations in the atmosphere under smoke/haze conditions are caused by other mechanisms.

In essence, this effect of a decrease of the electric field in smoke/haze is evidence of the fact that, as aerosol concentration in the atmosphere increases, the quantity of charged particles in smoke not only does not decrease (as it occurs under good weather conditions)—but significantly increases. This leads to the conclusion that either for some reason an intensive growth of the quantity of light ions, which leads to a decrease in field strength, occurs in the region of the fire, or else a fine aerosol is generated in the region of the fire and is initially charged, and the field dynamics is caused by some other mechanism.

As an alternative explanation of this effect, one could assume that the yield of radioactive emission increases during fire in soil that is drying out, and the processes

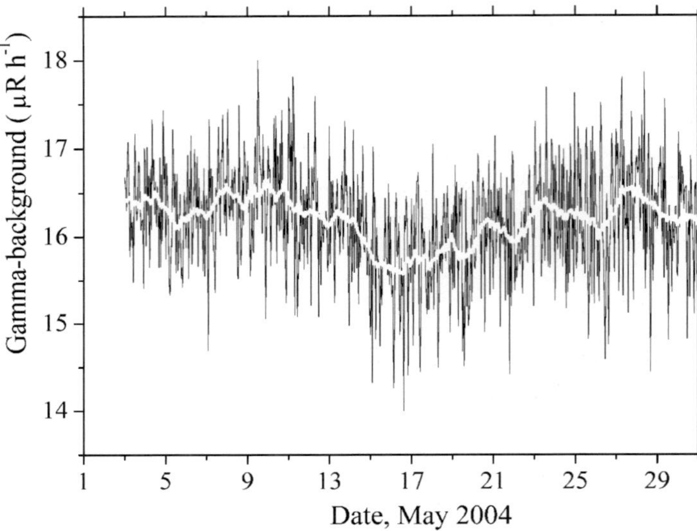

Figure 4.8. Temporal behavior of the gamma-ray background in the region obtained at the TO Station of the Institute for Atmosphere Optics, Siberian Branch of the RAS (the light curve is the smoothed temporal behavior of the gamma-ray background).

of ionization of air molecules become more intensive (Imyanitov and Shifrin, 1962; Smirnov, 1983). As already mentioned above, the gamma-ray background was measured in this experiment among many other parameters. The gamma-ray background varied during the entire period of measurements within the limits from 16 μroentgen to 17 μroentgen/hour while the measuring error is ~2 μroentgen/hour (Figure 4.8). This means that variations in gamma-ray background in the region of the fire could not affect the concentration of light ions in the atmosphere.

Assuming that the fine aerosol generated in the region of the fire is initially charged, one can consider the formation of a negative volume charge in the near-ground layer of the atmosphere, partially compensating the initial electrostatic field, as the most likely mechanism explaining the decrease of the electric field during smoke/haze (Imyanitov and Shifrin, 1962; Krechetov and Shamanskii, 2005).

4.6 CONCLUSIONS

This work has revealed a significant decrease of the electric field (from 200 $V m^{-1}$ to 30 $V m^{-1}$–60 $V m^{-1}$) with the increase in aerosol concentration during smoke/haze in the atmosphere. This obviously contradicts the known electro-optical relation and is evidence of the fact that, as the number density of aerosol particles increases, the quantity of light ions in the atmosphere not only does not decrease—but significantly increases. This leads to the suggestion that the fine photochemical aerosol generated at a fire is charged and forms a negative volume charge in the near-ground layer of the atmosphere and partially compensates the initial electrostatic field.

The results obtained provide evidence of a very strong effect of fires on the electro-optical characteristics of the lower troposphere. Due to the great number of fires on the surface of the Earth, it is necessary to take this into account in climatic models and in considering physical mechanisms of the effect of solar activity on weather and climate.

4.7 REFERENCES

Imyanitov I.M. and Shifrin K.S. (1962). Modern state of the study of atmospheric electricity. *Achievements of Physical Sciences*, **LXXVI**, 593–642 [in Russian].

Ivlev L.S. and Khvorostovskii S.N. (2000). Investigation into effect of cosmic radiation on microstructural parameters and optical properties of the lower atmosphere in middle and high latitudes, Part 2: Heterogeneous processes under the effect of a flow of high-energy particles. *Atmos. Ocean Optics*, **13**(12), 1000–1004 [in Russian].

Kondratyev K.Ya. and Grigoryev Al.A. (2004). Forest fires as a component of natural ecodynamics. *Atmos. Ocean Optics*, **17**(4), 245–255 [in Russian].

Kondratyev K.Ya. and Isidorov V.A. (2001). Effect of biomass burning on chemical composition of the atmosphere. *Atmos. Ocean Optics*, **14**(2), 93–101 [in Russian].

Krechetov A.A. and Shamanskii Yu.V. (2005). Gradient of the atmospheric electric field as an indicator of atmospheric pollution. *Atmos. Ocean Optics*, **18**(1/2), 128–130 [in Russian].

Pkhalagov Yu.A. and Uzhegov V.N. (1988). Statistical method for separation of the IR radiation extinction coefficients into components. *Atmos. Ocean Optics*, **1**(10), 3–11 [in Russian].

Pkhalagov Yu.A., Uzhegov V.N., and Shchelkanov N.N. (1992). Automated multiwave meter of spectral transmission of the ground layer of the atmosphere. *Atmos. Ocean Optics*, **5**(6), 423–424 [in Russian].

Pkhalagov Yu.A., Uzhegov V.N., Ovcharenko E.V., Genin V.N., Donchenko V.A., Kabanov M.V., and Shchelkanov N.N. (1999). Study of correlation between aerosol extinction of optical radiation and atmospheric electric field strength. *Atmos. Ocean Optics*, **12**(2), 99–102 [in Russian].

Pkhalagov Yu.A., Uzhegov V.N., Panchenko M.V., and Ippolitov I.I. (2006). Electro-optical relations in the atmosphere under smoke/haze conditions. *Atmos. Ocean Optics*, **19**(10), 774–777 [in Russian].

Smirnov V.V. (1983). Electric factors of air cleanness. *Proceedings Institute of Experimental Meteorology*, **30**(104), 64–106.

Struminskii V.I. (1981). *Sensor of Electrostatic Field*, Inventor's Certificate No. 830256. Novosibirsk State University, Novosibirsk.

5

Remote sensing of terrestrial chlorophyll content

Jadunandan Dash, Paul J. Curran, and Giles M. Foody

5.1 INTRODUCTION

Terrestrial chlorophyll content is a key environmental variable that is difficult to estimate accurately using remotely sensed data. Some of the pioneering studies in this field were undertaken by Kirill Kondratyev and co-workers in the 1970s and early 1980s. These studies paved the way for the operational global chlorophyll content maps of today. This chapter reviews some of Kondratyev's pioneering contributions to the development of the theory and practice for the remote sensing of terrestrial chlorophyll content. In particular, the chapter summarizes a third of a century of scientific research before concluding with a discussion of three contemporary applications of the remote sensing of terrestrial chlorophyll content.

Remote sensing may be used to monitor and evaluate terrestrial vegetation properties at a wide range of spatial and temporal scales. The type of remotely sensed data required depends on the ecological questions being asked (Curran, 2001). Three levels of questions may typically be posed about the vegetation of a study area: *"What is the type of vegetation there? How much vegetation is there? What is the condition of that vegetation?"* Kondratyev made major contributions to research that sought to answer each of these questions with many early and influential studies based on measurements acquired from space (Kozoderov and Dmitriev, 2008; Sushkevich, 2008). The first two questions can usually be answered using broadband remotely sensed data, whereas the third question usually requires data recorded in narrow wavebands. Specifically, data recorded in narrow visible/near-visible wavebands by a hyperspectral sensor can be used to estimate foliar biochemical content at local to global scales (Banninger, 1991; Curran *et al.*, 1997; Blackburn, 2007). Such information can, in turn, be used to quantify, understand, and support management of the vegetated environment.

Chlorophyll is one of the most important foliar biochemicals. The amount of chlorophyll within a vegetation canopy is related positively to both the productivity and the health of that vegetation. The chlorophyll concentration of leaves can be defined as the amount of chlorophyll per unit weight of leaf ($mg\,g^{-1}$) or as a percentage of leaf weight (% dry weight). It is often impractical to estimate chlorophyll concentration for a whole canopy using remote-sensing techniques because the mass of leaves per unit area can vary spatially. Therefore, unless the mass of leaves per unit area is constant, it is the chlorophyll content (concentration $\times$ leaf mass) rather than concentration alone that is most closely related to the remotely sensed response of the vegetation canopy (Curran *et al.*, 1995). For example, in a remote-sensing study of grassland the chlorophyll content (g) was defined as chlorophyll concentration ($mg\,g^{-1}$) $\times$ biomass (g) within an area covered by a pixel (Jago *et al.*, 1997). During the last third of a century many techniques have been developed to estimate canopy chlorophyll content remotely using data acquired from ground, airborne, and spaceborne sensors (Jago *et al.*, 1999). Compared with direct field-based sampling of chlorophyll content these remote-sensing approaches are less labor-intensive, are non-destructive, and offer the possibility of repetitive coverage. These estimates of chlorophyll content are now used widely to help understand and manage the vegetated portion of our environment. The origin of the remote-sensing approaches can be traced back to the early days of remote sensing, as researchers around the globe tried to unlock the information content of remotely sensed imagery. In the former Soviet Union, Kondratyev and his co-workers undertook pioneering research on the interaction of electromagnetic radiation with vegetation. Their work included developing a model to understand vegetation reflectance spectra over a wide range of spatial scales, developing techniques to estimate biochemical variables from remotely sensed data (Curran *et al.*, 1990b) and using remotely sensed estimates of, for example, chlorophyll concentration to map and monitor the status of vegetation in general and crops in particular. Kondratyev's early research in this field was explicitly inductive in that he and his colleagues collected spectra, measured vegetation variables, and described, qualitatively and quantitatively, what they found (Kondratyev and Smoktiy, 1973; Kondratyev and Fedchenko, 1979). Much of the remote-sensing community of the time was transfixed by innovations in, primarily space-based, sensing technology and were finding it difficult to link the large science questions of the day to the potential of this fledgling field.

Kondratyev was, by the mid-1970s, a leading figure in Soviet science, had an international reputation for research on the physical basis of environmental remote sensing (Kondratyev *et al.*, 1969), and was able to obtain relevant research resources, not least field and airborne spectrometers. Kondratyev, Fedchenko, and others focused their attention on soils and crops and devised a series of novel experiments to elucidate the spectral, spatial, and temporal characteristics of something as commonplace as a collection of fields (Fedchenko, 1982).

Between 1980 and 1982, and into a literature that was dominated by image classification and descriptions of one-off remote-sensing techniques, appeared a set of 18 books and papers that told the story of those fields (e.g., Kondratyev and Fedchenko, 1982a). These publications proved to be seminal in three respects.

- They were deductive and aimed to understand what we could and could not sense remotely.
- This understanding was used as a basis for answering all three levels of ecological question (above).
- One of the variables sensed was canopy chlorophyll content, and this was later to be crucial in understanding terrestrial ecosystem productivity.

The vast majority of the publications were published in Russian or when published in English focused on the first of the three questions (above); that is, *What is the type of vegetation there?* As Kondratyev and co-workers were assembling their studies for publication as a major book they also produced an English summary of their approach (Kondratyev *et al.*, 1986). Although this research was continued throughout the 1980s there is no evidence that their pioneering research on the remote sensing of chlorophyll content was reaching an international audience. However, the early 1990s were to see an opening up of this rich Soviet literature (Curran *et al.*, 1990b). This, along with improvements in technology and the awareness of climate change, opened the door to the sort of chlorophyll-based products that could be derived from remote sensing and used to understand anthropogenic changes to planet Earth.

5.2 SPECTRAL PROPERTIES OF VEGETATION

When incident solar radiation interacts with vegetation, some of it is reflected, some is absorbed, and the rest is transmitted. The intensity with which radiation is reflected at any particular wavelength is dependent upon both the spectral properties and also the area of the three main remotely sensed components of a vegetation canopy: leaves, substrate, and shadow (Dawson *et al.*, 1998, 1999). Leaves usually reflect weakly in the blue and red wavelengths as a result of absorption by pigments, and strongly in near-infrared wavelengths as a result of leaf structure (Curran, 1980; Kondratyev and Fedchenko, 1982b). The reflectance from a vegetation canopy usually comprises reflectance from live vegetation but also understory, senescent vegetation, and soil. Knowledge of the radiation interactions with a vegetation canopy allows estimation of vegetation properties from the remotely sensed response. The reflectance spectrum of a typical vegetation canopy (Figure 5.1) can be subdivided into three parts defined on the basis of the wavelength of the radiation. These are the visible (400 nm–700 nm), near-infrared (NIR) (701 nm–1,300 nm), and middle-infrared (1,301 nm–2,500 nm) spectral regions. Each part of the spectrum may be used to infer key vegetation properties. For example, the relations between multispectral reflectance and vegetation amount for six wavebands are summarized in Table 5.1. Moreover, Curran (1989) presented a list of 44 absorption features in the visible and NIR wavelengths which are related causally to foliar biochemical constituents.

5.2.1 Visible region

Chlorophyll is the major absorber of radiation in the visible region. Two types of chlorophyll occur, chlorophyll-a and chlorophyll-b; chlorophyll-a is the primary

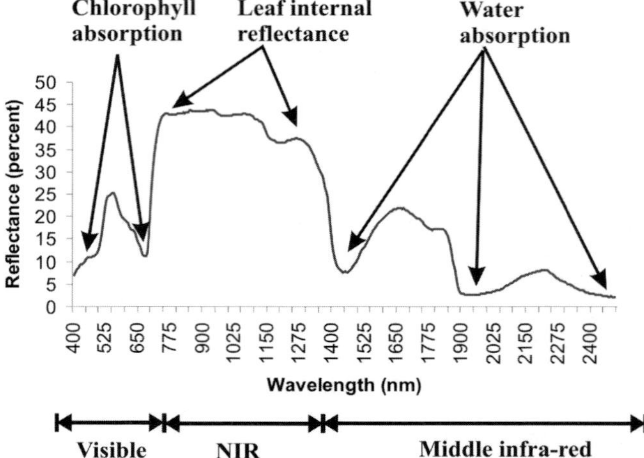

Figure 5.1. A typical leaf reflectance spectrum indicating major spectral features.

photosynthetic pigment while chlorophyll-b is an accessory pigment that collects energy to pass on to chlorophyll-a. The content of chlorophyll-a is usually between two and three times that of chlorophyll-b and dominates the absorption of radiation with wavelengths of 600 nm–700 nm (Lichtenthaler, 1987). Other leaf pigments also have an important effect on the visible spectrum. For example, the yellow to orange-red pigment, carotene, strongly absorbs radiation in the 350 nm–500 nm range and is

Table 5.1. Characteristic spectral features of some foliar biochemicals (adapted from Curran, 1980).

Waveband	*Waveband* (nm)	*Characteristics*	*Relation to vegetation amount*
Ultraviolet/blue	350–500	Strong chlorophyll and carotene absorption	Strong negative
Green	501–600	Reduced level of pigment absorption	Weak positive
Red	601–700	Strong chlorophyll absorption	Strong negative
Red edge	701–740	Transition between strong absorption and strong reflectance	Weak negative
Near-infrared	741–1,300	High vegetation reflectance	Strong positive
Middle-infrared	1,301–2,500	Water, cellulose and lignin absorption	Not specific

responsible for the color of some flowers and fruits, as well as leaves without chlorophyll. The red and blue pigment, xanthophyll, strongly absorbs radiation in the 350 nm–500 nm range and is responsible for the autumnal color of some leaves.

5.2.2 Near-infrared region

In the NIR spectral domain, structure explains the optical properties of vegetation. The NIR spectral region can be divided into two spectral sub-regions (Gausmann, 1974): first, between 701 nm and 1,100 nm, where reflectance from spongy mesophyll cells is high, except in two minor water-related absorption bands (960 nm and 1,100 nm), and, second, between 1,100 nm and 1,300 nm, which corresponds to the transition between high NIR reflectance and water-related absorption in the middle-infrared (Figure 5.1). The high reflectance between 701 nm to 1,100 nm is associated with scattering in the interior or back of a leaf where the radiation is reflected mainly at cell wall/air space interfaces. Leaf pigments and cellulose are transparent to NIR wavelengths, and therefore leaf absorption in the entire NIR region is very small and as a result there is typically a reflectance plateau. The level of this plateau is dependent on the internal structure of the leaf, with air spaces, water, and cells combining to provide a range of interfaces of different refractive indices. Leaf reflectance increases where cells are more heterogeneous in shape and size and where there are more cell layers and the leaf is thicker. The intensity of NIR reflectance is commonly greater than from most inorganic materials, so vegetation appears bright in NIR wavelengths.

5.2.3 Middle-infrared region

The middle-infrared region contains information about the absorption of radiation by leaf components such as water, cellulose, and lignin. For example, the nitrogen status of vegetation can be inferred indirectly by studying absorption features in that region (Baret and Fourty, 1997).

5.2.4 The red edge

The red edge is a region within the red–NIR transition zone of a vegetation reflectance spectrum and marks the boundary between absorption due to chlorophyll in the red region and scattering due to leaf internal structure in the NIR region (Horler *et al.*, 1983). The red-edge position (REP) can be defined as the position (wavelength) of the maximum of the first derivative of the reflectance spectrum of a leaf (Horler *et al.*, 1983; Curran *et al.*, 1990a). There is a negative exponential correlation between chemical concentration and absorption, and as a result an increase in chlorophyll concentration will increase absorption. This in turn will cause both broadening and deepening of the absorption feature (Filella and Peñuelas, 1994; Curran *et al.*, 1995) and a movement of the REP to longer wavelengths. This change in the REP can be used to estimate the amount of chlorophyll both in a leaf and over a canopy (Railyan and Korobov, 1993; Munden *et al.*, 1994; Pinar and Curran, 1996).

5.3 IMAGING SPECTROMETRY

To estimate the foliar biochemical content of vegetation canopies remotely a fundamental advance was required in instrumentation and techniques of analysis. This occurred in the 1970s, with the development of imaging spectrometry as this combined the spatial aspects of remote sensing with the analytical techniques of spectroscopy (Birth and Hecht, 1987; Curran, 2001). Spectrometry is the measurement of the interaction of electromagnetic radiation with matter and the use of these measurements to deduce the properties of material (Curran, 1989; Banwell, 1994). It is the extension of human vision and a non-destructive way of determining information about a material (Birth and Hecht, 1987). Spectrometers have been used in the laboratory (SF-18 spectrophotometer by Tarasov, 1968; Perstorp NIRSystem 6500 spectrometer by Kupiec and Curran, 1995), in the field ("mushroom" photometer by Kondratyev and Fedchenko, 1982b; Spectron SE590 Spectroradiometer by Blackburn, 1998; Geophysical Environmental Research IRIS Mark IV by Jago *et al.*, 1999), from the air (Compact Airborne Spectrographic Imager or CASI by Niemann, 1995; Airborne Visible/Infrared Imaging Spectrometer or AVIRIS by Zagolski *et al.*, 1996); and from space (Moderate Resolution Imaging Spectroradiometer or MODIS by Justice and Townshend, 2002; Medium Resolution Imaging Spectrometer or MERIS by Curran and Steele, 2005). The aim of these studies is to develop and modify techniques for estimating variables related to the type, amount, and condition of vegetation.

Laboratory spectrometers are standard instruments for estimating the chemical concentration of materials and are, in many cases, fully automated. Field spectrometers record a signal from a more complex surface and are becoming more convenient as user interfaces are improved and as the instruments become smaller and more efficient (Milton, 2000). Airborne imaging spectrometers provide a flexible operational and experimental remote-sensing tool with a fine spectral and spatial resolution (Wulder, 1998). One of the earliest systems to provide airborne imaging spectrometry for civilian use was the Airborne Imaging System (AIS) (Vane *et al.*, 1984). The AIS could record radiance in 128 contiguous wavebands over the spectral range of 1,200 nm–2,400 nm. Since the 1980s many airborne imaging spectrometers have been developed and used to acquire contiguous reflectance spectra over land and water (Treitz and Howarth, 1999) and today there are at least 15 types of imaging spectrometers that are either operational or at an advanced stage of development. Operational sensors include the Airborne Visible/Infrared Imaging Spectrometer (AVIRIS) (Vane *et al.*, 1993) and Compact Airborne Spectrographic Imager (CASI) (Anger *et al.*, 1996).

The increased demand for imaging spectrometry led to the development of spaceborne imaging spectrometers capable of providing global repetitive coverage (Table 5.2). The first civilian imaging spectrometer to be designed for use in orbit was the Shuttle Imaging Spectrometer (SIS) (Herring, 1987). The Space Shuttle Challenger accident brought an end to the SIS project but led to the development of the High Resolution Imaging Spectrometer (HIRIS). HIRIS was designed for use from the Space Station Polar Platform, part of the NASA Earth Observing System

Table 5.2. Measurement characteristics associated with three spectrometer locations for the measurement of spectra from Earth surface materials (adapted from Kupiec and Curran, 1995).

	Laboratory	*Field*	*Aircraft/satellite*
Radiation source	Strong, constant, controllable	Weak, variable, uncontrollable	Weak, variable, uncontrollable
Surface sensed	Homogenous	Heterogeneous	Heterogeneous
Distance from surface to detector	Centimeters	Meters	Kilometers
Measurement time	Long	Long	Short
Signal strength	Very strong	Strong	Weak
Signal-to-noise ratio	Very large	Large	Small

(EOS) program (Vane, 1987). HIRIS was similar to SIS in spatial and spectral design. It had 192 wavebands in the range of 400 nm–2,500 nm, a bandwidth of approximately 10 nm and a swathwidth of 24 km. Although it was never to fly, its development and the scientific thinking that underpinned that development paved the way for the launch of Hyperion on the Terra satellite in November 2000 (Curran, 2001).

The two operational spaceborne imaging spectrometers are the Moderate Resolution Imaging Spectroradiometer (MODIS) onboard NASA's Terra satellite, and the Medium Resolution Imaging Spectrometer (MERIS) onboard ESA's Envisat satellite. MODIS is a whiskbroom scanning imaging radiometer that can record radiation in 36 discontinuous wavebands with a variable bandwidth within the 400 nm–14,500 nm spectral range. MODIS has three spatial resolutions at nadir; two bands are imaged at a nominal resolution of 250 m, with five bands at 500 m and the remaining 29 bands at 1,000 m. A $\pm 55^\circ$ scanning pattern at the Terra orbit of 705 km achieves a 2,330 km swath and provides global coverage every one to two days. MERIS records radiation in 15 discontinuous wavebands with variable bandwidth (2.5 nm–20 nm) within the 390 nm–1,040 nm spectral range. Unlike other imaging spectrometers, the band centers and bandwidths are programmable in-flight. This allows users to define their bands for a specific application. A detailed description of the development of MERIS bands is given by Curran and Steele (2005). MERIS is a pushbroom sensor with five optical modules arranged symmetrically about nadir, each containing a two-dimensional charge-coupled device (CCD) array. These optical modules are positioned in a fan shape so that the viewing aperture converges towards each other. The field of view (FOV) of each module is 14° and the FOV of the whole instrument is 68.5°, thus allowing slight overlap between adjacent modules. From the platform altitude of 799 km it has a swath of 1,150 km and complete global coverage within 3 days.

MERIS has dual spatial resolution: full resolution (FR) (300 m at nadir) and reduced resolution (RR) (1,200 m at nadir). Full spatial resolution data are used mainly for coastal and land applications, while reduced spatial resolution data are used mainly for large-area ocean and atmospheric applications.

Several spaceborne imaging spectrometers are currently under development (Table 5.3). For example, the Surface Process and Ecosystem Changes Through Response Analysis (SPECTRA) mission, to be launched by ESA in the near future, is based around an imaging spectrometer and will be used to further develop our understanding of the interaction of terrestrial ecosystems with the atmosphere (Tobehn *et al.*, 2003). SPECTRA will acquire data in over 200 selectable wavebands in visible and NIR wavelengths with a spatial resolution of 50 m.

5.4 METHODS USED TO ESTIMATE CHLOROPHYLL CONTENT USING REMOTELY SENSED DATA

Various methods have been developed to estimate the chlorophyll concentration of leaves (Kondratyev *et al.*, 1982a, b, c; Gitelson and Merzlyak, 1998; Mariotti *et al.*, 1996) and content of canopies (Daughtry *et al.*, 2000; O'Neill *et al.*, 2002). The most obvious method, based on the magnitude of red-light absorption, is inadequate if the measurement conditions are not well controlled. Popular methods that have been found to be successful are those based on colorimetry and use of the REP.

5.4.1 Colorimetric method

Colorimetry, in this context, uses color rather than spectral reflectance alone to characterize remotely sensed data. This is based on three laws proposed by Grassmann in the 19th century (Bouma, 1971) and is discussed in detail in Curran *et al.* (1990b). The basis of this method is that any color can be obtained by mixing, in the correct proportion, three mutually independent colours: red (R), green (G), and blue (B). Three basic systems are used to express color: (i) $B_{\lambda p}$; (ii) RGB; and (iii) XYZ.

The $B_{\lambda p}$ system is based on the principle that any color can be produced by adding or subtracting a color from white; however, this does not allow for color calculation. The RGB system is based on the principle that the three independent colors of red (R), green (G), and blue (B) form the apex of a color triangle. Colors obtained from mixing these colors lie inside the triangle and others lie outside the triangle (Wright, 1969) and the position of a color within the triangle depends on the relative proportion of the independent colors. However, the use of this system is complicated by the possibility of obtaining negative color coordinates. The XYZ system (Hunt, 1987) also uses the three independent colors (red, green, and blue) but it assumes (i) the existing colors are inside the color triangles; (ii) the quantitative estimate of color is determined from the Y coordinate (the light flux for this coordinate corresponds to 0.68 μm) and zero brightness corresponds to X and Z; and (iii) to obtain white the colority coordinates must satisfy the equation $x = y = z = 1/3$.

Table 5.3. Eight spaceborne imaging spectrometers designed for environmental research.

Name	*Satellite*	*Launch date*	*No. of spectral channels*	*Spectral coverage* (nm)	*Spatial resolution* (m)
HYPERION	Earth Observation-1	November 24, 2000	220	400–2,500	30
CHRIS (Compact High Resolution Imaging Spectrometer)	Proba (Project for On-Board Autonomy)	October 22, 2001	19 at a time	425–1,050	20
MODIS (Moderate resolution Imaging Spectroradiometer)	Earth observing system (EOS), Terra, and Aqua	December 18, 1999	36	620-14,385	250 m (Bands 1–2) 500 m (Bands 3–7) 1,000 m (Bands 8–36)
MERIS (Medium Resolution Imaging spectrometer)	Envisat	March 1, 2002	15 (programmable)	390–1,040	300 (full resolution) 1200 (reduced resolution)
GLI (Global Imager)	ADEOS-II (Advanced Earth Observing Satellite -II)	December 14, 2002	36	375–1,250	250
ARIES (Australian Resource Information and Environmental Satellite)	NA	To be launched	32 32	400–1,100 2,000–25,000	30
SPECTRA	Not yet decided	To be launched	200 (selectable)	450–2,350	50
Veμus	Not yet decided	To be launched in 2008	12	395–920	5.3

For the XYZ system the basic colors can be expressed in terms of the RGB system as

$$\left.\begin{aligned} X &= a_1R + a_2G + a_3B \\ Y &= a_4R + a_5G + a_6B \\ Z &= a_7R + a_8G + a_9B \end{aligned}\right\} \tag{5.1}$$

where a_1 to a_9 are constants and any color C can be expressed in the XYZ system as

$$C = xX + yY + zZ$$

where

$$x = \frac{X}{X+Y+Z}; \qquad y = \frac{Y}{X+Y+Z}; \qquad z = \frac{Z}{X+Y+Z}$$

with $x + y + z = 1$. Therefore, the determination of any color using the XYZ system requires only two values, with a third used to ensure that the sum is 1.

The technique used by Kondratyev to estimate the chlorophyll concentration of an individual leaf is based on color coordinates estimated using weighted coordinates in the XYZ system. In using this method it is assumed that the entire spectrum is divided into n narrow intervals λ_i ($i = 1, 2, \ldots, n$) of equal width $\Delta\lambda$ and that within each of these intervals the radiation and reflectivity are constant. Using this method the X, Y, Z coordinates can be derived as

$$\left.\begin{aligned} X &= k_x \sum_{i=1}^{n} S(\lambda_i)r(\lambda_i)\bar{x}(\lambda_i)\Delta\lambda \\ Y &= k_y \sum_{i=1}^{n} S(\lambda_i)r(\lambda_i)\bar{y}(\lambda_i)\Delta\lambda \\ Z &= k_z \sum_{i=1}^{n} S(\lambda_i)r(\lambda_i)\bar{z}(\lambda_i)\Delta\lambda \end{aligned}\right\} \tag{5.2}$$

where $k_x = \sum S(\lambda_i)\bar{x}(\lambda_i)\Delta\lambda$, etc.; $S(\lambda_i)$ is the spectral distribution of the incident radiation; $r(\lambda_i)$ is the reflectivity at wavelength $\lambda_i \cdot \bar{x}(\lambda_i)$; $\bar{y}(\lambda_i)$ and $\bar{z}(\lambda_i)$ are adding functions with standard values at different wavelengths recommended by the International Luminance Council and for which $\bar{x}(\lambda_i) + \bar{y}(\lambda_i) + \bar{z}(\lambda_i) = 1$.

For any reflectance spectra, color coordinates can be calculated using the equation above and the sum of all color coordinates is

$$W = X + Y + Z$$

The spectral reflectance of leaves can then be used to estimate color coordinates which in turn can be related to chlorophyll concentration. For example, Kondratyev *et al.* (1982a) reported a strong negative relationship between the sum of color coordinates and chlorophyll concentration for individual leaves (Figure 5.2). Most of the studies conducted by Kondratyev and his co-workers used the colorimetry method to quantify chlorophyll concentration (Kondratyev *et al.*, 1982b, c); although the technique is not widely used now it did enhance our understanding of the link

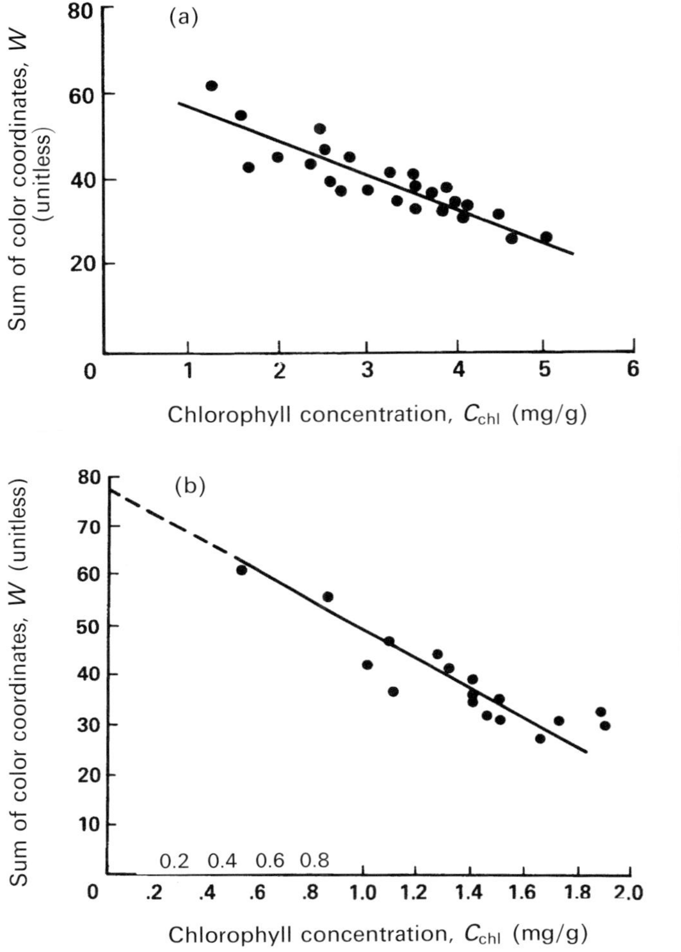

Figure 5.2. The relationship between the sum of the color coordinates and the chlorophyll concentration of (a) potato leaves and (b) buckwheat leaves (Curran *et al.*, 1990b).

between remotely sensed data and the chlorophyll concentration and later content of vegetation.

5.4.2 Red-edge position

Mathematically, the REP is the position (i.e., the wavelength) of the maximum of the first derivative spectrum in the red-edge region. The derivative spectrum can be

Table 5.4. Performance of six techniques used to locate the REP in vegetation spectra.

Requirements	*Maximum of first derivative*	*High-order curve fitting*	*Inverted Gaussian*	*Linear interpolation*	*Lagrangian interpolation*	*Rational function*
Spectral resolution	High	High	Medium	Medium	Low	High
Number of bands in red edge	Many	Many	3	4	3/4	3
Band continuity	High	High	Low	Low	Low	Low
Need for a derivative spectra	High	Low	Low	Low	High	High
Need for a modeled relationship	Low	Medium	High	High	Medium	Medium
Signal-to-noise ratio	Medium	Medium	Medium	Medium	High	Medium

estimated by

$$D_{\lambda_i} = \frac{R_{\lambda_i} - R_{\lambda_{i-1}}}{\Delta\lambda} \tag{5.3}$$

where R_{λ_i} and $R_{\lambda_{i-1}}$ are the reflectances at wavelength i and $i-1$, respectively. The REP estimated using the maximum of the first derivative method is to a first approximation independent of the influence of background reflectance in the reflectance spectra of a vegetation canopy (Horler *et al.*, 1983; Demetriades-Shah *et al.*, 1990). However, an accurate estimation of the REP using this method requires both spectral continuity and fine spectral resolution of the reflectance spectra. To overcome this dependence on spectral continuity, researchers suggested different techniques for REP estimation. Most commonly used techniques include (i) high-order curve fitting (Demetriades-Shah *et al.*, 1990); (ii) inverted Gaussian interpolation (Bonham-Carter, 1988); (iii) linear interpolation (Guyot *et al.*, 1988; Danson and Plummer, 1995); (iv) Lagrangian interpolation (Dawson and Curran, 1998); and (v) rational function interpolation (Baranoski and Ronke, 2005, see Table 5.4).

5.4.2.1 *High-order curve fitting*

Baret *et al.* (1992) proposed a polynomial equation for estimating the REP using three spectral bands. They used model simulations to select optimal spectral bands at 672 nm (R_1), 710 nm (R_2), and 780 nm (R_3) and subsequently fit a polynomial equation to the maximum of the second derivative, see Equation (5.4)

$$\mathrm{REP} = C_0 + C_1 R_1 + C_2 R_2 + C_3 R_3 + C_4 R_1^2 + C_5 R_2^2 + C_6 R_3^2 + C_7 R_1 R_2 + C_8 R_1 R_3 + C_9 R_2 R_3 + C_{10} R_1 R_2 R_3 \tag{5.4}$$

where C_0 to C_{10} are the constants to be determined through iteration.

Broge and Leblanc (2001) proposed a sixth-order polynomial function for estimating REP. The polynomial was described by

$$R(\lambda) = C_0 + C_1\lambda + C_2\lambda^2 + C_3\lambda^3 + C_4\lambda^4 + C_5\lambda^5 + C_6\lambda^6 \tag{5.5}$$

REP is determined by finding the value of λ for which the second derivative of the polynomial is zero, where λ is considered to be close to 720 nm, depending upon the nature of the curve.

Computationally these higher order curve-fitting techniques are complex. However, they will capture the potential asymmetry of the red edge unlike methods based on an inverted Gaussian function (Broge and Leblanc, 2001).

5.4.2.2 *Inverted Gaussian*

Hare *et al.* (1984) suggested that the shape of the spectral reflectance curve for vegetation in the red-edge region could be approximated by an inverted Gaussian function. The inverted Gaussian model (IGM) includes the central wavelength (λ_0), and the reflectance minimum (R_0) of the chlorophyll absorption region that occurs at approximately 680 nm (Figure 5.3). The IGM describes variation of reflectance, R, as a function of wavelength (λ)

$$R(\lambda) = R_s - (R_s - R_0)\exp - \left[\frac{(\lambda - \lambda_0)^2}{2\sigma^2}\right] \tag{5.6}$$

where R_s is the "shoulder" reflectance (at approximately 800 nm); R_0 is the minimum reflectance at the chlorophyll absorption feature; and σ is the Gaussian shape parameter.

For N data points in the red-edge region, the problem is to determine R_s, R_0, σ, and λ_0 in a way that minimizes the sum of the squared deviation of the fitted curve, $R'(\lambda)$ from the observed reflectance, $R(\lambda)$

$$\sum_{i=1}^{N}[R_i(\lambda) - R'_i(\lambda)]^2 \rightarrow \text{minimum} \tag{5.7}$$

Then the REP can be defined as the

$$\mathrm{REP} = \lambda_0 + \sigma \tag{5.8}$$

Miller *et al.* (1990) suggested a linear fitting approach which was computationally more efficient than (i) the iterative optimisation fitting procedure as described above and (ii) non-linear fitting methods; for example, the Gauss–Newton differential–correlation technique (Bonham-Carter, 1988). Despite the lack of a theoretical basis, the IGM has a shape that effectively characterizes the shape of the red edge and has

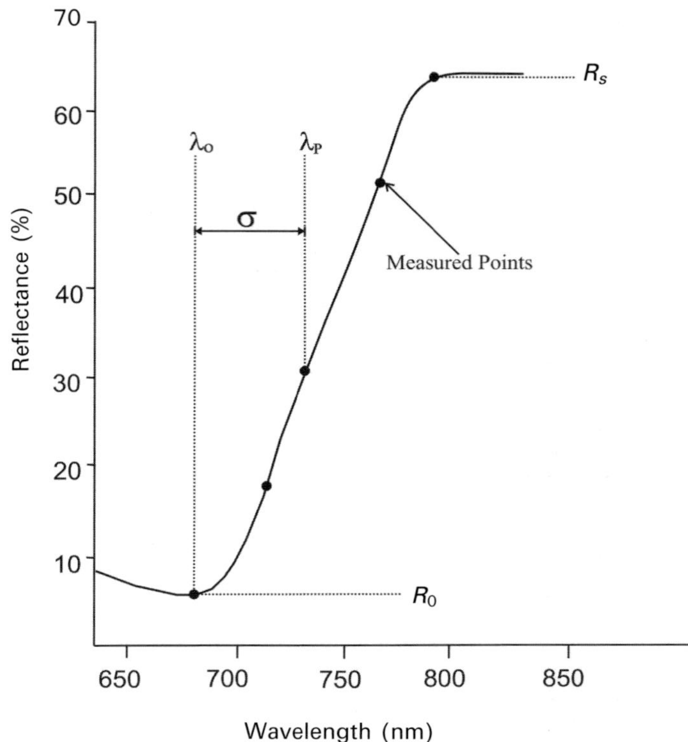

Figure 5.3. Inverted Gaussian technique, with red-edge curve-fit parameters indicated: the reflectance maximum (R_s), the reflectance minimum (R_0), the spectral position of the reflectance minimum (λ_0), the spectral position of the curve inflection (λ_p), and the Gaussian curve width parameter σ.

been fitted to laboratory and field measurements of vegetation spectra (Zarco-Tejada and Miller, 1999; Lucas *et al.*, 2000).

Vegetation canopy spectra contain information on shadow, soil, and understory reflectance, which affect the accuracy with which the REP can be estimated for the canopy alone. These effects, coupled with the need to know predetermined points (e.g., R_0 and R_s) weakens the argument for using IGM for the estimation of the REP (Dawson and Curran, 1998).

5.4.2.3 Linear interpolation

Guyot *et al.* (1988) proposed a linear interpolation technique for estimating the REP. They assumed that the red edge could be represented as a straight line on a spectrum between reflectance at 670 nm and 780 nm. The reflectances measured at 670 nm and 780 nm were then used to calculate the reflectance of the inflection point, and a linear interpolation technique was used to calculate the wavelength of this inflection point

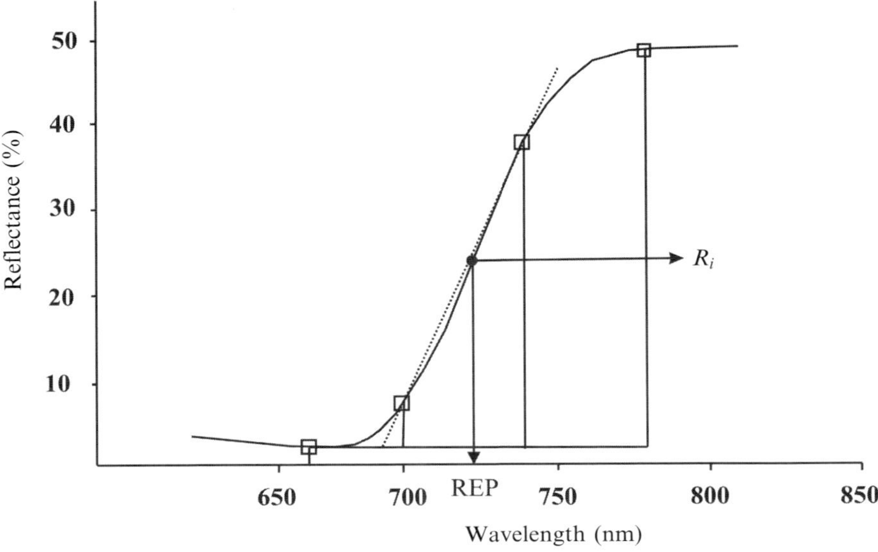

Figure 5.4. Linear interpolation technique, indicating the red-edge inflection point (R_i) and red edge position (REP).

(Figure 5.4). So there are two steps: first, calculation of reflectance at the inflection point and, second, calculation of the REP

$$R_i = \frac{(R_{670} + R_{780})}{2} \tag{5.9}$$

where R_i is the reflectance at wavelength i. Then

$$\text{REP} = 700 + 40\frac{(R_i - R_{700})}{(R_{740} - R_{700})} \tag{5.10}$$

Danson and Plummer (1995) used slightly different wavelengths and the following equation for estimating the reflectance at the REP, see Equation (5.11), and calculation of the REP, see Equation (5.12)

$$R_i = \frac{(R_{673} + R_{780})}{2} + R_{673} \tag{5.11}$$

where R_i is the reflectance at wavelength i. Then

$$\text{REP} = 700 + 40\frac{(R_i - R_{700})}{(R_{740} - R_{700})} \tag{5.12}$$

The linear interpolation technique is conceptually and computationally simple (Guyot *et al.*, 1988); however, in some cases this can result in a loss of useful spectral detail.

5.4.2.4 Lagrangian interpolation

Dawson and Curran (1998) proposed a technique based on three-point Lagrangian interpolation (Jeffrey, 1985) for the estimation of the REP. This method uses a second-order polynomial fit to the first derivative vegetation reflectance spectrum and reflectance in three wavebands: the band with maximum first-derivative reflectance and two adjoining bands. REP is

$$\text{REP} = \frac{A(\lambda_i + \lambda_{i+1} + B(\lambda_{i-1} + \lambda_{i+1}) + C(\lambda_{i-1} + \lambda_i)}{2(A + B + C)} \tag{5.13}$$

where

$$A = \frac{\mathrm{D}\lambda_{i-1}}{(\lambda_{i-1} - \lambda_i)(\lambda_{i-1} - \lambda_{i+1})}, \qquad B = \frac{\mathrm{D}\lambda_i}{(\lambda_i - \lambda_{i-1})(\lambda_i - \lambda_{i+1})},$$
$$C = \frac{\mathrm{D}\lambda_{i+1}}{(\lambda_{i+1} - \lambda_{i-1})(\lambda_{i+1} - \lambda_i)}$$

In this case $\mathrm{D}\lambda_{i-1}, \mathrm{D}\lambda_i, \mathrm{D}\lambda_{i+1}$ are the first-derivative reflectances corresponding to wavebands $\lambda_{i-1}, \lambda_i, \lambda_{i+1}$, respectively ($\lambda_i$ is the band with maximum first-derivative reflectance with λ_{i-1} and λ_{i+1} representing the bands either side of it, see Figure 5.5).

The advantages of Lagrangian interpolation are (i) wavebands used for the estimation of the REP need not be spaced equally; (ii) the use of a first-derivative spectrum minimizes interpolation errors; and (iii) it is computationally one of the simpler curve-fitting techniques. However, Clevers *et al.* (2002) reported a "jumping" feature in a nonlinear REP/chlorophyll content relationship derived using Lagrangian interpolation, and this has yet to be explained.

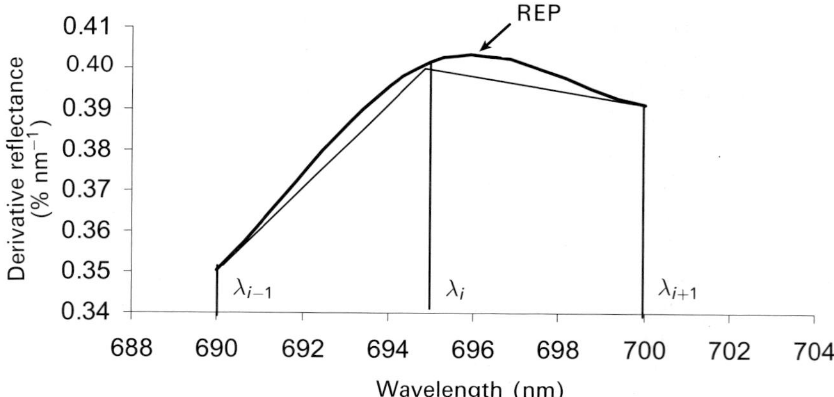

Figure 5.5. The three-point Lagrangian interpolation technique to determine the red edge position, indicating the REP along the derivative curve.

5.4.2.5 *Rational function*

Recently, Baranoski and Ronke (2005) proposed a red-edge estimation technique based on a rational function. They assumed that reflectance in the vicinity of the red edge could be classified into three regions: (i) region of low and relatively low reflectance; (ii) red-edge region; and (ii) region of high and relatively constant reflectance. These features could be approximated by a rational function

$$f(\lambda) = \frac{a + b\lambda + c\lambda^2}{1 + c\lambda + d\lambda^2} \tag{5.14}$$

Four conditions could be used to estimate the four parameters a, b, c, and d as

$$f(\lambda_l) = Y_l$$

$$f'(\lambda_l) = 0$$

$$f(\lambda_h) = Y_h$$

$$f'(\lambda_h) = 0$$

where λ_l and Y_l are the wavelength and reflectance at the lower bound; and λ_h and Y_h are the wavelength and reflectance at the upper bound of the red edge. The center of the red edge (λ_c, Y_c) is required to estimate the REP.

The accuracy of the estimation of the REP depends on the accuracy of picking the three points mentioned above. The REP estimated using the rational function has two advantages: (i) rational functions tend to spread the error more evenly over the approximation interval; and (ii) this technique could be automated using the fixed values for three input wavelengths (680 nm, 725 nm, 770 nm). Baranoski and Ronke (2005) found a relative error of less than 1% when comparing the REP estimated using the rational function and actual REP for 80 spectra. However, this technique needs to be compared with the REP estimated by other techniques.

Three problems remain with the use of the REP for estimation of foliar chlorophyll content at a regional to global scale. First, there is no generally accepted technique for estimating REP, and each technique produces a different value of the REP from the same set of data (Table 5.4, Figure 5.6). Second, the methods used to estimate the REP have been designed for use on continuous spectra without thought for standardization or automation (Dawson and Curran, 1998). Third, the REP is not an accurate indicator of chlorophyll content at high chlorophyll contents because of the asymptotic relationship between REP and chlorophyll content (Munden *et al.*, 1994; Jago *et al.* 1999). The launch of MERIS on Envisat with five discontinuous wavebands in red and near-infrared (NIR) wavelengths with band centers at 665 nm, 681.25 nm, 708.75 nm, 753.75 nm, and 760.623 nm provided an opportunity to develop an index to estimate canopy chlorophyll content at a regional to global scale. An index called the MERIS Terrestrial Chlorophyll Index (MTCI) (Dash and Curran, 2004) was developed for this purpose.

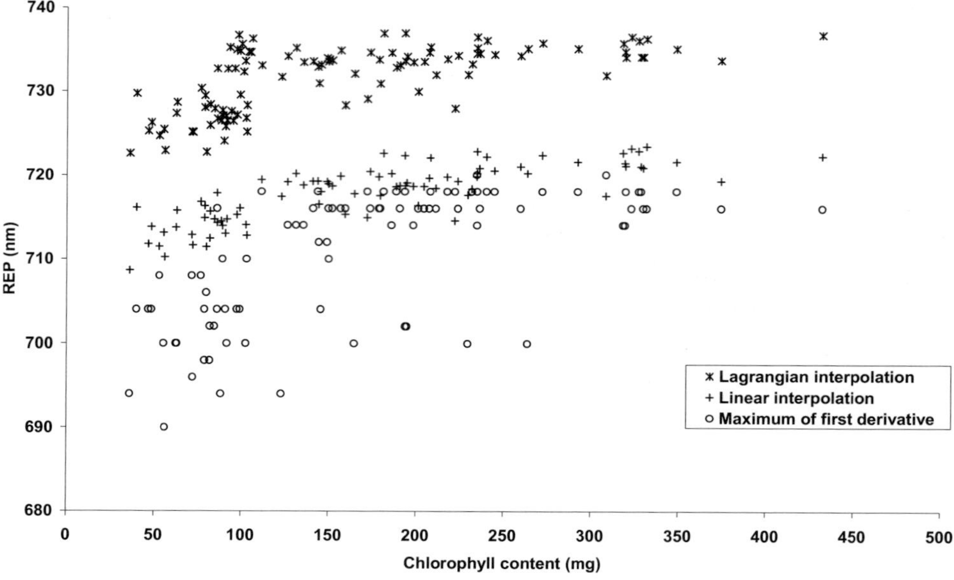

Figure 5.6. Relationship between chlorophyll content and REP estimated using three methods.

5.4.2.6 MERIS Terrestrial Chlorophyll Index (MTCI)

The MTCI is a ratio of the difference in reflectance between Band 10 and Band 9 and the difference in reflectance between Band 9 and Band 8 of the MERIS standard band setting

$$\text{MTCI} = \frac{R_{\text{Band 10}} - R_{\text{Band 9}}}{R_{\text{Band 9}} - R_{\text{Band 8}}} = \frac{R_{753.75} - R_{708.75}}{R_{708.75} - R_{681.25}} \tag{5.15}$$

where $R_{753.75}$, $R_{708.75}$, $R_{681.25}$ are the reflectances for the center wavelengths of each MERIS band.

The MTCI may be used to derive an estimate of the relative location of the reflectance "red edge" of vegetation and is more sensitive than red-edge position to canopy chlorophyll content, notably at high chlorophyll contents. This product effectively combines information on leaf area index and the chlorophyll concentration of leaves to produce an image of chlorophyll content. The MTCI has been applied to several species using data from the laboratory (Boyd *et al.*, 2007; Dash *et al.*, 2007a), field, and even at MERIS spatial resolution (Dash and Curran, 2008) (Figure 5.7). For each of these data there was a strong positive relationship between MTCI and chlorophyll content. In April 2004 MTCI became an operational ESA Level 2 land product, and currently MTCI weekly and monthly global composites are produced in near-real time (Curran *et al.*, 2007b).

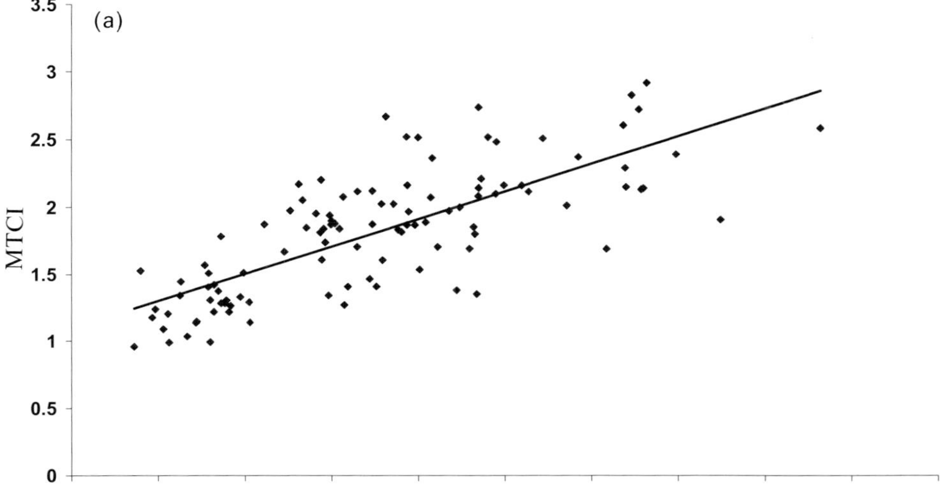

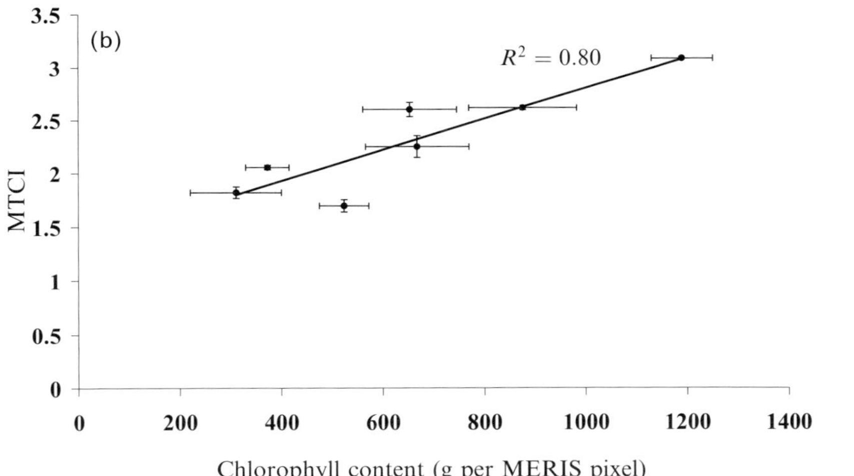

Figure 5.7. Relationship between MTCI and chlorophyll content for (a) spinach (greenhouse experiment) and (b) seven agricultural fields in southern England (field experiment).

5.5 APPLICATIONS OF REMOTELY SENSED CHLOROPHYLL CONTENT DATA

An increase or decrease in the amount of chlorophyll will affect the resulting vegetation canopy spectra. Using the estimated chlorophyll amount, all three ecological questions can be posed (Section 5.1). For example, the type of vegetation can be identified, its cover mapped, and information on physiological and biochemical condition derived. Information on the amount and spatial distribution of canopy chlorophyll content is of importance for the study of vegetation productivity and health, nutrient cycling, crop stress, crop yield, and most recently for driving ecosystem simulation models at regional scales (Ustin *et al.*, 2004; Treitz *et al.*, 2008). Therefore, although techniques for the remote sensing of chlorophyll amount have progressed since the 1970s and 1980s their applications remain at the heart of Kondratyev's latter-day research on anthropogenic effects on the state of planet Earth. Three particular applications capture well the major ecodynamic topics he pursued: vegetation productivity, vegetation stress, and land cover mapping.

5.5.1 Vegetation productivity

According to Reeves *et al.* (2005), the estimation of vegetation productivity using remotely sensed data has generally made use of two approaches: (i) use of the measured spectral reflectance to estimate the amount of absorbed photosynthetically active radiation (APAR) (Choudhury, 1987); and (ii) establishment of an empirical relationship between spectral reflectance and productivity (Wylie *et al.*, 1995). As chlorophyll is one of the ingredients for photosynthesis the estimation of chlorophyll amount can also give information on productivity. In a well-managed cereal crop the concentration of chlorophyll is related directly to yield (Reeves *et al.*, 1995), so the remote sensing of chlorophyll concentration offers the possibility of the estimation of crop yield. A remotely sensed measure of reflectance at the absorption wavelength of chlorophyll has limited suitability for estimating yield, however, as it couples with the effect of chlorophyll concentration, leaf mass, and other effects like ground reflectance and Sun-sensor geometry (Curran, 1983). Gitelson *et al.* (2006) demonstrated that canopy chlorophyll content was related closely to gross primary productivity in cereal crops. They have also demonstrated that the remotely sensed estimates of chlorophyll content have a stronger correlation with day-to-day variation in gross primary productivity than LAI alone (Gitelson *et al.*, 2006). More recently the MTCI has been used to estimate crop productivity (Dash and Curran, 2007).

5.5.2 Vegetation stress

Vegetation stress may be defined as any factor that reduces productivity below the optimum value (Steven *et al.*, 1990). Stress may result from a change in the physico-chemical condition of the environment, pests, or pathogens. Detection of vegetation stress from remotely sensed data usually involves studying the change of reflection spectra of a vegetation canopy over space or over time (Figure 5.8) (Adams *et al.*, 1999). The effect of vegetation stress on reflectance spectra has been studied in relation to heavy metals (Horler *et al.*, 1980), arsenic and selenium (Milton *et al.*, 1989), *Phylloxera* infestation (Johnson, 1999), powdery mildew disease (Carter, 1993), and water deficiency (Yang and Su, 2000). Horler *et al.* (1980) detected a shift in the long wavelength edge of chlorophyll absorption to shorter wavelengths as a result of pollutant stress. Similar effects have been observed as a result of disease (Johnson, 1999) and water deficiency (Yang and Su, 2000). Milton *et al.* (1989) noted

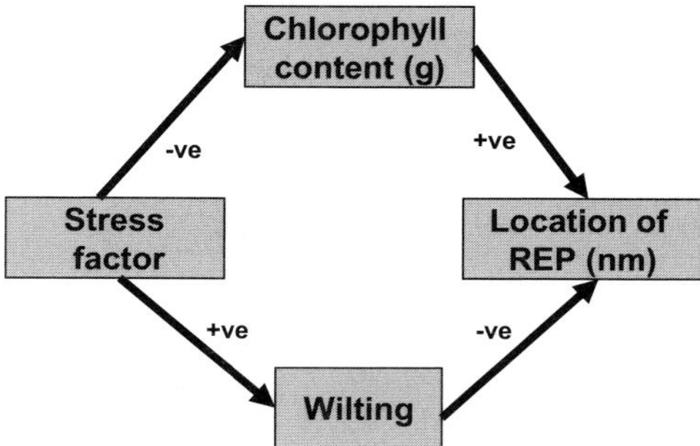

Figure 5.8. Relationship between stress factors, chlorophyll content, wilting, and red-edge location.

the same red-edge effects as Horler *et al.* (1980) for plants stressed with arsenic. Carter (1993) in an experiment with various stressed plants noted an increase in visible reflectance, particularly in wavebands centered near 510 nm and 710 nm in response to stress, regardless of stress agents and vegetation type. The relationship between stress factors, chlorophyll content, wilting, and location of the REP is indicated in Figure 5.8.

The shift in the red-edge position and the resulting change in the shape of the reflectance curve has been quantified using indices; for example, the red-edge vegetation stress index (RVSI) (Merton, 1998), the tricolor algorithm (Clark *et al.*, 1995), and yellowness index (Adams *et al.*, 1999).

Several studies have reported the use of the red edge to identify vegetation stress due to a release of natural gas (Smith *et al.*, 2004; Li *et al.*, 2005). Pre-visual stress was detected using the ratio of derivative reflectances in the red-edge region for wavelengths of 725 nm and 705 nm as this ratio was low for vegetation affected by natural gas. Li *et al.* (2005) reported a 20 nm shift in the REP of AVIRIS data for vegetation exposed to an oil spill at the Jornada experimental range in New Mexico. In their study of the effect of hydrocarbon contamination of grassland soil, Llewellyn *et al.* (2001) found that grassland on highly contaminated soil had an REP at a shorter wavelength than those estimated from grassland with a low level of soil contamination. More recently, the MTCI was used to identify areas having a low level of chlorophyll content in the forests of southern Vietnam, as these coincided with areas where large amounts of herbicide had been sprayed during the Vietnam War (Dash and Curran, 2006). MTCI has also been used to detect salt stress in coastal vegetation affected by the Indian Ocean tsunami in 2004 (Curran *et al.*, 2007a) and to monitor vegetation response to short-term changes in growing season (Almond *et al.*, 2007).

5.5.3 Land cover mapping

Land cover has been a fundamental variable in many parts of Kondratyev's research, notably in relation to work on ecodynamics (Kondratyev *et al.*, 2004). Remote sensing has also been widely used to map land cover types and condition. Indeed, land cover mapping is one of the most common applications of remote sensing. Given the importance of land cover and land cover change on the environment (Feddema *et al.*, 2005), the ability to derive accurate maps from remote sensing continues to be a major research priority. For example, remote sensing provides the only feasible means to monitor major processes such as deforestation that impact on a suite of environmental concerns such as climate change, flooding, soil erosion, or biodiversity.

Much attention in thematic mapping from remotely sensed data has focused on vegetation. Maps of vegetation type provide valuable information on vegetation distribution and change. Maps of vegetation properties, notably biophysical properties such as biomass or biochemical properties such as chlorophyll, convey considerable information on issues such as vegetation health and productivity. These various maps differ greatly in their general nature, from nominal/categorical level products showing unordered classes through to continuous products that illustrate gradations in the mapped property. All of the maps, however, require the property of interest to impact, directly or indirectly, on the remotely sensed response to allow the desired information to be extracted from the imagery. Fortunately, the variables determining the spectral response of vegetation allow the discrimination of many classes and estimation of major biophysical and biochemical properties (Jensen and Binford, 2004).

While mapping commonly makes use of spectral data acquired in a range of wavebands and other information (e.g., texture) considerable use is made of information related to the chlorophyll content of the vegetation. The latter provides valuable information on vegetation properties but also provides a variable with which different vegetation classes may be identified. The information used may be direct estimates of chlorophyll content but more commonly indices strongly influenced by it such as many vegetation indices or red-edge variables. While various vegetation indices have been developed to indicate mainly biophysical and biochemical properties, which may be useful in mapping properties such as vegetation biomass and productivity (Wessman, 1991), they may also be useful in mapping vegetation classes. Vegetation indices have, for example, been the basis of many major mapping programs. For example, the Normalized Difference Vegetation Index (NDVI) has been used extensively in mapping global land cover (Loveland *et al.*, 2000). Many other studies have also used products that are a function of chlorophyll content in mapping. For example, Zarco-Tejada and Miller (1999) show how red-edge variables, which are a function of the chlorophyll content of vegetation, may be used to map land cover classes as accurately as, or more accurately than, some other standard approaches to thematic mapping. Similarly, Dash *et al.* (2007b) show that the MERIS MTCI may be used to map broad land cover classes. The variation in leaf properties such as chlorophyll may help in discriminating detailed vegetation classes such as tree species (Castro-Esau *et al.*, 2006) and reveal differences between sites contaminated with radionuclides (Boyd *et al.*, 2006). Unfortunately, however, the degree of intra-class variation in chlorophyll content may sometimes act to reduce class separability and hence degrade thematic map accuracy. This within-class variability is, for example, a source of end-member variability that degrades the accuracy of sub-pixel estimation (Song, 2005).

5.6 CONCLUSION

Kondratyev was a pioneer in the remote sensing of chlorophyll content. In the past few decades this research has evolved from the development of empirical relationships between chlorophyll content and spectral reflectance in individual wavebands to the generation of an operational product: weekly global terrestrial chlorophyll content maps derived from MERIS data. With recent evidence that canopy chlorophyll content is closely related to gross primary productivity and the current availability of data from MERIS (and future missions like Sentinel 2, Sentinel 3) to estimate chlorophyll content remotely and at a global scale, it will soon be possible to produce a method which is an inexpensive yet accurate tool for estimating primary productivity. This information in turn could be used to improve the understanding of the carbon cycle, in general, and the impact of climate change on the carbon budget, in particular. Many other applications, such as those focused on yield estimation or impacts of environmental stress, can be envisaged. In short, however, Kondratyev's pioneering research helped lay the foundations of current techniques and applications that are important for fulfilling the potential of remote sensing as a source of information on the environment.

5.7 REFERENCES

Adams M.L., Philpot W.D., and Norvell W.A. (1999). Yellowness Index: An application of spectral second derivatives to estimate chlorosis of leaves in stressed vegetation. *Int. J. Remote Sensing*, **20**, 3663–3675.

Almond S.F., Boyd D.S., Curran P.J., and Dash J. (2007). The response of UK vegetation to elevated temperatures in 2006: Coupling Envisat MERIS Terrestrial Chlorophyll Index (MTCI) and mean air temperature. *Challenges for Earth Observation: Technical and Commercial*. Remote Sensing and Photogrammetry Society, Nottingham, U.K. (CD- ROM).

Anger C.D., Achal S., Ivanco T., Mah S., and Price R. (1996). Extended operational capabilities of CASI. *Proc. Second Int. Airborne Remote Sensing Conf., San Francisco, California*, pp. 124-133. Environmental Research Institute, Ann Arbor, MI.

Banninger C. (1991). Phenological changes in the red edge shift of Norway spruce needles and their relationship to needle chlorophyll content. *Proc. Fifth Int. Colloquium: Physical Measurements and Signatures in Remote Sensing, Courcheval, France*, pp. 155–158. European Space Agency, Noordwijk, ESA SP-319.

Banwell C.N. (1994). *Fundamentals of Molecular Spectroscopy*. McGraw-Hill, London.

Baranoski G.V.G. and Ronke J.G. (2005). A practical approach for estimating the red edge position of plant leaf reflectance. *Int. J. Remote Sensing*, **26**, 503–521.

Baret F. and Fourty T. (1997). Radiometric estimates of nitrogen status of leaves and canopies. In G. Lemaire (ed.), *Diagnosis of the Nitrogen Status in Crops*, pp. 207–227. Springer-Verlag, Berlin.

Baret F., Jacquemoud S., Guyot G. and Leprieur C. (1992). Modelled analysis of the biophysical nature of spectral shifts and comparison with information content of broad band. *Remote Sensing of Environment*, **41**, 133–142.

Birth G.S. and Hecht H.G. (1987). The physics of near infrared reflectance. In P. Williams and K. Norris (eds.), *Near-Infrared Technology in Agricultural and Food Industry*, pp. 1–6. American Association of Cereal Chemists, St. Paul, MN.

Blackburn G.A. (1998). Quantifying chlorophyll and carotenoids at leaf and canopy scales: An evaluation of some hyperspectral data. *Remote Sensing of Environment*, **66**, 273–285.

Blackburn G.A. (2007). Hyperspectral remote sensing of plant pigments. *J. Experimental Botany*, **58**, 855–867.

Bonham-Carter G.F. (1988). Numerical procedures and computer program for fitting an inverted Gaussian model to vegetation reflectance data. *Computers and Geosciences*, **14**, 339–356.

Bouma P.J. (1971). *Physical Aspects of Colour*, second edition. Macmillan, London.

Boyd D.S., Entwistle J.A., Flowers A.G., Armitage R.P., and Goldsmith P.C. (2006). Remote sensing the radionuclide contaminated Belorussian landscape: A potential for imaging spectrometry? *Int. J. Remote Sensing*, **27**, 1865–1874.

Boyd D.S., Almond S.F., Dash J., and Curran P.J. (2007). Investigating the factors affecting the relationship between the Envisat MERIS Terrestrial Chlorophyll Index (MTCI) and chlorophyll content: Preliminary findings. *Challenges for Earth Observation: Scientific, Technical and Commercial.* Remote Sensing and Photogrammetry Society, Nottingham, U.K. (CD-ROM).

Broge N.H. and Leblanc E. (2001). Comparing prediction power and stability of broadband and hyperspectral vegetation indices for estimation of green leaf area index and canopy chlorophyll density. *Remote Sensing of Environment*, **76**, 156–172.

Carter G.A. (1993). Responses of leaf spectral reflectance to plant stress. *Amer. J. Botany*, **80**, 239–243.

Castro-Esau K.L., Sanchez-Azofeifa G.A., Rivard B., Wright S.J., and Quesada, M. (2006). Variability in leaf optical properties of Mesoamerican trees and the potential for species classification. *Amer. J. Botany*, **93**, 517–530.

Choudhury B.J. (1987). Relationship between vegetation indices, radiation absorption and net photosynthesis evaluated by a sensitivity analysis. *Remote Sensing of Environment*, **22**, 209–233.

Clark R.N., King T.V.V., Ager C., and Swayze G.A. (1995). Initial vegetation species and senescence/stress indicator mapping in the San Luis valley, Colorado using imaging spectrometer data. *Proc. AVIRIS Airborne Geoscience Workshop, Pasadena, Callifornia*, pp. 35-38. NASA Jet Propulsion Laboratory, Pasadena, CA.

Clevers J.G.P.W., De Jong S.M., Epema G.F., Van Der Meer F., Bakker W.H., Skidmore A.K., and Scholte K.H. (2002). Derivation of the red edge index using the MERIS standard band setting. *Int. J. Remote Sensing*, **23**, 3169–3184.

Curran P.J. (1980). Multispectral remote sensing of vegetation amount. *Progress in Physical Geography*, **4**, 315–341.

Curran P.J. (1983). Multispectral remote sensing for the estimation of green leaf area index. *Philos. Trans. Royal Society of London, Series A*, **309**, 257–270.

Curran P.J. (1989). Remote sensing of foliar chemistry. *Remote Sensing of Environment*, **30**, 271–278.

Curran P.J. (2001). Imaging spectrometry for ecological applications. *Int. J. Applied Earth Observation and Geoinformation*, **3**, 305–312.

Curran P.J. and Steele C.M. (2005). MERIS: The re-branding of an ocean sensor. *Int. J. Remote Sensing*, **26**, 1781–1798.

Curran P.J., Dungan J.L., and Gholz H.L. (1990a). Exploring the relationship between reflectance red edge and chlorophyll content in slash pine. *Tree Physiology*, **7**, 33–48.

Curran P.J., Foody G.M., Kondratyev K.Ya., Kozoderov V.V., and Fedchenko P.P. (1990b). *Remote Sensing of Soils and Vegetation in the USSR*. Taylor & Francis, London.

Curran P.J., Windham W.R., and Gholz H.L. (1995). Exploring the relationship between reflectance red edge and chlorophyll concentration in slash pine leaves. *Tree Physiology*, **15**, 203–206.

Curran P.J., Kupiec J.A., and Smith G.M. (1997). Remote sensing the biochemical composition of a slash pine canopy. *IEEE Trans. Geoscience and Remote Sensing*, **35**, 415–420.

Curran P.J., Dash J., and Llewellyn G.M. (2007a). The Indian Ocean Tsunami: Use of the MERIS (MTCI) data to detect salt stress on near coastal vegetation. *Int. J. Remote Sensing*, **28**, 729–735.

Curran P.J., Dash J., Lankester T., and Hubbard S. (2007b). Global composites of the MERIS Terrestrial Chlorophyll Index. *Int. J. Remote Sensing*, **28**, 3757–3758.

Danson F.M. and Plummer S.E. (1995). Red edge response to forest leaf area index. *Int. J. Remote Sensing*, **16**, 183–188.

Dash J. and Curran P.J. (2004). The MERIS Terrestrial Chlorophyll Index. *Int. J. Remote Sensing*, **25**, 5003–5013.

Dash J. and Curran P.J. (2006). Relationship between herbicide concentration during the 1960s and 1970s and the contemporary MERIS Terrestrial Chlorophyll Index (MTCI) for southern Vietnam. *Int. J. Geographical Information Science*, **20**, 929–939.

Dash J. and Curran P.J. (2007). Relationship between the MERIS vegetation indices and crop yield for the state of South Dakota, USA. *Second Envisat Symposium*. European Space Agency, Noordwijk, ESA SP-636 (CD-ROM).

Dash J. and Curran P.J. (2008). The relationship between MTCI and terrestrial chlorophyll content. *Int. J. Remote Sensing*, **29** (in press).

Dash J., Curran P.J., Tallis M.J., Llewellyn G.M., Taylor G., and Snoeij P. (2007a). The relationship between the MERIS Terrestrial Chlorophyll Index and chlorophyll content, *Second Envisat Symposium*. European Space Agency, Noordwijk, ESA SP-636 (CD-ROM).

Dash J., Mathur A., Foody G.M., Curran P.J., Chipman J., and Lillesand T.M. (2007b). Landcover classification using multi-temporal MERIS vegetation indices. *Int. J. Remote Sensing*, **28**, 1137–1159.

Daughtry C.S.T., Walthall C.L., Kim M.S., Brown de Colstoun E., and McMurtrey III J.E. (2000). Estimating corn leaf chlorophyll concentration from leaf and canopy reflectance. *Remote Sensing of Environment*, **74**, 229–239.

Dawson, T.P. and Curran P.J. (1998). A new technique for interpolating the reflectance red edge position. *Int. J. Remote Sensing*, **19**, 2133–2139.

Dawson T.P., Curran P.J., and Plummer S.E. (1998). LIBERTY: Modelling the effects of leaf biochemical concentration on reflectance spectra. *Remote Sensing of Environment*, **65**, 50–60.

Dawson T.P., Curran P.J., North P.R.J., and Plummer S.E. (1999). The propagation of foliar biochemical absorption features in forest canopy reflectance: A theoretical analysis. *Remote Sensing of Environment*, **67**, 147–159.

Demetriades-Shah T.H., Steven M.D., and Clark J.A. (1990). High resolution derivative spectra in remote sensing. *Remote Sensing of Environment*, **33**, 55–64.

Fedchenko P.P. (1982). *A Remote Sensing Technique to Assess the Green Crop Canopy Mass*, Invention Bulletin A.C. No. 969204. U.S.S.R. Technical Press, Moscow.

Feddema J.J., Oleson K.W., Bonan G.B., Mearns L.O., Buja L.E., Meehl G., and Washington W.M. (2005). The importance of land-cover change in simulating future climate. *Science*, **5754**, 1674–1677.

Filella I. and Peñuelas J. (1994). The red edge position and shape as indicator of plant chlorophyll content, biomass and hydric status. *Int. J. Remote Sensing*, **15**, 1459–1470.

Gausmann H.W. (1974). Leaf reflectance of near-infrared. *Photogrammetric Engineering*, **40**, 183–191.

Gitelson A.A. and Merzlyak M.N. (1998). Remote sensing of chlorophyll concentration in higher plant leaves. *Advances in Space Research*, **22**, 689–692.

Gitelson A.A., Viña A., Verma S.B., Rundquist D.C., Arkebauer T.J., Keydan G., Leavitt B., Ciganda V., Burba G.G., and Suyker A.E. (2006). Relationship between gross primary production and chlorophyll content in crops: Implications for the synoptic monitoring of vegetation productivity. *J. Geophys. Res.*, **111**, D08S11.

Guyot G., Baret F., and Major D.J. (1988). High spectral resolution: Determination of spectral shifts between the red and the near infrared. *Int. Archives of Photogrammetry and Remote Sensing*, **11**, 740–760.

Hare E.W., Miller J.R., and Edwards G.R. (1984). Studies of the vegetation red reflectance edge in geobotanical remote sensing in eastern Canada. *Proc. Ninth Canadian Symp. on Remote Sensing, Newfoundland*, pp. 433–440. Canadian Astronautics and Space Institute, Ottawa.

Herring M. (1987). *Shuttle Imaging Spectrometer Experiment (SISEX): Imaging Spectroscopy II*. SPIE, Bellingham, WA [*Proc. SPIE*, **384**, 181–187].

Horler D.N.H., Barber J., and Barringer A.R. (1980). Effect of heavy metals on the absorption and reflectance spectra of plants. *Int. J. Remote Sensing*, **1**, 121–136.

Horler D.N.H., Dockray M., and Barber J. (1983). The red edge of plant leaf reflectance. *Int. J. Remote Sensing*, **4**, 273–288.

Hunt R.W.G. (1987). *Measuring Colour*. Ellis Horwood, Chichester, U.K.

Jago R.A., Curran P.J., and Cutler M.E.J. (1997). Estimating canopy chlorophyll concentration from both field and airborne spectra. In G. Guyot and T. Phulpin (eds.), *Physical Measurements and Signatures in Remote Sensing*, pp. 517–521. Balkema, Rotterdam.

Jago R.A., Cutler M.E., and Curran P.J. (1999). Estimation of canopy chlorophyll concentration from field and airborne spectra. *Remote Sensing of Environment*, **68**, 217–224.

Jeffrey A. (1985). *Mathematics for Engineers and Scientists*. Van Nostrand Reinhold, Wokingham, U.K.

Jensen R.R. and Binford M.W. (2004). Measurement and comparison of leaf area index estimators derived from satellite remote sensing techniques. *Int. J. Remote Sensing*, **25**, 4251–4265.

Johnson L.F. (1999). *Response of Grape Leaf Spectra to Phylloxera Infestation*, NASA Report #CR-208765. NASA Center for AeroSpace Information, Hanover, MD.

Justice C. and Townshend J. (2002). Special issue on the moderate resolution imaging spectroradiometer (MODIS): A new generation of land surface monitoring. *Remote Sensing of Environment*, **83**, 1–2.

Kozoderov V.V. and Dmitriev E.V. (2008). Remote sensing of soils and vegetation: Regional aspects. *Int. J. Remote Sensing*, **29**, 2733–2748.

Kondratyev K.Ya. and Fedchenko P.P. (1979). Spectral reflectance of some weeds. *Doklady U.S.S.R. Academy of Sciences*, **248**, 1318–1320.

Kondratyev K.Ya and Fedchenko P.P. (1982a). Remote sensing of areas under damaged and dead winter crops. *Meteorology and Hydrology*, **8**, 102–108.

Kondratyev K.Ya. and Fedchenko P.P. (1982b). *Spectral Reflectivity and Recognition of Some Vegetation Types*. Hydrometeoizdat, Leningrad.

Kondratyev K.Ya. and Smoktiy O.I. (1973). On the determination of the spectral transfer function for the brightness of natural formations and their contrasts in spectrophotometry of the atmosphere–surface system from space. *Trudy MGO*, **295**, 24–25.

Kondratyev K.Ya., Badinov I.Ya., Ivlev L.S., and Nikolsky G.A. (1969). Aerosol structure of the troposphere and stratosphere. *Physics of the Atmosphere and Ocean*, **5**, 480–493.

Kondratyev K.Ya., Fedchenko P.P., and Barmina Yu.M. (1982a). An experience in determining the chlorophyll content in the leaves of plants from colour coordinates. *Doklady U.S.S.R. Academy of Sciences*, **262**, 1022–1024.

Kondratyev K.Ya., Kozoderov V.V., and Fedchenko P.P. (1982b). Possibilities of determining the chlorophyll content in plants from their reflectance spectra. *Studies of the Earth from Space*, **3**, 63–68.

Kondratyev K.Ya., Kozoderov V.V., Fedchenko P.P., and Barmina Yu.M. (1982c). On the determination of the chlorophyll content in the leaves of plants from data of spectral measurements. *Doklady U.S.S.R. Academy of Sciences*, **256**, 1508–1510.

Kondratyev K.Ya., Kozoderov V.V., and Fedchenko P.P. (1986). Remote sensing of the state of crops and soils. *Int. J. Remote Sensing*, **7**, 1213–1235.

Kondratyev K.Ya., Krapivin V.F., Savinykh V.P., and Varotsos C.A. (2004). *Global Ecodynamics*. Springer/Praxis, Chichester, U.K.

Kupiec J. and Curran P.J. (1995). Decoupling effect of the canopy and foliar biochemicals in AVIRIS spectra. *Int. J. Remote Sensing*, **16**, 1731–1739.

Li L., Ustin S.L., and Lay M. (2005). Application of AVIRIS data in detection of oil-induced vegetation stress and cover change at Jornada, New Mexico. *Remote Sensing of Environment*, **94**, 1–16.

Lichtenthaler H.K. (1987). Chlorophyll and carotenoids: Pigments of photosynthetic biomembranes. *Methods of Enzymology*, **184**, 350–382.

Llewellyn G.M., Kooistra L., and Curran P.J. (2001). The red-edge of soil contaminated grassland. *Proceedings Eighth Int. Symp.: Physical Measurements and Signature in Remote Sensing*. International Society for Photogrammetry and Remote Sensing and Centre National D'Etudes Spatiales, Aussois, France, pp. 381–386.

Loveland T.R., Reed B.C., Brown J.F., Ohlen D.O., Zhu Z., Yang L., and Merchant J.W. (2000). Development of a global land cover characteristics database and IGBP DISCover from 1 km AVHRR data. *Int. J. Remote Sensing*, **21**, 1303–1330.

Lucas N.S., Curran P.J., Plummer S.E., and Danson F.M. (2000). Estimating the stem carbon production of a coniferous forest using an ecosystem simulation model driven by the remotely sensed red edge. *Int. J. Remote Sensing*, **21**, 619–631.

Mariotti M., Ercoli L., and Masoni A. (1996). Spectral properties of iron-deficient corn and sunflower leaves. *Remote Sensing of Environment*, **58**, 282–288.

Merton R. (1998). Monitoring community hysteresis using spectral shift analysis and the red-edge vegetation stress index. *AVIRIS Airborne Geosciences Workshop Proceedings*, pp. 32–38, JPL Publication 97-21, NASA Jet Propulsion Laboratory, Pasadena, CA.

Miller J.R., Hare E.W., and Wu J. (1990). Quantitative characterization of the vegetation red edge reflectance, 1: An inverted-Gaussian reflectance model. *Int. J. Remote Sensing*, **11**, 1755–1773.

Milton E.J. (2000). Practical methodologies for the reflectance calibration of CASI data. *Airborne Remote Sensing Facility Annual Meeting, Keyworth, Nottingham, U.K.*, pp. 1–14. NERC EPFS, Southampton, U.K.

Milton N.M., Ager C.M., Eiswerth B.A., and Power M.S. (1989). Arsenic and selenium induced changes in spectral reflectance and morphology of soyabean plants. *Remote Sensing of Environment*, **30**, 263–269.

Munden R., Curran P.J., and Catt J.A. (1994). The relationship between red edge and chlorophyll concentration in the Broadbalk winter wheat experiment at Rothamsted. *Int. J. Remote Sensing*, **15**, 705–709.

Niemann K.O. (1995). Remote sensing of forest stands age using airborne spectrometer data. *Photogrammetric Engineering and Remote Sensing*, **61**, 1119–1127.

O'Neill A.L., Kupiec J.A., and Curran P.J. (2002). Biochemical and reflectance variation throughout a Stika spruce canopy. *Remote Sensing of Environment*, **80**, 134–142.

Pinar A. and Curran P.J. (1996). Grass chlorophyll and the reflectance red edge. *Int. J. Remote Sensing*, **17**, 351–357.

Railyan Y. and Korobov R.M. (1993). Red edge structure of canopy reflectance spectra of triticale. *Remote Sensing of Environment*, **46**, 173–182.

Reeves D.W., Mask P.L., Wood C.W., and Delaney D.P. (1993). Determination of wheat nitrogen status with a hand-held chlorophyll meter influence of management-practices. *J. Plant Nutrition*, **16**, 781–796.

Reeves M.C., Zhao M., and Running S.W. (2005). Usefulness and limits of MODIS GPP for estimating wheat yield. *Int. J. Remote Sensing*, **26**, 1403–1421.

Smith K.L., Steven M.D., and Colls J.J. (2004). Use of hyperspectral derivative ratios in the red-edge region to identify plant stress responses to gas leaks. *Remote Sensing of Environment*, **92**, 207–217.

Song C.H. (2005). Spectral mixture analysis for subpixel vegetation fractions in the urban environment: How to incorporate endmember variability? *Remote Sensing of Environment*, **95**, 248–263.

Steven M.D., Malthus T.J., Demetriades-Shah T.H., Danson F.M., and Clark J.A. (1990). High-spectral resolution indices for crop stress. In M.D. Steven and J.A. Clark (eds.), *Applications of Remote Sensing in Agriculture*, pp. 201–228. Butterworth, London.

Sushkevich T.A. (2008). Pioneering remote sensing in the USSR, 1: Radiation transfer in the optical wavelength region of the electromagnetic spectrum. *Int. J. Remote Sensing*, **29**, 2585–2597.

Tarasov K.I. (1968). *Spectral Instruments.* Mashinostroenie, Moscow [in Russian].

Tobehn C., Kassebohm M., and Schmälter E. (2003). SPECTRA–ESA candidate Earth explorer core mission: Feasibility, results and outlook. *Proc. Fourth IAA Symposium on Small Satellites for Earth Observation, Berlin, Germany*. Wissenschaft und Technik Verlag, Berlin.

Treitz P.M. and Howarth P.J. (1999). Hyperspectral remote sensing for estimating biophysical parameters of forest ecosystems. *Progress in Physical Geography*, **23**, 359–390.

Treitz P., Thomas V., Zarco-Tejada P., Gong P., and Curran P.J. (2008). Hyperspectral remote sensing for forestry. In E.A. Clostis (ed.), *Manual of Remote Sensing*, third edition. American Society for Photogrammetry and Remote Sensing, Bethesda, MD (in press).

Ustin S.L., Roberts D.A., Gamon J.A., Asner G.P., and Green R.O. (2004). Using imaging spectroscopy to study ecosystem processes and properties. *Bioscience*, **54**, 523–534.

Vane G. (1987). *Earth Observing System—A Platform for Imaging Spectrometers: Imaging Spectroscopy II*. SPIE, Bellingham, WA [*Proc. SPIE*, **834**, 176–180].

Vane G., Goetz A.F.H., and Wellman J.B. (1984). Airborne imaging spectrometer: A new tool for remote-sensing. *IEEE Trans. Geoscience and Remote Sensing*, **22**, 546–549.

Vane G., Green R.O., Chrien T.G., Enmark H.T., Hansen E.G., and Porter W.M. (1993). The Airborne Visible/infrared Imaging Spectrometer (AVIRIS). *Remote Sensing of Environment*, **44**, 127–143.

Wessman C.A. (1991). Remote-sensing of soil processes. *Agriculture Ecosystems and Environment*, **34**, 479–493.

Wright W.D. (1969). *The Measurement of Colour*. Adam Hilger, London.

Wulder M. (1998). Optical remote sensing techniques for the assessment of forest inventory and biophysical parameters. *Progress in Physical Geography*, **22**, 449–476.

Wylie B.K., Denda I., Pieper R.D., Harrington J.A., Reed B.C., and Southward G.M. (1995). Satellite based herbaceous biomass estimates in the pastoral zone of Niger. *J. Range Management*, **48**, 159–164.

Yang C. and Su M. (2000). Analysis of spectral characteristics of rice canopy under water deficiency. *Proc. 21st Asian Conf. on Remote Sensing, Taipei, Taiwan*. National Central University of Taiwan, Taipei. Available at *http://www.gisdevelopment.net/aars/acrs/2000/ts1/agri003.asp*

Zagolski F., Pinel V., Romier R., Alcayde D., Fontanari, J., Gastellu-Etchegorry J.P., Giordano G., Marty G., Mougin E., and Joffre R. (1996). Forest canopy chemistry with high spectral resolution remote sensing. *Int. J. Remote Sensing*, **17**, 1107–1128.

Zarco-Tejada P.J. and Miller J.R. (1999). Land cover mapping at BOREAS using red edge spectral parameters from CASI imagery. *J. Geophys. Res.*, **104**, 921–933.

6

Regarding greenhouse explosion

Vyacheslav I. Zakharov

6.1 INTRODUCTION

Can our planet Earth become as hot as Venus as a result of runaway accumulation of carbon dioxide in the atmosphere and explosive increasing of the greenhouse effect? This question is reasonable because the Earth's reservoirs, such as the oceans, biota, and carbonates in the Earth's crust, contain approximately the same total amount of CO_2 as the atmosphere of Venus (Nicholls, 1967). On the other hand, simulations of the radiation regime of the atmosphere of Venus (Kondratyev and Moskalenko, 1985) confirm that the temperature of the lower atmosphere of the planet may be very hot even if the incoming solar flux inside its atmosphere is weaker than the incoming solar flux inside the Earth's atmosphere. For example, according to observations, the surface temperature of Venus is about 730 K (Kondratyev, 1990), in comparison with the annual mean surface temperature of the Earth which is about 288.2 K. However, the flux of solar radiation at the top of the atmosphere of Venus is mainly reflected by clouds (due to the high value of the albedo of the clouds, about 0.75); thus, the flux of solar radiation incoming to the atmosphere of Venus and heating the surface is about 165 $W\,m^{-2}$ (Kondratyev, 1990; Gorshkov, 1995) by comparison with the incoming flux of solar radiation to the atmosphere of the Earth of about 240 $W\,m^{-2}$ (Bach, 1987). The reason for the large difference between their surface temperatures is the very strong greenhouse effect on Venus due to the high content of greenhouse gases, mainly carbon dioxide (pressure about 90 atmospheres), in the atmosphere of Venus.

The present-day classical models of the greenhouse effect (Budyko, 1980; Bach, 1987; McGuffie and Henderson-Sellers, 1997) basically assume that the thermal balance of the Earth is mainly regulated by the variation of the downward thermal radiation of the atmosphere in the wings of the 15 μm fundamental absorption band of CO_2, because absorption (emission) in the center of the band is saturated (Figures 6.1 and 6.2). For changes in CO_2 alone, radiative forcing can be

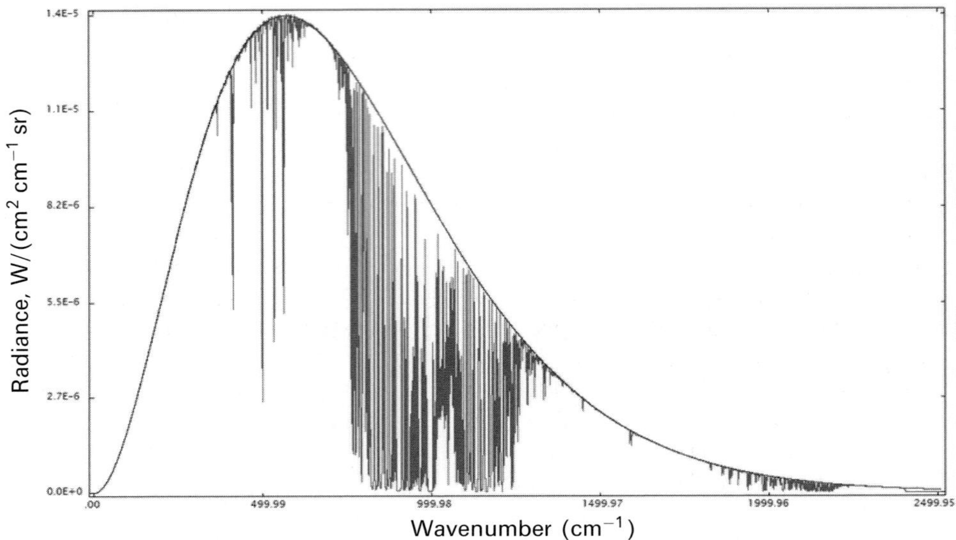

Figure 6.1. Downward thermal atmospheric radiance at the surface of the Earth in the spectral range of 0 cm^{-1}–2,500 cm^{-1} simulated with FIRE-ARMS using the U.S. standard model of the atmosphere. The envelope line is the Planck emission of the surface at 288.2 K.

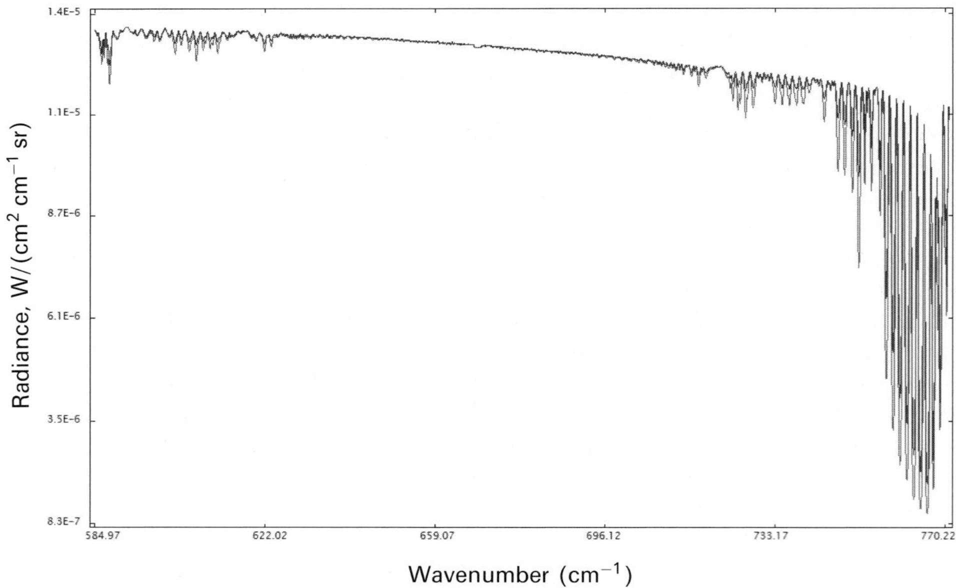

Figure 6.2. Effect of doubling of CO_2 in the atmosphere on the downward thermal atmospheric radiance at the surface of the Earth in the range of the CO_2 fundamental absorption band 15 μm. The gray line is the present concentration of CO_2, and the dark line is $2CO_2$. The change of downward radiance in the range of the right wing of the CO_2 band is shown.

approximated by a natural logarithmic function of CO_2 concentration with a climate sensitivity parameter determining the change in temperature (McGuffie and Henderson-Sellers, 1997). This means that a process of an unlimited accumulation of carbon dioxide in the atmosphere should increase the Earth's surface temperature only by a few degrees as a result of the absorption saturation of thermal radiation. However, it takes place until the surface temperature of the Earth and concentration of CO_2 in the atmosphere are lower than threshold magnitudes (Zakharov *et al.*, 1991a, 1992, 1997). Direct measurements of the temperature profile of the atmosphere of Venus using descent modules confirm that a strong greenhouse effect takes place in the high-pressure CO_2 atmosphere on Venus (Kondratyev, 1990, pp. 245–250), and the latter was also modeled by Kondratyev and Moskalenko (1985). In order to discuss the radiation balance model of the Earth within the surface temperature in the range from 288 K to about 1,000 K, the climate component of the model is intended to be more general and to be able to capture changes in global average surface temperature due to changes in albedo, solar flux, and the optical thickness of key greenhouse gases (CO_2, H_2O, and CH_4) in the thermal infrared. Several papers considering the possibility of a super-hot thermal regime of the atmosphere of a planet like Venus have been published since 1967. For example, Komabayashi (1967, 1968) considered discrete equilibrium temperatures of a hypothetical planet with the atmosphere and the hydrosphere of a one-component two-phase system under constant solar radiation. Ingersoll (1969) discussed the runaway greenhouse, the history of water on Venus. Abe and Matsui (1988) considered the evolution of an impact-generated H_2O–CO_2 atmosphere and the formation of a hot proto-ocean on Earth. Kasting (1988) discussed questions regarding runaway and moist greenhouse atmospheres and the evolution of Earth and Venus. Nakajima *et al.* (1992) performed a study on the runaway greenhouse effect with a one-dimensional radiative–convective equilibrium model. These publications also support the point of view that there is no saturation of the greenhouse effect resulting from an unlimited accumulation of greenhouse gases in the atmosphere of a planet.

So, the question "Can the Earth's atmosphere become as hot as Venus' atmosphere as a result of unlimited accumulation of carbon dioxide and increase of the greenhouse effect?" is quite reasonable. The main reason is that there is surface heating and positive feedback between the absorption of outgoing radiation in the atmosphere in the thermal infrared bands of CO_2 and H_2O (e.g., in the middle of the atmospheric transparency window 8 μm–13 μm). For example, thermal radiation is strongly absorbed by carbon dioxide in the vibrational thermal infrared bands (100)–(001) and (020)–(001), near the regions of 943 cm^{-1} and 1,064 cm^{-1}, respectively, and by water vapor over the entire ranges of transparency windows.

The mechanism for this positive feedback is based on the exponential temperature dependence of absorption in the thermal infrared bands as well as on the exponentially increasing CO_2 equilibrium concentration in the atmosphere due to emission from the oceans, from the Earth's crust, or as a result of exhalation of biota, on the one hand, and the increasing of the equilibrium concentration of water vapor in the atmosphere, on the other hand. Also, there is an additional positive feedback between surface temperature and the concentration of methane in the Earth's atmosphere (Gorham, 1991).

In the Earth's climate system there are also two mechanisms of negative feedback. The first one is due to photosynthesis leading to the non-linear dependence of the CO_2 concentration in the atmosphere on change in temperature within the range of 288 K–315 K (Bach, 1987; Bolin, 1987), probably tending to stabilize the concentration of carbon dioxide in the atmosphere. The second one is the dependence of the albedo on surface temperature within the range of 288 K–1,000 K. It expects a positive feedback, at the first stage of heating of the surface, several degrees higher than 288 K (Karol, 1988; Matveev, 1991), mainly due to thawing of glaciers and decreasing size of the mean area of the cryosystem. When the surface temperature becomes higher, negative feedback may dominate, resulting in the growth of evaporation and increase of the density and area of clouds over the planet. Obviously, if the negative feedback is dominant in the climate system the Earth's surface temperature is stable, otherwise global thermal instability can be expected to take place. Moreover, we should expect threshold conditions for this process, because a physical system with nonlinear positive feedback usually has threshold conditions for values of the parameters governing the system (Haken, 1984).

6.2 RADIATION BALANCE AT THE SURFACE WITHIN THE FRAMEWORK OF A MODEL OF A GRAY ATMOSPHERE; SEVERAL STATIONARY THERMAL STATES OF THE HYPOTHETICAL EARTH

To start with, it is better to consider the simplest model of radiation balance at the surface of the Earth within the framework of the concept of a gray atmosphere to clarify the physics and show the possibility of the existence of different stationary states of thermal balance of the planet. The model of a gray atmosphere for calculation of the radiation balance of a planet at the surface level is a useful approximation in order to obtain analytical solutions for the greenhouse effect and investigate the main qualitative features of the behavior of the surface temperature of the planet, T_s, as a function of the (equivalent gray) vertical optical thickness (opacity) of the greenhouse atmosphere τ and albedo of the planet A (see, for example, Chamberlain, 1980;, Matveev, 1991; McKay *et al.*, 1999; Lenton, 2000). A general gray atmosphere approach gives the following expression for the surface temperature T_s:

$$T_s = \left[\frac{I_s(1-A)}{4\sigma}\left(1+\tfrac{1}{2}\tau\right)\right]^{1/4}. \tag{6.1}$$

In the case of the Earth's atmosphere, $I_s \approx 1{,}370\,\mathrm{W\,m^{-2}}$ is the solar constant, and $\sigma = 5.67\times10^{-8}\,\mathrm{W\,m^{-2}\,K^{-4}}$ is the Stefan–Boltzmann constant. For the current state of the atmosphere $A = A(T_s^{(0)}) = 0.3$, $T_s^{(0)} = 288.2\,\mathrm{K}$, and $\tau = \tau_e^{(0)} = 1.26$.

In the general case we can expect a dependence of the albedo, A, of our planet and the opacity, τ, of its atmosphere vs. surface temperature (i.e., $A = A(T_s)$ and

$\tau = \tau(T_s)$). For simplicity we approximate the temperature dependence of the opacity of the atmosphere by an exponential function, for example, $\tau = \tau(T_s) = \tau_e^{(0)} a \exp(-b/T_s)$, where $a = \exp(b/T_s^{(0)})$. The evaluation of b gives $b = 2{,}275$ K, based on the assumption that the maximum value of the opacity of the Earth's atmosphere is the same as the opacity of the atmosphere of Venus (i.e., $\tau_{\max} = 216$), as a result of the unlimited accumulation of CO_2 and other greenhouse gases in the atmosphere. In this case Equation (6.1) can be written:

$$T_s = f(T_s) = \left[\frac{I_s(1 - A(T_s))}{4\sigma}\left\{1 + \tfrac{1}{2}\tau_e^{(0)} \exp\left(\frac{b}{T_s^{(0)}} \frac{T_s - T_s^{(0)}}{T_s}\right)\right\}\right]^{1/4}. \tag{6.2}$$

Solutions of Equation (6.2) for T_s are stationary states of the thermal regime of the surface of the planet. As an example, Figures 6.1 and 6.2 show possible stationary thermal states of the Earth's gray atmosphere following from Equation (6.2) at constant albedo, which is equal to the present-day value of the albedo of the Earth (i.e., $A(T_s) = A(T_s^{(0)}) = 0.3$). For this case Equation (6.2) can be written:

$$T_s = 255.2\left[1 + 0.63 \exp\left(7.9 \frac{T_s - 288.2}{T_s}\right)\right]^{1/4}.$$

This model leads to three stationary temperature states of the Earth's surface: the present-day stable thermal state with surface temperature 288.2 K, an unstable warm state with a temperature of the surface of about 350 K, and a stable hot state with the temperature of the surface about 840 K. Figure 6.2 demonstrates that the thermal state with surface temperature 288 K is stable, but the nearer warm thermal state with surface temperature 350 K is unstable. As for a hot thermal state with surface temperature 840 K (Figure 6.13a), it is stable as well.

Of course, how many stationary thermal states of the Earth are possible in the temperature range of 288 K–1,000 K and their locations in the temperature scale depends on the real temperature behavior of the albedo of the Earth $A = A(T_s)$ and the opacity of the atmosphere $\tau = \tau(T_s)$. The impact of the albedo can change both the number and the positions of these stationary states within the range of 288 K–1,000 K. For instance, in the beginning of warming of the present state of the Earth the value of the albedo will decrease because of the thawing of the glaciers and the decreasing area of the cryosystem (Karol, 1988; Matveev, 1991), until the process of growth of evaporation and cloud formation increase the cloud albedo and dominate the contribution of the cryosystem to the total albedo of the planet. In this case it should be expected that the unstable warm stationary state is located at a lower temperature than is shown in Figure 6.1.

It is important that the simple model of radiation balance at the surface of the planet described above in Equation (6.2) makes it possible to investigate analytically a threshold condition for a transition from the present stable thermal state with surface temperature 288.2 K to the hot thermal state at 840 K. The threshold condition of the radiation balance can be written in the form:

$$\left[\frac{I_s(1-A(T_s))}{4\sigma T_s^4}\left\{1+\tfrac{1}{2}\tau_e^{(0)}\exp\left(\frac{b}{T_s^{(0)}}\frac{T_s-T_s^{(0)}}{T_s}\right)\right\}\right] > 1. \tag{6.3}$$

This condition gives the magnitude of the threshold temperature of the surface T_s which depends on three parameters, which are the initial temperature of the surface $T_s^{(0)}$, the initial opacity of the atmosphere $\tau_e^{(0)}(T_s^{(0)})$, and the initial value of the albedo $A(T_s^{(0)})$ of the planet. These are currently 288.2 K, 1.26, and 0.3, respectively, for the Earth. The model gives the following magnitudes of the threshold parameters: $A(T_s^{(0)}) \approx 0.27$ for current values of $\tau_e^{(0)} = 1.26$ and $T_s^{(0)} = 288.2$ K, and $\tau_e^{(0)} \approx 1.39$ for current values of $A(T_s^{(0)}) = 0.3$ and $T_s^{(0)} = 288.2$ K. That is around a 10% variation of albedo as well as around a 10% variation of the opacity of the atmosphere. So, the qualitative model considered above gives a magnitude of threshold surface temperature $T_s \approx 350$ K at initial values of the albedo $A(T_s^{(0)}) = 0.3$ and opacity of the atmosphere $\tau_e^{(0)} = 1.26$.

Figures 6.3 and 6.4 show examples of the location of stationary states of surface temperature vs. magnitudes of the initial opacity of the atmosphere and albedo of the planet; above-critical and sub-critical thermal balances are presented. Herein the above model gives the following: if the initial albedo of the planet is greater than 0.46 only one stationary thermal regime with surface temperature around 288 K may exist (Figure 6.3). But if the initial opacity of the atmosphere is greater than the threshold magnitude, one hot thermal state with stationary surface temperature around 840 K occurs just as if initial albedo is lower than the threshold magnitude (Figure 6.4).

More accurate modeling of the Earth's surface radiation balance is possible if the equivalent vertical opacity of the gray atmosphere τ is assumed to be the sum of the opacities of the key greenhouse gases $\tau_{CO_2}(T_s)$, $\tau_{H_2O}(T_s)$, and $\tau_{CH_4}(T_s)$ (Lenton, 2000), aerosol $\tau_a(T_s)$, and other molecular constituents of atmosphere τ_m; that is,

$$\tau = \tau_{CO_2}(T_s) + \tau_{H_2O}(T_s) + \sigma_{CH_4}(T_s) + \tau_a(T_s) + \tau_m. \tag{6.4}$$

Positive feedback is taken into account here by the temperature dependences of the opacities of the key greenhouse gases and cloud aerosol, but with the temperature behavior of the albedo of the Earth as negative feedback. The following model of hypothetical temperature dependences of the albedo of our planet can be used:

$$A(T_s) = A(T_s^{(0)})\left(1 - a\frac{T_s - T_s^{(0)}}{2T_s - T_s^{(0)}}\right)\exp\left(c\frac{T_s - T_s^{(0)}}{T_s}\right), \tag{6.5}$$

where parameters a and c are within the range of 1–3. Figure 6.4 illustrates the modeled temperature behaviors of the albedo of the Earth in the range of 288 K–840 K.

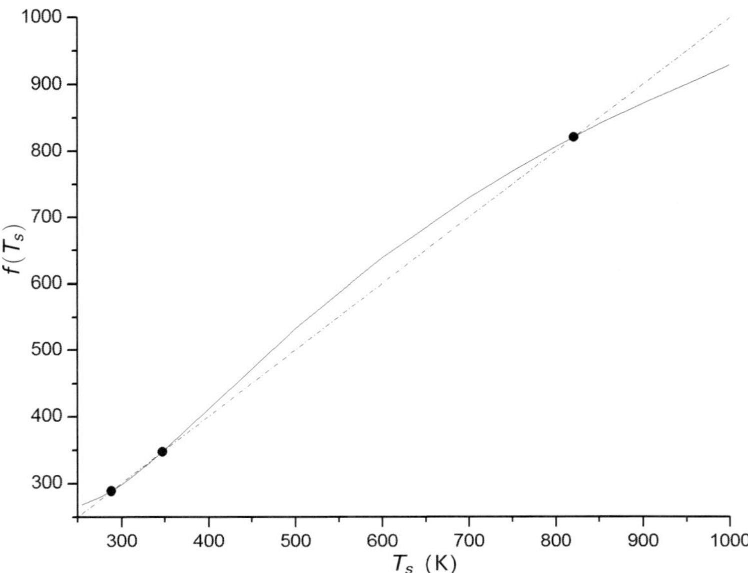

Figure 6.3a. Possible stationary states of surface temperature of the Earth within the framework of the model of gray atmosphere, current state with surface temperature 288.2 K, warm state with surface temperature around 350 K, and hot state with surface temperature around 840 K. Solid line is $f(T_s)$ described by the right part of Equation (6.2), the dashed-dot line is $f(T_s) = T_s$.

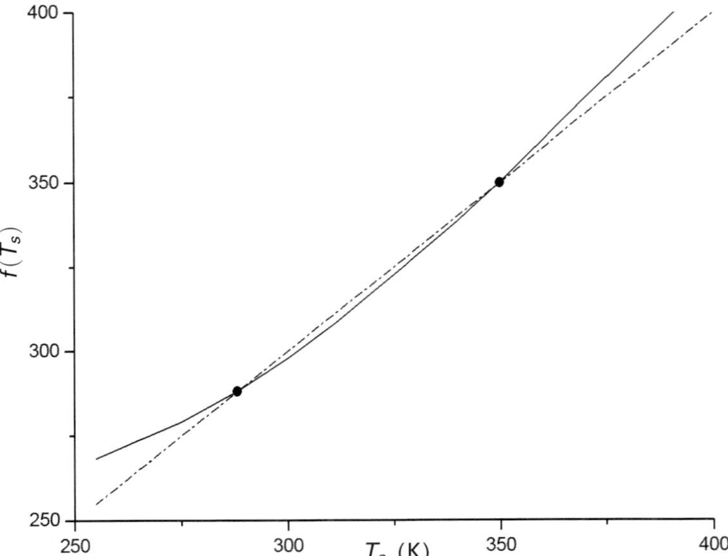

Figure 6.3b. Extended fragment of Figure 6.3a, which shows locations of present stable thermal state and possible nearest warm unstable thermal state of the Earth.

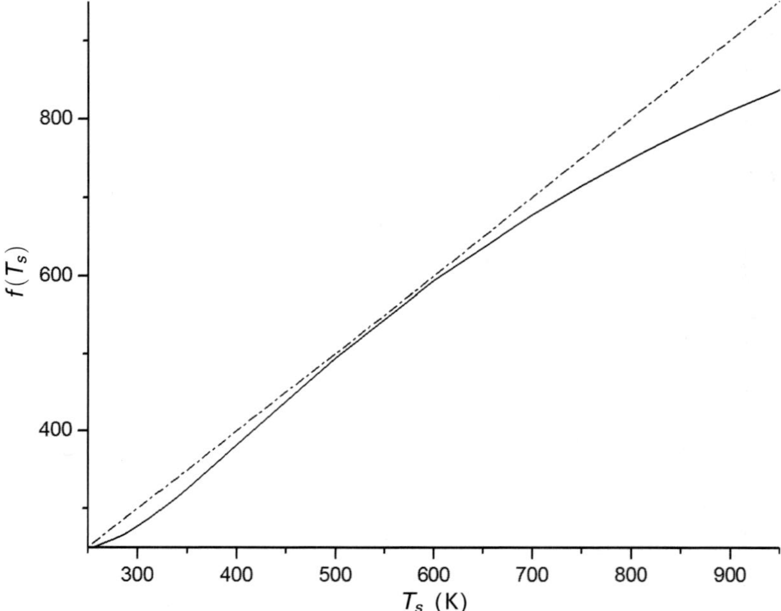

Figure 6.3c. Sub-critical thermal regime of the planet.

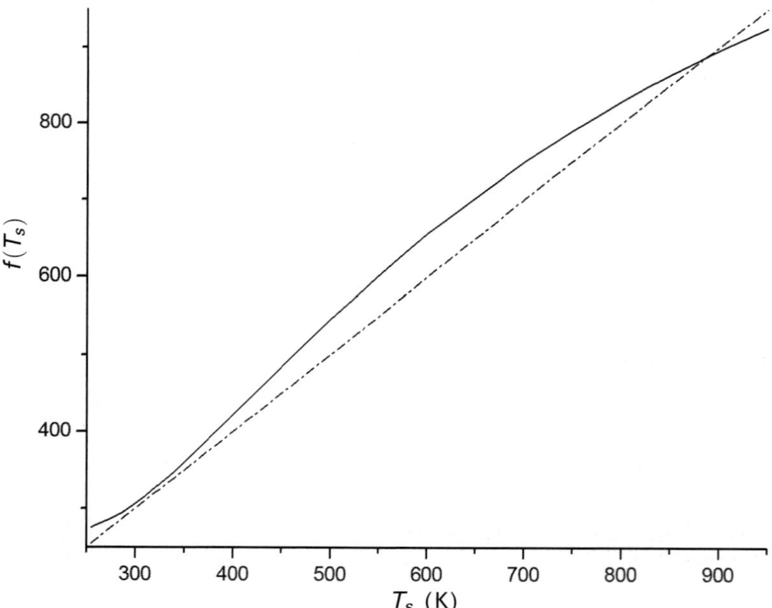

Figure 6.3d. Above-critical thermal regime of the planet.

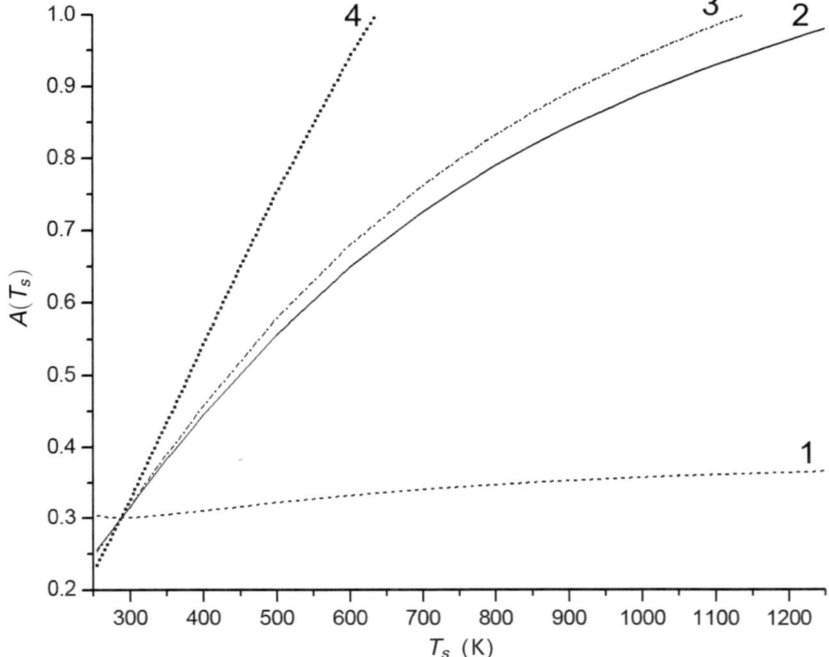

Figure 6.4. A model of the temperature behavior of albedo according to formula (6.5). Here: 1, $a = 1, c = 1$; 2, $a = 1.2, c = 2.5$; 3, $a = 1.3, c = 3$; and 4, $a = 1, c = 3$.

In this case Equation (6.1) becomes:

$$T_s = F(T_s) = \left[\frac{I_s(1 - A(T_s))}{4\sigma} \left[1 + \frac{1}{2} \left\{ \tau_m + \tau_a^{(0)} \right. \right. \right.$$
$$\left. \left. \left. \times \exp\left(b_a \frac{T_s - T_s^{(0)}}{T_s} \right) + \sum_i \tau_e^{(0)}(i) \exp\left(b_i \frac{T_s - T_s^{(0)}}{T_s} \right) \right\} \right] \right]^{1/4}, \qquad (6.6)$$

where i denotes CO_2, H_2O, and CH_4. The temperature dependence of the albedo $A(T_s)$ is described by the expression (6.5). According to direct simulations of atmospheric transmittances within the range of $0\,\mathrm{cm}^{-1}$–$2{,}500\,\mathrm{cm}^{-1}$ using the spectroscopic parameters of the HITRAN database (Rothman *et al.*, 2003) and the U.S. standard model of the atmosphere (Anderson *et al.*, 1986) we have $\tau_e^{(0)}(CO_2) \approx 0.15$, $\tau_e^{(0)}(H_2O) \approx 0.73$, $\tau_e^{(0)}(CH_4) \approx 0.02$. For other molecular constituents of the atmosphere it gives $\tau_m \approx 0.07$. As for the current opacity of atmospheric aerosol (mainly cloud aerosol), it is $\tau_a^{(0)} \approx 0.29$ as the balance of the total value of the opacity of the present Earth's atmosphere 1.26.

The change of concentration of carbon dioxide in the atmosphere for a surface temperature higher than 300 K due to emission from the oceans and mainly from carbonates in the Earth crust, can be approximated by the expression

$$n_{CO_2}(T_s)/n_{CO_2}(T_s^{(0)}) \approx \exp\left\{19.1\frac{T_s - T_s^{(0)}}{T_s}\right\}$$

(Bach, 1987; Nicholls, 1967). Here $T_s^{(0)} = 288.2\,K$ is the present surface temperature of the Earth. As for the concentration of water vapor in the atmosphere, it can be derived by the exponential temperature dependence of the partial pressure of water vapor

$$n_{H_2O}(T_s)/n_{H_2O}(T_s^{(0)}) \approx \exp\left(18.3\frac{T_s - T_s^{(0)}}{T_s}\right)$$

(Matveev, 1984; Lenton, 2000). The accumulation of methane in the atmosphere due to surface temperature increase within the range of 288 K–840 K is assumed here to occur at the same rate as the accumulation of carbon dioxide, and so is approximated by the same exponential expression:

$$n_{CH_4}(T_s)/n_{CH_4}(T_s^{(0)}) \approx \exp\left(19.1\frac{T_s - T_s^{(0)}}{T_s}\right).$$

Both direct simulation of atmospheric transmittance with different atmospheric models and literature data give the values $b_{CO_2} \approx 5.02$ and $b_{H_2O} \approx 9.20$. As for methane, direct simulation of atmospheric transmittances gives the value $b_{CH_4} \approx 7.79$ (Lenton, 2000). According to literature data (Lenton, 2000), the above temperature approximations of the opacities of carbon dioxide and water vapor are valid for the temperature range 288 K–320 K. In this study we use these approximations over the entire temperature range 288 K–840 K. As for the temperature dependence of the opacity of cloud aerosol in the thermal infrared $\tau_a(T_s)$, the same relative temperature dependence as the equilibrium water vapor content in the atmosphere has been assumed based on another assumption, namely that the ratio of the number of H_2O molecules condensed in clouds to the number of H_2O molecules of water vapor in the atmosphere is approximately constant at different annual mean surface temperatures; thus, it gives $b_a = 9.20$.

Figure 6.5 shows solutions of Equation (6.6) describing the positions of stationary states of surface temperature of the Earth obtained in the case of albedo functions presented in Figure 6.4. It is necessary to stress that the improved model of a gray atmosphere also leads to three possible stationary states of the thermal balance for constant albedo, even in the case that we completely neglect the temperature change of cloud opacity in the thermal infrared and the opacity of methane. It confirms that positive feedback between opacities of the greenhouse gases H_2O and CO_2 and temperature of the surface of the Earth is very strong.

This model gives three stationary states of the temperature of the surface of the Earth in a wide range of temperature behaviors of albedo, which is inside the area limited by lines 1–4 (Figure 6.4). All the thermal regimes obtained include the present-day stable thermal state with surface temperature $T_s^{(0)} = 288.2$ K, Models 1–3 include additionally one unstable and one stable thermal state at higher temperatures. In the

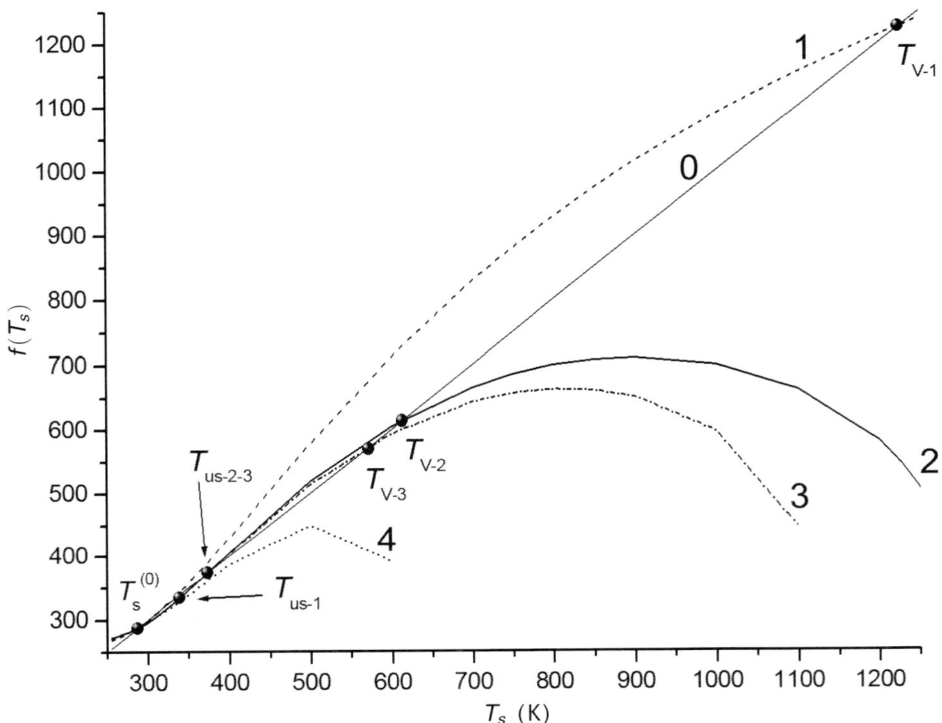

Figure 6.5. Different thermal regimes of Earth (6.6) depending on the hypothetical behavior of albedo (Figure 6.4). In the straight line, 0 is $f(T_s) = T_s$, and in the dotted line, 4 describes a sub-critical thermal regime of the Earth.

case of the most probable temperature behavior of albedo similar to the albedo of Venus (i.e., if $T_s = 730$ K then $A(T_s) = 0.75$, see Curve 2 in Figure 6.4), this model gives the present-day stable temperature state with surface temperature 288.2 K, a warm unstable state T_{us-2} with surface temperature around 365 K, and a hot stable state T_{V-2} with surface temperature around 610 K (Figure 6.5). As for Model 4 of the temperature behavior of the albedo (Figure 6.4), in this sub-critical case only a stable thermal regime of the Earth is possible in the range of 288 K–610 K, which is located around a stationary point with temperature 288.2 K. It also should be mentioned that taking into account the possible behavior of the albedo of our planet at a lower temperature than the current surface temperature 288.2 K we can expect the existence of a cold stationary state with temperature lower than 288 K, corresponding to past ice age climates, and probably a so-called "Snowball Earth" (Budyko, 1968; Crowley and Hyde, 2001; Karol, 1988).

The goal of the following sections of this chapter is consideration of the main spectroscopic features of the transmittance of the atmosphere of the hypothetical warming of the Earth in the range of 0 cm^{-1}–4,000 cm^{-1} near the most probable stationary states of surface temperatures that we have obtained above (namely,

288.2 K, 365 K), and discussing a possibility of transition from the present thermal regime of the Earth to a hot stable state 610 K like the atmosphere of Venus. This study is based on the line-by-line modeling of the wavenumber dependence of the molecular atmospheric transmittance function and outgoing atmospheric thermal radiance within the range of $0\,cm^{-1}$–$4{,}000\,cm^{-1}$ (Zakharov *et al.*, 1997). The U.S. standard model and the Tropical model of the atmosphere are used as basic models where temperature, CO_2, H_2O, and CH_4 vertical profiles are given functions of the surface temperature.

6.3 MOLECULAR TRANSMITTANCE FUNCTIONS OF THE EARTH'S ATMOSPHERE IN THE REGION FROM $0\,CM^{-1}$ TO $4{,}000\,CM^{-1}$ AT THE STATIONARY STATES OF SURFACE TEMPERATURES: 288.2 K, 365 K

The successful application of spectroscopic methods to study the Earth's atmosphere, weather, and climate requires the use of reliable information on high-resolution molecular spectra. The HITRAN database is usually used in Earth and planetary atmospheric applications and has been selected to simulate the change in transmittances of the Earth's atmosphere as a result of increasing the temperature of the surface (Rothman *et al.*, 2003). The software and the graphic package for personal computers FIRE-ARMS (Fine Infrared Explorer of Atmospheric Radiation Measurements, *http://remotesensing.ru*) (Gribanov *et al.*, 2001) designed to manage large-scale spectroscopic databases has been used for the present simulations.

Calculation of the atmospheric transmittance function in the spectral interval $0\,cm^{-1}$–$4{,}000\,cm^{-1}$ can be performed by a *line-by-line method* for different climatological situations of the Earth's atmosphere, assuming a constant gradient of the temperature profile in the troposphere according to the model of radiative–convective adjustment (Manabe and Stouffer, 1993; Nakajima *et al.*, 1992). An example of the temperature profile of the Earth's atmosphere used corresponding to the value of the stationary surface temperatures 365 K obtained above is shown in Figure 6.6.

The initial altitude profiles of $T(h)$, $n_{CO_2}(h)$, $n_{H_2O}(h)$, $n_{CH_4}(h)$, and other molecular constituents for the present atmosphere are taken from a paper by Anderson *et al.* (1986). Two sets of altitude profiles of key greenhouse gases at surface temperatures 288.2 K and 365 K are used for modeling the atmospheric transmittances up to altitudes of 120 km. The first model is the standard U.S. model describing the present thermal state of the Earth's atmosphere, and the other has been composed according to modeling of the possible thermal evolution of the Earth's atmosphere as a result of surface temperature increase and changes in the concentrations of CO_2, H_2O, and CH_4. The mixing ratios of carbon dioxide, water vapor, and methane are calculated at a given surface temperature T_s and this multiplies the profiles taken from the U.S. standard atmosphere by the values of the relative change of their concentrations in the atmosphere as a result of surface temperature

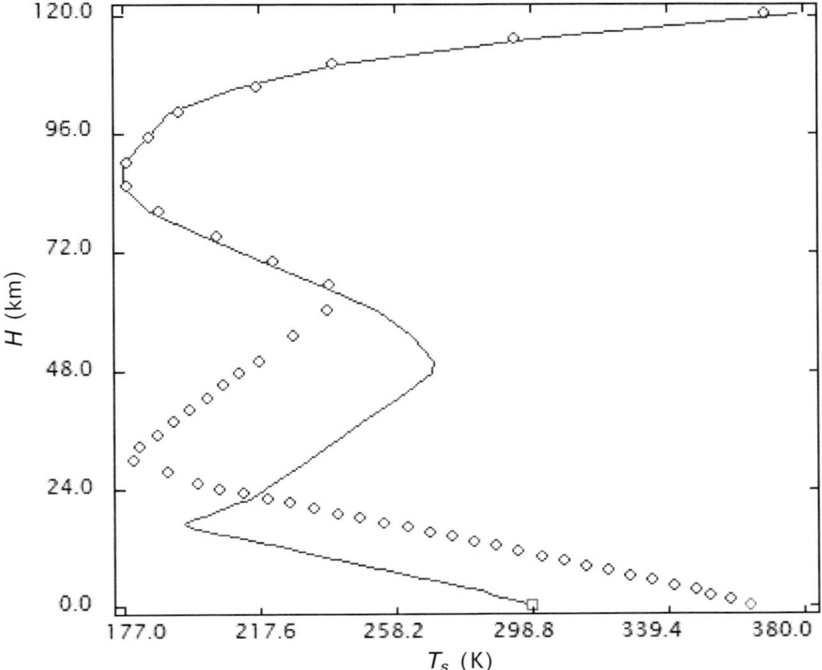

Figure 6.6. Squares show example of temperature profile of atmosphere corresponding to temperature of the surface of the Earth of 365 K; the solid line is the Tropical standard atmospheric model.

increase:

$$n_{CO_2}(T_s)/n_{CO_2}(T_s^{(0)}) \approx \exp\left\{19.1\frac{T_s - T_s^{(0)}}{T_s}\right\},$$

$$n_{H_2O}(T_s)/n_{H_2O}(T_s^{(0)}) \approx \exp\left(18.3\frac{T_s - T_s^{(0)}}{T_s}\right),$$

$$\text{and} \quad n_{CH_4}(T_s)/n_{CH_4}(T_s^{(0)}) \approx \exp\left(19.1\frac{T_s - T_s^{(0)}}{T_s}\right),$$

respectively.

Calculation of high-resolution spectra for radiation going through a whole non-uniform atmosphere in the entire frequency range ($0\,\text{cm}^{-1}$–$4{,}000\,\text{cm}^{-1}$) is a time-consuming task. To simplify its implementation, only molecular constituents have been selected for the calculations. Aerosol absorption and scattering are not taken into account in this study. Only the temperature dependences of concentrations of water vapor, carbon dioxide, and methane in the atmosphere are taken into account. As for the contribution of clouds to the transmittance of the atmosphere in the thermal infrared it is evaluated by using a factor >1 (between 1 and 2) and here the equilibrium water vapor concentration profile is multiplied by this factor. The accurate Voigt line shape is used for line-by-line calculation of absorption coefficients. The line-by-line calculations were performed with a grid step of 0.01 cm^{-1} and then convoluted with a Gauss ILS function of 0.25 cm^{-1} HWHM (half-width at half-maximum). It is sufficient to provide good accuracy of the calculations for water, carbon dioxide, and methane molecules. The wing extension for each line was 20 half-widths from the line center. This extension suffices in order to understand the climatological aspect. The water vapor continuum is taken into account for the atmospheric model 288.2 K; as for the surface temperature of 365 K, the problem of continuous absorption was beyond the scope of the paper by Clough *et al.* (1989).

Since LTE (local thermodynamic equilibrium) conditions are valid in the troposphere, the temperature dependence of the line intensities taken from the HITRAN database is described by the well-known Boltzmann dependence. The pressure and temperature dependence of half-widths is taken into consideration according to the information available in the HITRAN database.

Figures 6.7a and 6.7b show a function of the transmittance of the Earth's atmosphere in the range of 0 cm^{-1}–4,000 cm^{-1} for the U.S. standard atmospheric model (surface temperature $T_s = 288.2$ K) and for the warm stationary state of the atmosphere with surface temperature 365 K, respectively. From these simple examples one can show how the surface temperature and spectral dependence of the atmospheric transmittance function could provide a significant change in the Earth's radiation balance. The total flux of outgoing heat radiation in the upper layers of the atmosphere is considered to be mainly a sum of two contributions: the Planck radiation flux from the Earth's surface, going out mainly through the 8 μm–13 μm window, and atmospheric thermal emission. If the Earth's atmosphere accumulates carbon dioxide as on Venus, the 8 μm–13 μm window would be closed. In this case, the radiation budget would be the sum consisting of, certainly, thermal atmospheric emission itself and, probably, the Planck radiation of the Earth's surface, going out through the 8 μm–13 μm window. The former must be essentially greater than the latter, even without the consideration of the influence of the hot bands of CO_2 in the 3 μm–4 μm region, due to the increase in surface temperature.

If the atmospheric 8 μm–13 μm window is closed, the temperature of the lower atmosphere would increase up to the value at which the addition to the infrared emission of the atmosphere is the same as the Earth's heat radiation going out through this window. This process has been called the explosive greenhouse effect (Zakharov *et al.*, 1991a, 1992, 1997).

6.4 REGARDING THE RADIATION BALANCE OF THE EARTH AT THE TOP OF THE ATMOSPHERE

A basic concept of energy balance (Budyko, 1969; Sellers, 1969; North *et al.*, 1981) is applied here to the analysis of radiation balances at the top of the atmosphere for surface temperature corresponding to the stationary point 365 K of the unstable

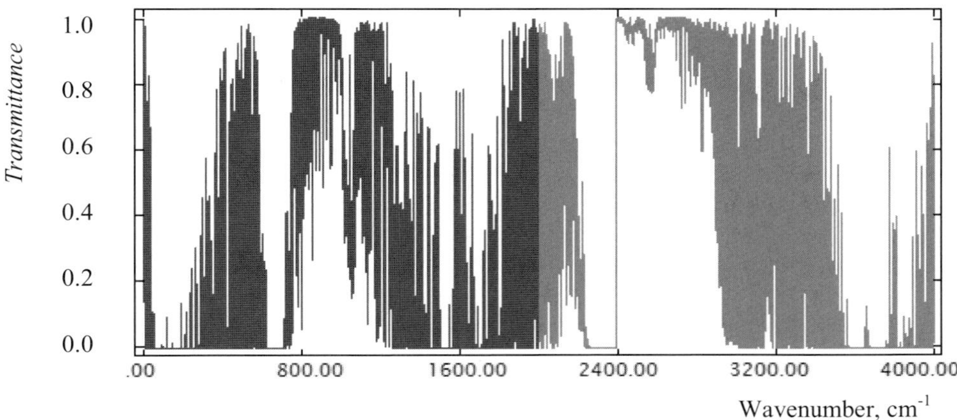

Figure 6.7a. Transmittance of the Earth's atmosphere in the thermal infrared using the U.S. standard atmospheric model with a surface temperature of $T_s = 288.2$ K.

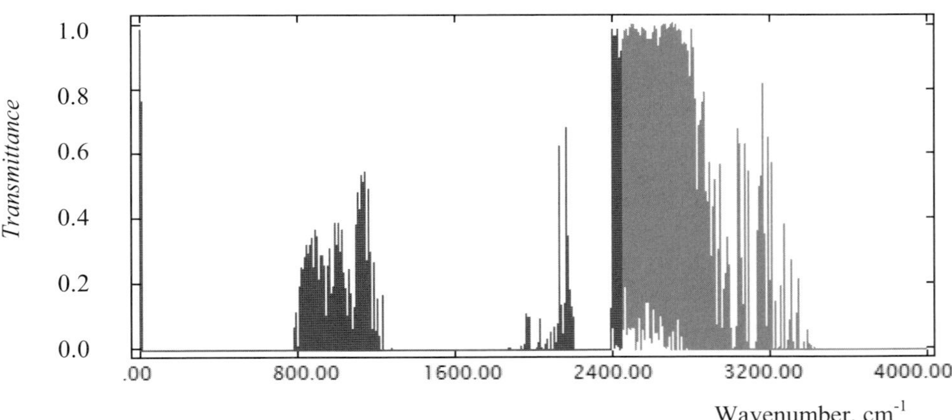

Figure 6.7b. Transmittance of the Earth's atmosphere in the thermal infrared. The surface temperature of the Earth is equal to the critical value $T_s = 365$ K.

thermal regime of the Earth obtained above. The radiation balance at the top of the Earth's atmosphere $\Delta Q_{\text{top}}(T_s)$ as a function of surface temperature T_s can be represented in the general form:

$$\left.\begin{aligned} \Delta Q_{\text{top}}(T_s) &= Q^+(T_s) - Q^-(T_s), \\ Q^+(T_s) &= \frac{I_s}{4}[1 - A(T_s)], \\ Q^-(T_s) &= Q_m^-(T_s)\{1 - \alpha(T_s)\} + Q_c^-(T_s)\alpha(T_s). \end{aligned}\right\} \qquad (6.7)$$

where $Q^+(T_s)$ and $Q^-(T_s)$ are the incoming solar flux heating the atmosphere and the outgoing thermal radiation flux of the Earth cooling the atmosphere, respectively;

$Q_m^-(T_s)$ is the mean outgoing flux of the cloud-free atmosphere; $Q_c^-(T_s)$ is the mean outgoing flux from the top of the cloud system of the atmosphere; $\alpha(T_s)$ is a fraction of the cloudy atmosphere depending on surface temperature T_s; I_s is the solar constant; and $A = A(T_s)$ is the albedo of the planet which can be described by the expression (6.5). For the Earth's atmosphere, meteorological observations demonstrate that the flux $Q_c^-(T_s)$ is around $180\,\mathrm{W\,m^{-2}}$, which can be calculated also by using the expression σT_c^4 for equivalent blackbody emission of a cloud surface, assuming a mean temperature of the top of the Earth's cloud system $T_c \approx 237.4\,\mathrm{K}$ (Matveev, 1984).

As for the flux $Q_m^-(T_s)$, it can be calculated with the help of the radiative transfer equation in the thermal infrared. The following forward radiative transfer model included into FIRE-ARMS software is used to simulate the outgoing radiance of the Earth W_ν^{out} (Gribanov *et al.*, 2001; Kondratyev and Timofeyev, 1970). Aerosol scattering and aerosol absorption of thermal infrared radiation in the atmosphere are not taken into account in the framework of this study:

$$W_\nu^{\mathrm{out}} = B_\nu(T_0)\exp\left(-\int_0^H k_\nu\,dh\right) + \int_0^H k_\nu B_\nu \exp\left(-\int_h^H k_\nu\,dh\right)dh \tag{6.8}$$

where W_ν^{out} is the atmospheric outgoing radiance at spectral frequency ν; $B_\nu(T_0)$ is the Planck radiance of the surface of the Earth assumed here as a blackbody; $B_\nu(T(h))$ is the Planck radiance of the atmosphere at temperature $T(h)$; h is the altitude in kilometers; H is the upper altitude of the atmosphere; $k_\nu(h)$ is the absorption coefficient of gaseous constituents of the atmosphere calculated using line-by-line summation and spectral parameters of atmospheric molecules obtained from the HITRAN-2000 database (Rothmann *et al.*, 2003). Since LTE conditions are valid in the troposphere, the temperature dependence of vibration–rotation populations for the above-mentioned molecular components of the Earth's atmosphere is determined by a Boltzmann function.

In order to reduce calculations, the outgoing flux $Q_m^-(T_s)$ of the heat radiation of the Earth, a result of integration over all zenith angles, is approximated by the expression:

$$Q_m^-(T_s) = \pi \int_0^{4000} W_\nu^{\mathrm{out}}(T_s, \theta_0)\,d\nu, \tag{6.9}$$

where W_ν^{out} is calculated at zenith angle $\theta_0 = 0$, and integration is over the entire spectral range $0\,\mathrm{cm^{-1}}$–$4{,}000\,\mathrm{cm^{-1}}$. This provides sufficient accuracy for this kind of climatological study.

The mixing ratios of carbon dioxide, water vapor, and methane is calculated at a given surface temperature T_s in the same manner as in Section 6.4, and the temperature profile presented in Figure 6.6 is used. Figure 6.8a shows the outgoing radiance W_ν^{out} for a cloud-free Earth atmosphere in the case of its surface temperature $T_s = 354\,\mathrm{K}$, calculated in the framework of the above-mentioned assumptions and approximations. For the flux of $Q_m^-(T_s)$, Equation (6.9) gives a value of $493.6\,\mathrm{W\,m^{-2}}$.

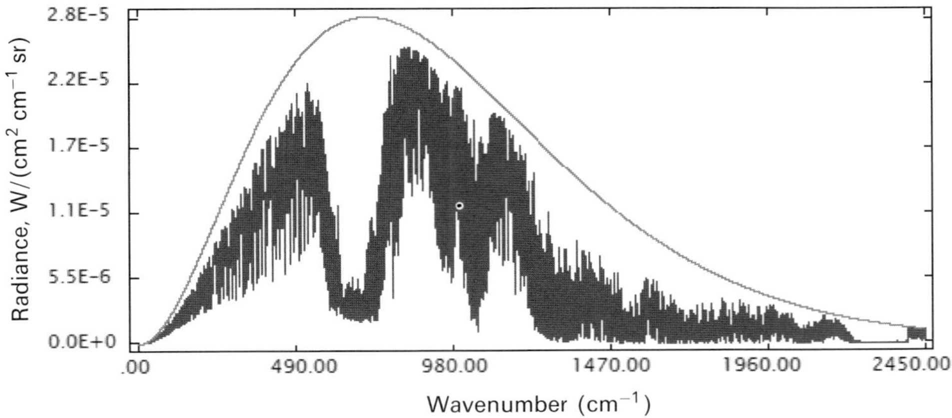

Figure 6.8a. Thermal outgoing radiance W_ν^{out} corresponding to a cloud-free Earth atmosphere at a surface temperature of 365 K. Solid envelope line is the Planck radiance of the surface at 365 K.

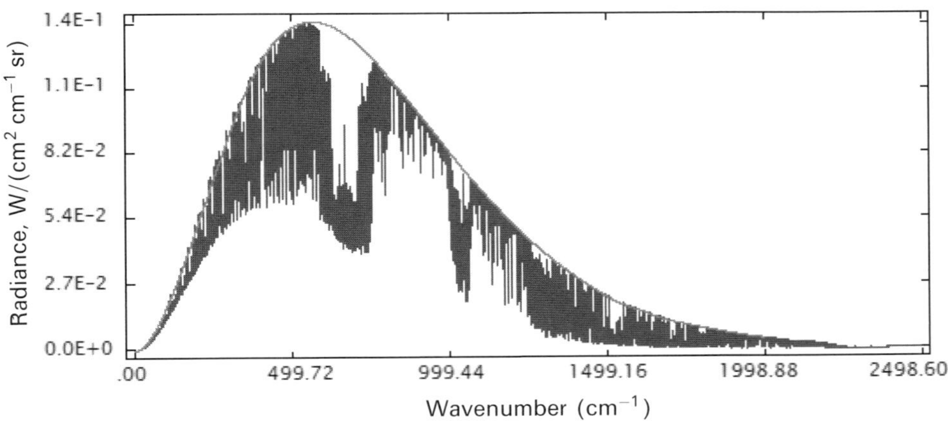

Figure 6.8b. Thermal outgoing radiance W_ν^{out} corresponding to a cloud-free Earth atmosphere at a surface temperature of 288.2 K (U.S. standard atmosphere). The solid envelope line is the Planck radiance of the surface at 288.2 K.

Unfortunately, the task of calculating the radiance of the Earth's atmosphere at a surface temperature of 610 K is very complex. It is especially difficult to discuss, because the pressure of water vapor is greater than 100 atm and the pressure of carbon dioxide is about 10 atm in the case of such a model of the atmosphere.

Cloud-free atmospheric radiance for the U.S. standard atmosphere with a surface temperature of 288.2 K is shown in Figure 6.8b for comparative purposes. Because the energy balance at the top of the atmosphere is equal to zero for both stationary thermal states 288.2 K and 365 K, the question of how the entropy balance

Table 6.1. Energy, entropy, and free-energy balances (incoming flux minus outgoing flux) at the top of the atmosphere of the Earth for the two possible stationary thermal states.

Stationary state of surface temperature (K)	*Energy balance*	*Entropy balance* ($W\,m^{-2}\,K^{-1}$)	*Free-energy balance* ($W\,m^{-2}$)
$T_{\text{present}} = 288.2$	0	−1.03	59.85
$T_{\text{us}} = 365$	0	−0.98	55.66

and free energy balance change as a result of a transition from one stationary thermal regime to another one is significant. The fraction of a cloudy atmosphere $\alpha(T_s)$ in the case of a surface temperature $T_s = 365$ K becomes about 0.9. This evaluation is made by using the calculated radiance (Figure 6.8a) expressions (6.7) for $Q^+(T_s)$, $Q^-(T_s)$ and (6.9) for $Q_m^-(T_s)$, and assuming a constant value of the flux $Q_c^-(T_s) \approx 180\,W\,m^{-2}$ within the temperature range of 288.2 K–365 K. Based on the data obtained it is not difficult to evaluate the entropy balance at the top of the atmosphere using the method discussed by Stephens and O'Brien (1993) and Goody and Abdou (1996) and the free-energy balance at the top of the atmosphere using the method discussed by Zakharov *et al.* (2005, 2008) for each stationary state of surface temperatures 288.2 K and 365 K. Nevertheless, the energy balance is zero for the two stationary states obtained of the Earth, but the entropy balance and free-energy balance are different. Table 6.1 presents the energy, entropy, and free-energy balance at the top of the atmosphere for the present-day thermal state of the Earth and a possible unstable warm thermal state. Comparison of the entropy and free-energy balances between these two stationary states shows that both the export of entropy and the import of free energy by the Earth are lower in the case of the stationary state with surface temperature 365 K than in the case of the present-day thermal state.

6.5 DISCUSSION REGARDING GREENHOUSE EXPLOSION ON THE EARTH

Transition between the two stable states of the Earth's surface T_{present} and T_{V-E} can be provoked by the positive feedback between the accumulation of carbon dioxide in the atmosphere and the absorption of outgoing thermal radiation as a result of the physical mechanism of closing the atmospheric transparency window (8 μm–13 μm) discussed above. This transition is an explosive greenhouse effect (or greenhouse explosion) because it develops according to the general equation of a thermal explosion by Franck-Kamentskii (1987) and Shmelev *et al.* (1889).

Due to the acceleration of the rate of burning of fossil fuels, large-scale agriculture, cement production, and the development of other technological processes giving rise to pollution, the concentration of greenhouse gases in the atmosphere can be expected to increase significantly, and this could lead to extremely strong global

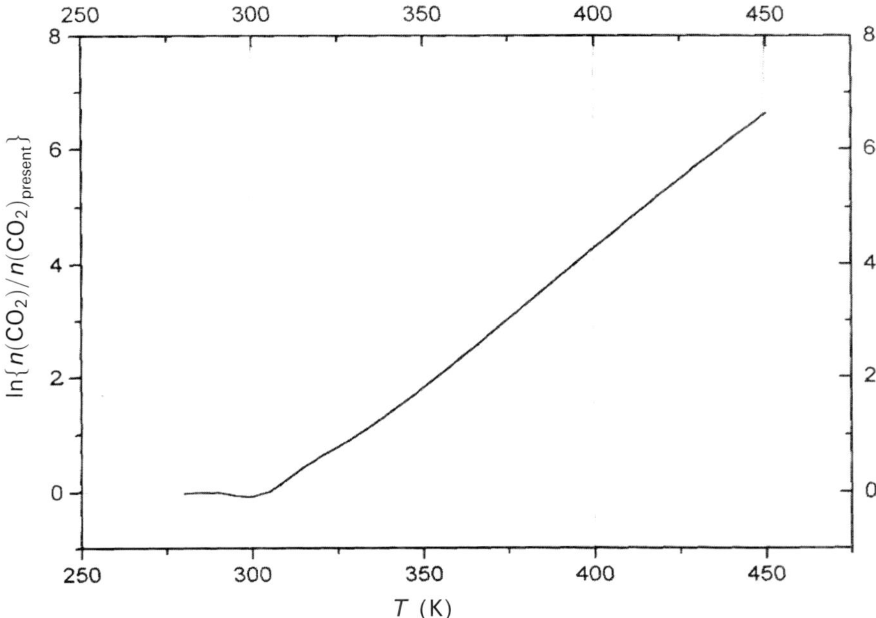

Figure 6.9. Possible temperature behavior of the relative carbon dioxide concentration in the atmosphere within the temperature range of 280 K–450 K (Zakharov *et al.*, 1997).

climate change and an irreversible transition from the present comfortable thermal state to a hot state like the atmosphere of Venus. The methods of satellite monitoring of the greenhouse gases in the atmosphere over the globe and of observing the thermal regime of the Earth are developing now in order to accumulate knowledge and understand this complex problem to predict and mitigate global warming before it becomes irreversible (Chedin *et al.*, 1994). For example, Figure 6.9 illustrates the temperature behavior of $n_{CO_2}(T_s)$ in the range of 280 K–450 K (see p. 119), and the negative and positive feedbacks dominating in the regions of 288 K–300 K and $T_s > 300$ K, respectively (Zakharov *et al.*, 1997). Photosynthesis is the only natural negative feedback mechanism which can control the stability of the concentration of CO_2 in the atmosphere. Otherwise, we can expect that the increasing anthropogenic greenhouse effect will initiate, in turn, the runaway emission of carbon dioxide from the oceans and from the Earth's crust, which contains a comparable amount of carbon dioxide to the atmosphere of Venus. The threshold radiation balance models presented in this chapter confirm that for the development of the greenhouse explosion the positive feedback temperature of the surface of the Earth and concentration of water vapor in the atmosphere are very important in addition to the temperature–carbon dioxide positive feedback. And a huge amount of liquid water on the Earth could make possible the process of a greenhouse explosion on the Earth.

Possible steady thermal condition developments at different initial concentrations of CO_2 obtained within the framework of a model of the thermal balance at the

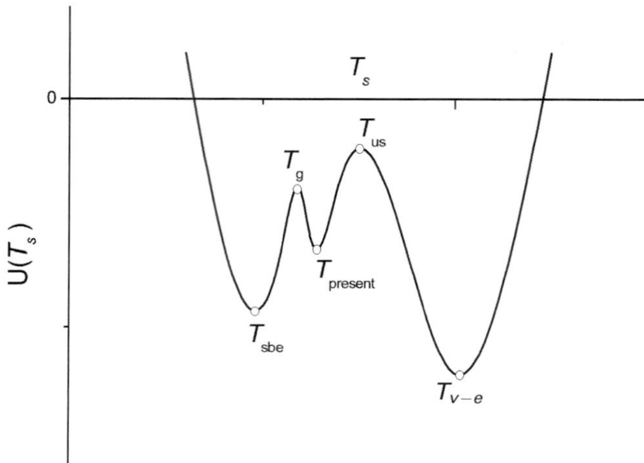

Figure 6.10. A qualitative picture of the behavior of the kinetic potential function describing several stationary thermal states of the Earth. It describes three possible stable ($T_{\rm sbe}$, $T_{\rm present}$, and T_{v-e}) and two unstable (T_g, $T_{\rm us}$) stationary thermal regimes of the Earth.

surface of the Earth are shown in Figure 6.10 (Zakharov *et al.*, 1991a, 1992, 1997). There would be only a very hot thermal state, once the CO_2 concentration in the atmosphere is about 12 times greater than the present one (Figure 6.11). These conditions can lead to the development of a thermal explosion and an explosive greenhouse effect. In this case the final thermal regime is a hot stable state of the Earth like the state of Venus. The duration of the induction time t_i of the thermal explosion is determined by the integral (Franck-Kamentskii, 1987; Shmelev *et al.*, 1989):

$$t_i = \int_{T_1}^{T_{V-E}} \frac{c_e\,dT}{Q^+ - Q^-} \tag{6.10}$$

where c_e is the heat capacity of the Earth's surface (mainly ocean); T_1 is the initial temperature of the surface of the Earth, which is higher than the temperature of the unstable state (i.e., $T_1 > T_{\rm us}$).

According to some extreme scenarios of carbon dioxide accumulation in the atmosphere the ten-fold increase in CO_2 amount could be reached in 300–500 years (Bach, 1987). The magnitude of t_i (i.e., the characteristic time of the exponential increase in surface temperature) is estimated as about $\sim 10^4$ years in this case.

In conclusion, let us discuss an approach combining a global climate model with the presented radiation balance models. The former is based on the albedo feedback in the temperature range of the Earth's surface lower than 288 K (Budyko, 1968; Golitsyn and Mokhov, 1978; Vinnikov, 1986; Nicolis, 1992), which might lead to glaciation cycles and even lead to a so-called snowball state of the Earth. Generally, the existence of five stationary thermal states of the Earth should be expected (including two stationary states at lower temperatures than the present temperature). A combined model is based on the kinetic potential function $U(T)$, which is defined

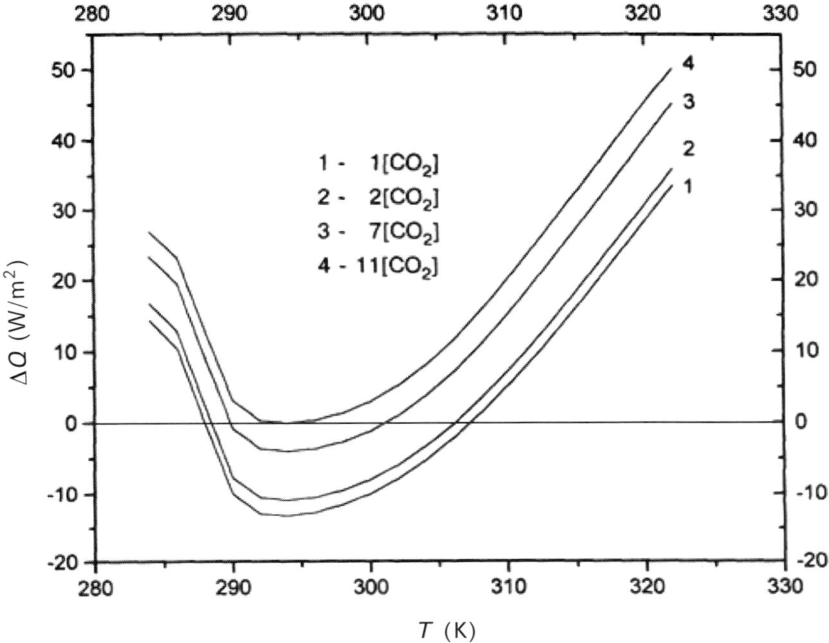

Figure 6.11a. Curves for the heat balance at surface ΔQ illustrating possible stationary thermal states (the points of crossing ΔQ and the zero line) of the Earth's surface at different initial concentration n_{CO_2} of carbon dioxide in the atmosphere: Curve 1 for $n = n_{CO_2}/n_{CO_2}^{(0)} = 1$; Curve 2 for $n = 2$; Curve 3 for $n = 7$; Curve 4 for $n = 11$. The value $n_{CO_2}^{(0)}$ is the present-day concentration of CO_2 (Zakharov *et al.*, 1997).

as (Haken, 1984):

$$U(T_s) = -\int dT\{Q^+(T_s) - Q^-(T_s)\}. \tag{6.11}$$

This combined qualitative model (Figure 6.10) produces three stable climates which correspond to the three minima of potential holes. One of them T_{sbe} corresponds to a low temperature, and it can be taken to represent a "snowball Earth" climate (Crowley and Hyde, 2001; Karol, 1988; Matveev, 1991), while the temperature $T_{present}$ describes the present thermal state. The third temperature point T_{V-E} corresponds to a hypothetical thermal state of the Earth's atmosphere similar to the state of Venus' atmosphere. These three stable climate states are separated by two intermediate unstable states: T_g (probable past glaciation climate) and T_{us}.

It should be stressed that the possibilities of the existence of several stationary thermal states of the Earth's atmosphere in the range of surface temperature higher than the present (threshold conditions, explosive development of greenhouse effect, and catastrophic change of the Earth's climate as a result of the accumulation of greenhouse gases in the atmosphere) have been discussed in the literature for about 40

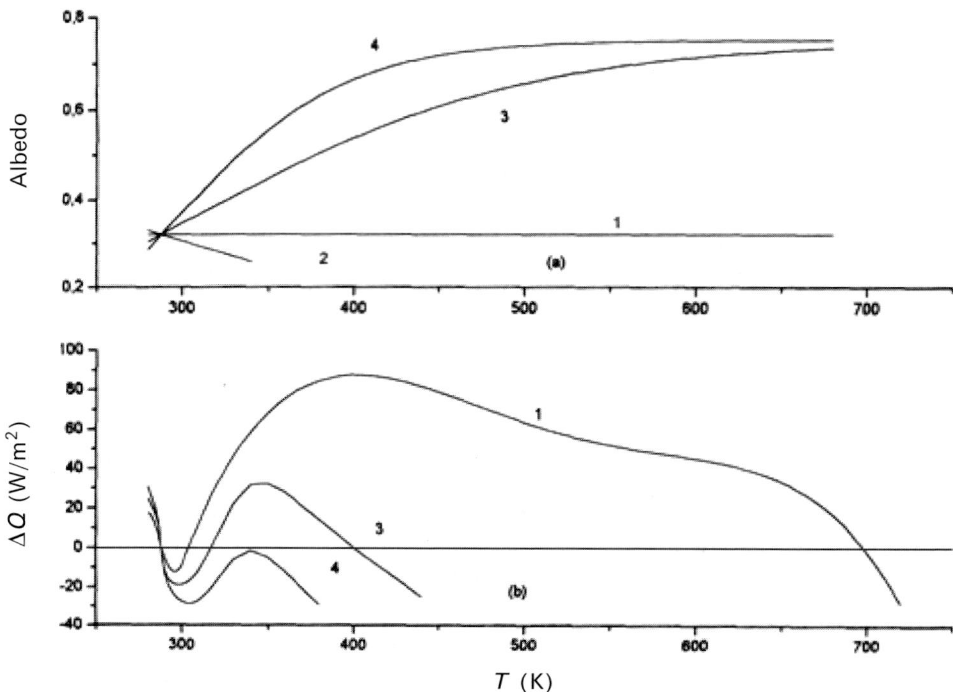

Figure 6.11b. Influence of the Earth's albedo on the thermal state of the Earth (the points of crossing ΔQ and the zero line). Panel (a) shows approximations of the temperature dependence of albedo, which are used for the simulation. Panel (b) shows the temperature dependence of the heat balance ΔQ (Curves 1, 3, and 4) at the initial present-day concentration of carbon dioxide; these correspond to the temperature approximations 1, 3, and 4 of albedo in panel (a) (Zakharov *et al.*, 1997).

years (Komabayashi, 1967, 1968; Ingersoll, 1969; Abe and Matsui, 1988; Kasting, 1988; Zakharov *et al.*, 1991a, 1992–1994, 1996, 1997; Nakajima *et al.*, 1992; Gribanov and Zakharov, 1994; Karnaukhov, 1994; Gorshkov, 1994, 1995; Makarieva and Gorshkov, 2001; Lovelock, 2004). Some attempts at experimental observations of threshold features of the explosive greenhouse effect have been undertaken as well. For example, a first laboratory registration of the explosive absorption of radiation at 10.6 μm in atmospheric air with an admixture of carbon dioxide was made by Asiptsov *et al.* (2000).

Of course, prediction of the real behavior of the greenhouse on the Earth in the future is a very complex problem. There are still uncertainties in the atmospheric models for different surface temperatures as well as in the temperature dependence of the albedo of the Earth. The models of radiation balance considered above, which predicted the possible existence of more than one stationary thermal state of a planet like our Earth for a temperature range greater than 288.2 K, are not perfect in a quantitative sense and improvement of them is needed. But the problem is very important and is a real challenge to humankind (Lovelock, 2004). The main question now is whether there are limits to the global stability of the current thermal regime of our Earth to perturbations of the content of greenhouse gases in the atmosphere and variations in Earth's surface temperature and, if so, what are the values of these limits? There is a hope that such investigations of the global stability of the present thermal regime of the Earth will continue and helpful quantitative results will be obtained soon.

Let us now try to formulate the most important problems for further research to improve the threshold models of the greenhouse effect in order to specify the number and positions of possible stationary states of the Earth's surface temperature and quantify threshold conditions for the greenhouse explosion. These include

- Investigation of the temperature dependence of the Earth's albedo and model development within the range of 288 K–850 K of the Earth's surface temperature.
- Investigation of the temperature dependence of cloud coverage of the Earth and model development.
- Improving models of temperature dependences of concentrations of key greenhouse gases in the atmosphere within the range of 288 K–850 K of the Earth's surface temperature.
- Improving models of atmospheric temperature profiles within the range of 288 K–850 K of the Earth's surface temperature.
- Development of fast and accurate models of radiative transfer in the molecular atmosphere taking into account aerosol absorption and scattering.

Acknowledgements. The author would like to thank V.I. Prokop'ev, V.M. Shmelev, Vl.G. Tyuterev, V.G. Gorshkov, T. Aoki, R. Imasu, B.A. Fomin, A. Khain, S.F. Borisov, and S.A. Beresnev for fruitful discussions on the concept and results of models of the explosive greenhouse effect; special thanks go to K.G. Gribanov, V.F. Golovko, and A. Chursin for joint research in this field and computer simulations of atmospheric transmittances, radiances, and threshold regimes of the radiation balance of the Earth's atmosphere; O.I. Asiptsov for experimental confirmation of the regime of explosive absorption of radiation at 10.6 μm in atmospheric air with an admixture of carbon dioxide; and finally S.V. Zakharov for help in preparation of the figures.

This study was partially supported by RFBR grants No. 06-01-00669 and No. 07-07-00269a.

6.6 REFERENCES

Abe Y. and Matsui T. (1988). Evolution of an impact-generated H_2O–CO_2 atmosphere and formation of a hot proto-ocean on earth. *J. Atmos. Sci.*, **45**, 3081–3101.

Anderson G.P., Clough S.A., Kneizys F.X., Cherwynd J.H., and Shettle E.P. (1986). *AFGL Atmospheric Constituents Profiles (0–120 km)*, AFGL-TR-86-0110, Environmental

Research Papers No. 954. Air Force Geophysics Laboratory, Hanscom Air Force Base, MA, 46 pp.

Asiptsov O.I., Zakharov V.I., and Gribanov K.G. (2000). Observation of explosive absorption phenomenon of CO_2 laser radiation 10.6 μm in atmospheric air with admixture of carbon dioxide. *Atmospheric and Ocean Optics*, **13**(11), 905–909.

Bach W. (ed.) (1987). *Carbon Dioxide in the Atmosphere*. Mir, Moscow [in Russian].

Bolin B. (ed.) (1987). *Greenhouse Effect, Climate Variation and Ecosystems*. Mir, Moscow, 231 pp. [in Russian].

Budyko M.I. (1968). Glaciation ages origin. *Meteorology and Hydrology*, **11**, 3–12 [in Russian].

Budyko M.I. (1969). The effect of solar radiation variations on the climate of the earth. *Tellus*, **21**(5), 611–619.

Budyko M.I. (1980). *Climate in the Past and Future*. Hydrometeoizdat, Leningrad, 320 pp. [in Russian].

Chamberlain J.W. (1980). Changes in the planetary heat balance with chemical changes in air. *Planetary and Space Science*, **28**, 1011–1018.

Chedin A., Chahine M.T., and Scott N.A. (eds.) (1994). *High Spectral Resolution Infrared Remote Sensing for Earth's Weather and Climate Studies*, NATO ASI Series, Series I: Global Environmental Change, Vol. 9. Springer-Verlag, Berlin, pp. 264–271.

Clough S.A., Kneizis F.X., and Davies R.W. (1989). Line shape and the water vapour continuum. *Atmospheric Research*, **23**, 229–241.

Crowley T.J. and Hyde W.T. (2001). CO_2 levels required for deglaciation of a "Near-Snowball" Earth. *Geophysical Research Letters*, **28**(2), 283–286.

Franck-Kamenetskii D.A. (1987). *Diffusion and Heat Transfer in Chemical Kinetics*. Science, Moscow, 265 pp. [in Russian].

Golitsyn G.S. and Mokhov I.I. (1978a). Stability and external properties of climate models. *Proceedings of USSR Academy of Sciences, Physics of Atmosphere and Ocean*, **14**(8), 271–277 [in Russian].

Golitsyn G.S. and Mokhov I.I. (1978b). Evaluation of sensitivity and role of cloudiness in simple climate models. *Proceedings of USSR Academy of Sciences, Physics of Atmosphere and Ocean*, **14**(8), 803–814.

Goody R. and Abdou W. (1996). Reversible and irreversible sources of radiation entropy. *Q. J. Roy. Meteorol. Soc.*, **122**, 483–494.

Gorham E. (1991). Northern peatlands: Role in the carbon cycle and probable responses to climatic warming. *Ecol. Appl.*, **1**, 182–195.

Gorshkov V.G. (1994). Thermal stability of climate. *Proceedings of Russian Geographical Society*, **216**(3), 26–35 [in Russian].

Gorshkov V.G. (1995). *Physical and Biological Principles of Stability of Life*. ARISTI, Moscow, 470 pp. [in Russian].

Gribanov K.G. and Zakharov V.I. (1994). Radiation regimes of Earth's atmosphere taking into account threshold absorption of thermal radiance in the range of atmospheric transparency window 8–13 μm. *Computational Technologies*, **3**(8), 62–71 [in Russian].

Gribanov K.G., Zakharov V.I., Tashkun S.A., and Tyuterev V.G. (2001). A new software tool for radiative transfer calculations and its application to IMG/ADEOS data. *J. Quant. Spectrosc. Radiative Transfer*, **68**(4), 435–451.

Haken H. (1984). *Advanced Synergetics*. Springer-Verlag, Berlin, 289 pp.

Ingersoll A.P. (1969). The runaway greenhouse: A history of water on Venus. *J. Atmos. Sci.*, **26**, 1191–1198.

Karnaukhov A.V. (1994). Regarding the stability of chemical composition of the atmosphere and thermal balance of the Earth. *Biophysics*, **39**(1), 148–152 [in Russian].

Karol I.L. (1988). *Introduction to the Earth's Climate Dynamics*. Hydrometeoizdat, Leningrad, 215 pp. [in Russian].

Kasting J.F. (1988). Runaway and moist greenhouse atmospheres and the evolution of Earth and Venus. *Icarus*, **74**, 472–494.

Komabayashi M. (1967). Discrete equilibrium temperatures of a hypothetical planet with the atmosphere and the hydrosphere of a one component–two phase system under constant solar radiation. *J. Meteor. Soc. Japan*, **45**, 137–139.

Komabayashi M. (1968). Conditions for the coexistence of the atmosphere and the oceans. *Shizen*, **23**(2), 24–31.

Kondratyev K.Ya. (1990). *Planet Mars*. Hydrometeoizdat, Leningrad, 367 pp. [in Russian].

Kondratyev K.Ya. and Moskalenko N.I. (1985). *Greenhouse Effect of the Atmospheres of Planets*. ARISTI, Moscow, 157 pp. [in Russian].

Kondratyev K.Ya. and Timofeyev Yu.M. (1970). *Thermal Sounding of the Earth from Space*. Hydrometeoizdat, Leningrad, 421 pp. [in Russian].

Lenton T.M. (2000). Land and ocean carbon cycle feedback effects on global warming in a simple Earth system model. *Tellus*, **52B**, 1159–1188.

Lovelock J. (2004). Something nasty in the greenhouse. *Atmos. Sci. Lett.*, **5**, 108–109.

Makarieva A.M. and Gorshkov V.G. (2001). Greenhouse effect and problem of stability of annual mean surface temperature of the globe. *Proceedings of RAS*, **346**(6), 810–814 [in Russian].

Manabe S. and Stouffer R.J. (1993). Century-scale effects of increasing atmospheric CO_2 on the ocean–atmosphere system. *Nature*, **364**, 215–218.

Matveev L.T. (1984). *The Course of General Meteorology (Physics of Atmosphere)*. Hydrometeoizdat, Leningrad, 751 pp. [in Russian].

Matveev L.T. (1991). *Theory of Atmospheric Circulation and Climate of the Earth*. Hydrometeoizdat, Leningrad, 340 pp. [in Russian].

McGuffie K. and Henderson-Sellers A. (1997). *A Climate Modelling Primer*. John Wiley & Sons, Chichester, U.K., 178 pp.

McKay C.P., Lorenz R.D., and Linine J.I. (1999). Analytic solutions for the antigreenhouse effect: Titan and the early Earth. *Icarus*, **137**, 56–61.

Nakajima S., Hayashi Y., and Abe Y. (1992). A study on the "Runaway Greenhouse Effect" with a One-Dimensional Radiative–Convective Equilibrium Model. *J. Atmos. Sci.*, **49**(23), 2256–2266.

Nicholls G.D. (1967). *In Mantles of the Earth and Terrestrial Planets*. Intersience, New York, 285 pp.

Nicolis C. (1992). *Long Term Climate Transitions and Stochastic Resonance*. Institut Royal Météorologique de Belgique, Brussels, 54 pp.

North G.R., Cahalan R.F., and Coackley J.A. (1981). Energy balance climate models. *Rev. Geophys. Space Phys.*, **19**(1), 91–121.

Rothmann L.S., Barbe A., Chris Benner D., Brown L.R., Camy-Peyret C., Carleer M.R., Chance K., Clerbaux C., Dana V., Devi V.M. *et al.* (2003). The HITRAN molecular spectroscopic database: Edition of 2000 including updates through 2001. *J. Quant. Spectrosc. Radiative Transfer*, **82**, 5–44.

Sellers W.D. (1969). A global climatic model based on the energy balance of the Earth–atmosphere system. *J. Appl. Met.*, **8**, 392–398.

Shmelev V.M., Zakharov V.I., and Nesterenko A.I. (1989). Explosive absorption of power CO_2 laser beams in the atmosphere. *Atmospheric Optics*, **2**(6), 489–496.

Stephens G.L. and O'Brien D.M. (1993). Entropy and climate, I: ERBE observations of the entropy production of the earth. *Q. J. Roy. Meteorol. Soc.*, **119**, 121–152.

Vinnikov K.Ya. (1986). *Climate Sensitivity*. Hydrometeoizdat, Leningrad, 224 pp. [in Russian].

Zakharov V.I., Prokop'ev V.E., Shmelev V.M, and Gribanov K.G. (1991a). *Stability of the Present Thermal State of the Earth* (Preprint No. 7). Tomsk Science Center of Siberian Branch of Academy of Science of U.S.S.R, 15 pp. [in Russian].

Zakharov V.I., Shmelev V.M., and Nesterenko A.I. (1991b). Explosive absorption of CO_2 laser radiation 10.6 μm in the atmosphere. *J. de Phys.*, **IV**, 775–781.

Zakharov V.I., Gribanov K.G., Prokop'ev V.E., and Shmelev V.M. (1992). Influence of atmospheric transparency window 8–13 μm on stability of the thermal state of the Earth. *Atomic Energy*, **72**(1), 98–102 [in Russian].

Zakharov V.I., Shmelev V.M., Gribanov K.G., and Prokop'ev V.E. (1993). Influence of atmospheric transparency window 8–13 micron on thermal stability of the Earth atmosphere. *Proceedings of International ASA Colloquium, September 8–10, 1993, Reims, France*, pp. 39–42.

Zakharov V.I., Gribanov K.G., Shmelev V.M., Chursin A.A., Husson N., Golovko V.F., and Tyuterev Vl.G. (1994). Temperature dependence of atmospheric transparency function in field of 100–5000 cm^{-1} and model of explosive greenhouse effect. *Proceedings of the Fifth International Workshop on ASS/FTS, November 30–December 2, 1994, Tokyo, Japan*, pp. 419–445.

Zakharov V.I., Gribanov K.G., Falko M.V., Golovko V.F., Chursin A.A., Husson N., Scott N.A., and Tyuterev Vl.G. (1996). Temperature dependence of molecular atmospheric transmission function in field of 2–400 micron and the Earth radiation balance. *Proceedings of the Seventh Global Warming International Conference, April 1–3, 1996, Vienna, Austria*, pp. 234–240.

Zakharov V.I., Gribanov K.G., Falko M.V., Golovko V.F., Chursin A.A., Nikitin A.V., and Tyuterev Vl.G. (1997). Molecular atmospheric transmittance function in the range of 2–400 micron and Earth radiation balance. *J. Quant. Spectrosc. Radiat. Transfer*, **57**(1), 1–10.

Zakharov V.I., Imasu R., and Gribanov K.G. (2005). Net free energy of the Earth and its monitoring from space concept. *SPIE*, **5655**, 540–547.

Zakharov V.I., Imasu R., Gribanov K.G., and Zakharov S.V. (2008). Free energy balance at top of the atmosphere. *Atmospheric and Ocean Optics*, **21**, 240–247.

7

Model-based method for the assessment of global change in the nature–society system

Vladimir F. Krapivin and John J. Kelley

7.1 INTRODUCTION

Numerous problems arising from the interaction between nature and society are considered by various authors (Adamenko and Kondratyev, 1999; Bartsev *et al.*, 2003; Degermendzhi and Bartsev, 2003; Gorshkov *et al.*, 2000; Kondratyev, 1990, 1992, 2002, 2004a; Kondratyev *et al.*, 2003a–c, 2004a, b, 2006b). The growing number of published works dedicated to global environmental change leads to the realization that protection of the natural environment has become an urgent problem. The question of working out the principles underlying coevolution of human beings and nature is being posed with ever-increasing persistence. Scientists in many countries are making attempts to find ways of formulating laws governing human processes acting on the environment. Numerous national and international programs of biosphere and climate studies contribute to the quest for means of resolving the conflict between human society and nature. However, attempts to find efficient methods of regulating human activity on the global scale encounter many difficulties. The major difficulty is the absence of an adequate knowledge base pertaining to climatic and biospheric processes as well as the largely incomplete state of the databases concerning global processes occurring in the atmosphere, in the ocean, and on land. Another difficulty is the inability of modern science to formulate the requirements that must be met by the global databases necessary for reliable evaluation of the state of the environment and forecasting its development for sufficiently long time intervals.

Many scientists are trying to find answers to the above questions. The majority suggest the creation of a unified planetary-scale adaptive Geoinformation Monitoring System (GIMS) as one of the efficient ways of resolving the conflicts between nature and human beings (Kondratyev *et al.*, 2000). Based on regenerated knowledge bases and global datasets, the adaptive nature of such a system should be provided through correction of the data-acquisition mode as well as by varying the parameters and structure of the global model. The main idea of this approach to studying the nature–society system (NSS) was developed in detail by Kondratyev *et al.* (2002).

This chapter gives a schematic description of a simulation model, the Global Model of the Nature–Society System (GMNSS), considering different aspects of global ecodynamics and globalization processes. The main idea of this model is based on the interactivity concept developed by Kondratyev (1998). The model is constructed of blocks parameterizing natural and anthropogenic processes. The various blocks describe biogeochemical cycles of greenhouse gases; the global hydrologic cycle in liquid, gaseous, and solid phases; productivity of soil–plant formations with numerous types defined; photosynthesis in ocean ecosystems, taking into account ocean depth and surface heterogeneity; demographic processes and anthropogenic changes. The model makes it possible to compute the dynamics of industrial CO_2 distribution between the ocean, terrestrial biota, and the atmosphere. The ocean is described by a spatial four-layer model with due regard for water chemistry. The model is designed for connection in turn to a global climate model. Input model data are combined from existing global databases, and model-oriented environmental monitoring is proposed to be adapted to simulation model input. Examples are presented of using the model to estimate the state of the NSS and its sub-systems. The respective roles of vegetation and of the global ocean in climate change are evaluated. Different hypotheses of global change causes are considered.

7.2 A NEW TYPE OF GLOBAL MODEL

Approaches to the synthesis of a global model include the need to describe all aspects of human interactions with the environment and with its physical, biological, and chemical systems. One such application has its origin in the studies of the Computer Center of the Russian Academy of Sciences in Moscow (Krapivin *et al.*, 1982). This type of global model is formulated on the basis of a detailed description of the climate system with the consideration of a small set of biospheric components. Such a strategy of global modeling is adhered to in the Potsdam Institute for Climate Impact Research studies (Boysen, 2000), where Moscow Global Model prototypes are developed. More than 30 climate models are being developed in different countries in an attempt to generate new trends in the science of global change (Demirchian and Kondratyev, 1999). Unfortunately, studies conducted on a global and regional scale using this approach to assess the processes and impacts of global change have not produced sufficiently acceptable results (Kondratyev, 2004b; Kondratyev and Krapivin, 2003). That is why another approach to the global modeling problem, known as evolutionary modeling, has been developed by many authors (Kondratyev *et al.*, 2004a; Krapivin, 1993; Sellers *et al.*, 1996).

Traditional approaches to building a global model entail some difficulties of algorithmic description with respect to many socio-economic, ecological, and climatic processes (Kondratyev, 1999b), so that one has to deal with information uncertainty. These approaches to global modeling simply ignore such uncertainty, and consequently the structure of the resultant models does not adequately reflect the real processes. Evolutionary modeling makes it possible to remove this drawback by the synthesis of a combined model whose structure allows for adapting the background history of a system to the biosphere and climate components. The implementation of such a model can also be combined in various classes of models using conventional software and hardware and special-purpose processors of the evolutionary type. The form of such combination is diverse, depending on the spatio-temporal completeness of the databases (Rochon *et al.*, 1996).

Experiences in global modeling abound in examples of insoluble problems which one encounters when looking for ways to describe scientific and technological advances and human activity in their diverse manifestations. No less difficulty arises in modeling climate described by a superimposition of processes with different temporal variability rates. As to the inclusiveness of description in the global model, it is impossible to delineate clearly the bounds of information availability and the extent of the required spatial and structural detail. Therefore, without going into a natural–philosophical analysis of global problems and skirting the issue of the ultimate solution to global modeling, we will confine ourselves to the discussion of only one of the possible approaches. This approach will demonstrate the way in which evolutionary modeling developed on special processors can help overcome some of the difficulties of global modeling, such as those inherent in computing and algorithmic variations. All of this does imply that a search for effective models of the traditional type can well be of value. At present, the building of global biogeocenotic models is not seen as difficult. Many such models have been created (Alexandrov and Oikawa, 2002; Alexandrov *et al.*, 2005; Bartsev *et al.*, 2003), and the gathering of information to support them is underway. The history of the interaction of the biosphere with the climate system and human society is not sufficiently understood, which is one of the obvious hurdles in describing climatic cycles. That is why an evolutionary approach is essential to building a global model that accounts for the interaction of the biosphere, climate system, magnetosphere, etc. Such an approach helps to overcome the uncertainties in describing such a complex interaction. As a result of adjusting such a model to the history of the prescribed cycle, we shall obtain a model implicitly tracing various regularities of the dynamics of the biosphere in the past and allowing for forecast assessments to be made in the same temporal cycle. A special processor version of this model completely removes all the existing algorithmic and computing hurdles arising from the large dimensionality of the global model and the conditions of irreducible non-parametrical uncertainty.

Figure 7.1 shows the key elements of this new type of global model. The data archive is formed here as two structures. Data of the first type for the computer models of the biosphere processes are stored as climatic maps and as tables of the model equation coefficients. It is necessary to fill in all the cells of the schematic maps. Data of the second type are represented as fragments recorded disparately (possibly irregularly) in time and space (i.e., CO_2 concentration, temperature, precipitation, pressure, population numbers, availability of resources, etc.). Data of this type are used to adjust the evolutionary processor to the given class of models (e.g., finite automata). As a result of this procedure the model is adapted to the history of the prescribed time cycle. As has been shown by Kondratyev *et al.* (2002) a stable forecast

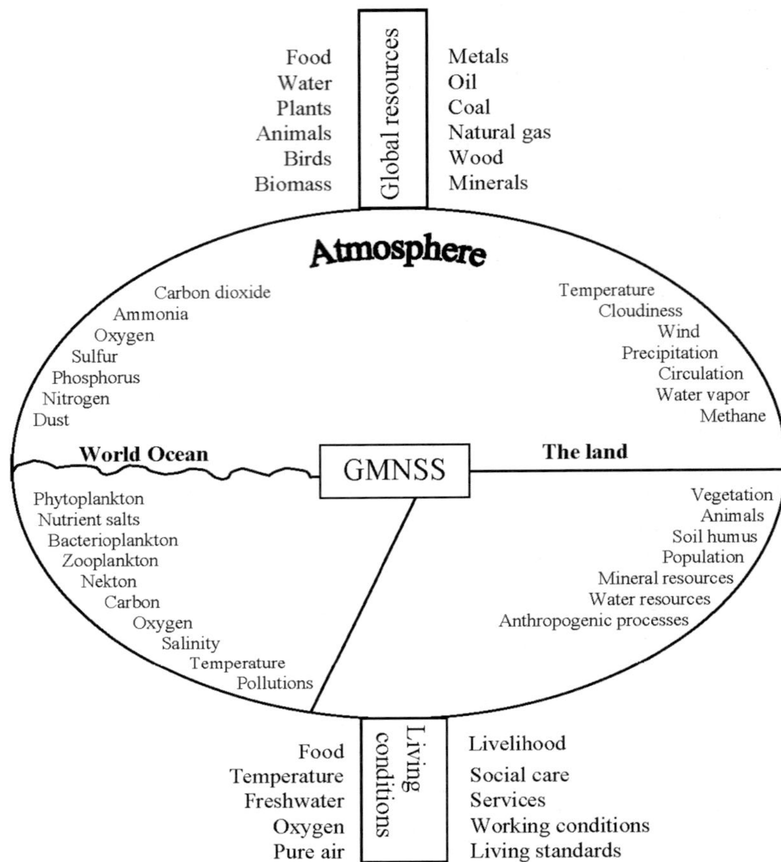

Figure 7.1. Key elements of the nature–society system and energy components to be taken into account in the global ecodynamics forecast within the framework of global model use.

is produced with 75%–95% reliability covering several temporal steps. The extent of a forecast is determined by the length of its history. Given the need for a forecast under the conditions of change in the trends of human economic activities, an evolutionary processor is adjusted to the assigned scenario, thus automatically providing for simulation of the corresponding response of the biosphere to this change.

The suggested structure of the global model thus ensures a flexible combination of models of the traditional and evolutionary types. The proposed approach helps to avoid the need to model non-stationary processes (climatic, socio-economic, demographic, etc.) and provides for overcoming uncertainty. A model of this new type makes it possible to go from learning experiments to the assessment of the viability of the biosphere with regard to actual trends of anthropogenic stresses in all regions of the globe.

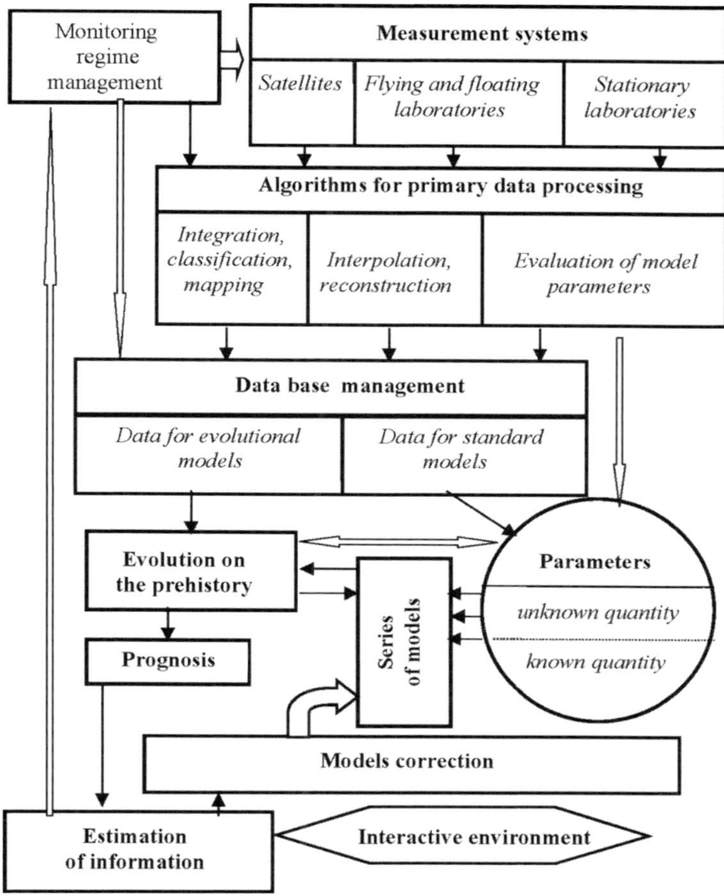

Figure 7.2. The scheme of GMNSS interactive adjustment and control of the geoinformation monitoring regime.

A departure from the established global modeling techniques based on new information technology makes it possible to proceed to creating a global system of monitoring with the global model as a portion of the support for the system. A conceptual block diagram of geoinformation monitoring and the use of a global model is represented in Figure 7.2. Application of the evolutionary computer technology provides for categorization of the whole system by a class of sub-systems with variable structure and for making it adaptable to changes in the natural process or entity under observation. Furthermore, it becomes possible to detail heterogeneously the natural systems under study in the space of phase variables and to select non-uniform geographical grids in a sampling analysis of the planetary surface (i.e., arbitrary insertion of significant regularities at the regional level becomes possible).

The automatic system for processing global information is aimed at the acquisition of combined models reflected in the real-time scale of the climatic and anthropogenic changes in the biosphere and is based on its known history (or, rather, its simulation). The system relies on a set of models of processes in the biosphere and, using software of other units with the help of the scenario of anthropogenic behavior formulated at input, provides for prompt assessment of the environmental state and for forecast assessments within the framework of this scenario. The automatic system for processing global information consists also of a further advantage in that it formulates the initial and boundary conditions for particular built-in models in the study of regional systems and virtually substitutes for field measurements of those conditions.

The first version of the global model (Krapivin *et al.*, 1982) was oriented towards rigid spatio-temporal detailing and therefore required a large quantity of information. The subsequent development of an automatic system for processing global information has made it possible, owing to the evolutionary technology, to discard the generally accepted regular geophysical grid in archive development (Rosen, 2000) and to solve this problem using algorithms for the recovery of spatio-temporal information.

7.3 MATHEMATICAL MODEL OF NATURE–SOCIETY SYSTEM (NSS) DYNAMICS

7.3.1 General description of the global model

In connection with the different aspects of environmental change taking place during the last few decades, experts have put forward numerous conceptions of nature–society system (NSS) global description, and models of various complexity have been developed to parameterize the dynamics of the characteristics of the biosphere and the climate. The availability of a large database of these characteristics enables one to consider and estimate the consequences of possible realization of different scenarios of the development of NSS sub-systems. Traditional approaches to the synthesis of global models are based on the consideration of a totality of balance equations, which include parameters $\{x_i\}$ in the form of functions, arguments, coefficients, and conditions of transition between parametric descriptions of processes taking place in the environment. Also, other approaches are applied, which use evolutionary and neuronet algorithms. Organization of the global model of NSS functioning is presented as a conceptual scheme in Figure 7.3. This scheme is realized by introducing a geographical grid $\{\varphi_i, \lambda_j\}$ with step progression in sampling the land surface and the oceans by $\Delta\varphi_i$ and $\Delta\lambda_j$ in latitude and longitude, respectively, so that within a pixel $\Omega_{ij} = \{(\varphi, \lambda) : \varphi_i \leq \varphi \leq \varphi_i + \Delta\varphi_i, \lambda_j \leq \lambda \leq \lambda_j + \Delta\lambda_j\}$ all the processes and elements of the NSS are considered as homogeneous and parameterized by point models. The choice of the pixel size is determined by several conditions governed by the spatial resolution of satellite measurements and the availability of a needed global database. In the case of the water surface, in pixel Ω_{ij} the water mass

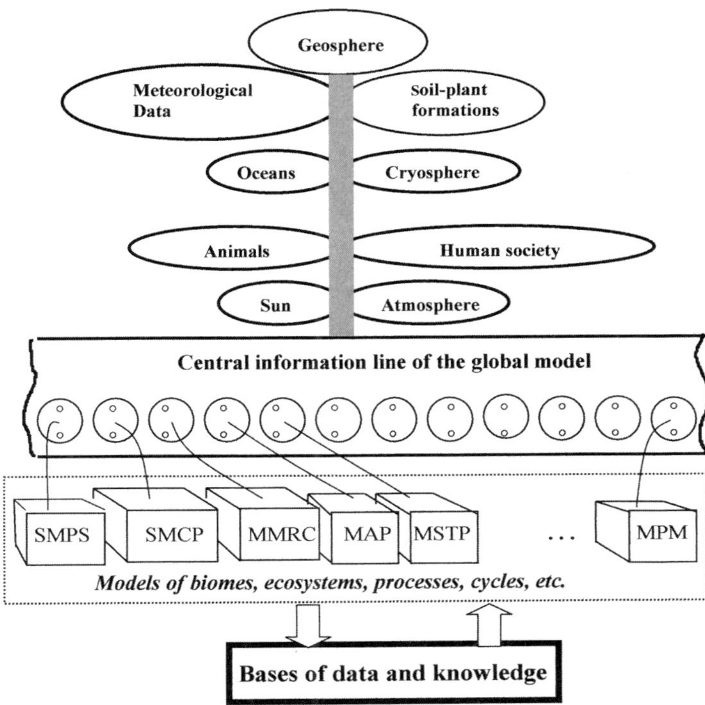

Figure 7.3. The information–functional structure of the global NSS model. The notation is given in Table 7.1.

is divided into layers by depth z (i.e., 3-D volumes $\Omega_{ijk} = \{(\varphi, \lambda, z) : (\varphi, \lambda) \in \Omega_{ij}, z_k \leq z \leq z_k + \Delta z_k\}$ are selected and all elements are distributed uniformly within them). Finally, the atmosphere over pixel Ω_{ij} is digitized by altitude h either by atmospheric pressure levels or by characteristic layers of altitude Δh_s.

It is clear that the development of a global model is only possible using knowledge and data on the international level. Among numerous models, the most adequate is that described in Kondratyev *et al.* (2004a). The block scheme construction of this model is shown in Figure 7.4. The synthesis of a model of such scale requires careful consideration of existing models of various partial processes derived from information on climatology, ecology, hydrology, geomorphology, etc.

An adaptive procedure of introducing the global model into a system of geoinformation monitoring has been proposed by Kondratyev *et al.* (2003b). This procedure is shown schematically in Figure 7.5.

Note that since the NSS is a part of the Earth system, it can be considered as a closed object of the energy exchange with space as well as part of the Earth system that includes the core and mantle as sources of planetary energy formed due to the process of gravitational differentiation and radioactive decay. The GMNSS should be improved on this methodical basis, which has certain mechanisms for (V, W)-

Table 7.1. Characteristics of GMNSS units.

Identifier of the unit in Figure 7.4	*Characteristics of unit functions*
SMPS	A set of models of population size dynamics with regard to age structure (Logofet, 2002)
SMCP	A set of models of climatic processes with differently detailed consideration of parameters and their correlations (Kiehl and Gent, 2004)
MMRC	Model of mineral resources control (Nitu *et al.*, 2004)
MAP	Model of agricultural production (Kondratyev and Krapivin, 2001)
MSTP	Model of scientific–technical progress
CGMU	Control of global model units and the database interface
AGM	Adjustment of the global model to simulate experimental conditions and global model control
PSR	Preparation of simulation results to visualization or other forms of account
MBWB	Model of the biospheric water balance (Krapivin and Kondratyev, 2002)
MGBC	Model of the global biogeochemical cycle of carbon dioxide (Kondratyev and Krapivin, 2004; Kondratyev *et al.*, 2003b)
MGBS	Model of the global biogeochemical cycle of sulfur compounds (Krapivin and Nazaryan, 1987)
MGBO	Model of the global biogeochemical cycle of oxygen and ozone (Krapivin, 2000a)
MGBN	Model of the global biogeochemical cycle of nitrogen (Krapivin, 2000b)
MGBP	Model of the global biogeochemical cycle of phosphorus (Krapivin, 2000c)
SMKP	A set of models of the kinetics of some types of pollutants in different media (Kondratyev *et al.*, 2006b; Krapivin and Potapov, 2006)
SMWE	A set of models of water ecosystems in different climatic zones (Degermendzy, 1987)
MHP	Model of hydrodynamic processes (Kondratyev *et al.*, 2002)
SMSF	A set of models of soil–plant formations (Kondratyev *et al.*, 2004a)
MPM	Model of processes in the magnetosphere (Korgenevsky *et al.*, 1989)

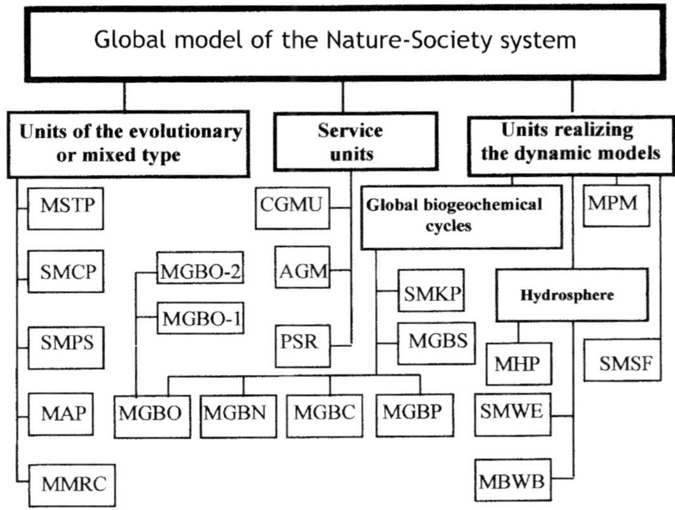

Figure 7.4. The block scheme of the GMNSS. The notation is given in Table 7.1.

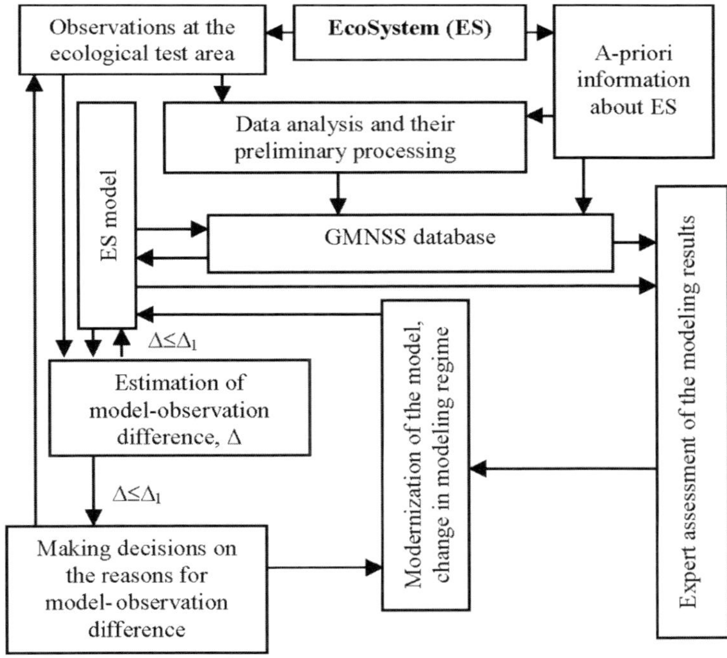

Figure 7.5. A principal scheme of the organization of ecological monitoring using an adaptive modeling regime. Notation: Δ = integral or subject estimate of difference between the modeling and observational data; Δ_1 = a permissible level of difference for Δ estimation.

exchange regulation via consideration of correlations of biogeochemical cycles of carbon, methane, ozone, water, oxygen, nitrogen, sulfur, and phosphorus. Interdisciplinary studies under development enable elaboration of this series and make it possible to establish functional connections between all spheres of the Earth system. Here an attempt has been made to synthesize the GMNSS elements responsible for simulation of a complex of biogeochemical cycles of some important elements.

7.3.2 Model of the global biogeochemical cycle of carbon dioxide

Focus has been directed over recent decades on the problem of a global carbon cycle (the carbon cycle) in terms of numerous, often speculative, explanations of the role of CO_2 in future climate change. Unfortunately, an objective assessment of this role is still absent. Recently published studies (Kondratyev, 2000b, 2004a, b; Kondratyev and Krapivin, 2004; Krapivin and Chukhlantsev, 2004) have summed up the first results of development of a formalized technology assessing the greenhouse effect due to CO_2 with due regard to the role of land and ocean ecosystems. An interactive connection has been demonstrated between the global carbon cycle in the form of CO_2 and climate change. The formalization of this connection is based on synthesis of GMNSS functioning, with the spatial distribution of the elements of this system taken into account, which makes it possible to reduce into a single correlated scheme the cause-and-effect connections of carbon fluxes between its different biospheric and geospheric reservoirs.

An objective formalization of biospheric sources and sinks of CO_2 as functions of environmental parameters and consideration of the actual role of anthropogenic processes becomes possible due to the recent studies of many experts who have developed models with varied degrees of detailed description of space-distributed carbon fluxes and their interaction with NSS components (Kondratyev *et al.*, 2004a).

In a recently published first report of an international project on the global carbon cycle, GCP (Global Carbon Project) (Canadel *et al.*, 2003), a strategy has been initiated of interdisciplinary cooperation within a broad spectrum of environmental problems considered in the context of a global system of nature–society interaction, with special emphasis on the need to develop methods and information technologies to analyze the carbon–climate–society system (CCSS). The central goal is the consideration of the following five aspects of the general problem of a global carbon cycle:

- Study of the carbon cycle which should be based on the integration of natural and anthropogenic components by the interactive analysis of interactions between energy systems based on fossil fuels, the biogeochemical carbon cycle, and the physical climate system.
- Development of new methods of analysis and numerical modeling of an integrated carbon cycle.
- Global studies of the carbon cycle carried out with due regard to the results of national and regional research programs on studies of the carbon cycle between its reservoirs.

- One of the strategic problems of the international project should be a search of means in regional development of achieving a stabilized concentration of CO_2 in the atmosphere.
- A separate aspect of the carbon cycle should be a classification of all countries into developed and developing status in order that the production technologies are respectively divided into industrial, economic, and energy sectors of the NSS by their significance as sources of anthropogenic CO_2 emissions.

In addition to this, we enumerate the key directions of developments within the CCSP program (*Our Changing Planet*, 2004):

- What are the specific features of spatio-temporal variability of the sources and sinks of carbon on the continent of North America on time scales from seasonal to centennial, and what processes are attributable to the prevailing effect on carbon cycle dynamics?
- What are the respective features of variability and their determining factors in the case of ocean components (sources and sinks) of the carbon cycle?
- How do local, regional, and global processes on the land surface (including land use) affect the formation of carbon sources and sinks in the past, present, and future?
- How do the sources and sinks of carbon vary on land, in the ocean, and in the atmosphere on time scales from seasonal to centennial, and how can the respective information be used to obtain a better understanding of the laws of the carbon cycle formation?
- What future changes can be expected in concentrations of atmospheric CO_2, methane, and other carbon-containing greenhouse gases as well as changes in sources and sinks of carbon on land and in the ocean?
- How will the Earth system and its components respond to various choices of strategy for regulating the carbon content in the environment, and what information is needed to answer this question?

The carbon cycle is closely connected with climate, water, and nutrient cycles and with photosynthesis production on land and in the ocean. Therefore, all studies of the carbon cycle that overlook such connections are inevitably doomed to failure and hence cannot give even approximately adequate estimates of the consequences of anthropogenic emissions of carbon to the environment. For this reason, many international projects on the study of the greenhouse effect and its impact on climate have failed, such as the case of the Kyoto Protocol intention of regulating CO_2 emissions. The GCP is hopeful of making progress in this sense by planning interdisciplinary studies of the carbon cycle. Such studies can be divided into three directions:

- Formation of a strategy for carbon cycle studies and evaluation of carbon cycle variability.

- Analysis of connections between causes and consequences in studies of mechanisms of environmental interaction with natural and anthropogenic CO_2 sources and sinks.
- Identification and quantitative estimation of evolutionary processes in the carbon–climate–society system (CCSS).

The first GCP report (Canadel *et al.*, 2003) formulates the goal of the coming decadal period of the carbon cycle study, which ideologically combines the earlier isolated programs: IGBP (the International Geosphere–Biosphere Program), IHDP (the International Human Dimensions Program), and WCRP (the World Climate Research Program). The authors of this report substantiated the detailed scheme of the cause-and-effect connections between components of the climate–biosphere system and pointed to the necessity of their joint consideration in order to raise the level of reliability of estimates and forecasts for the climatic impact of CO_2. All of these problems had been discussed earlier in other publications (Kondratyev, 2004a; Kondratyev and Krapivin, 2003; Kondratyev *et al.*, 2003b). In the first GCP report, the role of other greenhouse gases remains underestimated, unfortunately, even though its contribution in the most immediate future can conceivably exceed that of CO_2. The list of greenhouse gases, such as methane, nitric oxide, hydrofluorocarbons, perfluorocarbons, and hydrofluoroethers, increases with time. Moreover, these gases, molecule for molecule, have much higher global-warming potential than that of CO_2. Their total equivalent emissions in 1990 constituted 3.6 $GtCO_2$, and by the year 2010 the level of 4.0 $GtCO_2$ will be exceeded (Bacastow, 1981). At the same time, anthropogenic CO_2 emissions are estimated at 6 $GtC\,yr^{-1}$ (EPA, 2001; Houghton *et al.*, 2001; IPCC, 2007). Table 7.2 compares the potential impact of various greenhouse gases on climate change. According to EPA (2001), historical data from the beginning of the industrial revolution of the relative levels of anthropogenic emissions of certain greenhouse gases contributing to enhancement of the greenhouse effect consisted of: CO_2 55%, CH_4 17%, O_3 14%, N_2O 5%, and others 9%.

All of this testifies to the fact that continuing with inadequate descriptions (primitive in most cases) of the carbon cycle, together with even as much as a parameterization for other greenhouse gases, cannot lead to reliable estimates of possible future climate changes due to anthropogenic activity within the NSS. The idea of identifying the locations and impact potential of CO_2 sources and sinks on land and in the oceans as declared in the GCP report (Canadel *et al.*, 2003) has not been supported with serious and substantial motivation in the development of new information technologies for a comprehensive analysis of the Earth's radiation budget.

Nevertheless, it should be recognized that the fundamental postulate of the GCP promotes a better understanding of the carbon cycle, basing its underlying studies on the concept of combined natural and anthropogenic components with the application of established analytic methods, algorithms, and models. The main structure of the carbon cycle is determined by its fluxes between the basic reservoirs, including carbon in the atmosphere (mainly in the form of CO_2), the oceans (with division into surface,

Table 7.2. Potentials of relative global warming due to various greenhouse gases (EPA, 2001). Global warming potential (GWP) is a measure of how much a given mass of greenhouse gas is estimated to contribute to global warming. The GWP is defined as the ratio of time-integrated radiative forcing from the instantaneous release of 1 kg of a trace substance relative to that of 1 kg of a reference gas. Carbon dioxide has a GWP of exactly 1 (since it is the baseline unit with which all other greenhouse gases are compared).

Gas	*Potential*	*Gas*	*Potential*
CO_2	1	HFC-227es	2,900
CH_4	21	HFC-236fa	6,300
N_2O	310	HFC-4310mee	1,300
HFC-23	11,700	CF_4	6,500
HFC-125	2,800	C_2F_6	9,200
HFC-134a	1,300	C_4F_{10}	7,000
HFC-143a	3,800	C_6F_{14}	7,400
HFC-152a	140	SF_c	23,900

intermediate, and deep layers and bottom deposits), and terrestrial ecosystems (vegetation, litter, and soil), rivers and estuaries, and fossil fuels. All of these reservoirs should be studied with due regard for their spatial heterogeneity and dynamics as influenced by natural and anthropogenic factors according to such bases of accumulated knowledge as the following observations:

- Anthropogenic carbon emissions have been growing constantly from the beginning of industrial development, reaching levels of 5.2 GtC in 1980 and 6.3 GtC in 2002 (IPCC, 2007).
- The content of the main greenhouse gases CO_2, CH_4, and N_2O in the atmosphere has increased since 1750 by 31%, 150%, and 16%, respectively. About 50% of the CO_2 emitted to the atmosphere due to the burning of fuel was assimilated by vegetation and the oceans.
- The observed distribution of atmospheric CO_2 and the oxygen/nitrogen relationship show that a land sink of carbon prevails in northern and middle latitudes over the oceanic sink. In tropical latitudes, emissions of CO_2 to the atmosphere are substantially due to the use of the Earth's resources.
- Inter-annual oscillations of CO_2 concentration in the atmosphere follow changes in the use of fossil fuels. The intra-annual variability of atmospheric CO_2 concentration correlates more closely with the dynamics of land ecosystems rather than with that of ocean ecosystems.
- The regional flow of carbon in 2000, due to production and commercial trade of crops, timber, and paper, constituted 0.72 GtC yr^{-1}. The global pure carbon flux

at the atmosphere–ocean boundary observed in 1995 was estimated at 2.2 GtC (−19%–+22%), with an intra-annual variation of about 0.5 GtC. Maximum amplitudes of CO_2 flux oscillations in the atmosphere–ocean system are observed in the equatorial zone of the Pacific Ocean.

- An approximate picture of CO_2 distribution in ocean sources and atmospheric sinks is known: the tropical basins of the oceans are sources of CO_2, and high-latitude water basins are CO_2 sinks. The role of rivers is reduced mainly to the transport of carbon to the coastal zones of the oceans ($\sim$1 GtC yr^{-1}).

The most important section of the GCP is global environmental monitoring with the accumulation of detailed information on land life zone production, CO_2 fluxes at the atmosphere–ocean boundary, and volumes of anthropogenic emissions. The spaceborne sounding of CO_2 with the use of AIRS (Atmospheric Infrared Sounder) on the spaceborne laboratory EOS-Aqua launched by NASA on March 4, 2002 to an altitude of 705 km and IASI (Infrared Atmospheric Sounder Interferometer) carried by the satellite METOP (Meteorological Operational Polar) plays a special role (Nishida *et al.*, 2003). Other space vehicles, either presently functioning or planned to be launched, will be used to evaluate CO_2 fluxes from the data of indirect measurements of environmental characteristics. In particular, these are aims of the satellite TIROS-N (Television Infrared Observational Satellite-N) and instrumentation SCIAMACHY (Scanning Imaging Absorption Spectrometer for Atmospheric Chartography). The latter spectrometer launched in 2002 provides a high spectral resolution within the absorption bands of greenhouse gases such as CO_2, CH_4, H_2O (accuracy 1%) and N_2O, CO (accuracy 10%) with the surface resolution ranging from 30 km to 240 km depending on latitude.

Traditional ground measurements will be continued with the particular goal of substantiating national strategies for compatible use of the Earth's resources, including the development of forestry and agriculture, stock-breeding, and cultivation of field crops.

The GCP program foresees an extensive study of physical, biological, biogeochemical, and ecophysiological mechanisms involved in the formation of environmental carbon fluxes. A deeper understanding of these mechanisms and their parameterization will make it possible to specify carbon cycle models and related climate changes. Broadening the respective base of knowledge will make it possible to specify the following information about these mechanisms that has accumulated:

- Atmosphere–ocean carbon exchange is controlled mainly by physical processes, including mixing between surface and deep layers of the ocean through the thermocline. Biological processes promote the transport of carbon from the ocean surface to deeper layers and further to bottom deposits. A biological pump functions due to phytoplankton photosynthesis.
- A complex of feedbacks control the interactive exchange of energy, water, and carbon between the atmosphere and land surface, causing a response of these fluxes to such disturbances as transformation of land covers or oil pollution of

the oceans. Plant communities respond physiologically to changes in temperature and humidity of the atmosphere and soil.

- The carbon sink in the northern hemisphere depends on forest growth, climate change, soil erosion, fertilization, and the accumulation of carbon in freshwater systems. Unfortunately, the processes taking place in the northern hemisphere have been poorly studied, and factual information is practically absent. The significance of the terrestrial carbon sink can increase with certain dimensions of climate change. When the atmospheric CO_2 concentration exceeds a level of 550 ppm, many processes in land ecosystems become short of nutrients and water, and therefore the photosynthetic accumulation of carbon by terrestrial vegetation becomes physiologically saturated.
- The extent of key factors that determine directions and amplitudes of CO_2 fluxes between the atmosphere and land ecosystems is limited by several factors:
  - extreme climatic phenomena such as droughts, serious drifts of seasonal temperatures, solar radiation change due to a large-scale input of aerosol to the atmosphere (e.g., from volcanic eruptions or from large-scale fires like those which took place in Iraq in connection with recent military operations);
  - forest wildfires and other fires which introduce large-scale and long-term changes in carbon cycle characteristics (about 5%–10% of pure primary production, estimated at 57 GtC yr^{-1} globally, is emitted to the atmosphere by the burning of wood);
  - land use leading to a change in the boundaries of biomes and a change of their types (from evergreen forests to coniferous stands, forests changed to pastures, meadows becoming built-up areas);
  - reduction of biodiversity and change of the structures of communities, which changes the character of their impact on nutrient, carbon, and water cycles.
- Phenomena of the El Niño type or thermohaline circulation in the North Atlantic lead to global instability in processes of energy–matter exchange, which should be reflected in the parameterization of non-linear feedbacks.

Future dynamics of carbon exchange in the NSS will be determined by the strategy of managing the interaction of natural and anthropogenic factors, which, on the one hand, is apparently obvious, but yet raises doubts, on the other hand, since the problem of the greenhouse effect within the GCP cannot be simply solved. The GCP as a program is isolated from other investigative directions of effort concerning global ecodynamics. Although a broader approach to this problem has been introduced (Kondratyev *et al.*, 2002, 2003c, 2004b), these studies have unfortunately been neglected by the GCP authors.

In a number of recent studies (Bartsev *et al.*, 2003; Kondratyev *et al.*, 2002, 2003a, 2004a), it has been proposed that the GCP be considered in the context of its interaction with other processes in the NSS. As shown in Figure 7.6, the carbon cycle correlates with a multitude of natural and anthropogenic factors whose interaction forms the dynamics of the key processes in the NSS. For CO_2, such processes are exchanges at the boundaries of the atmosphere with land surfaces and sea and

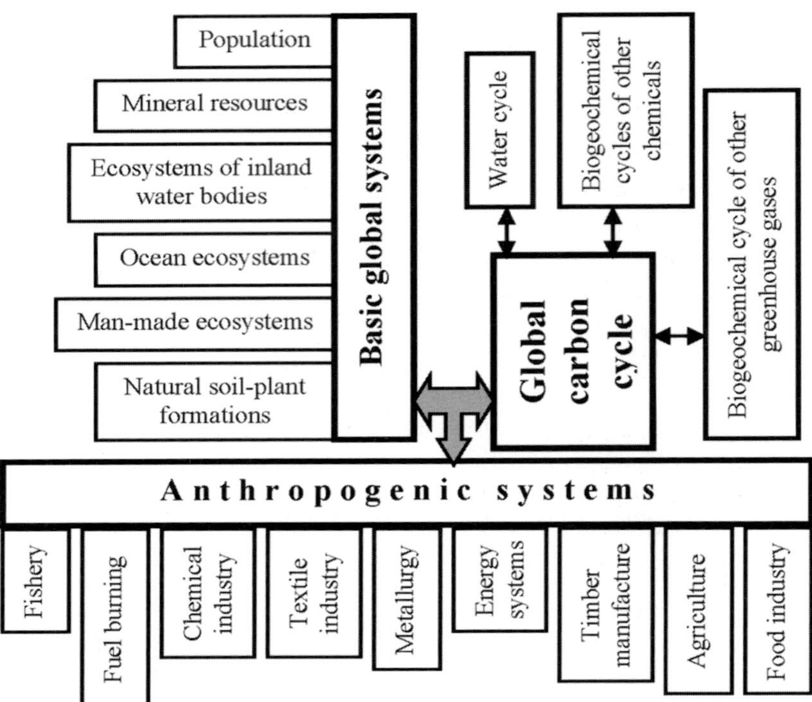

Figure 7.6. The conceptual scheme of the carbon cycle in the environment, and the place of the biogeochemical carbon cycle in the global system of energy exchange.

ocean basins. It is clear that CO_2 dynamics in the biosphere can be analyzed with available data concerning the spatial distribution of sinks and sources. The present level of knowledge makes it possible to specify and solve the problem of the impact of the greenhouse effect on climate, and thereby to decrease the level of uncertainty in estimates of future climate change. However, the applied carbon cycle model should reflect not only the spatial mosaic of its reservoirs, sinks, and sources, but should also provide a dynamic calculation of the respective influences. Earlier calculations using carbon cycle models have not adequately taken into account information on the status and classification of land cover and have considered even less variability of in oceans' basins. Therefore, the scheme in Figure 7.7 and Table 7.3 is aimed at compensating for these shortcomings of other models. The system of balance equations for such a scheme is written as

$$\frac{\partial \alpha_S^i(\varphi, \lambda, z, t)}{\partial t} + V_\varphi \frac{\partial \alpha_S^i(\varphi, \lambda, z, t)}{\partial \varphi} + V_\lambda \frac{\partial \alpha_S^i(\varphi, \lambda, z, t)}{\partial \lambda} + V_z \frac{\partial \alpha_S^i(\varphi, \lambda, z, t)}{\partial z} = \sum_{j \in \Omega_S} H_{jS} - \sum_{m \in \Omega_S} H_{Sm} \quad (i = 1, \ldots, N); \quad (7.1)$$

where S is the carbon reservoir in the ith cell (pixel) of spatial digitization; φ is the

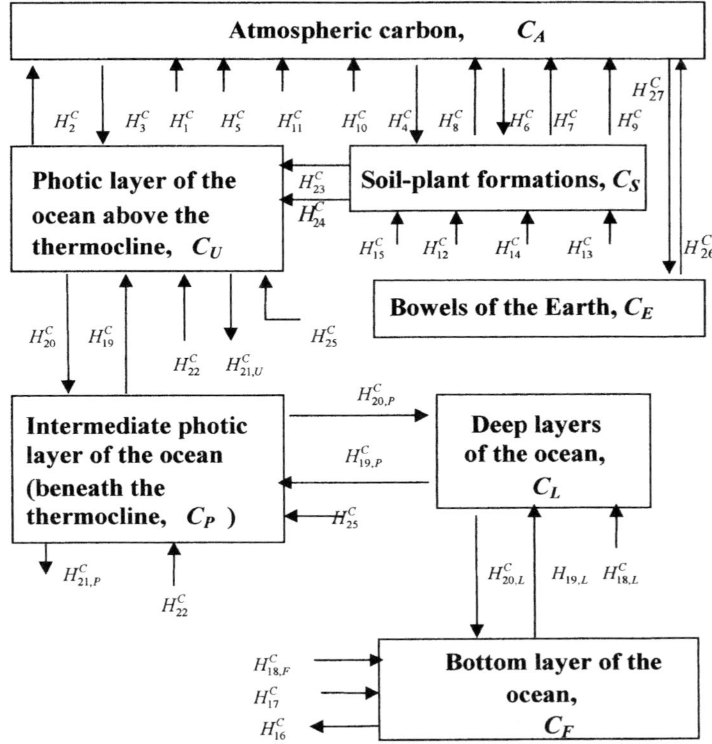

Figure 7.7. The block scheme of the global biogeochemical cycle of carbon dioxide (MGBC unit of the GMNSS) in the "atmosphere–land–ocean" system. The CO_2 reservoirs and fluxes are described in Table 7.3.

latitude; z is the longitude; z is the depth; t is the time, H_{jS} is the carbon sink from the jth reservoir to the reservoir S; H_{Sm} is the carbon sink from the reservoir S to the mth reservoir; Ω_S is the multitude of carbon reservoirs bordering the reservoir S; N is the number of carbon reservoirs; and $V(V_\varphi, V_\lambda, V_z)$ is the rate of exchange between reservoirs.

In Equation (7.1), the rate V and fluxes H are non-linear functions of environmental characteristics. These functions have been described in detail by Krapivin and Kondratyev (2002) and are only specified here. Mainly, the elements of the biogeocenotic unit of the global model shown in Figures 7.6 and 7.7 should be designated. This can be visualized by covering the whole land surface Σ with a homogeneous grid of geographic pixels $\Sigma_{ij} = \{(\varphi, \lambda) : \varphi_{i-1} \le \varphi < \varphi_i; \lambda_{j-1} \le \lambda < \lambda_j\}$ with boundaries in latitude $(\varphi_{i-1}, \varphi_i)$ and longitude $(\lambda_{j-1}, \lambda_j)$ and area σ. The number of pixels is determined by the available database (i.e., by the choice of grid size $(\Delta\varphi, \Delta\lambda) : i = 1, \ldots, n; n = [180/\Delta\varphi]; j = 1, \ldots, k; k = [180/\Delta\lambda]$). Each pixel can contain N types of surfaces, including the types of soil–plant formations, water basins, and other objects. The dynamics of the vegetation cover of the sth type

Table 7.3. Reservoirs and fluxes of carbon as CO_2 in the biosphere considered in the simulation model of the global biogeochemical cycle of carbon dioxide shown in Figure 7.7.

Reservoirs and fluxes of carbon dioxide	*Identifier*	*Estimate of reservoir* (10^9 t) *and flux* (10^9 t/yr)
Carbon		
atmosphere	C_A	650–750
photic layer of the ocean	C_U	580–1,020
deep layers of the ocean	C_L	34,500–37,890
soil humus	C_S	1,500–3,000
Emission in burning		
vegetation	H_8^C	6.9
fossil fuel	H_1^C	3.6
Desorption	H_2^C	97.08
Sorption	H_3^C	100
Rock weathering	H_4^C	0.04
Volcanic emanations	H_5^C	2.7
Assimilation by land vegetation	H_6^C	224.4
Respiration		
plants	H_7^C	50–59.3
humans	H_{10}^C	0.7
animals	H_{11}^C	4.1
Emission		
soil humus decomposition	H_9^C	139.5
plant roots	H_{15}^C	56.1
Vital activity		
population	H_{12}^C	0.3
animals	H_{13}^C	3.1
Plants dying off	H_{14}^C	31.5–50
Bottom deposits	H_{16}^C	0.1–0.2
Solution of marine deposits	H_{17}^C	0.1
Detritus decomposition		
photic layer	H_{22}^C	35
deep layers of the ocean	H_{18}^C	5

Reservoirs and fluxes of carbon dioxide	*Identifier*	*Estimate of reservoir* (10^9 t) *and flux* (10^9 t/yr)
Rising with deep waters	H_{19}^{C}	34
Lowering with surface waters and due to gravitational sedimentation	H_{20}^{C}	40
Photosynthesis	H_{21}^{C}	69
Groundwater runoff	H_{23}^{C}	0,5
Surface runoff	H_{24}^{C}	0.5–0.6
Respiration of living organisms in the ocean	H_{25}^{C}	25
Degasation processes	H_{26}^{C}	21.16
Sink to the Earth's bowels	H_{27}^{C}	1.3

follows the law:

$$\frac{dB_s}{dt} = R_s - M_s - T_s, \tag{7.2}$$

where R_s is photosynthesis; M_s and T_s are losses of biomass B_s due to die-off and evapotranspiration, respectively.

The components shown on the right-hand side of Equation (7.2) are functions of environmental characteristics: illumination, temperature, air and soil humidity, and atmospheric CO_2 concentration. There are several methods and forms of parameterizing these functions. An example is the model of Collatz *et al.* (2000), which provided the basis for developing the global biospheric model SiB2 (Sellers *et al.*, 1996). Temperature, humidity, and rate of evaporation in the vegetation cover and soil depend on the biospheric parameters and energy fluxes in the atmosphere–plant–soil system. By analogy with electrostatics, the notion of "resistance" is introduced, and fluxes are calculated from a simple formula: flux = potentials difference/resistance. The model SiB2 takes into account the fluxes of sensible and latent heat through evaporation of water vapor in plants and soil, and CO_2 fluxes are divided into classes C_3 and C_4, which substantially raises the accuracy of parameterization of the functions on the right-hand side of Equation (7.2). According to Collatz *et al.* (2000), three factors regulate the function R_s: the efficiency of the photosynthetic enzymatic system, the amount of photosynthetically active radiation (PAR) absorbed by cellulose chlorophyll, and the ability of plant species to assimilate and transmit the products of photosynthesis to the outside medium. Application of the Libich principle (Kondratyev *et al.*, 2002; Nitu *et al.*, 2000), and consideration of the data on the distribution of the types of vegetation cover by pixels $\{\Sigma_{ij}\}$, on partial pressures of CO_2 and O_2, the temperature and density of the atmosphere, and the level of illumination makes it possible to calculate fluxes H in Equation (7.1) for all pixels on land.

A model of the carbon cycle in the atmosphere–ocean system has been described in detail by Tarko (2005). It is based on the same grid of geographic pixels, but is combined with the zonal principle according to classification by Tarko (2001, 2005). The ocean thickness is considered a single biogeocenosis in which the main binding factor is the flux of organic matter produced in surface layers and then penetrating down to the deepest layers of the ocean. In this medium the carbonate system, a parametric description of which has been given by Kondratyev *et al.* (2004b), is a regulator of carbon fluxes.

One of the principal questions concerning CO_2 atmosphere–ocean exchange is the role of hurricanes, which has not been studied in detail. Perrie *et al.* (2004) conducted a study of hurricane influence on the local rates of air–sea CO_2 exchange. Hurricanes are shown to affect the thermal and physical structure of the upper ocean. Air–sea gas transfer includes processes such as upper-ocean temperature changes and the upwelling of carbon-rich deep water. Observations show that sea surface temperature and CO_2 partial pressure can decrease by 4°C and 20 μatm, respectively, due to the effect of hurricane activity. Perrie *et al.* (2004) proposed a model to parameterize CO_2 flux H_3 with the following formula:

$$H_3 = k_L \alpha \, \Delta[CO_2], \tag{7.3}$$

where α is the solubility of CO_2; and $\Delta[CO_2]$ is the difference between its partial pressure in the atmosphere and upper layer of the sea. Parameter k_L (cm h^{-1}) is determined with one of the following correlations depending on the wind speed:

$$k_L = \begin{cases} 0.31 U_{10}^2 (S_c/660)^{-0.5} & \text{for hurricanes 1–3 category;} \\ 0.0283 U_{10}^3 (S_c/660)^{-0.5} & \text{for hurricanes 4–5 category;} \end{cases} \tag{7.4}$$

where S_c is the Schmidt number (Hasegawa and Kasagi, 2001, 2005); U_{10} is the wind speed at an altitude of 10 m (m s^{-1}).

By introducing the wave spectrum peak frequency ω_p, the air-side friction velocity u^*, and the kinematic viscosity ν, parameter k_L can be calculated by the formula (Perrie *et al.*, 2004):

$$k_L = 0.13 \left[\frac{u_*^2}{\nu \omega_p} \right]^{0.63}. \tag{7.5}$$

Parameter k_L is actually formed from two components: $k_L = k_{L1} + k_{L2}$, where k_{L1} and k_{L2} are the wave-breaking and the interfacial terms, respectively. The terms

k_{L1} and k_{L2} are calculated with the use of the following formulas:

$$\left.\begin{aligned} k_{L1} &= u_*^{-1}[\sqrt{\rho_w/\rho_a}(h_w S_{cw}^{0.5} + \kappa^{-1}\ln\{z_w/\delta_w\}) \\ &\quad + \alpha(h_a S_{ca}^{0.5} + c_d^{-0.5} - 5 + 0.5\kappa^{-1}\ln S_{ca})]; \\ k_{L2} &= fV\alpha^{-1}[1 + (e\alpha S_c^{-0.5})^{-1/n}]^{-n}, \end{aligned}\right\} \quad (7.6)$$

where $f = 3.8 \times 10^{-6} U_{10}^3$; α is gas solubility; subscript $a(w)$ denotes the air (water) side; ρ is density; z is measurement depth; δ is the turbulent surface layer thickness; κ is the von Kármán constant; c_d is the drag coefficient; $h \equiv \Lambda\varphi^{-1}R_r^{0.25}$; Λ is an adjustable constant; R_r is the roughness Reynolds number; φ is an empirical function that accounts for buoyancy effects on turbulent transfer in the ocean; and V, e, and n are empirical constants equal to $14\,\text{cm}\,\text{h}^{-1}$, $1.2\,\text{cm}\,\text{h}^{-1}$, and $4{,}900\,\text{cm}\,\text{h}^{-1}$ in the GasEx-1998 field experiment (Perrie *et al.*, 2004), and may need readjustment for other datasets.

The principal significance is the fact that hurricane activity initiates an upwelling zone where air–water gas exchange takes on another character. Hales *et al.* (2005) studied atmospheric CO_2 uptake in a coastal upwelling system located off the Pacific coast of Oregon using high-resolution measurements of the partial pressure of CO_2 and nutrient concentrations in May to August 2001. Results showed that the dominance of low-CO_2 waters over the shelf area renders the region a net sink during the upwelling season due to

- the presence of upwelled water rich in preformed nutrients;
- complete photosynthetic uptake of these excess nutrients and a stoichiometric proportion of CO_2;
- moderate warming of upwelled waters.

It is estimated that:

- The eastern boundary area of the North Pacific can constitute a sink of atmospheric CO_2 that is 5% of the annual North Pacific CO_2 uptake.
- By mid-August, the partial pressure of CO_2 in sub-surface waters increases 20%–60%, corresponding to an increase of 1.0%–2.3% total dissolved CO_2 due to respiration of settling biogenic debris.

Many parameters of the global carbon cycle model are measured in the satellite-monitoring regime, which makes it possible to apply an adaptive scheme of calculation of greenhouse effect characteristics (Figure 7.8). This scheme makes it possible to add information to the model of the continuous regime by correction of its structure and parameters. Satellite measurements in the visible and near-IR regions provide operational estimates of photosynthetically active radiation and vegetation characteristics such as canopy greenness, area of living photosynthetically active elements, soil humidity and water content in the elements of vegetation cover, CO_2 concentration on the surface of leaves, etc. The regime of prediction of the

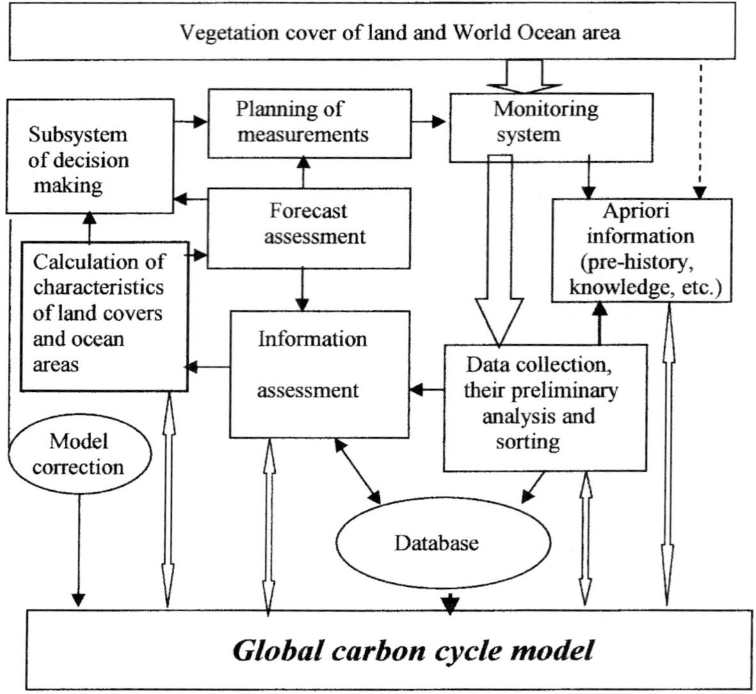

Figure 7.8. An adaptive regime of greenhouse effect monitoring with assessment of the role of vegetation cover of land and ocean areas.

vegetation cover biomass in each pixel Σ_{ij} and comparisons with satellite measurements enables one to correct some fragments of the model; for instance, using the doubling of its units or their parametric adjustment to minimize discrepancies between prediction and measurements (Figure 7.8). In particular, to calculate primary production, there are some semi-empirical models that can be used by a sample criterion in different pixels. There is a certain freedom in the choice for estimating evaporation from vegetation cover (Wange and Archer, 2003).

The key component of the global CO_2 cycle is anthropogenic emissions to the environment. The main problem studied in this connection by most scientists is an assessment of the ability of the biosphere to neutralize an excess amount of CO_2. Table 7.4 and Figures 7.9 and 7.10 illustrate the modeling results. It is seen that 41.3% of the 6.3 GtC emitted to the atmosphere by industry remains in the atmosphere, while the oceans and land vegetation absorb 20.2% and 38.5%, respectively. Taking as a basis the dependence of air temperature changes on CO_2 variation (Mintzer, 1987):

$$\Delta T_{\mathrm{CO}_2} = -0.677 + 3.019 \ln[C_a(t)/338.5], \tag{7.7}$$

for the realistic scenario in Figure 7.9, we obtain $\Delta T_{\mathrm{CO}_2} \leq 2.4°\mathrm{C}$. This substantially

Table 7.4. Model estimates of excessive CO_2 assimilation over Russia. A more detailed classification of soil–plant formations is given in Table 7.5.

Soil–plant formation	*Flux of assimilated carbon as* CO_2 (10^6 tC/year)
Arctic deserts and tundras, sub-arctic meadows and marshes	2.2
Tundras	3.3
Mountain tundras	3.6
Forest tundra	2.8
North-taiga forests	10.8
Mid-taiga forests	31.2
South-taiga forests	22.9
Broad-leaved–coniferous forests	4.8
Steppes	3.6
Alpine and sub-alpine meadows	1.1
Deserts	2.2

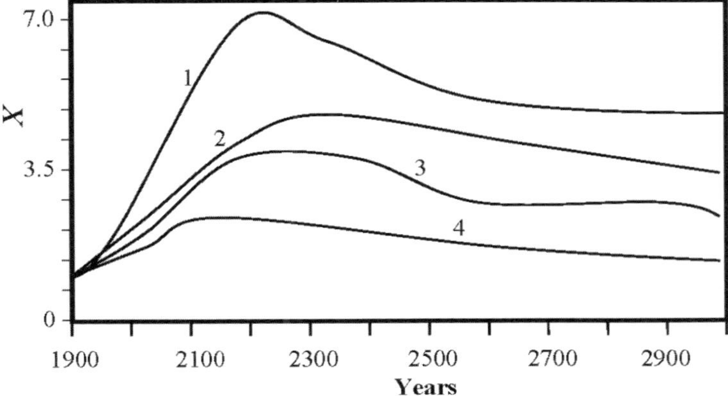

Figure 7.9. Forecast of CO_2 concentration in the atmosphere with different scenarios of mineral resources expenditure: 1, pessimistic scenario (Bacastow, 1981); 2, optimistic scenario (Björkstrom, 1979); 3, scenario of IPCC (Intergovernmental Panel on Climate Change) (Dore *et al.*, 2003); 4, realistic scenario (Demirchian and Kondratyev, 2004). $X = C_a(t)/C_a(1900)$.

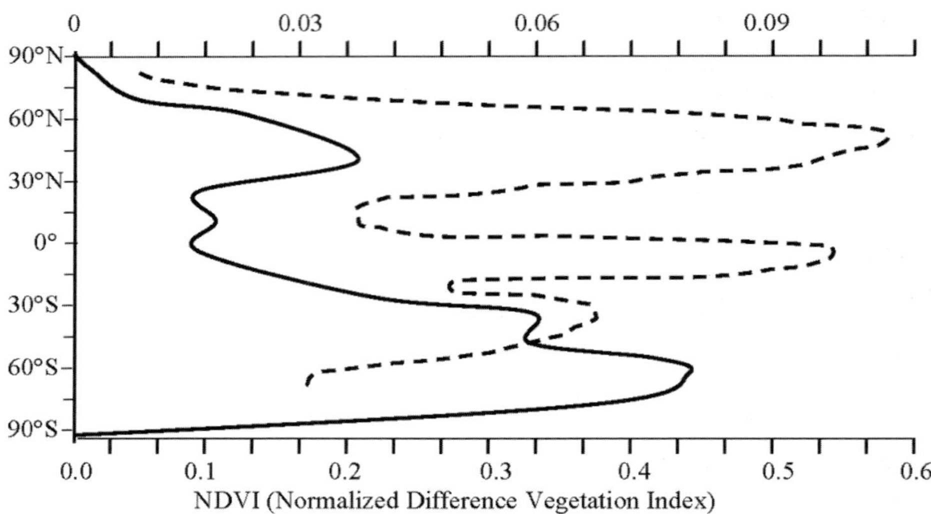

Figure 7.10. The latitudinal distribution of rate F (GtC yr^{-1} deg^{-1}) of carbon absorption (solid curve) from the atmosphere and vegetation index (dashed curve). Types and spatial distribution of soil–plant formations are determined in Table 7.5 and Figure 7.11. Industrial emissions of CO_2 are assumed to be 6.3 GtC yr^{-1}.

supports the estimate $\Delta T_{CO_2} \leq 4.2°C$ published by many authors and assumed in the Kyoto Protocol.

As can be seen from Figure 7.10, the discrepancy between the forms of CO_2 distribution in the absorption curve and the vegetation index suggests the possibility that in the southern hemisphere the structure of model pixels and their correspondence with observational data should be specified. Nevertheless, an introduction to the carbon cycle model of pixel mosaic has made it possible to evaluate the role of some types of ecosystems and regions of Russia in the regulation of the greenhouse effect. Table 7.4 demonstrates the role of taiga in the territory of Russia in this regulation. On the whole, the model enables one to consider various scenarios of land cover changes and study the dependence of CO_2 partial pressure in the atmosphere on their structure. For instance, if by 2050 the forest areas are reduced only by 10% with respect to 1970 (~42 million km^2), then by the end of the 21st century the content of atmospheric CO_2 can increase by 46.7%, with stable anthropogenic emissions of carbon about 6 GtC yr^{-1}. On the contrary, broadening of the forest areas in the northern hemisphere by 10% will reduce the anthropogenic impact on the greenhouse effect by 14.8%.

The problem of global warming due to the growth of greenhouse gas concentration is the problem of sustainable development of civilization. The approach proposed here enables one to synthesize the accumulated data and knowledge of the carbon cycle and other greenhouse gases into a single monitoring system.

Table 7.5. Identifier of the types of soil–plant formations following the classification of Bazilevich and Rodin (1967).

Type of soil–plant formation	*Symbol*
Arctic deserts and tundras	*A*
Alpine deserts	*B*
Tundras	*C*
Mid-taiga forests	*D*
Pampas and grass savannahs	*E*
North-taiga forests	*F*
South-taiga forests	*F*
Sub-tropical deserts	*G*
Sub-tropical and tropical grass–tree thickets of tugai type	*I*
Tropical savannahs	*J*
Saline lands	*K*
Forest tundra	*L*
Mountain tundra	*M*
Tropical xerophytic open woodlands	*N*
Aspen–Birch sub-taiga forests	*O*
Sub-tropical broad-leaved and coniferous forests	*P*
Alpine and sub-alpine meadows	*Q*
Broad-leaved coniferous forests	*R*
Sub-boreal and saltwort deserts	*S*
Tropical deserts	*T*
Xerophytic open woodlands and shrubs	*U*
Dry steppes	*V*
Moderately arid and arid (mountains included) steppes	*W*
Forest steppes (meadow steppes)	*X*
Variably humid deciduous tropical forests	*Y*
Humid evergreen tropical forests	*Z*
Broad-leaved forests	+
Sub-tropical semi-deserts	&
Sub-boreal and wormwood deserts	@
Mangrove forests	#
Lack of vegetation	*

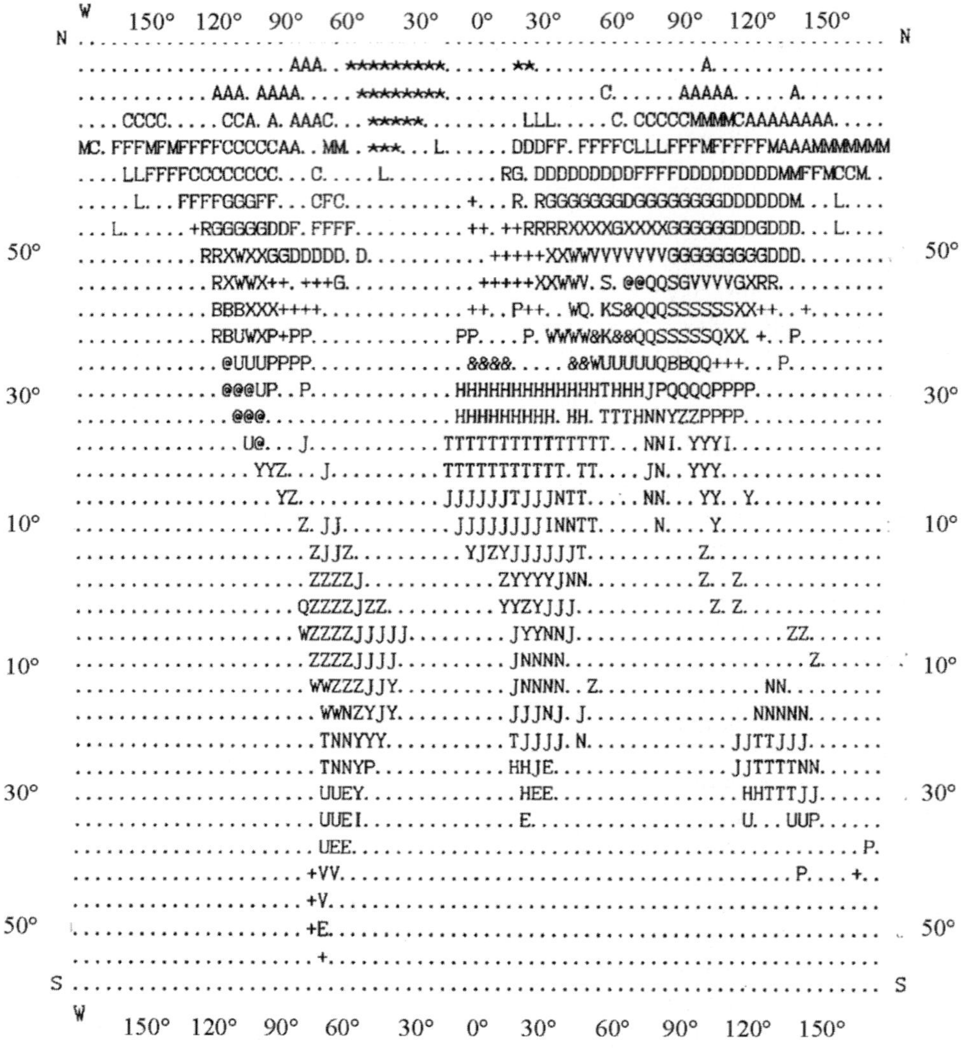

```
     W  150°  120°   90°   60°    30°    0°   30°    60°    90°   120°   150°
N  ...................................................................................  N
   ....................AAA..*********......**...............A...............
   ............AAA.AAAA.....********..............C......AAAAA.....A........
   ....CCCC.....CCA.A.AAAC...*****.........LLL.....C.CCCCCMMMMCAAAAAAAAA.....
   MC.FFFMFMFFFFCCCCCAA..MM.***...L......DDDFF.FFFFCLLLFFFMFFFFFMAAAMMMMMMM
   ....LLFFFFCCCCCCCC...C.....L..........RG.DDDDDDDDDFFFFDDDDDDDDDMMFFMCCM..
   .....L...FFFFGGGFF...CFC...........+...R.RGGGGGGGDGGGGGGGGDDDDDDM...L....
   ...L......+RGGGGGDDF.FFFF..........++.++RRRRXXXXGXXXXGGGGGGDDGDDD...L....
50°............RRXWXXGGDDDDD.D...........+++++XXWWVVVVVVVGGGGGGGGGDDD.........  50°
   ............RXWWX++.+++G............+++++XXWWV.S.@@QQSGVVVVGXRR..........
   ............BBBXXX++++..............++..P++..WQ.KS&QQQSSSSSSXX++..+........
   ............RBUWXP+PP.............PP....P.WWWW&K&&QQSSSSSQXX.+..P........
   .............@UUUPPPP...............&&&&.....&&WUUUUUQBBQQ+++...P.........
30°.............@@@UP..P.............HHHHHHHHHHHHHHTHHHJPQQQQPPPP............  30°
   ..............@@@.................HHHHHHHHH.HH.TTTHNNYZZPPPP.............
   ...............U@...J............TTTTTTTTTTTTTTT...NNI.YYYI..............
   ................YYZ...J..........TTTTTTTTTTT.TT....JN..YYY...............
   ..................YZ.............JJJJJTJJJNTT.....NN...YY..Y............
10°....................Z.JJ..........JJJJJJJJINNTT.....N....Y................  10°
   .....................ZJJZ..........YJZYJJJJJJT..........Z...............
   .....................ZZZZJ............ZYYYYJNN..........Z..Z.............
   ....................QZZZZJZZ..........YYZYJJJ............Z.Z.............
   ....................WZZZZJJJJJ.........JYYNNJ...................ZZ.......
10°.....................ZZZZJJJJ..........JNNNN......................Z......  10°
   .....................WWZZZJJY..........JNNNN..Z................NN.........
   ......................WWNZYJY..........JJJNJ.J...............NNNNN.......
   ......................TNNYYY...........TJJJJ.N.............JJTTJJJ.......
   ......................TNNYP............HHJE................JJTTTTNN......
30°......................UUEY.............HEE..................HHTTTJJ......  30°
   ......................UUEI.............E..................U...UUP......
   ......................UEE............................................P.
   .....................+VV..................................P....+..
   .....................+V..................................................
50°.....................+E..................................................  50°
   ......................+..................................................
S  ...................................................................................  S
     W  150°  120°   90°   60°    30°    0°   30°    60°    90°   120°   150°
```

Figure 7.11. Distribution of the types of soil–plant formations by pixels of the GMNSS spatial structure. The notation is given in Table 7.5.

Unfortunately, the initiated international program on carbon cycle study (Canadel *et al.*, 2003), like other similar programs of global character, has not been aimed at developing a constructive information technology able to raise substantially the reliability of prognostic estimates of future climate change. Nevertheless, the ideas and approaches of Russian specialists published recently (Bartsev *et al.*, 2003; Kondratyev *et al.*, 2002, 2003a, 2004b; Krapivin and Kondratyev, 2002; Tarko,

2005), as well as models developed by American scientists (Collatz *et al.*, 2000; Sellers *et al.*, 1996) will make it possible, though not within the GCP, to overcome the existing isolation of carbon cycle studies and create a global model able in the operational regime of satellite monitoring to give reliable estimates of the role of regions in the greenhouse effect dynamics. Such a model will be a tool to work out an efficient strategy of land use and will lead to making well-informed international decisions (in contrast to the Kyoto Protocol).

Revealing the key factors of global change by modeling global ecodynamics faces some problems connected with the choice of the form and methods of modeling. Prevailing problems have been discussed in studies by Bartsev *et al.* (2003), Kondratyev *et al.* (2003c), and others. The main problem here is a combination of the parameters used in models developed with available data and knowledge bases, as well as the choice of a compromise accepted between the complicated structures of these models and their semi-empirical realizations. An example is provided by the case of soil and vegetation cover and the spatial averaging that is necessary in this case; see Figure 7.11 where the notation is explained in Table 7.5. It is clear that processes of the choice of technology used in modeling and interpretation of the results obtained are similar. In this process one can select for important parameters of the NSS which affect global ecodynamics (Kondratyev *et al.*, 2004b). However, as follows from publications of Barenbaum (2002, 2004) and Yasanov (2003), the description of the carbon cycle lacks an important fact connected with carbon buried in geological structures and its intake from space. Therefore, in perspective, when synthesizing the global model of the carbon cycle, it is necessary to consider a more detailed combined description of the biospheric and lithospheric parts of this cycle. Clearly, the lifetime of carbon in each sub-cycle should be estimated more accurately. The lithospheric part of the carbon cycle includes its transformation in the process of long interactions and conversions, including the transformation to methane, oil, coal deposits, etc. Depending on temperature and pressure, hydrocarbons can be oxidized and become the main component of underground fluids and magmas. In this way the carbon cycle correlates with the cycles of methane and water.

The various global biogeochemical cycles have the same common uncertainties the overcoming of which is possible with use of new data and modified models. A perspective modeling method is described by Degermendzhy and Bartsev (2003). This method is based on small-scale models that minimize their requirements from the global database.These models oversimplify the formalization of climatic, biotic, geochemical, economic, and social processes. The level of adequacy of these studies is determined by the extent to which real processes are simplified in the models.

7.3.3 Global model units for other biogeochemical cycles

7.3.3.1 Sulfur unit

Taking into account the designations in Figure 7.12 and Table 7.6, the equations of the sulfur unit of the GMNSS are written in the form of balance correlations (Krapivin and Nazaryan, 1997):

$$
\left.
\begin{aligned}
\frac{dAH2SL}{dt} &= C_1 + C_2 + C_3 + C_{21} - C_4 \\
\frac{dASO2L}{dt} &= C_4 + C_5 + C_6 - C_7 - C_8 - C_9 \\
\frac{dASO4L}{dt} &= C_9 + C_{13} + C_{20} - C_{11} - C_{12} \\
\frac{dS}{dt} &= C_{17} - C_{16} - C_{19} \\
\frac{dSO4L}{dt} &= C_{10} + C_{11} + C_{12} + C_{16} - C_3 \\
&\quad - C_{13} - C_{14} \\
\frac{dFIX}{dt} &= C_7 + C_{15} + C_{22} - C_{17} \\
\frac{dH2SO4L}{dt} &= C_8 - C_{18} - C_{21} - C_{22} \\
\frac{dAH2SO}{dt} &= H_1 + H_3 + H_4 + H_{26} - H_2 \\
\frac{dASO2O}{dt} &= H_2 + H_5 + H_6 - H_7 - H_8 \\
&\quad - H_{24} \\
\frac{dASO4O}{dt} &= H_8 + H_9 + H_{12} - H_{10} - H_{11} \\
\frac{\partial SO4OU}{\partial t} + v_z \frac{\partial SO4OU}{\partial z} + k_z \frac{\partial^2 SO2OU}{\partial z^2} &= H_7 + H_{10} + H_{11} + H_{20} + H_{22} \\
&\quad + H_{27} + C_{14} - H_{12} - H_{13} \\
\frac{\partial H2SOU}{\partial t} + v_z \frac{\partial H2SOU}{\partial z} + k_z \frac{\partial^2 H2SOU}{\partial z^2} &= H_{21} + H_{23} - H_4 - H_{22} \\
\frac{\partial H2SOD}{\partial t} + v_z \frac{\partial H2SOD}{\partial z} + k_z \frac{\partial^2 H2SOD}{\partial z^2} &= H_{17} - H_{18} - H_{21} \\
\frac{\partial SO4OD}{\partial t} + v_z \frac{\partial SO4OD}{\partial z} + k_z \frac{\partial^2 SO4OD}{\partial z^2} &= H_{18} - H_{19} - H_{20} \\
\frac{\partial DU}{\partial t} + v_z \frac{\partial DU}{\partial z} + k_z \frac{\partial^2 DU}{\partial z^2} &= H_{14} - H_{15} - H_{23} \\
\frac{\partial DD}{\partial t} + v_z \frac{\partial DD}{\partial z} + k_z \frac{\partial^2 DD}{\partial z^2} &= H_{15} - H_{16} - H_{17} \\
\frac{\partial FI}{\partial t} + v_z \frac{\partial FI}{\partial z} + k_z \frac{\partial^2 FI}{\partial z^2} &= H_{13} - H_{14} \\
\frac{dBOT}{dt} &= H_{16} + H_{19}
\end{aligned}
\right\}
\tag{7.8}
$$

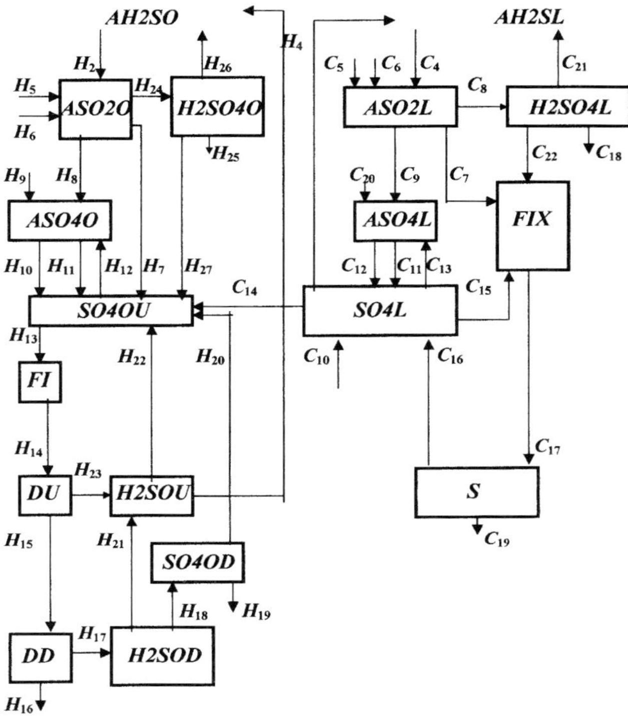

Figure 7.12. The scheme of sulfur fluxes in the environment (MGBS unit of the GMNSS). The notation is given in Table 7.6.

where v_z is advection velocity (m da^{-1}); and k_z is the coefficient of turbulent mixing (m^2 da^{-1}). Reservoir designations are given in Table 7.7. Functional representations of the sulfur flows are given by Krapivin and Nazaryan (1997).

The discharge speed of H_2S to the atmosphere due to humus decomposition is described by a linear function $C_3 = \mu_1(pH) \cdot SO4L \cdot T_L$, where μ_1 is the proportionality coefficient depending on soil acidity (i.e., pH) (da^{-1} K^{-1}), and T_L is soil temperature (K).

It is supposed that flow H_4 is a function of the rates of alignment for H_2S oxidation in the photic layer with the vertical velocity of water rising. Therefore, for the description of flow H_4 the parameter t_{H2SU} which reflects the lifetime of H_2S in the water is used: $H_4 = H2SU/t_{H2SU}$, where t_{H2SU} is a function of the velocity of vertical advection u_z and of the oxygen concentration $O2$ in the upper layer having the thickness Z_{H2S}:

$$t_{H2SU} = H2SOU \cdot O2^{-1} u_z(\theta_2 + O2)(\theta_1 + u_z)^{-1}. \tag{7.9}$$

The constants θ_1 and θ_2 are defined empirically, while the value of $O2$ is estimated by the oxygen unit of the GSM. Flows H_2 and C_4 reflect the correlation between the

Table 7.6. Characteristics of land and hydrospheric fluxes of sulfur in the structure of Figure 7.12. Numerical estimates of fluxes (mg m^{-2} da^{-1}) are obtained by averaging over the respective territories (Krapivin and Nazaryan, 1997).

Sulfur flux	*Land*		*Hydrosphere*	
	Identifier	*Estimate*	*Identifier*	*Estimate*
Volcanic invasions				
H_2S	C_1	0.018	H_3	0.0068
SO_2	C_5	0.036	H_5	0.0073
SO_4^{2-}	C_{20}	0.035	H_9	0.0074
Anthropogenic emissions				
H_2S	C_2	0.072	H_1	0.00076
SO_2	C_6	0.92	H_6	0.038
SO_4^{2-}	C_{10}	0.47		
Oxidation of H_2S to SO_2	C_2	1.13	H_2	0.3
Oxidation of SO_2 to SO_4^{2-}	C_9	1.35	H_8	0.16
Dry sedimentation of SO_4^{2-}	C_{12}	0.37	H_{11}	0.11
Fallout of SO_4^{2-} with rain	C_{11}	1.26	H_{10}	0.38
Biological decomposition and emission of H_2S to the atmosphere	C_3	1.03	H_4	0.31
Assimilation of SO_4^{2-} by biota	C_{15}	0.41	H_{13}	1.09
Biological decomposition and formation of SO_4^{2-}	C_{16}	1.13	H_{17}	0.43
			H_{23}	0.12
Sedimentation and deposits	C_{18}	0.22	H_{15}	0.98
	C_{19}	0.11	H_{16}	0.55
			H_{19}	0.0076
			H_{25}	0.036
Wind-driven return to the atmosphere	C_{13}	0.25	H_{12}	0.33
Replenishing sulfur supplies due to dead biomass	C_{17}	0.86	H_{14}	1.1
Assimilation of atmospheric SO_2	C_7	0.46	H_7	0.18
Washing out of SO_2 from the atmosphere	C_8	0.27	H_{24}	0.061
River runoff of SO_4^{2-} to the ocean	C_{14}	1.17		
Transition of gas-phase H_2SO_4 to H_2S	C_{21}	0.018	H_{26}	0.0076

Sulphur flux	*Land*		*Hydrosphere*	
	Identifier	*Estimate*	*Identifier*	*Estimate*
Assimilation of the washed-out part of atmospheric SO_2 by biota	C_{22}	0.036	H_{27}	0.015
Oxidation of H_2S to SO_2 in water medium			H_{18} H_{22}	0.045 0.19
Advection of SO_2			H_{20}	0.38
Advection of H_2S			H_{21}	0.37

Table 7.7. Initial data taken into account under simulation experiments.

Reservoir	*Identifier of the GSM*	*Preliminary estimation of reservoir* ($mg\,m^{-2}$)
Atmosphere above the ocean		
H_2S	*AH2SO*	10
SO_2	*ASO2O*	5.3
SO_4^{2-}	*ASO4O*	2
Atmosphere above land		
H_2S	*AH2SL*	36.9
SO_2	*ASO2L*	17.9
SO_4^{2-}	*ASO4L*	12.9
Land		
SO_4^{2-}	*SO4L*	11.2
biomass	*FIX*	600
soil	*S*	5,000
Ocean photic layer		
H_2S	*H2SOU*	1.9
SO_4^{2-}	*SO4OU*	19×10^7
phytomass	*FI*	*66.5*
DOM	*DU*	730
Deep ocean layers		
H_2S	*H2SOD*	2×10^6
SO_4^{2-}	*SO4OD*	3.4×10^9
DOM	*DD*	13,120

sulfur and oxygen cycles: $C_4 = AH2SL/t_{H2SA}$, $H_2 = AH2SO/t_{H2SA}$, where t_{H2SA} is the lifetime of H_2S in the atmosphere.

The mechanism of SO_2 removal from the atmosphere is described by flows H_7, H_8, H_{27}, C_7, and C_9. These flows are characterized by typical parameters t_{SO2L} and t_{SO2A1}, which are the lifetimes of SO_2 above the land and water surface, respectively. SO_2 is absorbed from the atmosphere by minerals, vegetation, and soil. Dry absorption of SO_2 by vegetation from the atmosphere is described by the model $C_7 = q_2 RX$, where $q_2 = q_2' \cdot ASO2L/(r_{tl} + r_s)$, r_{tl} is the atmospheric resistance to SO_2 transport over the vegetation of lth type (da m^{-1}), r_s is surface resistance to SO_2 transport over the surface of sth type (da m^{-1}), RX is the production of X-type vegetation (mg m^{-2} da^{-1}), and q_2' is the proportionality coefficient. Production RX is calculated by the biogeocenotic unit of the GSM.

The process of washing out SO_2 from the atmosphere is described by the model: $C_8 = q_{11} W \cdot ASO2L$, where q_{11} is the characteristic parameter for the surface of lth type ,and $W(t, \varphi, \lambda)$ is precipitation intensity. The interaction of acid rain with the land surface was reflected in Figure 7.12 by means of flows C_{18}, C_{21}, C_{22}, H_{25}, H_{26} and H_{27}. These flows are parameterized by models: $C_{18} = h_1 \cdot H2SO4L$, $C_{22} = h_2 \cdot RX \cdot H2SO4L$, $C_{21} = h_3 T_a \cdot H2SO4L$, $H_{25} = h_6 \cdot H2SO4O$, $H_{26} = h_4 T_a \cdot H2SO4O$, $H_{27} = h_5 \cdot RFI \cdot H2SO4O$, where $T_a(t, \varphi, \lambda)$ is atmosphere temperature, $h_1 + h_2 \cdot RX + h_3 T_a = 1$, $h_4 T_a + h_5 \cdot RFI + h_6 = 1$, and RFI is the production of phytoplankton.

Similarly, the flows H_8, C_9, H_7, and H_{24} are simulated by the following models: $H_8 = ASO2O/t_{SO2A1}$, $C_9 = ASOL/t_{SO2L}$, $H_7 = ASO2O/t_{SO2A2}$, and $H_{24} = q_{11} W \cdot ASO2O$.

The physical mechanisms of sulfate transportation in the environment are described by the models of Bodenbender *et al.* (1999), Luecken *et al.* (1991), Krapivin (1993), Park *et al.* (1999): $H_{10} = \mu W \cdot ASO4O$, $H_{11} = \rho v_0 \cdot ASO4O$, $C_{11} = b_3 W \cdot ASO4L$, $C_{12} = d_1 v_a \cdot ASO4L$, where v_0 and v_a are the rates of dry sedimentation of aerosols over the water surface and land, respectively.

For the flows C_{13}, H_{12}, C_{14}, and C_{16} we consider the following models: $C_{13} = d_2 \cdot RATE \cdot SO4L$, $H_{12} = \theta \cdot RATE \cdot SO4L$, $C_{16} = b_2 ST_L$, and $C_{14} = d_3 W \cdot SO4L + (C_{11} + C_{12})\sigma$, where $RATE(t, \varphi, \lambda)$ is wind velocity over the surface (m s^{-1}), and coefficient b_2 reflects the sulfur content in dead plants.

The terrestrial part of the sulfur cycle correlates with the water part through flows in the atmosphere–hydrosphere–land system. We have $H_{13} = \gamma \cdot RFI$, $H_{14} = b \cdot MFI$, $H_{15} = f \cdot DU$, $H_{16} = p \cdot DD$, $H_{17} = q \cdot DD$, $H_{18} = H2SOD/t_{H2SOD}$, $H_{19} = u \cdot SO4D$, $H_{20} = a_1 v_D \cdot SO4D$, $H_{21} = b_1 v_D \cdot H2SOD$, $H_2 = H2SOU/t_{H2SOU}$, $H_{23} = g \cdot DU$, where MFI is the mass of dead phytoplankton, t_{H2SOU} and t_{H2SOD} are the characteristic times for H_2S total oxidation in the photic layer and deep waters, respectively.

7.3.3.2 *Nitrogen unit*

The MGBN unit simulating the fluxes of nitrogen in the environment is necessary in the global biospheric model for several indisputable reasons: nitrogen compounds can affect the environmental conditions, change the food quality, affect the climate, and transform the hydrospheric parameters. An abundant use of nitrates leads to water pollution and reduces the quality of food products. It is well known that intensive exploitation of soils without taking into account the consequences of the misuse of nitrogen fertilizers breaks the stability of agri-ecosystems and human health. Moreover, nitrous oxide (N_2O), nitrogen dioxide (NO_2), and nitric oxide (NO), being minor gas components of the atmosphere, substantially affect the formation of the processes of optical radiation absorption in the atmosphere. Small deviations in their concentrations can cause significant climatic variations near the Earth's surface.

The nitrogen cycle is closely connected with the fluxes of hydrogen, sulfur, and other chemicals. The global cycle of nitrogen as one of the nutrient elements is a mosaic structure of local processes of its compounds formed due to water migration and atmospheric processes. The present-day nitrogen cycle is especially vulnerable to anthropogenic impacts manifested through interference with the nitrogen cycle both directly and via the influence on related processes. Therefore, the construction of an adequate model of the nitrogen cycle in nature at present should be based on a description of the whole complex of natural processes and those initiated by humans.

The natural sources of nitrogen oxides are associated with the vital functions of bacteria, volcanic eruptions, as well as several atmospheric phenomena (e.g., lightning discharges). The biogeochemical cycle of nitrogen includes processes such as fixation, mineralization, nitrification, assimilation, and dissimilation. The structural schemes of these processes have been described in detail by many authors (Ehhalt, 1981; Ronner, 1983). Their complexity level is determined by the goal of studies, the availability of data on the rates of transformation of the nitrogen-containing compounds and their supplies, by the level of detailing, etc.

Nitrogen transport in the biosphere is driven by a complicated meandering structure of fluxes, including a hierarchy of cycles at various levels of life organization. From the atmosphere, nitrogen enters the cells of microorganisms, from where it goes to soil and then passes to higher plants, animals, and humans. The survivability of living organisms results in the return of nitrogen into the soil, from which it either again goes to plants and living organisms or is emitted to the atmosphere. Approximately the same scheme of nitrogen oxide cycling prevails in the hydrosphere. The characteristic feature of these cycles is their accessibility to available processes of nitrogen removal from the biospheric balance and subsequent migration into rock formations, from where it returns to the biosphere at a much slower rate than its outbound flux. Taking into account the nature of the nitrogen cycle in the biosphere and its reservoir structure enables one to formulate a global scheme of nitrogen fluxes.

A comparative analysis of the model schemes for representing the flux of nitrogen compounds in nature as proposed by various experts makes it possible to construct the block scheme presented in Figure 7.13. Here the atmosphere, soil, lithosphere, and hydrosphere are considered as nitrogen reservoirs. The first three reservoirs are described by 2-D models, and the hydrosphere is described by a 3-D multi-layer

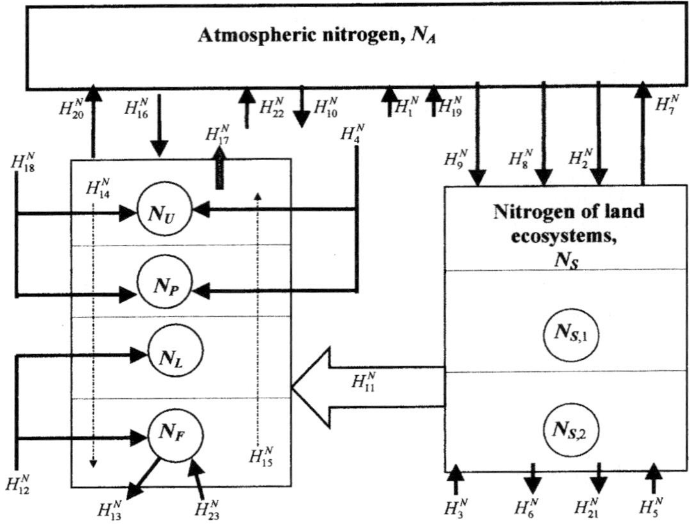

Figure 7.13. The scheme of nitrogen fluxes in the GMNSS. The notation is given in Table 7.8.

model. The characteristics of nitrogen fluxes between these reservoirs are given in Table 7.8. The equations of the model are written as

$$
\left.
\begin{aligned}
\frac{\partial N_A}{\partial t} + V_\varphi \frac{\partial N_A}{\partial \varphi} + V_\lambda \frac{\partial N_A}{\partial \lambda} &= H_1^N + \\
&\quad \begin{cases} H_{20}^N - H_{16}^N, & (\varphi,\lambda) \in \Omega_O \\ H_7^N + H_{19}^N - H_8^N - H_9^N + H_{22}^N - H_2^N - H_{10}^N, & (\varphi,\lambda) \in \Omega/\Omega_O \end{cases} \\
\frac{\partial N_{S_1}}{\partial t} &= H_8^N + H_6^N - H_3^N \\
\frac{\partial N_{S_2}}{\partial t} &= H_2^N + H_3^N + H_5^N + H_9^N + H_6^N + H_7^N \\
&\quad + H_{11}^N - H_{21}^N \\
\frac{\partial N_U}{\partial t} + v_\varphi \frac{\partial N_U}{\partial \varphi} + v_\varphi \frac{\partial N_U}{\partial \lambda} &= H_{16}^N + H_{4,U}^N + H_{18,U}^N + H_{11}^N - H_{17,U}^N - H_{20}^N \\
&\quad - H_{14,UP}^N - H_{15,UP}^N \\
\frac{\partial N_P}{\partial t} + v_\varphi \frac{\partial N_P}{\partial \varphi} + v_\lambda \frac{\partial N_P}{\partial \lambda} &= H_{18,P}^N + H_{4,P}^N + H_{14,UP}^N + H_{15,PL}^N - H_{17,P}^N \\
&\quad - H_{14,PL}^N - H_{15,UP}^N \\
\frac{\partial N_L}{\partial t} &= Q_L + H_{12,L}^N + H_{14,PL}^N + H_{15,LF}^N - H_{14,LF}^N - H_{15,PL}^N \\
\frac{\partial N_F}{\partial t} &= Q_F + H_{12,F}^N + H_{23}^N + H_{14,LF}^N - H_{13}^N - H_{15,LF}^N
\end{aligned}
\right\}
\tag{7.10}
$$

Table 7.8. Characteristics of reservoirs and fluxes of nitrogen in the biosphere (Figure 7.13).

Reservoirs (Gt) *and fluxes* (10^6 t yr^{-1})	*Identifier*	*Estimate*
Nitrogen supplies		
atmosphere	N_A	39×10^5
soil	N_S	280
photic and intermediate layer of the ocean	$N_U + N_P$	2,800
deep and bottom layer of the ocean	$N_L + N_F$	36,400
Natural sources of the hydrosphere	H_1^N	0.392
Technogenic accumulation		
fuel burning	H_2^N	22.8
fertilizer production	H_9^N	41.8
Input due to dead organisms		
on land	H_3^N	42.2
in upper layers of the oceans	H_{18}^N	5
in deep layers of the oceans	H_{12}^N	7.8
Input due to organisms functioning		
on land	H_5^N	0.1
in the oceans	H_4^N	0.3
Biological fixation		
on land	H_6^N	20.3
in the oceans	H_{17}^N	10
in the atmosphere	H_{10}^N	40
Denitrification		
on land	H_7^N	52
in the oceans	H_{20}^N	49.8
Atmospheric fixation		
over land	H_8^N	4
over the oceans	H_{16}^N	3.6
Runoff from land into the oceans	H_{11}^N	38.6
Precipitation	H_{13}^N	0.5
Vertical exchange processes in the oceans		
descending	H_{14}^N	0.2
lifting	H_{15}^N	7.5
Anthropogenic emissions to the atmosphere	H_{19}^N	15
Removal of nitrogen from the cycle due to sedimentation	H_{21}^N	0.2
Input of nitrogen to the atmosphere during rock weathering	H_{22}^N	0.217
Input of nitrogen to the water medium with dissolving sediments	H_{23}^N	0.091

where $V(V_\varphi, V_\lambda)$ is the wind speed; $v(v_\varphi, v_\lambda)$ is the current velocity in the ocean; and Q_L and Q_F are functions describing the mixing of the deep waters of the ocean.

To simplify the calculation scheme shown in Figure 7.13, advection processes in the equations can be described by superposition of fluxes H_{14}^N and H_{15}^N. The computer realization of the equations of the nitrogen unit introduces into its equations some corrections for the agreement between the dimensionalities of the variables in conformity with the spatial digitization of Ω. Therefore, the estimates of fluxes H_i^N given in Table 7.8 should be corrected according to this criterion.

7.3.3.3 Phosphorus, oxygen, and ozone units

Biogeochemical cycles are characterized by a high level of interactivity with other environmental processes. The global cycles of phosphorus, oxygen, and ozone have been described in detail by Kondratyev *et al.* (2004a). In contrast to nitrogen, the main reservoir of phosphorus in the biosphere is not the atmosphere but rather the rocks and other deposits formed in past geological epochs, which, being subject to erosion, release phosphates. There are other mechanisms by which phosphorus is returned to the biospheric cycle, but, as a rule, they are not very efficient. One of these mechanisms is fish harvesting, which returns about $60 \cdot 10^3$ tP yr^{-1} to land from the hydrosphere, as well as the extraction of phosphorus-containing rocks at an estimated rate of $1\text{–}2 \times 10^6$ tP yr^{-1}. The present cycle of phosphorus terminates by its fluxes to bottom deposits in the oceans, where it combines with sewage, or it may become involved as well with coast and river runoff.

The oxygen cycle in nature is composed of characteristic biogeochemical transitions among the various reservoirs of basic constituents circulating in the biosphere. The block scheme of oxygen exchange is similar, therefore, to those of sulfur, nitrogen, carbon, and phosphorus. However, oxygen occurs in various constituents which are spread very widely all over the globe; this makes it one of the most substantial components of biogeochemical cycles. The proportion of oxygen in the Earth's crust, including the hydrosphere, is about 49% by mass. The lithosphere (without considering the ocean and atmosphere) contains 47.2% oxygen, and pure water contains 88.89% oxygen. Oxygen constitutes 85.82% of ocean water, and marine biota account for 65% of oxygen by mass. These estimates testify to the dominant significance of oxygen in the biosphere, whose very appearance and existence are determined by oxygen. Presently, about 39×10^{14} tO_2 circulate in the biosphere.

Oxygen is present in the biosphere in the form of molecular oxygen (O_2), ozone (O_3), atomic oxygen (O), and as a constituent of various oxides. On one hand, oxygen maintains life on Earth through the process of respiration and formation of the ozone layer, and yet oxygen is itself the product of organismic functioning. This fact confounds any simple description of an oxygen cycle, since it requires a synthesis of the descriptions of various processes. A detailed description of the model of the global oxygen cycle (MGOC) was made by Kondratyev *et al.* (2003b).

Many authors believe that in the imminent future nothing threatens the stability of the global biogeochemical cycle of oxygen. Such a statement is not valid for ozone. In relation to atmospheric oxygen and oxygen bound in oxides, etc. in the ground, the mass of the ozone in the atmosphere is negligible. However, the importance of this ozone—for human beings and many other lifeforms—is out of all proportion to its mass. This is because of its role in blocking a large proportion of the solar ultraviolet radiation. The depletion of stratospheric ozone in recent years and the spectacular reduction in the Antarctic spring (the "ozone hole") are well known. According to Kondratyev and Varotsos (2000), available observations of the vertical profile of atmospheric ozone show a very complicated spatio-temporal variability that depends on many characteristics of the nature–society system. The MGOC unit as a parameterization of ozone fluxes follows a numerical model by Aloyan (2004), with a necessary correlation taken into account. This correction consists of the substitution of certain functional dependencies for scenarios reflecting the dynamics of change in concentrations of chemicals not described in the global model of the carbon cycle.

7.3.4 The oceans' bioproductivity unit

Ocean ecosystems are represented by three trophic structures characterizing (1) tropical pelagic zones (long trophic chains), (2) tropical latitude shelf zones and mid-latitude aquatic zones (medium-length trophic chains), and (3) Arctic latitudes (short trophic chains). In each of these structures the water column is considered as a single biogeocenosis. The major factor ensuring this unity is the flow of organic matter, which is produced in the surface layers and subsequently reaches maximum depths.

The functioning of the trophic pyramid is characterized by the consumption intensity for the sth food variety at the ith level:

$$C_{is} = k_{is}\overline{B}_s \Big/ \sum_{j \in S_i} k_{ij}\overline{B}_j, \tag{7.11}$$

where $\overline{B}_j$ is the effective biomass of the sth level; S_i is the food spectrum of the ith level; and k_{is} is the Ivlev coefficient used in the formula for the ith component ration:

$$R_i = k_i\left[1 - \exp\left(-\sum_{j \in S_i} k_{ij}\overline{B}_j\right)\right]. \tag{7.12}$$

The equations used in describing the bioproduction process in the water column have the following forms:

$$\left.\begin{aligned}
\frac{\partial p}{\partial t} &= P_p - \tau_p - M_p - \beta \frac{\partial p}{\partial z} + \frac{\partial (A\, \partial p/\partial z)}{\partial z} \\
&\quad - \sum_{i \in \Gamma_p} k_{ip} \bar{p} R_Z^i \Big/ \left(k_{ip}\bar{p} + k_{id}\bar{d} + \sum_{s \in S_i} k_{is}\overline{Z}_s \right) \\
\frac{\partial Z_i}{\partial t} &= (1 - h_Z^i) R_Z^i - \tau_Z^i Z_i - M_Z^i - \beta_Z^i \frac{\partial Z_i}{\partial z} \\
&\quad - \sum_{j \in \Gamma_i} k_{ji} \overline{Z}_i R_Z^i \Big/ \left(k_{jp}\bar{p} + k_{jd}\bar{d} + \sum_{s \in S_j} k_{js}\overline{Z}_s \right) \\
\frac{\partial d}{\partial t} + \beta \frac{\partial d}{\partial z} + A \frac{\partial^2 d}{\partial z^2} &= M_p + \sum_{i=1}^{m} M_Z^i - \mu_d + \sum_{i=1}^{m} h_Z^i R_Z^i \\
\frac{\partial n}{\partial t} + \beta \frac{\partial n}{\partial z} + A \frac{\partial^2 n}{\partial z^2} &= \lambda_0 d - \delta P_p + \rho \sum_{i=1}^{m} \tau_Z^i Z_i
\end{aligned}\right\} \tag{7.13}$$

where $M_\omega^i = \mu_\omega \max\{0, \omega_i - \omega_{i,\min}\}^{r_Z^i}$ is the mortality velocity of element $\omega(p, Z_1, \ldots, Z_m)$; p is the phytoplankton biomass; Z_i $(i = 1, \ldots, m)$ is the biomass of the ith component of zooplankton; d and n are the concentrations of detritus and nutrients, respectively; τ_ω is the index of energy inputs of component ω; A is the turbulent diffusion coefficient; β_Z^i is the mobility index of the ith component of zooplankton in vertical migrations; and β is the upwelling velocity of the water.

7.3.5 Units of biogeocenotic, hydrologic, and climatic processes

As shown in Figures 7.11 and 7.14, the GMNSS comprises 30 models (or fewer) for soil–plant formations. In synthesizing these models, use was made of results obtained by Friend (1998), Holmberg *et al.* (2000), Krapivin and Kondratyev (2002), Papakyriakou and McCaughey (1991), Peng (2000), Wirtz (2000), Yokozawa (1998). All of these are based on the equation for the balance of the biomass $X(t, \varphi, \lambda)$:

$$\partial X/\partial t = \xi - \omega_X - \tau - \Sigma, \tag{7.14}$$

where ξ is the actual plant productivity; ω_X and τ are the quantities of mortality and the outlays for energy exchange with the environment; and Σ are biomass losses due to anthropogenic reasons. These functions are described in detail by many authors (including those cited above, for instance). In the Global Simulation Model (GSM) the value of ξ is approximated as follows:

$$\xi = \delta_c \delta_o (1 + \alpha_T \cdot \Delta T/100) \exp(-\beta_1/X) \cdot \min\{\delta_e, \delta_Z, \delta_W, \delta_N, \delta_S, \delta_P\}, \tag{7.15}$$

where α_T and β_1 are indices corresponding to the dependence of production on temperature and biomass, respectively; δ_e is the index of production limitation by

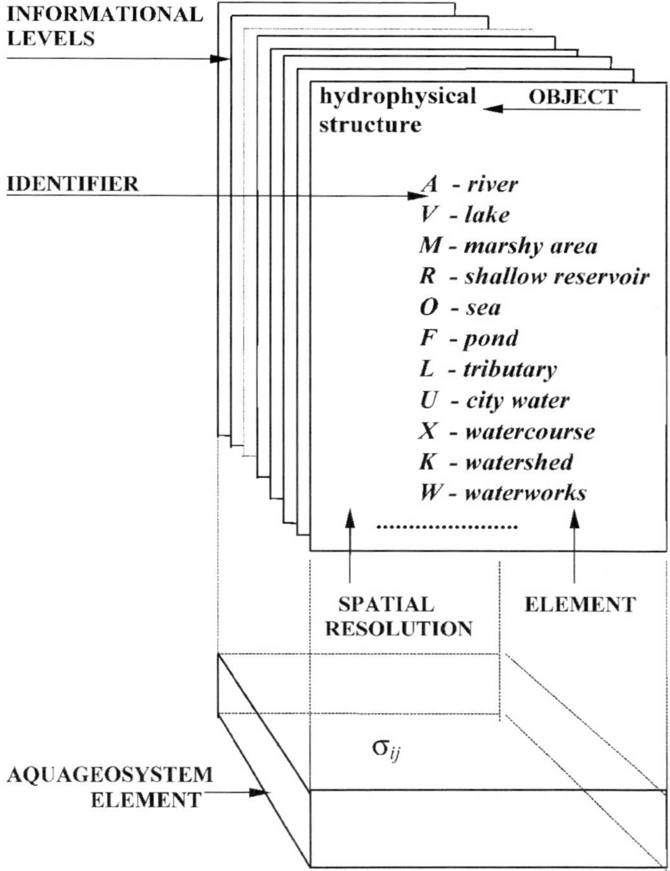

Figure 7.14. Cartographic identification and formation scheme of the GMNSS database.

the θ factor (e = illumination, Z = pollution, W = soil moisture, N, S and P are the nitrogen, sulfur, and phosphorus concentrations in the soil, respectively).

Formula (7.3) was chosen after performing numerous computational experiments taking into account various options for the limiting factor dependence of plant productivity. The δ_θ functions actually used were calculated based on data published in the literature. Thus, the role played by C_A in photosynthesis is described by the relation $\delta_c = bC_A/(C_A + C_{0.5})$, where $C_{0.5}$ is the CO_2 concentration for which $\delta_c = b/2$. The influence of the solar radiation intensity $e(t, \varphi, \lambda)$ on photosynthesis is parameterized by the relation $\delta_e = \delta^* \exp(1 - \delta^*)$, where $\delta^* = e/e^*, e^*$ is optimal illuminance. In the soil–plant formations unit for which the maximum photosynthesis value d_1 and the initial slope of the photosynthesis curve m_1 are known, use was made of the relation $\delta_e = d_1 e/(d_1/m_1 + e)$. The limiting of photosynthesis by pollution is defined by the exponential dependence $\delta_Z = \exp(-fZ)$, where f is a constant. The effect of soil moisture on photosynthesis is expressed by the function $\delta_W = 1 - \exp(-gW)$, where g is a constant. The biogenic element dependence of plant production is represented in the form of $\delta_\theta = \theta/(\theta + \theta_A)$, where θ_A is the θ element concentration in soil for which $\delta_\theta = 0.5$.

Water is responsible for channels of interrelations between natural systems in the biosphere. The water cycle in the biosphere includes the exchange of water in its various phase states between the hydrosphere, atmosphere, and living organisms. The reserves of water in their various forms are described in the literature at great length; therefore, there is every possibility of constructing a mathematical model of the global water cycle. Such a model was suggested in the study by Krapivin *et al.* (1982) in conformity with the diagram of the water balance which is described in that paper. In this version of the model, atmospheric water circulation is simulated by a simplified diagram of stable transports. In reality the process of atmospheric circulation is far more complex in space and time. It is characterized only roughly by alternation of zonal and meridian motions.

Satellite systems for measuring environmental parameters allow rapid acquisition of data pertaining to water content in various biospheric reservoirs, and, in particular, of atmospheric moisture content. This information may be obtained simultaneously with synoptic data on the temperature, velocity, and direction of wind, atmospheric pressure, content of pollutants in the atmosphere, and the ground surface heat flux balance. Such measurements and published data on biospheric water distribution have made it possible to construct a flow chart of global water balance. These flow charts are based on balance equations. The form in which the latter are written is illustrated by the following example:

$$dW_{iH}/dt = W_{SiH} - W_{iHG} - \sum_k W^k_{HiO} \quad (i = 1, \ldots, n), \tag{7.16}$$

where W_{iH} is the level of underground waters; W_{SiH} is infiltration; W_{iHG} is irrigation waters; W_{HiO} is runoff to the oceans; and n is the number of land regions.

The precipitation formation regime is represented by a threshold algorithm: namely, rainfall for $T > T^*$, snowfall for $T < T^*$, and snow thawing for $T > T_W$, where $T^* = 0°\text{C}$, and $T_W = 5.5°\text{C}$.

The most important climatic factor responsible for the character of human activity in the various regions appears to be atmospheric temperature. A change in atmospheric temperature leads to changes in the intensity of biological processes on land and causes disturbances in biogeochemical cycles (Chen *et al.*, 2000; Power, 2000).

Atmospheric temperature is formed as a function of carbon dioxide C_A and water vapor W_A content in the near-surface layer of the Earth: $T = T_D + \Delta T(C_A, W_A)$, where T_D is temperature in the pre-industrial period. Estimations of ΔT are calculated by $\Delta T = \Delta T_C + \Delta T_W$, where ΔT_C and ΔT_W reflect changes in T caused by fluctuations of the CO_2 and W_A concentrations in the atmosphere, respectively. The value of ΔT_C has two components: $\Delta T_C = 0.5(\Delta T_{C_1} + \Delta T_{C_2})$. The values of ΔT_{C_1} and ΔT_W are calculated using the climate model. The spatial distributions of ΔT_{C_1} and ΔT_W are calculated as functions of C_A and W_A, respectively. The value of ΔT_{C_2} is calculated using the equation:

$$\Delta T_{CO_2} = \begin{cases} L_1 & \text{when } X \geq 1 \\ L_2 & \text{when } X < 1, \end{cases} \tag{7.17}$$

where $X = C_A(t)/C_A(1900)$; $L_1 = -0.847 + 4.528 \ln X - 1.25 \exp\{-0.82(X - 1)\}$; and $L_2 = -2.63X^2 + 6.27X + 1.509 \ln X - 3.988$.

7.3.6 Demographic unit

The effect of numerous environmental and social factors on population dynamics in the ith region G_i is reflected in the birth rate R_{Gi} and death rate M_{Gi}:

$$dG_i/dt = R_{Gi} - M_{Gi} \quad (i = 1, \ldots, K). \tag{7.18}$$

In each of the K regions, the birth and death rates depend on food supply and quality, environmental contamination, gas composition of the atmosphere, the standard of living, power resource sufficiency, and population density as follows:

$$\begin{aligned} R_{Gi} &= (1 - h_{Gi})K_{Gi}G_i \min\{H_{GVi}, H_{GGi}, H_{GMi}, H_{GAi}\}, \\ M_{Gi} &= \mu_{Gi}G_i \max\{H_{\mu Ai}, H_{\mu Gi}, H_{\mu Vi}\}, \end{aligned} \tag{7.19}$$

where h_{Gi} is the quality coefficient representing the lack of nutrition in the food consumed by the population ("food inassimilability"); $H_{Gi} = 0$ for ideal conditions; K_{Gi} and μ_{Gi} are the constants of birth rate and death rate, respectively; and the functions

$$H_{GVi}(H_{\mu Vi}), \quad H_{GGi}(H_{\mu Gi}), \quad H_{GMi}(H_{\mu Mi}), \quad H_{GAi}(H_{\mu Ai}) \tag{7.20}$$

indicate the effect on birth rate (death rate) of various factors (namely, food supply, population density, standard of living, and environmental quality, respectively). Functional descriptions of these factors are related to the effects in human ecology. Thus, function H_{GVi} is represented in the form $H_{GVi} = 1 - \exp(-V_{Gi})$, where V_{Gi} is the effective food amount determined as a weighted sum of components in the *Homo sapiens* food spectrum:

$$\begin{aligned} V_{Gi} = {} & K_{Gpi}p + K_{GFi}\left(F_i + \sum_{j\neq i} a_{Fji}F_j\right) + K_{Gri}I_i(1 - \theta_{Fri} - \theta_{Uri}) \\ & + K_{GXi}\left[(1 - \theta_{FXi})X_i + (1 - \nu_{FXi})\sum_{j\neq i} a_{Xji}X_j\right], \end{aligned} \tag{7.21}$$

where coefficients K_{Gpi}, K_{GFi}, K_{Gri}, and K_{GXi} are defined following the method of Krapivin (1996); a_{Fji} and a_{Xij} are the portions of animal food and plant origin, respectively, for the ith region; θ_{FXi} and ν_{FXi} are the portions of food of plant origin used for cattle feeding produced in and imported to the ith region, respectively; and θ_{Fri} and θ_{Uri} are the portions of fishery I_i allotted for cattle feeding and fertilizer production, respectively.

It is assumed that with an increase in food supply, the population death rate diminishes at a rate ρ_{Gi} to a certain level determined by the constant $\rho_{\mu i}$ so that $H_{\mu Hi} = \rho_{\mu i} + \rho_{Gi}/F_{RGi}$, where $F_{RGi} = V_{Gi}/G_i$.

Similarly, it is assumed that birth rate dependency on the standard of living M_{SGi} is described by a saturating function, so that the birth rate is maximum for low M_{SGi} values and falls as a_{Gi} increases, down to a certain level a_{GMi}. The rate at which the transition between the maximum and minimum birth rate levels occurs is denoted by α_G and defined by the equation $H_{GMi} = a_{Gi} + a_{GMi} \exp(-\alpha_G M_{SGi})$.

The dependency of birth rate on population density is approximated by the relationship

$$H_{GGi} = g' + g^* \exp(-g'' G_i). \tag{7.22}$$

In general, the demographic unit has a branching structure permitting the use of different parameters describing population dynamics within the framework of a concrete computational experiment. The option used in the GMNSS is described in detail by Krapivin (1978) and Krapivin *et al.* (1982). The demographic unit includes a matrix model comprising three population age groups (0–14, 15–64, and 64 and older) and a group of disabled people. The unit structure also permits the use of different scenarios in describing both the population dynamics as a whole and its parts.

7.4 GLOBAL SIMULATION EXPERIMENTS

One of the special features of the present global ecodynamics is an intensified development of science and technology, and for this reason, the environment and human society are changing at a higher rate than they were 50 years ago, and the role of interdisciplinary studies of NSS with the use of accumulated knowledge in mathematics, ecology, sociology, medicine, chemistry, geophysics, etc. has grown. The global simulation experiment based on the use of the latest computer technologies has become a precursor of decision-making strategies in the field of natural resource management and, likewise, the conception and understanding of the experiment have changed as the process has evolved with real nature observations. Of course, problems arise in assessing the adequacy of the results of simulation experiments, and ecoinformatics solves these problems (Kondratyev *et al.*, 2002). The late 20th to early 21st centuries are characterized by a new mentality: the transition to experiments with models but not with natural components themselves. How, for instance, is it possible to carry out a field experiment with such unique systems as oceans, continental ecosystems, and the biosphere on the whole? Such experiments are obviously very dangerous and should be excluded.

The GMNSS makes it possible to realize a wide spectrum of simulation experiments within a multi-dimensional phase window of the NSS and to predict the development of NSS for this century. More long-term deeper predictions are not likely possible at the present level of knowledge and global databases. The apparent uncertainties have been analyzed, and they are closely connected with climate modeling. Unfortunately, there is no universal model, so far, which would reliably simulate the functioning of the Earth's climate system, reflecting the interaction of the atmosphere, oceans, land, and human society. Therefore, to use the GMNSS, it is necessary to make some *a priori* assumptions:

- Up to the the year 2100, anthropogenic emissions of CO_2 are described by the model of Demirchian *et al.* (2002), and emissions of other greenhouse gases until 2050 follow the pattern of 2000, after which they decrease by 25%.
- Extraction and consumption of non-renewable fossil fuels in the period up to 2050 grow by 0.5%/yr in developed countries and by 0.7%/yr in developing countries, after which time they become stabilized.
- The per capita gross domestic product increases in developed and developing countries at constant rates of 2.5% and 1.5% per annum, respectively.
- The level of medical service in developed and developing countries reduces mortality by 90% and 50%, respectively, due to environmental changes, and in 2020 a remedy for AIDS will be found.
- The atmospheric temperature regime is formed in accordance with the model of Mintzer (1987).
- Agricultural investments depend on providing the population with food (Krapivin *et al.*, 1982).
- The involvement of all countries with an effort against environmental pollution will grow by 0.2%/yr until 2050, after which it will increase no further.
- Processes of deforestation and replanting reach a balance in 2010 so that the total area of forests existing in 2000 is preserved, and the area of tropical forests decreases by no more than 5%.
- Oil pollution of the oceans in the 21st century follows the pattern of the last decade of the 20th century.
- Agricultural productivity grows steadily, reaching 80% in 2100.

Within the formulated assumptions, NSS dynamics are characterized by the following indicators:

- The ecodynamic pattern of NSS functioning as characterized by derivatives of the functions describing the state of NSS components is never equal to zero for time periods longer than the time step of the model.
- Population size by 2100 will reach 12.7 billion people, with the ratio between age groups characterized by the following average values: 0–15 years 32%; 16–65 years 53%, and >65 years 15%, and the disabled population will constitute 17%.
- Provision of the population with protein food will increase by 6.8% by the year 2030 (with the oceans contributing 12.1%) and then decrease by 3.2% with respect to 2000 (the share of the oceans increases up to 22.1%).
- The CO_2 concentration in the atmosphere will reach 486 ppm with a corresponding increase of 0.87°C in average planetary temperature.
- The dynamics of spatial distribution of atmospheric CO_2 sinks are characterized by the growing role of the oceans and land ecosystems, whose proportions in the

middle of the 21st century will constitute 31% and 19%, respectively. Then, by the end of the century, the proportion of the oceans will decrease to 26.7%, with the role of land ecosystems increasing up to 24.4%.

- The conversion of 5% of the tropical forest area to urban ecosystems will lead to a 1.2% reduction in total CO_2 sink.
- A complete succession of wet tropical forests to grass ecosystem by 2050 will later take its toll on global climate change due to interference with the regional moisture cycle on the "atmosphere–land" interface; increase in outgoing longwave radiation flux from deforested territory by 4.3 $W\,m^{-2}$; increase of this territory albedo by 5.7%; decrease of the absorbed shortwave radiation flux by 4.1%; increase of soil temperature by 1.3°C; reduction of rains and evaporation by 0.7 $mm\,da^{-1}$ and 0.9 $mm\,da^{-1}$, respectively.

For educational purposes, it is of interest to consider hypothetical versions of the man–nature interaction. Considering only some of the situations which would occur according to hypotheses of future dynamics of the most important anthropogenic factors, it can be noted that, in contrast to many other models, the GMNSS calculates the size of the population as a function of the environmental parameters.

Calculations show that the mechanisms of demographic process formation depend strongly on the strategies of natural resource use. So, it would seem a 50% increase of food provision, which is possible with the introduction of efficient technologies in agriculture and the development of aquaculture systems, should result in a 14% decrease of population density. On the contrary, a 10% decrease in the rate of mineral resource consumption causes a 3% increase of population density. The)NSS is very sensitive to possible changes of the global structure of soil–plant formations. So, for instance, if by 2050 the forest areas decrease steadily by only 10% with respect to 2000, the atmospheric CO_2 concentration in 2100 will reach 611 ppm (i.e., it will almost double with respect to 2000). But if the forest areas increase by 15% by 2050, then in 2100 the partial pressure of CO_2 will constitute 475 ppm.

These results of modeling the NSS response to possible anthropogenic changes of its parameters show that the suggested method enables one to assess the dynamics of various components of this system depending on hypotheses of possible rates of these changes. An application of this method to reconstruct the integral dynamics of NSS by some of its parameters for the period 1970–2000, on the basis of data published by Houghton *et al.* (2001), Kondratyev and Losev (2002), and Watson *et al.* (2000), has shown that the error of prediction for 10 and 20 years does not exceed 15% and 25%, respectively. Hence, GIMS technology can be recommended as one of the constructive approaches to solving the problems of global ecodynamics formulated by Kondratyev (1999b). Simulation experiments with the use of GMNSS, using available reliable data on the prehistory of the NSS, enable one to reveal conditions which constrain the development of civilization and to find mechanisms for its sustainable development. For this, it is necessary to concentrate the efforts of international and national scientific programs in synthesis of the GMNSS database. Kofi Annan, U.N. Secretary-General, emphasized this at his press conference in Johannesburg on the last day of the World Forum (September 4, 2002). He pointed out that governments came to an agreement to join efforts on biodiversity preservation and solution of global problems in such spheres as water supply, energy production, medical services, and agriculture. Speaking about possible ways to solve the global problems of the present civilization, Kofi Annan was fully confident that there is a way "that will reduce poverty while preserving the environment." GIMS technology shows this way. But its realization requires the concentration of accumulated knowledge in the development of GMNSS by improving its units, and primarily finding out and formalizing the factors and dynamic strategies of the formation of social evolution.

7.5 CONCLUDING REMARKS

Lomborg (2001, 2004) is of course right in rejecting apocalyptic predictions of global ecodynamics based on the exaggerated fear of limited natural resources and the environmental state. Such opinions and assessments are confirmed in particular by the data compiled by Holdren (2003), which characterize both real and potential global energy resources. Energy units are expressed here (in the case of non-renewable energy sources) in terawatt-years, which is equivalent to 31 exa-J ($1\,\text{TW} = 1\,\text{TW-yr}\,\text{yr}^{-1} = 31.5\,\text{exa-J}\,\text{yr}^{-1}$). It should be added that in 2000 global energy consumption constituted about 15 TW or $15\,\text{TW-yr}\,\text{yr}^{-1}$ with an assumed increase up to $60\,\text{TW-yr}\,\text{yr}^{-1}$ by 2100.

Despite the optimistic data, the absence of long-range prospects for development of the present consumption society illustrated by global ecodynamics estimates (Kondratyev *et al.*, 2003a, 2004a) raises no doubts. Therefore, at the World Summit on Sustainable Development held in Johannesburg in 2002, the necessity was emphasized of accomplishing 10-year programs in order to achieve stable production and consumption, which included the following recommendations (Starke, 2004):

- Developed countries should undertake the leading role in the provision of stable production and consumption.
- These goals should be achieved on the basis of common but differentiated responsibility.
- The problem of stable production and consumption should play the key role.
- The young generation must take part in solution of the problem of sustainable development.
- The "polluter pays" principle should be practiced.
- Control of the complete cycle of product evolution should be considered, from their production to consumption, as well as waste, in order to raise production efficiency.
- Support should be given to politics that favor the output of ecologically acceptable products and rendering of ecologically adequate services.
- Develop more ecological and effective methods of energy provision and liquidate energy subsidies.

- Support free-will initiatives of industry aimed at raising social and ecological responsibility.
- Study and introduce a practice of ecologically pure production, especially in developing countries, as well as in small and medium-scale businesses.

Although the enumerated recommendations are rather declarative, still they are clearly oriented toward the necessity of changing the paradigm of socio-economic development (this mainly refers to developed countries) from a consumption society to a conservation society. A concrete analysis of the means of such development requires the participation of specialists in the field of social sciences. Some related opinions have been expressed in Section 7.1 in the context of Earth Charter problems (Corcoran, 2005). Therefore, the question of whether humans can change climate requires further study (Borisov, 2005).

Finally, preliminary conclusions drawn from the above are as follows:

- Existing climate models cannot be used to make decisions and assess the risk of accomplishing anthropogenic scenarios.
- The level of uncertainty of climate forecasts can be reduced by means of a broader consideration in global models of interactive bonds in the NSS and mechanisms of biotic regulation of the environment as well as improvement of the global monitoring system.
- The use of hydrocarbon energy sources in the 21st century will not lead to catastrophic climate change if the Earth cover is preserved and the oceans are protected from pollution.

7.6 REFERENCES

Adamenko V.N. and Kondratyev K.Ya. (1999). Global climate change and its empirical diagnostics. *Anthropogenic Impact on Northern Nature and Its Consequences*. Kola Center of the Russian Acad. Sci., Apatity, pp. 17–35 [in Russian].

Alexandrov G. and Oikawa T. (2002). TsuBiMo: A biosphere model of the CO_2–fertilization effect. *Climate Res.*, **19**, 265–270.

Alexandrov G.A., Yamagata Y., Saigusa N., and Oikawa T. (2005). Long-term carbon exchange at the Takayama, Japan Forest re-calibration TsuBiMo with eddy-covariance measurements at Takayama. *Agricultural and Forest Meteorology*, **134**(1/4), 135–142.

Aloyan A.E. (2004). Numerical modelling of minor gas constituents and aerosols in the atmosphere. *Ecological Modelling*, **179**, 163–175.

Bacastow R. (1981). Numerical evaluation of the evasion factor. *Carbon Cycle Modeling, SCOPE-16*. John Wiley & Sons, New York, pp. 95–101.

Barenbaum A.S. (2002). *Galaxy. Solar System. The Earth: Subordinate Processes and Evolution*. Geos, Moscow, 393 pp. [in Russian].

Barenbaum A.S. (2004). Mechanism for the formation of gas and oil accumulation. *Annals of Acad. Sci.*, **399**(6), 1–4 [in Russian].

Bartsev S.I., Degermendzhy A.G., and Erokhin D.V. (2003). Global generalized models of carbon dioxide dynamics. *Problems of the Environment and Natural Resources*, **12**, 11–28 [in Russian].

Bazilevich N.I. and Rodin L.E. (1967). The map-schemes of productivity and of the biological cycle of the most significant types of land vegetation. *Bull. of the All-Union Geographical Soc.*, **99**(3), 190–194 [in Russian].

Björkstrom A. (1979). A model of CO_2 interaction between atmosphere, ocean, and land biota. *Global Carbon Cycle, SCOPE-13*. John Willey & Sons, New York, pp. 403–458.

Bodenbender J., Wassmann R., Papen H., and Rennenberg H. (1999). Temporal and spatial variation of sulfur-gas-transfer between coastal marine sediments and the atmosphere. *Atmospheric Environment*, **33**(21), 3487–3502.

Borisov P.M. (2005). *Can Humankind Change Climate?* Science, Moscow, 270 pp. [in Russian].

Boysen M. (ed.) (2000). *Biennial Report 1998 & 1999*. Potsdam Institute for Climate Impact Research, Potsdam, Germany, 130 pp.

Canadel I.G., Dickinson R., Hibbard K., Raupach M., and Young O. (eds.) (2003). *Global Carbon Project: The Science Framework and Implementation*, Report No. 1. Earth System Science Partnership, Canberra, 69 pp.

Chen W., Chen J., and Cihlar J. (2000). An integrated terrestrial ecosystem carbon-budget model based on changes in disturbance, climate, and atmospheric chemistry. *Ecological Modelling*, **135**(1), 55–79

Collatz G.J., Bounoua L., Los S.O., Randall D.A., Fung I.Y., and Sellers P.J. (2000). A mechanism for the influence of vegetation on the response of the diurnal temperature range to changing climate. *Geophys. Res. Lett.*, **27**(20), 3381–3384.

Corcoran P.P. (ed.) (2005). *The Earth Charter in Action: Toward a Sustainable World.* Koninklijk Instituut voor de Tropen, Amsterdam, The Netherlands, 192 pp.

Degermendzhy A.G. (1987). Objective laws for combined species formation under the modeling of water ecosystems (by the example of Krasnoyarsk's reservoir). Doctoral dissertation in biophysics, Institute of Biophysics, Siberian Branch of Russian Academy of Sciences, Krasnoyarsk, 510 pp.

Degermendzhy A.G. and Bartsev S.I. (2003). Global small-scale models of dynamics and stability of the biosphere. *Problems of the Environment and Natural Resources*, **7**, 32–49 [in Russian].

Demirchian K.S. and Kondratyev K.Ya. (1999). Scientific validity of predicted impacts of energetics on climate. *Bull. of Russian Acad. Sci.: Energetics*, **6**, 3–46 [in Russian].

Demirchian K.S. and Kondratyev K.Ya. (2004). Global carbon cycle and climate. *Proc. of the Russian Geographical Society*, **136**(1), 16–25 [in Russian].

Demirchian K.S., Demirchian K.K., Danilevich Ya.B., and Kondratyev K.Ya. (2002). Global climate warming, energetics and geopolicy. *Annals of RAS: Energetics*, **3**, 221–235 [in Russian].

Dore S.E., Likas R., Sadler D.W., and Karl D.M. (2003). Climate-driven changes to the atmospheric CO_2 sink in the subtropical North Pacific Ocean. *Nature (U.K.)*, **424**(6950), 754–757.

Ehhalt D.H. (1981). Chemical coupling of the nitrogen, sulphur, and carbon cycles in the atmosphere. In: G.E. Likens (ed.), *Some Perspectives of the Major Biogeochemical Cycles*. Elsevier, Amsterdam, pp. 81–91.

EPA (2001). *Non-CO_2 Greenhouse Gas Emissions from Developed Countries: 1990–2010*, EPA-430-R-01-007. U.S. Environmental Protection Agency, Washington, D.C., 79 pp.

Friend A.D. (1998) Parametrization of a global daily weather generator for terrestrial ecosystem modelling. *Ecological Modelling*, **109**(2), 121–140

Gorshkov V.G., Gorshkov V.V., and Makarieva A.M. (2000). *Biotic Regulation of the Environment: Key Issues of Global Change*. Springer/Praxis. Chichester, U.K., 367 pp.

Hales B., Takahashi T., and Bandstra L. (2005). Atmospheric CO_2 uptake by a coastal upwelling system. *Global Biogeochemical Cycles*, **19**(GB1009), doi:10.1029/2004GB002295, 1–11.

Hasegawa Y. and Kasagi N. (2001). The effect of Schmidt number on air–water interface mass transfer. *Proceedings of the Fourth International Conference on Multiphase Flow, New Orleans, May 27–June 1, 2001, University of Nottingham, New Orleans, Louisiana*, pp. 296–292.

Hasegawa Y. and Kasagi N. (2005). Turbulent mass transfer mechanism across a contaminated air–water interface. *Proceeding of the Fourth International Symposium on Turbulence and Shear Flow Phenomena (TSFP-4), Williamsburg, Virginia, June 27–29, 2005*, pp. 971–976.

Holdren J.P. (2003). Environmental change and human condition. *Bull. Amer. Acad. Arts. Sci., New York*, **57**(1), 25–31.

Holmberg M., Rankinen K., Johansson M., Forsius M., Kleemola S., Ahonen J., and Syri S. (2000). Sensitivity of soil acidification model to deposition and forest growth. *Ecological Modelling*, **135**(2/3), 311–325

Houghton J.T., Ding Y., Griggs D.J., Noguer M., Van der Linden P.J., Dai X., Masskell K., and Johnson C.A. (2001). *Climate Change 2001: The Scientific Basis* (contribution of WG1 to the third assessment report of the IPCC). Cambridge University Press, Cambridge, U.K., 881 pp.

IPCC (2007). *Climate Change 2007: The Physical Science Basis*. WMO/UNEP, Geneva, Switzerland, 18 pp.

Kiehl J.T. and Gent P.R. (2004). The Community Climate System Model, version 2. *J. Climate*, **17**, 3666–3682.

Kondratyev K.Ya. (1990). *Key Problems of Global Ecology*. ARISTI, Moscow, 454 pp. [in Russian].

Kondratyev K.Ya. (1992). *Global Climate*. Science, St. Petersburg, 359 pp. [in Russian].

Kondratyev K.Ya. (1998). *Multidimensional Global Change*. Wiley/Praxis, Chichester, U.K., 761 pp.

Kondratyev K.Ya. (1999a). *Climatic Effects of Aerosols and Clouds*. Springer/Praxis, Chichester, U.K., 264 pp.

Kondratyev K.Ya. (1999b). *Ecodynamics and Geopolitics, Vol. 1: Global Problems*. St. Petersburg State University, St. Petersburg, 1,040 pp. [in Russian].

Kondratyev K.Ya. (2000a). Earth researches from space: Scientific plane of the EOS system. *Earth Research from Space (Moscow)*, **3**, 82–91 [in Russian].

Kondratyev K.Ya. (2000b). Global changes on the verge of two millennia. *Herald of Russian Academy of Sciences*, **70**(9), 788–796 [in Russian].

Kondratyev K.Ya. (2002). Global climate change: Reality, hypotheses, and fiction. *Research of the Earth from Space*, **1**, 3–23 [in Russian].

Kondratyev K.Ya. (2004a). Global climate change: Observational data and numerical modeling results. *Research of the Earth from Space*, **1**, 3–25 [in Russian].

Kondratyev K.Ya. (2004b). Global climate change: Unsolved problems. *Meteorology and Hydrology*, **4**, 93–119 [in Russian].

Kondratyev K.Ya. and Krapivin V.F. (2001). Global dynamics of basic land ecosystems. *Research of the Earth from Space*, **4**, 3–12 [in Russian].

Kondratyev K.Ya. and Krapivin V.F. (2003). Global change: Real and possible in the future. *Research of the Earth from Space*, **4**, 1–10 [in Russian].

Kondratyev K.Ya. and Krapivin V.F. (2004). *Modeling the Global Carbon Cycle*. Physics-Mathematics, Moscow, 335 pp. [in Russian].

Kondratyev K.Ya. and Losev K.S. (2002). Present problems of global civilization development and its possible perspectives. *Earth Research from Space*, **2**, 3–23 [in Russian].

Kondratyev K.Ya. and Varotsos C.A. (2000). *Atmospheric Ozone Variability: Implications for Climate Change, Human Health, and Ecosystems*. Springer/Praxis, Chichester, U.K., 758 pp.

Kondratyev K.Ya., Krapivin V.F., and Pshenin E.S. (2000). Concept of the regional geoinformation monitoring. *Earth Research from Space (Moscow)*, **6**, 3–10 [in Russian].

Kondratyev K.Ya., Krapivin V.F., and Phillips G.W. (2002). *Global Environmental Change: Modelling and Monitoring*. Springer-Verlag, Heidelberg, 319 pp.

Kondratyev K.Ya., Krapivin V.F., and Savinykh V.P. (2003a). *Perspectives of Civilization Development: Multidimensional Analysis*. Logos, Moscow, 574 pp. [in Russian].

Kondratyev K.Ya., Krapivin V.F., and Varotsos C.A. (2003b). *Global Carbon Cycle and Climate Change*. Springer/Praxis, Chichester, U.K., 343 pp.

Kondratyev K.Ya., Losev K.S., Ananicheva M.D., and Chesnokova I.V. (2003c). Price of Russian ecological service. *Herald of Russian Academy of Sci.*, **73**(1), 3–10 [in Russian].

Kondratyev K.Ya., Krapivin V.F., Varotsos C.A., and Savinikh V.P. (2004a). *Global Ecodynamics: A Multidimensional Analysis*. Springer/Praxis, Chichester, U.K., 649 pp.

Kondratyev K.Ya., Losev K.S., Ananicheva M.D., and Chesnokova I.V. (2004b). *Stability of Life on Earth*. Springer/Praxis, Chichester, U.K., 152 pp.

Kondratyev K.Ya., Ivlev L.S., Krapivin V.F., and Varotsos C.A. (2006a). *Atmospheric Aerosol Properties: Formation, Processes and Impacts*. Springer/Praxis, Chichester, U.K., 572 pp.

Kondratyev K.Ya., Krapivin V.F., and Varotsos C.A. (2006b). *Natural Disasters as Interactive Components of Global Ecodynamics*. Springer/Praxis, Chichester, U.K., 620 pp.

Korgenevsky A.V., Krapivin V.F., and Cherepenin V.A. (1989). Modeling the global processes of the magnetosphere. In: E.P. Novitchikhin (ed.), *Methods of Informatics in Radio-physical Investigations of Environment*. Nauka, Moscow, pp. 25–43 [in Russian].

Krapivin V.F. (1978). *On the Theory of Complex System Survivability*. Science, Moscow, 248 pp. [in Russian].

Krapivin V.F. (1993). Mathematical model for global ecological investigations. *Ecological Modelling*, **67**(2/4), 103–127.

Krapivin V.F. (1996). The estimation of the Peruvian current ecosystem by a mathematical model of biosphere. *Ecological Modelling*, **91**(1), 1–14.

Krapivin V.F. (2000a). Biospheric balance of oxygen and its modeling. *Problems of Environment and Natural Resources*, **10**, 15–20 [in Russian].

Krapivin V.F. (2000b). The model of global nitrogen cycle. *Problems of Environment and Natural Resources*, **10**, 3–15 [in Russian].

Krapivin V.F. (2000c). Simulation model of biogeochemical cycle of phosphorus in the biosphere. *Problems of Environment and Natural Resources*, **10**, 26–30 [in Russian].

Krapivin, V.F. and Chukhlantsev A.A. (2004). Remote UHF radiometric sounding of soil and vegetation in the context of global environmental change. *Ecological Systems and Devices*, **9**, 37–45 [in Russian].

Krapivin V.F. and Kondratyev K.Ya. (2002). *Global Environmental Change: Ecoinformatics*. St. Petersburg State University, St. Petersburg, 724 pp. [in Russian].

Krapivin V.F. and Nazaryan N.A. (1997). Mathematical model for investigations of the global sulphur cycle. *Mathematical Modeling (Moscow)*, **9**(8), 36–50 [in Russian].

Krapivin V.F. and Potapov I.I. (2006). Monitoring of the chemical element cycles in the environment. *Problems of the Environment and Natural Resources*, **12**, 3–16.

Krapivin V.F., Svirezhev Yu.M., and Tarko A.M. (1982). *Mathematical Modeling of Global Biospheric Processes*. Science, Moscow, 272 pp. [in Russian].

Logofet D.O. (2002). Matrix population models: Construction, analysis and interpretation. *Ecological Modelling*, **148**(3), 307–310.

Lomborg B. (2001). *The Sceptical Environmentalist: Measuring the Real State of the World*. Cambridge University Press, Cambridge, U.K., 496 pp.

Lomborg B. (ed.) (2004). *Global Crisis, Global Solutions*. Cambridge University Press, Cambridge, U.K., 670 pp.

Luecken D.J., Berkowitz C.M., and Easter R.C. (1991). Use of a three-dimensional cloud-chemistry model to study the transatlantic transport of soluble sulfur species. *J. Geophys. Res. D.*, **96**(12), 22477–22490

Mintzer I.M. (1987). *A Matter of Degrees: The Potential for Controlling the Greenhouse Effect*. World Resources Institute Research Report No. 15, Washington, D.C., 70 pp.

Nishida K., Nemani R.R., Glassy J.M., and Running S.W. (2003). Development of an evapotranspiration index from Aqua/MODIS for monitoring surface moisture status. *IEEE Trans on Geosci. and Remote Sensing*, **41**(2), 493–501.

Nitu C., Krapivin V.F., and Bruno A. (2000). *Intelligent Techniques in Ecology*. Printech, Bucharest, 150 pp.

Nitu C., Krapivin V. F., and Pruteanu E. (2004). Ecoinformatics: Intelligent Systems in Ecology. Magic Print, Onesti, Bucharest, Rumania, 410 pp.

Our Changing Planet (2004). The U.S. Climate Change Science Program for Fiscal Years 2004 and 2005. DOE, Washington, D.C., 159 pp.

Papakyriakou T.N. and McCaughey J.H. (1991). An evaluation of evapotranspiration for a mixed forest. *Can. J. Forest. Res.*, **21**(11), 1622–1631

Park S.U., In H.J., and Lee Y.H. (1999). Parametrization of wet deposition of sulfate by precipitation rate. *Atmospheric Env.*, **33**(27), 4469–4475

Peng C. (2000). From static biogeographical model to dynamic global vegetation model: A global perspective on modelling vegetation dynamics. *Ecological Modelling*, **135**(1), 33–54.

Perrie W., Zhang W., Ren X., and Long Z. (2004). The role of midlatitude storms on air–sea exchange of CO_2. *Geophysical Research Letters*, **31**(L09306), doi:10.1029/2003GL019212, 1–4.

Power H.C. (2000). Estimating atmospheric turbidity from climate data. *Ecological Modelling*, **135**(1), 125–134

Rochon G.L., Krapivin V.F., Watson M., Fauria S., Tsang F.Y., and Fernandez M. (1996). Remote characterization of the landsea interface: A case study of the Viet-Nam/South China Sea coastal zone. *Proceedings of the Second HoChiMinh City Conference on Mechanics. HoChiMinh City, September 24–25, 1996*, pp. 72–74.

Ronner, U. (1983). *Biological Nitrogen Transformations in Marine Ecosystems with Emphasis on Denitrification*. Department of Marine Microbiology University of Goteborg, Goteborg, Sweden, 165 pp.

Rosen C. (2000). *People and Ecosystems: The Fraying Web of Life*. Elsevier, Washington, D.C., 250 pp.

Sellers P.J., Randall D.A., Collotz G.J., Berry J.A., Field C.B., Dazlich D.A., Zhang C., Collelo G.D., and Bounoua L. (1996). A revised land surface parameterization (SiB2) for atmospheric GCMs, Part 1: Model formulation. *J. of Climate*, **9**(4), 676–705.

Starke L. (ed.) (2004). *State of the World, 2004: Progress towards a Sustainable Society*. Earthscan, London, 246 pp.

Tarko A.M. (2001). Investigation of global biosphere processes with the aid of a global spatial carbon dioxide cycle model. *Sixth International Carbon Dioxide Conference, Tohoku University, Sendai, Japan*, Extended Abstract No. 2, pp. 899–902.

Tarko A.M. (2005). *Numerical Modeling of Anthropogenic Changes of Global Biospheric Processes*. Physics-Mathematics, Moscow, 278 pp. [in Russian].

Wange G. and Archer D.J. (2003). Evaporation of groundwater from arid playas measured by C-band SAR. *IEEE Trans. on Geosci. and Remote Sensing*, **41**(7), 1641–1650.

Watson R.T., Noble I.R., Bolin B., Ravindranath N.H., Verardo D.J, and Dokken D.J. (eds.). (2000). *Land Use, Land-use Change, and Forestry*. Cambridge University Press, Cambridge, U.K., 377 pp.

Wirtz K.W. (2000). Second order up-scaling: Theory and an exercise with a complex photosynthesis model. *Ecological Modelling*, **126**(1), 59–71.

Yasanov N.A. (2003). Climate of Phanerozoe and the greenhouse effect. *Herald of the Moscow State University, Ser. 4: Geology*, **6**, 3–11 [in Russian].

Yokozawa M. (1998). Effects of competition mode on spatial pattern dynamics in plant communities. *Ecological Modelling*, **106**(1), 1–16.

8

Self-learning statistical short-term climate predictive model for Europe

Oleg M. Pokrovsky

8.1 INTRODUCTION

Forecasting the weather from one month to one season ahead has become very important economically. A clear awareness of the scientific basis of long-term predictive skill began with the work of Walker and Bliss (1932) and Bjerknes (1969). Seasonal forecasts are possible whenever the chaotic atmospheric motion is perturbed in a predictable way by slowly varying boundary conditions, such as sea surface temperature (SST) or land conditions. The most important of these boundary conditions are the El Niño Southern Oscillation (ENSO) in the Pacific Ocean and the North Atlantic Oscillation (NAO) in the Atlantic Ocean. The El Niño Southern Oscillation is the strongest climate signal in inter-annual timescales (Rasmusson and Carpenter, 1982). It has quasi-periodic behavior with dominant periods of around 2–7 years. Many other similar features distributed around the world have been discovered in recent years. Although the weather is highly non-linear, perturbations to the average weather can often be taken as being proportional to the forcing plus an unpredictable weather noise. This means that simple, often linear, forecast models can be very useful in seasonal forecasting. In fact, statistical models based on the linear El Niño Southern Oscillation, North Atlantic Oscillation, and other teleconnections are used in many locations throughout the world (Peng and Whitaker, 1999; Wallace, and Gutzler, 1981; Wang, 2001).

The statistical climate may be considered as the statistical set of the daily weather for a given season over a specific geographical region. Thus, the timescale of seasonal climate goes far beyond the predictability that is defined in terms of sensitive dependence on initial atmospheric conditions. Instead, the predictability of seasonal climate is often connected with a forcing field such as sea surface temperature

(SST) or surface atmospheric pressure (SAP). The key to successful application of an empirical model lies in understanding the underlying physical mechanism for the relation between the predictor and the predicted fields. Unlike dynamical models that try to answer how a certain anomaly occurs by simulating the detailed processes that are necessary to produce the observed seasonal anomaly, statistical models directly try to determine the probability that a certain anomaly will occur over a specific geographical region under a known condition (in particular, the spacetime structure of a given forcing, such as the ENSO signal in the sea surface temperature field).

The effectiveness of empirical models, therefore, depends crucially on whether the relevant components (with respect to spatio-temporal scales) to be used as predictors are suitably incorporated in the prediction models, and whether the relationships between predictors and predictands are properly established. The above relationships may not necessarily be linear, particularly when mid and high latitudes are concerned. Nowadays, the use of a procedure, such as principal component analysis (Vautard *et al.*, 1999) or a rotated version of it, to extract coherent signals has enabled the compression of climatic variables into a few standard patterns.

The present chapter attempts to apply the idea of a non-linear model in low-dimensional phase space. In terms of a low-dimensional phase space approach, fuzzy patterns are used to determine non-linear metrics and the position of the initial time state as well as simulating state vectors with respect to the centers of action (e.g., centers of ocean–atmosphere interactions). Fuzzy set methodology is used to define an empirical rule to assign any current initial state to one of the atmospheric circulation regime sets determined by major low oscillations and represented in a few low-dimensional vector subsets. A specific meaning is assigned to the simulation trajectory in the phase space generated by a non-linear model to be close to the observing trajectory. A self-learning model is designed to approximate to the observing trajectory at the learning time interval and to evaluate deviations between model and measurement data at the verification stage.

The motivation for such an approach came from the following facts. First, a non-linear model might be efficient only in low-dimensional phase space because in high-dimensional space the model cannot be stable and an extremely long time is required to train such a model. Second, we have observed that the time evolution of a specific key sea surface temperature forcing region, relevant to season scales, can often be efficiently described in a low-dimensional phase space. The possibility for constructing a low-dimensional phase space from the predictor fields allows closer examination of their impact on the seasonal climate over a specific climate zone leading to the possibility of non-linear and more dynamically based statistical prediction in contrast to pure statistical approaches. Our preliminary application of this method to the prediction of the winter and spring surface air temperature (SAT) over Europe showed significant improvement in the skill score (Pokrovsky, 2006a). Third, the geographical distribution of the predictive skills, as well as their time behavior, varies from one key region of sea surface temperature forcing to another. Therefore, the final optimal prediction might be achieved through a linear or non-linear combination of the predicted results derived from different key forcing regions.

8.2 ATMOSPHERIC CIRCULATION IN THE ATLANTIC–EUROPEAN SYSTEM

Changes in stream flow patterns over Europe have serious consequences for a wide range of human activities in this densely populated region. An appearance of extreme events such as floods or droughts is caused in many cases by the persistence of some circulation type or types. The climate of the European–Atlantic sector exhibits considerable variability on a wide range of timescales. A substantial portion is associated with the North Atlantic Oscillation, a hemispheric meridional oscillation in atmospheric mass with centers of action near Iceland and over the sub-tropical Atlantic. The impact of the North Atlantic Oscillation on the winter climate extends from the East coast of the United States to Eurasia and from North Africa and the Middle East to the Arctic regions (Hurrell, 1995; James and James, 1989; Pokrovsky, 2006a, c; Trenberth *et al.*, 1998; Wallace and Thompson, 2002a, b). Analysis of proxy data of the North Atlantic Oscillation shows phases of enhanced (active) and reduced (passive) decadal variability (Appenzeller *et al.*, 1998).

The growing interest in the North Atlantic Oscillation is partly explained by the fact that the spatial signature of the observed climate warming over the last century (with a significant increase in the last three decades) resembles the surface temperature anomalies associated with the North Atlantic Oscillation (Pokrovsky, 2006b). The current climate trend could be partly explained by human activities and by the related increase of the concentrations of greenhouse gases (Raible *et al.*, 2001). The understanding of the mechanisms sustaining the North Atlantic Oscillation and their link to global climate change is thus crucial to the detection and identification of the signature of climate change (Halliwell, 1997; Pokrovsky, 2007; Raible and Blender, 2004). In addition, climate fluctuations on shorter timescales (from a week to a season) that are related to the North Atlantic Oscillation (in terms of temperature, precipitation anomalies, or preferred storm tracks) affect a large number of human activities, such as the management of energy and water resources, agriculture, and the fishing industry. Therefore, understanding the origin of the North Atlantic Oscillation and predicting its temporal fluctuations is necessary to meet social and economic demands.

One of our aims is to classify the atmospheric circulation patterns in the Atlantic–European sector and to reveal linkages between anomalies in the pressure field over the North Atlantic (e.g., the North Atlantic Oscillation) and the respective circulation pattern occurrence over the continent, on the one hand, and rain fields, on the other hand. Changes in atmospheric circulation over Europe during the past 50 years have been examined using both objective (modes of low-frequency variability by regression analysis and objective cluster classification of circulation types: fuzzy logic) and subjective (Hess–Brezowsky classification of weather types) methods. The Hess–Brezowsky catalog of large-scale circulation patterns (Gerstengarbe *et al.*, 1999; Hess and Brezowsky, 1952) recognizes three groups of circulations (zonal, half-meridional, and meridional) divided into 10 major types (Grosswettertypen, GWT) and 29 sub-types (Grosswetterlagen, GWL). Any circulation type (GWL) persists for several days (normally at least 3 days), which is the difference compared with the objective circulation types. For the description of individual GWLs see, for example, Gerstengarbe *et al.* (1999).

8.3 FORECASTING METHODOLOGY

Statistical analyses provide an empirical knowledge that can lead to more skillful forecasts in the absence of explicit physical understanding. Additionally, acquired information may provide guidance towards the identification of the physical process, contributing to or limiting the predictability. The choice to use an empirical approach reflects the fact that both simple and complex general circulation models (GCMs), either with prescribed boundary conditions or with actual oceanic coupling, currently do not adequately reproduce the processes of the real atmosphere in the mid and high latitudes for the lead times and averaging periods of concern here (Pokrovsky, 2004). It is not surprising that the seasonal skill score of GCMs may not be the best (Van den Dool, 1994). One of the main difficulties is actually to validate GCM forecasts since a large number of independent prediction cases (at least equal to the number of the independent grid variables in the model) is required to fully assess their skill. We hope and assume that eventually, with advances in physical understanding, dynamic prediction approaches will outperform statistical ones.

Prediction of the time-averaged surface climate has received considerable attention over the last two decades. First, the potentially predictable portion of the total variability of a given predictand has been empirically estimated using ratios of predictand variability at different frequencies (Trenberth, 1984). Second, direct attempts at forecasting and verification have been made using analog approaches (Barnston and Livezey, 1987), and linear statistical approaches with either several pre-selected predictor elements or whole predictor fields (Barnett, 1985).

It is known that the major contribution in the predictive skill for seasonal climate might be derived from sea surface temperature and surface air pressure fields (Barnett, 1985; Barnston, 1994; Barnston and Smith, 1996). However, it is difficult to embed grid field variability in low-dimensional space, because each key sea surface temperature forcing region has its own relative independent part of variability on the seasonal scale. Thus, to work in a low-dimensional space it is necessary to divide the oceans into a few key forcing regions, having impact on seasonal climate. The concept of phase space is based on dynamic system theory and associated time series analysis (Sauer *et al.*, 1991; Wallace *et al.*, 1993). As the original phase space is unknown, the first step is the reconstruction of phase space based on observed variables.

Empirical orthogonal functions (EOFs) and singular value decomposition (SVD) are the most commonly used techniques (Fraedrich and Wang, 1993; Vautard *et al.*, 1999) to build a phase space. EOFs are the eigenvectors of the covariance matrix obtained from calculating covariances of time series at different spatial points. EOFs are optimal in explaining as much total variance as possible with any specific number of spatial patterns. The first EOF explains most of the temporal variance in the dataset among all possible spatial fields. The subsequent EOFs are mutually orthogonal (in space and time) and successfully explain less variance. EOF analysis is non-local in that the loading values at two various spatial points in an EOF do not simply depend on the time series at those two points but depend on the whole dataset. This contrasts with the one-point correlation analyses used to define teleconnections, for which patterns can be interpreted locally.

The single-value decomposition approach provides the retrieval of both temporal and spatial modes, simultaneously (Cherry, 1996). The EOFs or single-value decomposition serve here only to prescribe a coordinate basis for phase space, containing all the observed states of the sea surface temperature or other fields. There is a difference in implication of these techniques. EOFs are used for the analysis of anomaly fields attributed to a given time index (e.g., a month or a season). The single-value decomposition might be applied to spatio-temporal observed fields distributed over some time window of the predictor (e.g., a few months, a season, or a year). Thus, any sea surface temperature or other predictor (predictand) field might be projected onto the m-dimensional phase space and represented as a state point in this vector space. When we use single-value decomposition phase space with a year time window, such a state point is equivalent to a coherent spacetime structure in physical space, describing the statistics for the past 12 months. However, leading EOF or single-value decomposition vectors usually tend to have the largest spatial scales, whereas fairly small anomalies may be predictable (Montroy, 1997).

A fuzzy set approach is more appropriate to approximate the temporal and spatial modes in low-dimensional phase space (Pokrovsky *et al.*, 2002). Certainly, there are atmosphere–ocean interactions generating a set of forward and feedback links. Some of them are non-linear and cannot be described by simplified statistical models based on linear regression. Therefore, a multivariate self-learning neural network model was developed to describe the predictive relationships between evolving large-scale patterns in northern hemisphere sea surface temperature, surface air pressure, and surface air temperature fields (predictors) and subsequent patterns in northern Europe surface air temperature and precipitation (predictands). A lead interval of varying length (from 1 to 6 months) is placed between a series of consecutive predictor periods and a single predictand period. Objective evaluation of the strength of such relationships is one of our primary aims.

The global monthly mean sea surface temperature, surface air temperature, and surface air pressure grid fields used in the present study were derived from the NCEP/NCAR (National Center for Environment Protection/National Center for Atmospheric Research) re-analysis data set. The original daily data were provided by NCEP and then averaged over monthly intervals. The dataset covers the period from January 1958 to December 1998. The re-analysis dataset includes all available satellite remote-sensing data updated after processing (Kondratyev *et al.*, 1996, 1997). The annual cycle and inter-annual linear trend were removed from predictor and predictand fields. The anomalies (departure from climate means) were used in all prediction model modifications. The data used were divided into training and verification sample sets. All calculations for subsequent model building were derived from the learning set only. The data contained in the verification set were used only for the evaluation of the predictive skill. It should be pointed out that the linear trend, calculated on each grid after the removal of the annual cycle, is related either to artificial factors (measurement errors) or to variability having a long timescale (equivalent to longer than a century), which is not relevant to the predictive problem considered here. The amplitude of the linear trend is very small. However, it may give rise to a trajectory shifting in phase space and thus affect the selection of the nearest fuzzy set activated in the non-linear model. Therefore, this filtering procedure might be considered as a necessary step in the present context.

Below we describe a numerical procedure which includes two major parts: fuzzy classification/declassification of atmospheric circulation regimes and meteorological field anomalies.

8.4 FUZZY ALGORITHM

Clustering analysis is a fundamental but important tool in statistical data analysis. In the past, clustering techniques have been widely applied in interdisciplinary scientific areas such as pattern recognition, information retrieval, clinical diagnosis, and microbiological analysis. In the literature, the k-means is a typical clustering algorithm, which partitions the input dataset $\{x_t\}_{t-1}^N$ that generally forms k^* true clusters into k categories (also simply called clusters without further distinction) with each represented by its center (Pokrovsky *et al.*, 2002). Although the k-means technique has been widely used due to its easy implementation, it has two major drawbacks:

(1) It implies that the data clusters are spherical because it performs clustering based on the Euclidean distance only;
(2) It needs to pre-assign the number, k, of clusters. Many experiments have shown that the k-means algorithm can work well when k is equal to k^*. However, in many practical cases, it is impossible to know the exact cluster number in advance. Under the circumstances, the k-means algorithm often leads to poor clustering performance.

Clustering based on k-means is closely related to a number of other clustering and location problems. A k-means algorithm is measured by two criteria: the intra-cluster criterion and the inter-cluster criterion. These include Euclidean k-medians, in which the objective is to minimize the sum of distances to the nearest center, and the geometric k-center problem, in which the objective is to minimize the maximum distance from every point to its closest center. The use of k-means is the most popular iterative centroid-based divisive algorithm. The specific fuzzy classification algorithm considered herein is now recalled and briefly discussed (Matoušek, 2000). In such algorithms the definition of the centroid will be used extensively; specifically, the centroid of M, say w, is given by

$$w = \frac{1}{N}\sum_{i=1}^{N} x_i, \tag{8.1}$$

where x_i is the ith column of matrix X. Similarly, the centroids of the sub-clusters X_l and X_r, say w_l and w_r, are given by

$$w_l = \frac{1}{N_l} \sum_{i=1}^{N_l} X_{l,i}; \qquad w_r = \sum_{i=1}^{N_r} X_{r,i}, \tag{8.2}$$

where $X_{l,i}$ and $X_{r,i}$ are the ith columns of X_l and X_r, respectively.

***k*-means algorithm**

Step 1. (Initialization). Randomly select a point, say $c_l \in R^p$; then compute the centroid w of M, see Equation (8.1), and compute $c_r = w - (c_l - w)$.

Step 2. Divide a set $M = \{x_1, x_2, \ldots, x_N\}$ into two sub-clusters M_l and M_r, according to the following rule:

$$\begin{cases} x_i \in M_l, & \text{if } \|x_i - c_l\| \le \|x_i - c_r\| \\ x_i \in M_r, & \text{if } \|x_i - c_r\| \le \|x_i - c_l\| \end{cases}.$$

Step 3. Compute the centroids of M_l and M_r: w_l and w_r, as in Equation (8.2).

Step 4. If $w_l = c_l$ and $w_r = c_r$, stop, else, let $c_l = w_l$; $c_r = w_r$ and go to Step 2.

8.5 LOW-OSCILLATION DYNAMIC AND PREDICTABILITY OF PRECIPITATION RATE

The North Atlantic Oscillation (NAO) exerts a dominant influence on wintertime temperature and precipitation across the North Atlantic basin and thus has major impacts on marine and terrestrial ecosystems. The NAO index is a normalized difference between sea level atmospheric pressures at Iceland minimum and Azores maximum. Linear regression analysis shows that a considerable portion of the climatic fluctuations in surface temperatures and sea surface temperatures is directly related to the North Atlantic Oscillation index. Changes of more than 1°C associated with a one standard deviation change in the North Atlantic Oscillation index occur over the northwest Atlantic and extend from northern Europe across much of Eurasia (Deser and Blackmon, 1993; Fraedrich *et al.*, 1993; James and James, 1989; Sutton and Allen, 1997). The changes in mean circulation patterns over the North Atlantic are accompanied by pronounced shifts in storm tracks and associated synoptic eddy activity (Sickmoller *et al.*, 2000; Walter *et al.*, 2001), which affect the transport and convergence of atmospheric moisture and can therefore be directly tied to changes in regional precipitation. Hurrell (1995) has shown that drier conditions during high North Atlantic Oscillation index winters occur over much of central and southern Europe and the Mediterranean, while wetter-than-normal conditions occur from Iceland to Scandinavia. This has been the case for much of the past two decades.

By contrast, increases in wintertime precipitation over Scandinavia may be related to recent positive mass balances in the maritime glaciers of southwest Norway, one of the few regions of the globe where glaciers are not retreating.

Beniston (1997) showed that snow depth and duration in Switzerland is correlated with the North Atlantic Oscillation. Beniston and Rebetez (1996) found that snow depth and duration over the past few winters have been among the lowest recorded this century, causing economic hardships to those industries dependent on winter snowfall. However, as 1996 was a low North Atlantic Oscillation index winter, Europe experienced a severe winter with record low temperatures and heavy snowfall in many parts of southern Europe. Let us consider some results of our correlation analysis aimed to reveal hidden linkages between the North Atlantic Oscillation and precipitation rate in southern Europe. We present here several figures that describe a case when the North Atlantic Oscillation is considered to be a predictor of precipitation rate with a lead time of two months.

We used monthly fields acquired from the NCEP/NCAR re-analysis dataset for 1948–1998. Seasonal prediction of surface air temperature from winter to spring is most reliable (Czaja and Frankignoul, 2002; Luksch, 1996). Therefore, it is not surprising that the winter North Atlantic Oscillation index values provide very high correlation (up to 0.9) with the precipitation rate in the following spring (see Figure 8.1). The highest correlation can be found in the Western Balkans and Italy as well as in the Southern Caucasus area. It is necessary to emphasize that the map in Figure 8.1 corresponds to the period 1995–2005, which might be attributed to the climate-warming time.

The North Atlantic Oscillation seems not to be a stationary stochastic (or deterministic) process on timescales that are common in climate research. Appenzeller *et al.* (1998) showed by means of wavelet analysis that in a 1,400-year simulation of the ECHAM3 GCM developed at the Max-Planck Institute in Hamburg, as well as in ice core data, the dominant frequencies of the North Atlantic Oscillation index change over time. One frequency in the North Atlantic Oscillation

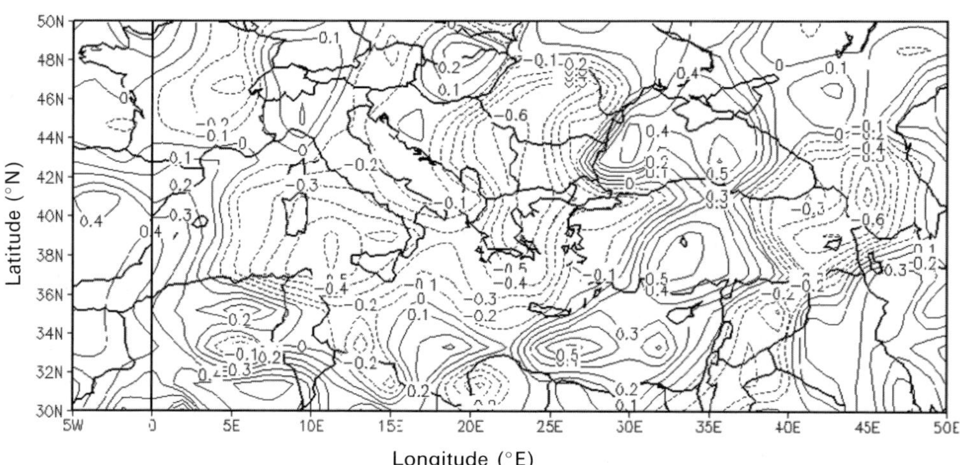

Figure 8.1. Seasonal correlation of April–May (1995–2005) precipitation rate with February–March NAO (index leads by 2 months).

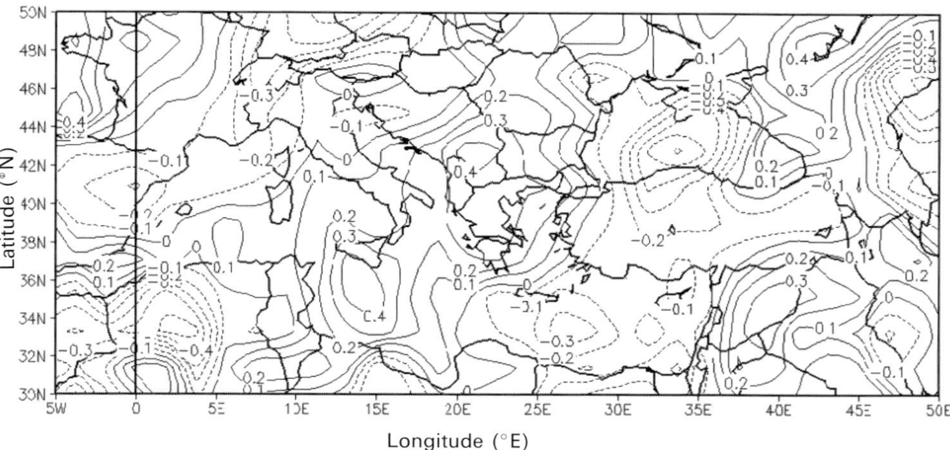

Figure 8.2. Seasonal correlation of April–May (1965–1975) precipitation rate with February–March NAO (index leads by 2 months).

index of ECHAM3 could be attributed to a coupled ocean–atmosphere mode, which projects into the North Atlantic Oscillation index. Another indication that the North Atlantic Oscillation can change its regime is the strong positive trend of the index since the late 1960s. During this latter part of the record, an 8-year oscillation may be observed (Latif, 1998; Stephenson and Xoplaki, 2001). This trend in the index may come from a very significant mechanism in a changing climate, which a GCM must reproduce if climate projections in the North Atlantic region are to have any meaning. Our calculations confirmed the major conclusion of past work that the North Atlantic Oscillation does not behave like a stationary process. Moreover, not only are its trends distinctive before and after 1975, which is considered as the beginning of global warming in the atmosphere, but also its correlation to other meteorological parameters changed sharply. In particular, the relationship with the precipitation rate became much stronger than previously (see Figure 8.2).

The North Atlantic Oscillation precipitation rate cross-correlation was rather weak even in the most promising double-season "winter–spring". The highest cross-correlation (0.5) was achieved in the Balkans. Reliable rain rate prediction for summer time is of practical value for most countries of southern Europe and northern Africa. Our study showed (Figure 8.3) that there is strong cross-correlation between spring North Atlantic Oscillation values and summer precipitation rate in some areas of the Mediterranean: Spain, western Africa, southern Italy, and the eastern area including the Southern Caucasus region. Autumn is the next important season for rain amount investigation. Its successful prediction depends mainly on the cross-correlation between the summer North Atlantic Oscillation (predictor) and the precipitation rate in September–October. Our calculations (see Figure 8.4) show that the most promising autumn rain seasonal forecast is related to France, northern Africa, the Balkans, southern Russia, and the eastern Mediterranean.

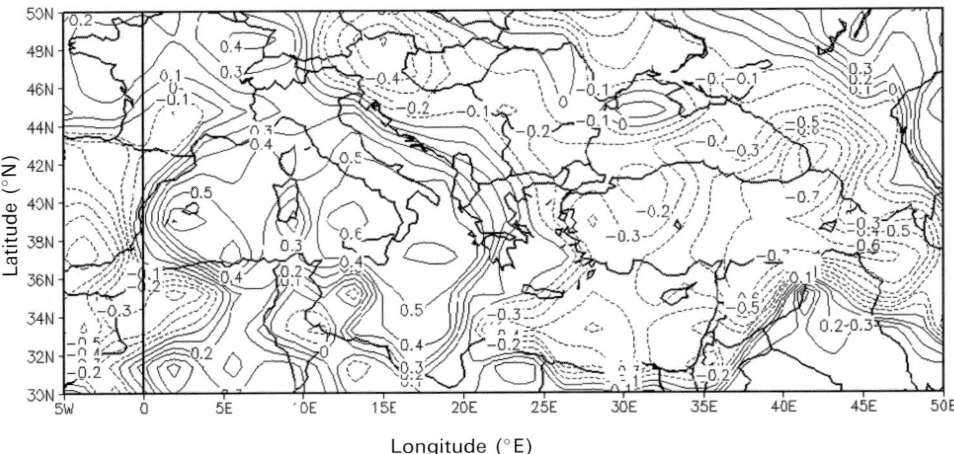

Figure 8.3. Seasonal correlation of June–July (1995–2005) precipitation rate with April–May NAO (index leads by 2 months).

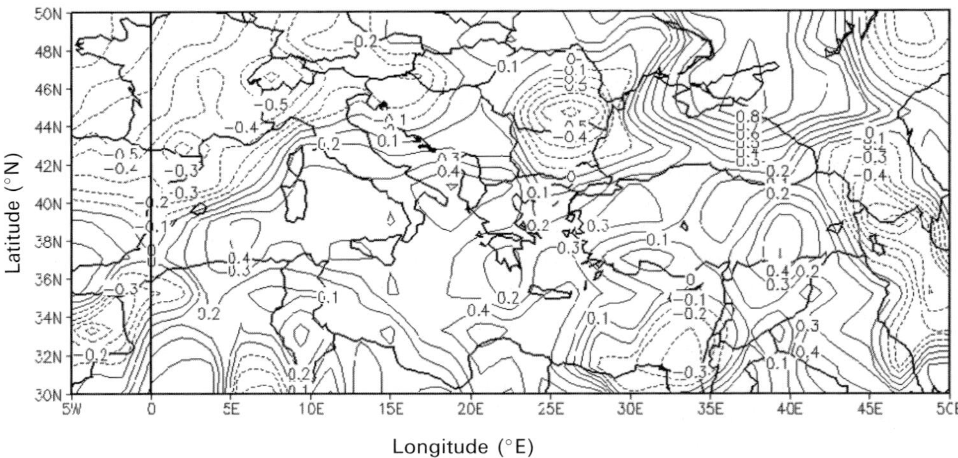

Figure 8.4. Seasonal correlation of September–October (1995–2005) precipitation rate with July–August NAO (index leads by 2 months).

Our study demonstrated also that the predictor explained variance based on the regression model provides a very high level of statistical confidence for forecast estimates. Finally, it is necessary to add here that very strong correlations were found with Arctic Oscillation (AO) index magnitudes obtained in a similar investigation mode. The AO index is a value of a coefficient corresponding to the first empirical orthogonal function (EOF) used in an expansion of the sea level pressure field in the high-latitude belt of the northern hemisphere.

8.6 FUZZY CLASSIFICATION OF REGIME CIRCULATION AND RAIN RATE SPATIAL DISTRIBUTION OVER EUROPE

It is well known that the wind field probability distribution function has a multimode signature (Trenberth, 1984). The vector character of the horizontal wind field complicates the task of approximating this probability distribution function by standard model functions. A more promising approach is related to splitting the multidimensional wind velocity phase space into several sub-domains in such a way that the wind velocity probability distribution function has a uni-modular structure within each of such sub-domains. Every one-modular probability distribution function obtained in the above way might be approximated by one of the standard models (Gaussian, log-normal, etc.).

Recently, Pokrovsky *et al.* (2002) successfully used fuzzy logic to classify spatial meteorological fields. In fuzzy set clustering, we are given a set of N points in a d-dimensional space, R^d, and we have to arrange them into a number of groups (called clusters). In *k-means* clustering (see Section 8.4), the groups are identified by a set of points that are called the cluster centers. The data points belong to the cluster whose center is closest.

Existing algorithms for k-means clustering suffer from three main drawbacks:

(i) the algorithms are slow and do not scale to a large number of data points;
(ii) these are restricted by low dimensionality, d, of phase space; and
(iii) they converge to different local minima based on the initializations.

To overcome these disadvantages we applied new recurrent algorithms recently developed by Pelleg and Moore (1998). This permits us to increase the dimensions d up to 800–900. Here we considered a joint distribution of three fields: zonal and meridional wind velocity at the standard atmospheric level of 850 mbar, U850, V850, and precipitation in a domain with 300 gridpoints.

Therefore, the general dimension d (see Section 8.4) was equal to 900. Joint analysis permits us to investigate not only atmospheric circulation patterns responding to major surface air pressure low oscillations in the Atlantic–Europe area, but also to find the related spatial anomaly of the rain rate fields during all seasons. The fuzzy set analysis of these fields revealed the major circulation regimes over the eastern North Atlantic and Europe. We found that there were three main regimes corresponding to seasons (see Figures 8.5–8.7): (1) winter and early spring; (2) summer, and (3) autumn. The first regime (Figure 8.5) is primary zonal and closely related to Type W in the Hess–Brezowsky classification. Therefore, major anomalies of rain rate might be found in mountain areas around Europe including the Alps, the Balkans and Apennine peninsulas in southern Europe. The summer airflow regime (Figure 8.6) substantially deviates from the zonal regime and is determined by three vorticity polar systems: (1) northwestern (Scandinavia), (2) western Mediterranean, and (3) the Caucasus. Non-zonal circulation in Middle Asia is closely related to a precipitation field anomaly in the Caucasus region. This circulation type has some similarity with Type E in the Hess–Brezowsky classification (dominated meridional

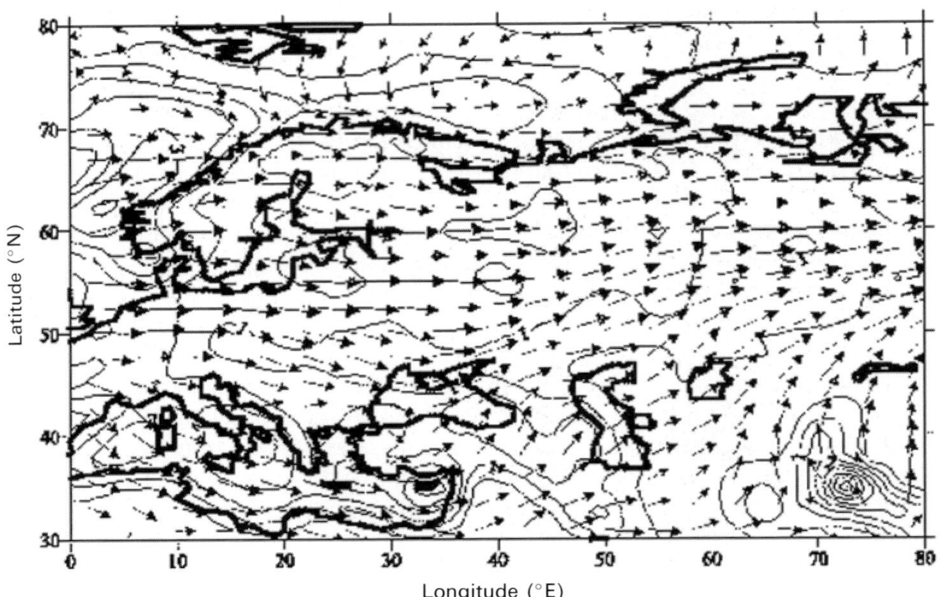

Figure 8.5. Monthly circulation regime 1 (winter and early spring): the joint pattern for U850–V850.

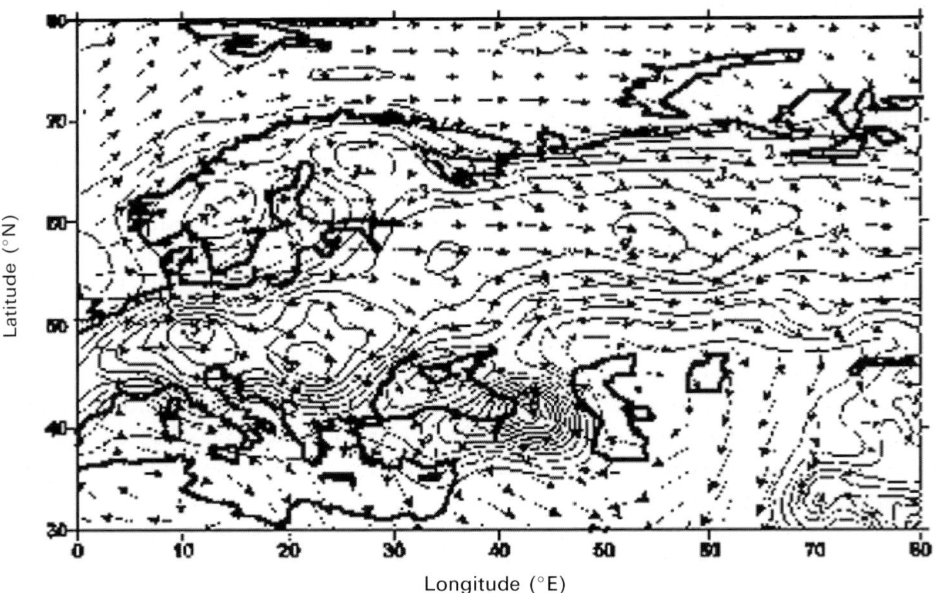

Figure 8.6. Monthly circulation regime 2 (summer): the joint pattern for U850–V850.

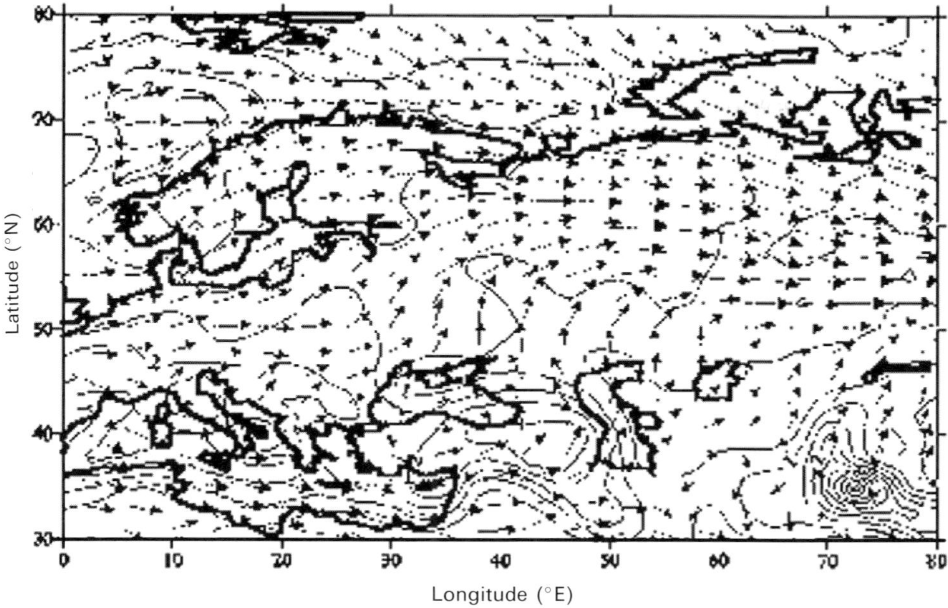

Figure 8.7. Monthly circulation regime 3 (autumn and early winter): the joint pattern for U850–V850.

circulation in eastern Europe). It is necessary to note that an anticyclone was found in the western part of the North Atlantic for both (warm and cold) seasons. The Scandinavia cyclone area explains rain rate maxima located in the 50°–60° latitude European area and lower rain rate in southern Europe because of hot and dry African air inflow. In late autumn and early winter (Figure 8.7) we found a vorticity system comprised of three polar systems: (1) northwestern, (2) northern Africa, and (3) northern Russia (Kara Sea).

The zonal circulation type dominates in Southern Europe and more precipitation is delivered from the Atlantic. The winter rain rate is more uniformly distributed in various latitude belts across Europe than in summer, but more intensive precipitation occurred in southern Europe because of moisture transport strengthening to this area from the Atlantic. Another important rain formation mechanism is non-zonal meridian airflow in the eastern Mediterranean and the Caucasus. North Atlantic Oscillation as well as Arctic Oscillation indexes substantially increased their magnitudes in the late 1980s and 1990s during global warming. The atmospheric circulation patterns have been moving northward during this time. As a consequence, the climate in southern Europe became drier, and respective rain amounts reduced primarily in the warm part of year. By contrast, the rain rate has been increasing here in the cold part of the year. It led to a wetter climate in winter and a drier one in summer.

The overall objective of this section was to propose a novel approach for classification of the atmospheric circulation patterns over Europe. The new method was validated and compared with the conventional low-oscillation approach based on a regression technique. It is known (Cherry, 1996) that the main problem with linear regression, the EOF, or single-value decomposition approach, in statistical prediction is the over-fitting problem. Fuzzy techniques permit us to reveal separate clusters in multi-dimensional phase space describing quasi-stationary circulation regimes. Points in phase space traced between them may be considered as transition states of a weather system subjected to stochastic factors. By contrast, points located within clusters describe the well-established circulation regimes discussed above. Each atmospheric regime generates particular air and moisture transport, and therefore determines rain probability distribution. The topology of each fuzzy set displays its low-dimensional structure and small diameter. Fuzzy metric introduction allowed us to establish links between the associated predictor and the predictand sets (instead of the single-value decomposition or EOF modes).

Therefore, rain prediction skill is found to be significant for winter, spring, and summer while it is much weaker for autumn. Both correlation and fuzzy analysis manifested that winter prediction provides a useful skill score over southern Europe. Climate change is a major cause of precipitation deficit in southern Europe which appeared in the 1980s and 1990s. We showed that this phenomenon is related to the northward displacement of the major route of the airflow transport of Atlantic moisture in summer, spring, and autumn.

8.7 MODEL DESCRIPTION

The use of a neural network (NN) is a powerful non-linear scheme based on black-box statistics, where one can tune the model parameters to arrive at a good prediction, but can see neither the phase relation between the predictands and predictors, nor the origin of skills. Therefore, we assume that the predictability of seasonal climate is connected with forcing fields such as sea surface temperature or others. The key to a truly successful application of a neural network model lies in the understanding of the underlying physical mechanism for the relation between predictor and predictand fields (Pokrovsky, 2000).

We used a comprehensive neural network model, which is based on a combination of fuzzy logic modules and neural network principle structures. The general scheme of our self-learning Fuzzy–Neural model is presented in Figure 8.8. Let us consider it from the left to the right. The left module defines the initial field assimilation as the input information. Further, this dataset should be classified in several fuzzy logic modules. It is necessary to emphasize that each input meteorological field is linked to several fuzzy sets. The classification procedure is related to diurnal or seasonal cycles or to various types of spatial distribution of meteorological parameters (e.g., type of atmospheric circulation). That means that each field should be evaluated by using a complex procedure and then should be attributed to one of the clusters. Nonetheless, the metric distances of each field to cluster centers are taken into account in the next model layer, which is called a hidden layer. A hidden layer is

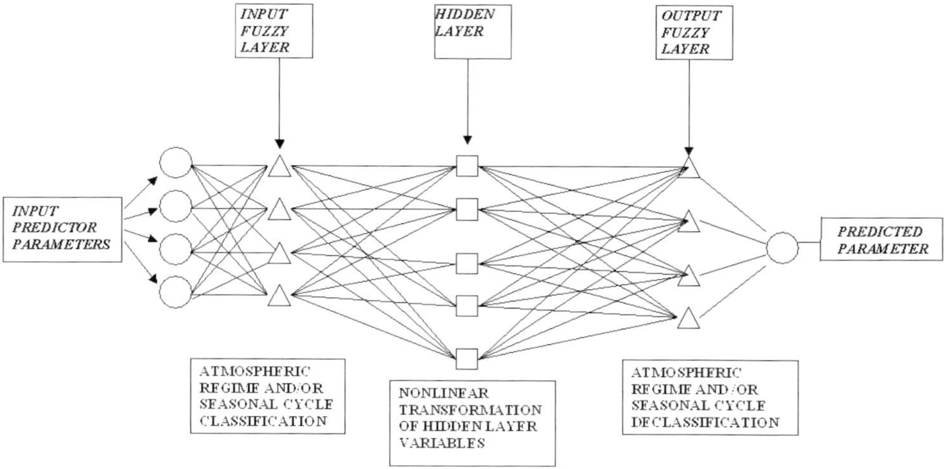

Figure 8.8. Scheme of the Fuzzy–Neural predictive model.

designed to perform non-linear transformation of distance variables into output variables.

A very important feature of a neural network is its ability to learn. This learning is based on the simultaneous analysis of some input and output samples, which are called learning samples. As a rule, the size of a learning sample is equal to one-third the general sample size. Therefore, two-thirds of the general sample is implemented to evaluate the efficiency and accuracy of a neural network as a predictive tool. The structure of the hidden layers is very complicated. Each of the hidden layer modules is linked to every one of the fuzzy logic modules. The first hidden layer output is connected to the second fuzzy logic layer. Its function is designed to declassify the results of non-linear transformations, which were carried out in the hidden layer. Hence, this declassification module provides an inverse transformation from the artificial inner variables to ordinary units of meteorological parameters. The final combination of output variables leads us to acquisition of the estimates of predictive variables (surface air temperature, etc.).

The decision to use the principal fuzzy patterns of surface atmospheric pressure and temperature as predictor fields is based on findings of other studies on field teleconnections (Lanzante, 1984; Namias, 1982). For this and other calculations described here the seasonal cycle has been removed by subtracting the calendar monthly means. The following teleconnection spatial areas (TCSAs) were selected (Barnston and Livezey, 1987) to derive principal fuzzy patterns for the Europe model: North Atlantic Oscillation, East Atlantic (EA), EA Jet (EAJ), East Atlantic/West Russia (EAWR), Scandinavia (S), Polar/Eurasian (PE). Another important issue is that each teleconnection spatial area is related to atmospheric low-oscillation regions, which regulate the circulation states and transition between them. It is known (Barnston, 1994; Wang, 2001) that the effectiveness of statistical models depends crucially on whether the relevant components (with respect to spacescales and timescales) to be used as predictors are suitably incorporated in the prediction model, and whether the relationships between predictors and predictands (which may not necessarily be linear, particularly, when mid and high latitude are concerned) are properly established. In this respect implementation of an optimal design technique (Pokrovsky and Roujean, 2003) makes it possible to determine an optimal set of predictors (principal fuzzy patterns) representing key low-oscillation patterns, which are most informative with respect to the predictand field for a prescribed lead interval. A five-layer neural network utilizes fuzzy classification input and output layers and radial basis functions for principal fuzzy patterns as activated units. In order to reduce the problem of artificial skill produced from over-fitting and thus receive a more representative estimate of real skill, we used a cross-validation method, in which the forecast model is developed using only part of the available dataset (learning sample) and then applied to the independent data (verification sample).

8.8 FORECAST SKILL EVALUATION

Time series of monthly magnitudes (re-analysis NCEP) for 1948–1998 were split into two parts: learning and verification samples. In contrast to the GCMs our self-learning model accumulates all past observation data in such a way that after training with data from 35 years it could provide very competitive prediction results for surface air temperature fields. It captured both the positive and negative phases of the NAO, EA, EAJ, EAWR, and AO climate indexes as well as transition periods in their relationships with predictand fields. Respective results for St. Petersburg's (northwest Russia) monthly air temperature time series are presented in Figure 8.9. Figure 8.9 demonstrates the approximation of observed time series of March

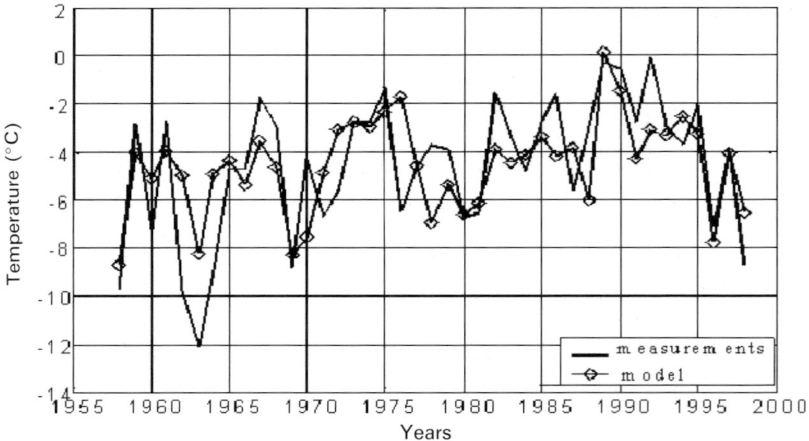

Figure 8.9. March monthly surface air temperature for St. Petersburg observed and explained by the model.

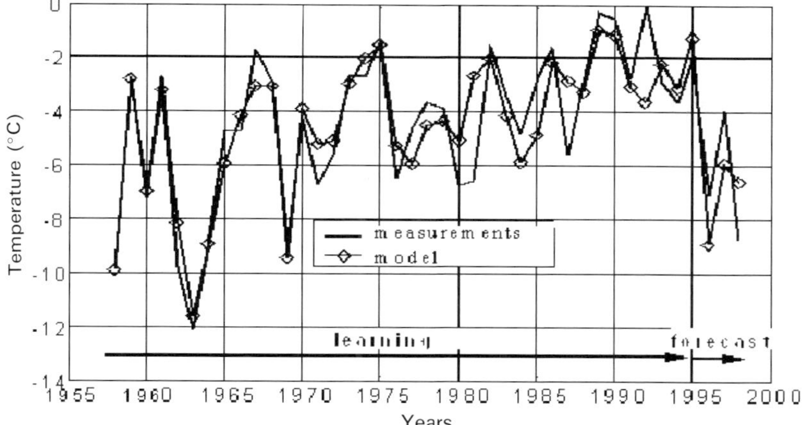

Figure 8.10. March monthly surface air temperature for St. Petersburg observed and predicted by the model.

surface air temperature for 41 years provided by our model, which implements the North Atlantic Oscillation and Arctic Oscillation principal fuzzy pattern components for January and February as predictors. In this case the approximation's root-mean-square deviation is equal to 0.3°C. When the predictor list includes other teleconnection spatial area patterns (East Atlantic, EA Jet, Scandinavia, East Atlantic/West Russia), the r.m.s. error might be reduced to 0.2°C. Those figures show that the model describes inter-annual temperature variability correctly. This means that low-oscillation teleconnection spatial area patterns contain all the necessary information for short-term climate forecasting.

Certainly, corresponding links have a non-linear feature. Fortunately, those might be efficiently described by a neural network. Therefore, these links might be identified and inferred at a training stage. Moreover, our experiments proved that the model developed at the learning stage is useful as a forecasting tool for subsequent years (Figure 8.10). In the case when the training process was terminated in 1990, the model temperature for 1989–1990 was underestimated by 1.6°C and as a consequence the predicted values for 1991–1992 were also underestimated. When the temperature estimate for the last year is in good agreement with the observed value, forecast predictions lie closer to the measured values for other years (Figure 8.10).

Our study showed that the r.m.s. temperature deviations amount to 0.2°C–0.3°C at the training stage and 0.7°C–0.9°C at the predictive stage. It is necessary to note that the standard deviation (STD) of inter-annual variability of the March monthly surface air temperature for 1958–1998 is equal to 2.6°C. Hence the *a priori* uncertainty of monthly temperature inter-annual variability might be decreased by three times due to this model implementation. A self-learning model permits us to obtain monthly temperature fields for various lead times.

The comparison of predictive and true fields for a two-month lead time allows us to conclude that the forecast reproduces the main features of temperature spatial

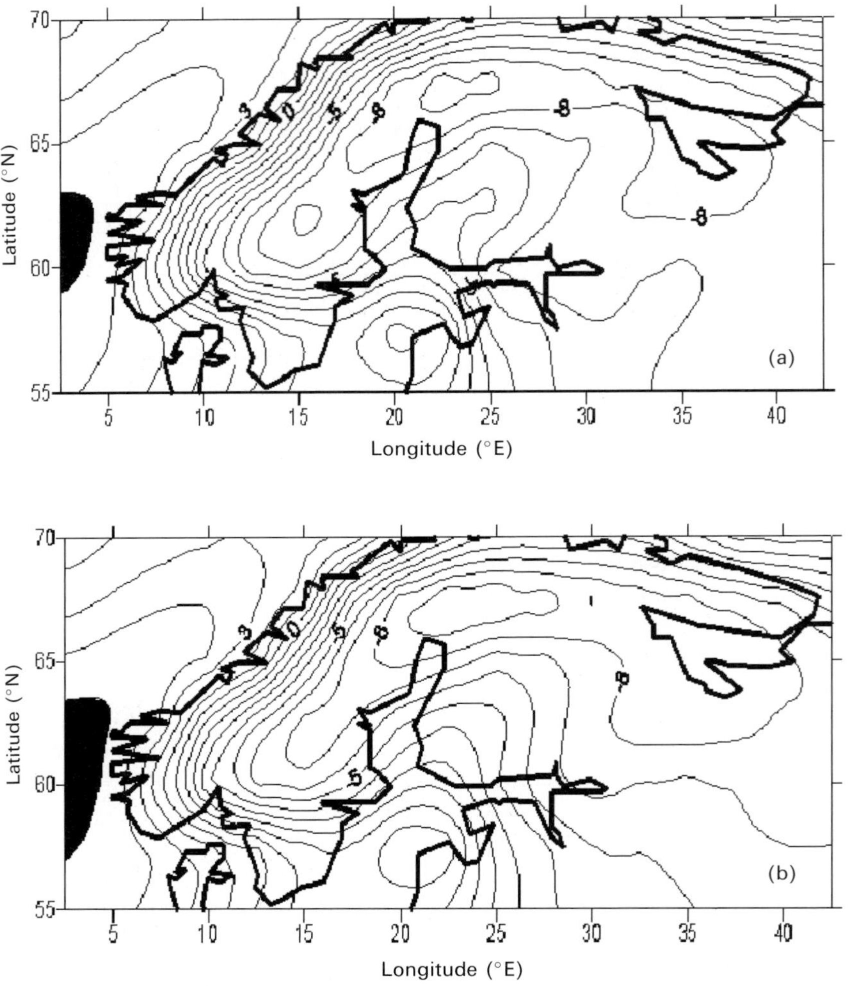

Figure 8.11. Comparison of (a, top) true field objective analysis and (b, bottom) forecast March SAT field of 1996: northern Europe (lead time of 2 months).

distribution correctly (see Figure 8.11). The r.m.s. deviation from the true field is equal to 0.8°C. When a lead time extends to four months, the r.m.s. deviation of the predictive field from the true field rises to 1.2°C (see Figure 8.12). In some cases (Figure 8.13) biases (systematic deviations) were revealed in the predictive fields. Nonetheless, predictive fields exhibit the right structure of spatial temperature distributions in northern Europe. The correlation skill score was used for examining the spatial behavior of a predictive model. The skill score for a grid site was defined as the correlation coefficient between the observed (re-analysis data) and the predicted surface air temperature time series values over all the verification periods.

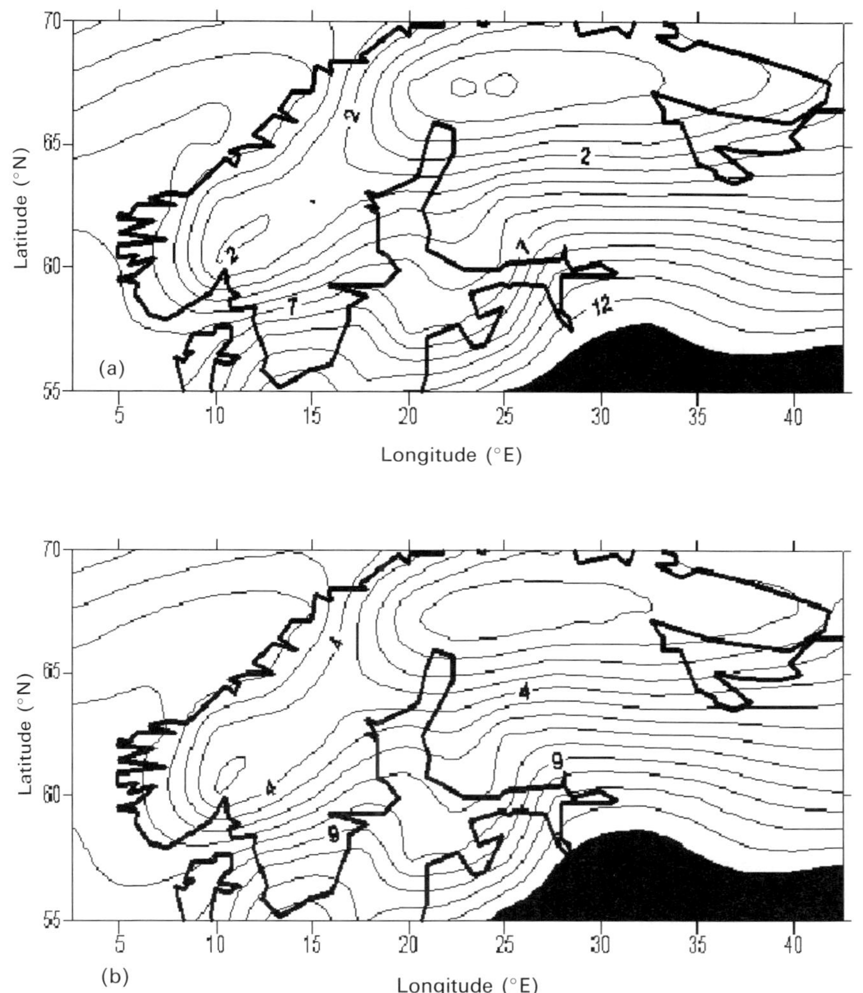

Figure 8.12. Comparison of (a, top) true field objective analysis and (b, bottom) forecast SAT field of May 1996: northern Europe (lead time of 4 months).

The spatial average value of the correlation coefficients was used as a predictive skill score. Predictive model efficiency was confirmed by the high level of skill score and correlation coefficient magnitudes (0.5–0.7) for predictive surface air temperature fields. It was not surprising as the general features of the surface air temperature fields were reproduced correctly by the model. Moreover, the low level of standard deviation magnitudes permits us to achieve a high 0.8–0.9th level for the explained variance of predictive surface air temperature fields, which is much higher than those obtainable for linear regression used in a recent study (Blender *et al.*, 2003).

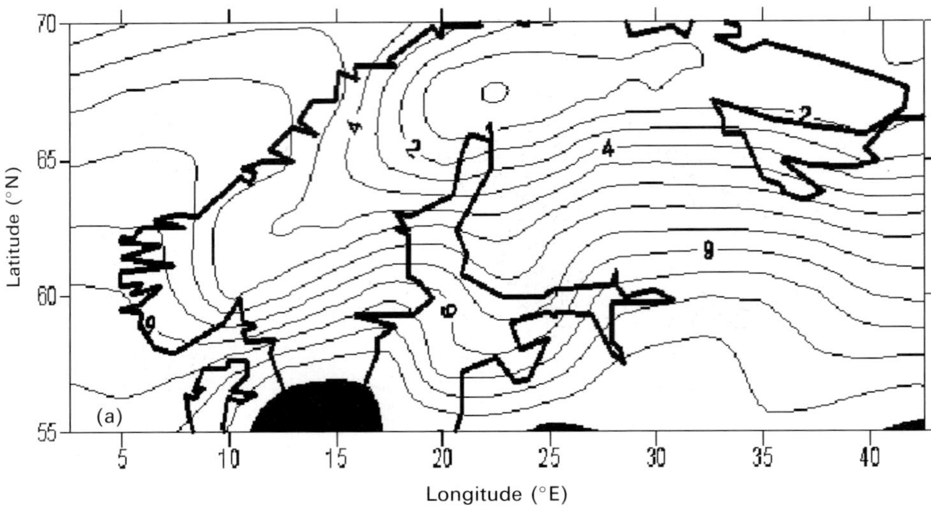

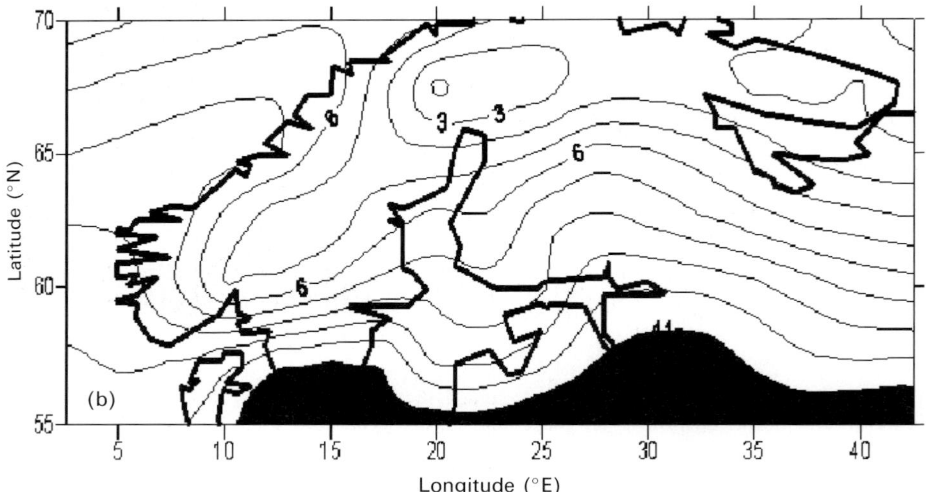

Figure 8.13. Comparison of (a, top) true field objective analysis and (b, bottom) forecast SAT field of May 1998: northern Europe (lead time of 4 months).

An important practical feature of our model is that it provides not only predictive, but also error fields, which permits us to acquire *a priori* information on forecast accuracy. These error fields were obtained as theoretical estimates supplied by the model. Therefore, we compared these estimates with the deviation of forecast from objective analysis fields. We consider the latter as the actual forecast

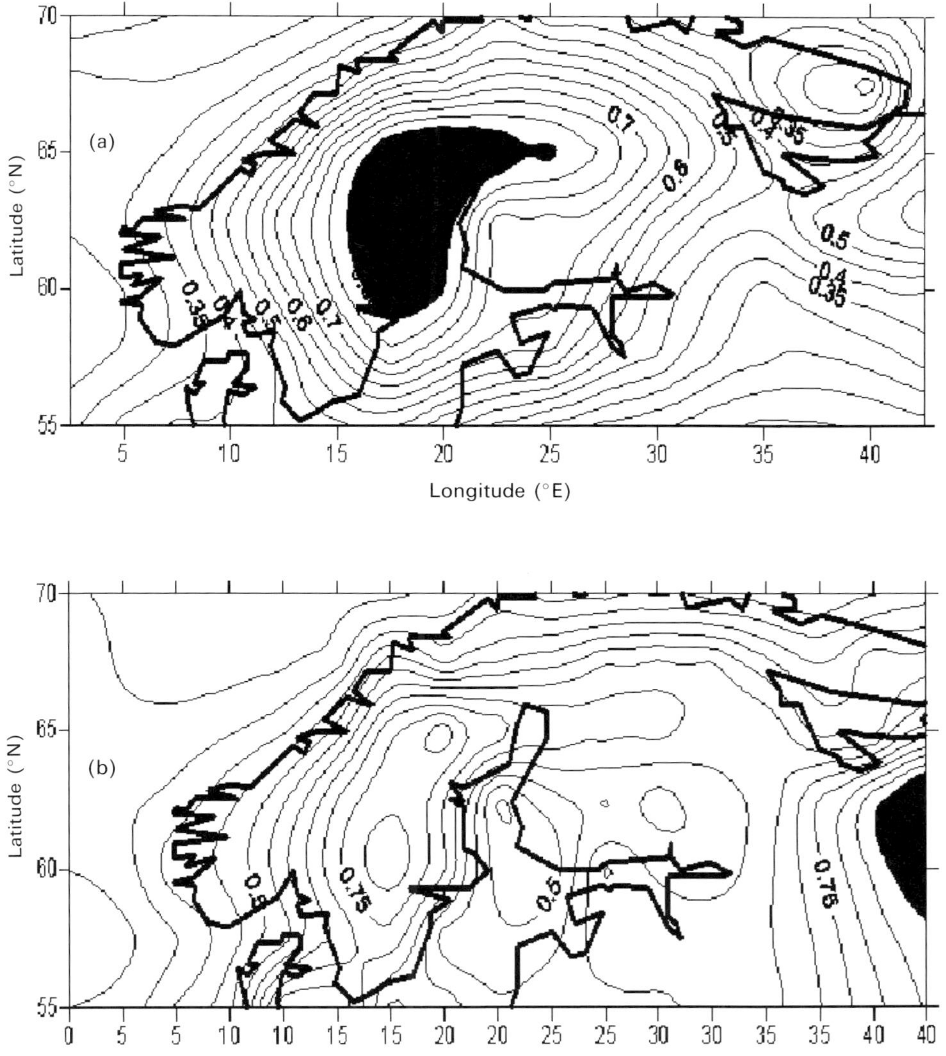

Figure 8.14. Comparison of (a, top) actual and (b, bottom) theoretical forecast error fields: SAT field of May 1998, northern Europe (lead time of 4 months).

error field (Figure 8.14a). The theoretical error field (Figure 8.14b) exhibits a similar structure with a maximum over Scandinavia and a minimum at the highest latitude on the map. But there are some differences. The theoretical field discloses another maximum at the east edge of the map. The theoretical values are 10% underestimated.

8.9 DISCUSSION

The overall objective of this chapter was to propose a self-learning model for seasonal prediction of surface air temperature over Europe. The model was validated using a cross-validation technique and was compared with the regression technique. It is known (Cherry, 1996) that the main problem with a regression or single-value decomposition approach in statistical prediction is the over-fitting problem. Pairs of most correlated predictor/predictand modes are sought over the training period and are used in prediction. When the training sample is not long enough, the associated correlations between these pairs of modes are largely overestimated. Since these correlations are precisely the linear prediction coefficients linking predictand modes to their corresponding predictor modes (entering in the regression model), the explained variance of forecast is also overestimated. In contrast, when a sample is long, the phase space points are rather scattered and the associated correlation coefficients become underestimated, while the explained variance of the forecast is also underestimated. The phase space points might be associated with the various circulation regimes. That is why these points comprise a large scattering pattern in phase space. A fuzzy set approach enables us to split this wide pattern into a number of more narrow sets of points. Each of these narrow sets is associated with a specific circulation regime.

8.10 REFERENCES

Appenzeller C., Stocker T.F., and Anklin M. (1998). North Atlantic oscillation dynamics recorded in Greenland ice cores. *Science*, **282**, 446–449.

Barnett T.P. (1985). Variations in the near-global sea level pressure. *J. Atmos. Sci.*, **42**, 478–501.

Barnston A. (1994). Linear statistical short-term climate predictive skill in the Northern Hemisphere. *J. Climate*, **7**, 1513–1564.

Barnston A. and Livezey G. (1987). Classification, seasonality and persistence of low-frequency atmospheric circulation patterns. *Mon. Weather Rev.*, **115**, 1083–1126.

Barnston A. and Smith T.M. (1996). Specification and prediction of global surface temperature and precipitation from global SST using CCA. *J. Climate*, **9**, 2660–2697.

Beniston, M. (1997). Variations of snow depth and duration in the Swiss Alps over the last 50 years: Links to changes in large-scale climatic forcings. *Climatic Change*, **36**, 281–300.

Beniston M. and Rebetez M. (1996). Regional behavior of minimum temperatures in Switzerland for the period 1979–1993. *Theor. Appl. Climatol.*, **53**, 231–243

Bjerknes J. (1969). Atmospheric teleconnections from Equatorial Pacific. *Mon. Weather Rev.*, **97**, 163–172.

Blender R., Luksch U., Fraedrich K., and Raible C. (2003). Predictability study of the observed and simulated European climate using linear regression. *Quarterly Journal of the Royal Meteorological Society*, **129**, 2299–2313.

Cherry, S. (1996). Singular value decomposition analysis and canonical correlation analysis. *J. Climate*, **9**, 2003–2009.

Czaja A. and Frankignoul C. (2002). Observed impact of Atlantic SST anomalies on the North Atlantic oscillation. *J. Climate*, **15**, 606–623.

Deser C. and Blackmon M.L. (1993). Surface climate variations over the North Atlantic ocean during winter: 1900–1989. *J. Climate*, **6**, 1743–1753.

Fraedrich K. and Wang, R. (1993). Estimating the correlating dimension from noisy and small data set base on re-embedding. *Physica D*, **65**, 373–398.

Fraedrich K., Bantzer C., and Burkhardt U. (1993). Winter climate anomalies in Europe and their associated circulation at 500 hPa. *Clim. Dyn.*, **8**, 161–175.

Gerstengarbe F.-W.,Werner P.C., and Rüge U. (1999). *Katalog der Grosswetterlagen Europas nach Paul Hess und Helmuth Brezowsky 1881–1998*. Deutscher Wetterdienst, Offenbach, Germany.

Hess P. and Brezowsky H. (1952). Katalog der Grosswetterlagen Europas. *Der Deutscher Wetterdienstes in der US-Zone*, **33**, 39.

Halliwell, G.R. (1997). Decadal and multidecadal North Atlantic SST anomalies driven by standing and propagating basin-scale atmospheric anomalies. *J. Climate*, **10**, 2405–2411.

Hurrell, J.W. (1995). Decadal trends in the North Atlantic oscillation: Regional temperatures and precipitation. *Science*, **269**, 676–679.

James, I.N. and James, P.M. (1989). Ultra-low-frequency variability in a simple atmospheric circulation model. *Nature*, **342**, 53–55.

Kondratyev K.Ya., Buznikov A.A., and Pokrovsky O.M. (1996). *Global Change and Remote Sensing*. Wiley/Praxis, Chichester, U.K., 370 pp.

Kondratyev K.Ya., Sumi A., and Pokrovsky O.M. (1997). *Global Change and Climate Dynamics: Optimization of Observing Systems*. Center for Climate System Research Rep. No. 3, University of Tokyo, Tokyo, 213 pp.

Latif M. (1998). Dynamics of interdecadal variability in coupled ocean–atmosphere models. *J. Climate*, **11**, 602–624.

Lanzante J.R. (1984). A rotated eigenvalue analysis of correlation between 700 mb heights and sea-surface temperature in the Pacific and Atlantic. *Mon. Weather Rev.*, **112**, 2270–2280.

Luksch U. (1996). Simulation of North Atlantic low-frequency SST variability. *J. Climate*, **9**, 2083–2092.

Matoušek, J. (2000). On the approximate geometric k-clustering. *Discrete and Computational Geometry*, **24**, 61–84.

Montroy D.L. (1997). Linear relation of central and eastern North American precipitation to tropical sea surface temperature anomalies. *J. Climate*, **10**, 541–558.

Namias J. (1982). Teleconnections of 700 mb height anomalies for the Northern Hemisphere, *Mon. Weather Rev.*, **110**, 824–828.

Pelleg D. and Moore A. (1998). Cached sufficient statistics for efficient machine learning with large datasets. *Journal of Artificial Intelligence Research*, **8**, 67–91.

Peng S. and Whitaker J.S. (1999). Mechanisms determining the atmospheric response to midlatitude SST anomalies. *J. Climate*, **12**, 1393–1408.

Pokrovsky O.M. (2000). Land surface energy exchange simulation based on combined "Fuzzy Sets and Neural Networks" approach. *Proceedings of Second Conference on Artificial Intelligence, January 17–21, 2000, Boston*. American Meteorological Society, Boston, MA, pp. 21–26.

Pokrovsky O.M. (2004). Optimization of Siberian RAOB network by maximization of information content. *Proceedings of Third CGC/WMO Workshop on the Impact of Various Observing Systems on Numerical Weather Prediction*. World Weather Watch Technical Rep. WMO/TD N1228, World Meteorological Organization, Geneva, pp. 270–282.

Pokrovsky O.M. (2006a). Climatic changes in air–sea interaction over Russian Arctic and its impact on extreme rain events occurred during monsoon in India and China. *Proceedings of the Third Annual Meeting Asia Oceania Geosciences Society (AOGS-2006)*. Interdisciplinary Working Group, Abstract 59-IWG-A0446. Asia Oceania Geosciences Society, Singapore.

Pokrovsky O.M. (2006b). The SST long-term trend features in North Atlantic currents and the climate change in the Eurasia. *Proceedings of the International Science Conference: Rapid Climate Change, October 24–27, 2006, Birmingham, U.K.*, p. 65.

Pokrovsky O.M. (2006c). Recent changes in atmospheric circulation regimes over northern Eurasia and suggestions to redesign the RAOB network. *Proceedings of the Second THORPEX International Scientific Symposium, December 4–8, 2006, Landshut, Germany*. WMO/TD N1355, World Meteorological Organization, Geneva, pp. 234–235.

Pokrovsky O.M. (2007). A causal link between the eastern Arctic ice extent reduction and changes in the atmospheric circulation regimes over northern Asia. *Proceedings of the Seventh International Conference on Global Change: Connection to Arctic (GCCA-7), February 19–20, 2007*. International Arctic Research Center, University of Alaska Fairbanks, AL, pp. 82–85.

Pokrovsky O.M. and Roujean J.L. (2003). Land surface albedo retrieval via kernel-based BRDF modeling: II. An optimal design scheme for the angular sampling. *Remote Sens. Environ.*, **84**, 120–142.

Pokrovsky O.M., Roger H.F. Kwok R.H., and Ng C.N (2002). Fuzzy logic approach for description of meteorological impacts on urban air pollution species: A Hong Kong case study. *Computers and Geosciences*, **28**, 119–127.

Raible C.C. and Blender R. (2004). Northern hemisphere midlatitude cyclone variability in GCM simulations with different ocean representations. *Clim. Dyn.*, **22**, 239–248.

Raible C.C., Luksch U., Fraedrich K., and Voss R. (2001). North Atlantic decadal regimes in a coupled GCM simulation. *Clim. Dyn.*, **17**, 321–330.

Rasmusson E.M. and Carpenter T.H. (1982). Variation in sea surface temperature and surface wind fields associated with the Southern Oscillation/El Nino. *Mon. Weather Rev.*, **110**, 354–384

Sauer T., Yorke J.A., and Gasdagli M. (1991). Embedology. *J. Stat. Phys.*, **65**, 579–616.

Sickmoller M., Blender R., and Fraedrich K. (2000). Observed winter cyclone tracks in the northern hemisphere in re-analysed ECMWF data. *Quarterly Journal of the Royal Meteorological Society*, **126**, 591–620.

Stephenson D.B. and Xoplaki E. (2001). North Atlantic oscillation: Concepts and studies. *Survey Geophys.*, **22**, 321–382.

Sutton R.T. and Allen M.R. (1997). Decadal predictability of the North Atlantic sea surface temperature and climate. *Nature*, **388**, 563–567.

Trenberth K.E. (1984). Some effect of finite sample size and persistence on meteorological statistics, Part 2: Potential predictability. *Mon. Weather Rev.*, **112**, 2369–2378.

Trenberth K.E., Branstator G.W., Karoly D., Kumar A., Lau N.-C., and Ropolewski C. (1998). Progress during TOGA in understanding and modelling global teleconnections associated with tropical sea surface temperatures. *J. Geophys. Res.*, **103**, 14291–14324.

Van den Dool, H.M. (1994). Long-range weather forecast through numerical and empirical methods. *Dyn. Atmos. Ocean*, **20**, 247–270.

Vautard R, Plaut G., Wang R., and Brunet, G. (1999). Seasonal prediction of North American surface air temperatures using space-time principal components. *J. Climate*, **12**, 380–394.

Walker G.T. and Bliss W. (1932). World weather. *Mem. Roy. Met. Soc.*, **4**, 53–84.

Wallace J.M. and Gutzler D.S. (1981). Teleconnections in the geopotential height field during the Northern Hemisphere winter. *Mon. Wea. Rev.*, **109**, 782–812.

Wallace J.M. and Thompson, D.W.J. (2002a). Annular modes and climate prediction. *Phys. Today*, **55**, 28–33.

Wallace J.M. and Thompson, D.W.J. (2002b). The Pacific center of action of the Northern Hemisphere annular mode: Real or artifact? *J. Climate*, **15**, 1987–1991.

Wallace J.M., Panetta R.L., and Estberg J., (1993). Representation of the equatorial stratospheric quasi-biennial oscillation in EOF phase space. *J. Atmos. Sci.*, **50**, 1751–1762.

Walter K., Luksch U., and Fraedrich K. (2001). A response climatology to idealized mid-latitude thermal forcing experiments with and without a storm track. *J. Climate*, **14**, 467–484.

Wang R. (2001). Prediction of seasonal climate in low-dimensional phase space derived from the observed SST forcing. *J. Climate*, **14**, 77–97.

9

Theory of series of exponents and their application for analysis of radiation processes

Stanislav D. Tvorogov, Tatyana B. Zhuravleva, Olga B. Rodimova, and Konstantin M. Firsov

9.1 INTRODUCTION

Academician K.Ya. Kondratyev in one of his first monographs (Kondratyev, 1950) stressed the availability of the idea put forward by Academician V.A. Ambarzumyan (1968) which is associated now with the term "series of exponents". This involves computation of values integrated over the frequency spectrum necessary for analysis of radiation processes: in this case, a "palisade" of a great number of spectral lines gives rise to not purely technical difficulties.

The evolution of the idea includes a number of lines of investigation such as the use of absorption band models (Lacis and Oinas, 1991), invoking intuitive considerations (Goody *et al.*, 1989), and the treatment of series of exponents as only approximate expressions (Tarasova and Fomin, 2000). The list of papers devoted to elaboration of the different sides of the problem numbers tens of items. These variants may be acceptable as purely pragmatic ones. However, with all this work being done there still remain methodological questions, which can be referred to as "theorems of existence". They can be answered within the framework of the correct mathematical theory of the series of exponents, or Dirichlet series, as they are called in accepted mathematical terminology (Leont'ev, 1976, 1980, 1983). This allows one not only to identify the subtle details of the problem (Tvorogov, 1994, 1997, 1999), but also to obtain some important practical results (Nesmelova and Tvorogov, 1996; Nesmelova *et al.*, 1997, 1999; Tvorogov *et al.*, 1996, 2000; Firsov *et al.*, 1998; Zhuravleva and Firsov, 2004, 2005). We believe that combining exact formal results with physical aspects will provide an adequate mathematical climate for the development of appropriate methods and algorithms. Exact formulas for the coefficients of expansion of the radiation characteristics into series of exponents obtained with the use of the Dirichlet series theory constitute a qualitative forward step as compared with the purely quantitative improvements of the k-distribution method. Indeed, they enable one to introduce evaluated approximations, on the one hand, and, on the other hand, to open the correct approach to problems which are as yet unsolved. It is these points that determine the content of this chapter.

We shall discuss (in Section 9.2) two variants for obtaining series of exponents, by representing the transmission function $P(z)$ in terms of the Laplace transform $f(s)$ of the function $P(z)$ and in terms of the Laplace transform $g(s)$ of the function $P(z)/z$. The advantages of the second variant are highlighted from the viewpoint of both the feasibility and mathematical correctness in calculations of $g(s)$ through the absorption coefficient. The exact theoretical expressions for the coefficients of expansion of $g(s)$ in the series of exponents of the function under consideration are given in the cases of homogeneous and inhomogeneous media, overlapping spectra, and integrals with the source function.

We shall also (in Section 9.3) consider application of the series of exponents for derivation of the radiative transfer equation for radiation characteristics, integrated over the frequency spectrum. It will be shown that in the case of an aerosol–molecular medium the radiative transfer equation can be written through the function $s(g)$ (the function inverse of $g(s)$) under the condition of the smallness of the absorption coefficient gradient as compared with the aerosol extinction coefficient. We give examples of comparison of the solar radiation fluxes obtained with the use of this approach and the benchmark calculation values and the data of field measurements. In the case of the molecular medium the expressions of radiation fluxes in terms of $s(g)$ are derived without the use of the correlated k-distribution (CKD) approximation. We shall also present (in Section 9.4) some theoretical results which use the series of exponents to facilitate the calculations in the cases of small pressures, overlapping bands, and an inhomogeneous atmosphere.

The formal problem as applied to spectroscopy can be formulated as follows. Let us assume that there is an integral over frequency ω of the function $\Xi(\kappa(\omega))$, where $\kappa(\omega)$ is the spectral absorption coefficient, or, in other words, the function with a number of closely spaced maxima and minima (which is just what in fact creates computational difficulties). It is necessary to construct such a monotonic function $s(g)$, ensuring fulfillment of the equality

$$\frac{1}{\Delta\omega}\int_{\omega'}^{\omega''} d\omega\,\Xi(\kappa(\omega)) = \int_0^1 dg\,\Xi(s(g)), \qquad \Delta\omega = \omega'' - \omega' \tag{9.1}$$

with further application to $\int dg$ of an appropriate quadrature formula. It is clear that this operation diminishes radically the number of terms in the integral sum.

The theory of Dirichlet series convinces one that the construction of $s(g)$ can be made through the Laplace transform of $\Xi(\omega)$. Then, the existence of Equation (9.1) appears to be a simple consequence of the Parseval theorem (Tvorogov *et al.*, 2000).

9.2 EXACT EXPANSIONS OF THE TRANSMISSION FUNCTION IN A SERIES OF EXPONENTS

The transmission function $P(z)$ is the main characteristic of radiation processes in a molecular medium (it plays here the role of function Ξ in Equation (9.1)).

The mathematical foundation of the exact expansion of the transmission function in a series of exponents is completely illustrated by the example of transmission function $P(z)$ for a homogeneous medium, namely

$$P(z) = \frac{1}{\Delta\omega}\int_{\omega'}^{\omega''} e^{-\kappa(\omega)z}\,d\omega, \tag{9.2}$$

where z is a dimensionless length (for the convenience of calculations).

Let us introduce the function $f(s)$, the Laplace transform for $P(z)$

$$f(s) = \frac{1}{2\pi i}\int_{\varepsilon-i\infty}^{\varepsilon+i\infty} dz\, e^{sz} P(z), \qquad \varepsilon > 0. \tag{9.3}$$

Then

$$g(s) = \int_0^s f(s)\,ds = \frac{1}{2\pi i}\int_{\varepsilon-i\infty}^{\varepsilon+i\infty} \frac{dz}{z} P(z)\, e^{sz} \tag{9.4}$$

and

$$P(z) = \int_0^\infty ds\, f(s)\, e^{-sz} \tag{9.5}$$

$$= \int_0^1 dg\, e^{-zs(g)} \tag{9.6}$$

$$= \sum_\nu a_\nu\, e^{-zs_\nu} \tag{9.7}$$

$$= \sum_\nu b_\nu\, e^{-zs(g)}. \tag{9.8}$$

The series of exponents (9.7) and (9.8) are obtained as a result of application in Equations (9.5) and (9.6) of the relevant quadrature formula (s_ν and g_ν, a_ν and b_ν are the appropriate abscissas and ordinates of quadrature formulas).

The function $s(g)$ appearing in Equation (9.6) is the inverse function of $g(s)$; the latter is derived as a consequence of the substitution of Equation (9.2) in Equation (9.4) and the change in order of integration (Tvorogov, 1994; Tvorogov *et al.*, 2000):

$$g(s) = \frac{1}{\Delta\omega}\int_{\kappa(\omega)\le s} d\omega, \qquad \omega \in [\omega', \omega''], \tag{9.9}$$

$$g(s) = 1 - \frac{1}{\Delta\omega}\int_{\kappa(\omega)\ge s} d\omega. \tag{9.10}$$

Relations (9.9) and (9.10) are illustrated by Figure 9.1. The sum of the segments on the abscissa axis marked by thick lines corresponds to Equation (9.9); the sum of dotted segments at the s level refers to Equation (9.10). The monotonicity of $g(s)$, and hence of $s(g)$, necessary for the fulfillment of Equation (9.1) is quite obvious.

The task is to express $f(s)$ and $g(s)$ in terms of the absorption coefficient $\kappa(\omega)$. That means substitution of Equation (9.2) into Equation (9.3) or Equation (9.4) and changing the order of integration over z and ω in the resultant double integrals. One essential feature must be noted: after substitution of Equation (9.2) into Equation (9.3), changing the order of integrations is impossible because the condition of

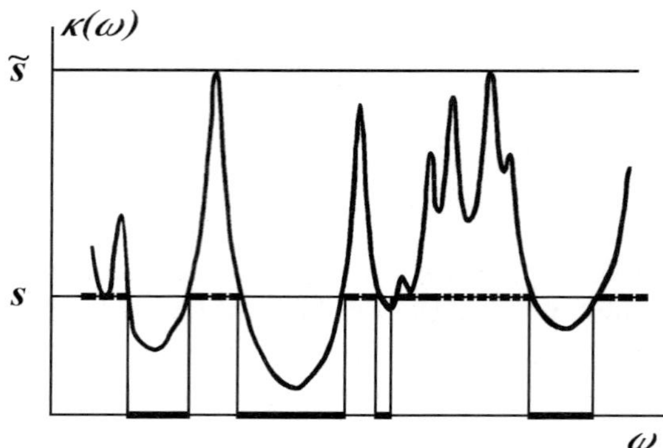

Figure 9.1. Function $g(s)$ is the sum of segments marked on abscissa axis for Equation (9.7a) and the sum of dotted segments at the level s for Equation (9.7b).

uniform convergence would be violated for the integrand obtained. Indeed, the integral

$$f(s) \Rightarrow (1/\Delta\omega) \int_{\omega'}^{\omega''} d\omega\, \delta(s - \kappa(\omega)),$$

obtained as a result of this integration reordering exists only if $\kappa'(\omega) \neq 0$ in the interval $\Delta\omega$, and this cannot apply to the physical problem under examination, with a number of maxima and minima in the real spectrum. This explains the difficulties of using this idea in computations (Lacis and Hansen, 1974; Liou and Sasamori, 1975). The problem of reordering the integration disappears when substitution of Equation (9.2) into Equation (9.4) is made. In this case, integration can be performed and the result will be Equations (9.9) and (9.10) (Tvorogov, 1994).

When expressions (9.3) and (9.4) are considered as a base for expansion into the series of exponents (9.7) and (9.8), the ordinates b_ν in Equation (9.8) are independent of the thermodynamic characteristics of the medium, which advantageously distinguishes them from a_ν in Equation (9.7). This fact allows one to treat the series structure of Equation (9.8) more rationally. Indeed, Equations (9.9) and (9.10) formalize the ordering $\kappa(\omega)$ according to their values (i.e., a procedure suggested in a number of papers and virtually based on qualitative considerations). Rigorous mathematical formulation enables one to extend this procedure to more complicated cases than that of a homogeneous medium (e.g., an inhomogeneous medium, overlapping bands, integrals with a source function). The fact that the coefficients of exponents do not depend on the thermodynamic characteristics of the medium plays a significant role in this case. We shall now consider briefly these extensions.

The transmission function for an inhomogeneous medium has the form

$$H = \frac{1}{\Delta\omega} \int_{\omega'}^{\omega''} e^{-\tau(\omega)}\, d\omega \tag{9.11}$$

with optical depth

$$\tau(\omega) = \int_{(l)} dl\kappa(\omega, l). \tag{9.12}$$

The integration in Equation (9.12) is performed over points l along the path of the ray; the absorption coefficient $\kappa(\omega, l)$ depends on l.

Formally, the transmission function for the inhomogeneous medium (9.11) is reduced to the transmission function for a homogeneous medium (9.2) using the virtual factor ζ: this is added to Equation (9.12), and then in the final formulas we let $\zeta \to 1$ (Tvorogov *et al.*, 2000). As a result we have a series

$$H = \sum_{\nu} b_{\nu}\, e^{-\tilde{s}(g_{\nu})}, \tag{9.13}$$

where $\tilde{s}$ is the inverse function to $\tilde{g}(s)$. The latter is defined by Equations (9.9) and (9.10) after substitution of τ for κ:

$$\tilde{g}(s, l) = \frac{1}{\Delta\omega} \int_{\tau(\omega,l) \le \tilde{s}, \omega \in [\omega', \omega'']} d\omega. \tag{9.14}$$

It is essential that Equation (9.14) is a rigorous formula. It implies again the ordering procedure, but now $\tau(\omega, l)$ is ordered as a function of ω in $\tilde{s}$ calculation for given $\int_{(l)} dl(\cdots)$.

The variant

$$H = \sum_{\nu} b_{\nu}\, e^{-\int_{(l)} s(g_{\nu}, l')\, dl'}, \tag{9.15}$$

where $s(g, l)$ is the inverse function of Equation (9.14) with $\kappa(\omega) \to \kappa(\omega, l)$, is very popular in the literature. In other words, $s(g, l)$ is obtained upon ordering the absorption coefficients in a given layer l. This is just the assumption of the correlation of absorption coefficients in different layers (the correlated k-distribution, or CKD, approximation). Numerical investigations show the validity of Equation (9.15) in most cases in the Earth's atmosphere. We think however that it is preferable to use the foolproof variant (9.13), especially considering that it requires virtually no extra computational effort.

A similar approach can be adopted with overlapping bands of different gases, when the absorption coefficient is equal to

$$\kappa(\omega) = \sum_{j} \kappa_j(\omega). \tag{9.16}$$

The subscript j numbers the components of the mixture. Equations (9.7)–(9.10) remain the same, and we just use Equation (9.16) instead of κ. In the case of a homogeneous medium it is possible to take a "dimension" length as a variable z by including the gas density into the definition of κ_j.

The example with overlapping spectra allows us to discuss one quite general question: Equation (9.5) and the relationship between s and κ lead us to interpret $f(s)$ as a probability density of the fact that the absorption coefficient takes a definite value s, and further to apply to Equation (9.4) the rules of probability theory.

This idea was used in a paper by Goody *et al.* (1989) to write function (9.4) for the mixture of two gases in the form

$$\Lambda(s) = \int_0^s ds' f_1(s') g_2(s - s'),$$

where subscripts 1 and 2 refer to different gases. Application of Equations (9.3) and (9.4) leads to

$$\Lambda(s) = \frac{1}{(\Delta\omega)^2} \int_{\kappa_1(\omega)+\kappa_2(\omega)\leq s} d\omega_1\, d\omega_2 \tag{9.17}$$

instead of the exact relation

$$\Lambda(s) = \frac{1}{\Delta\omega} \int_{\kappa_1(\omega)+\kappa_2(\omega)\leq s} d\omega.$$

The difference between the last two formulas is quite understandable. The mathematical reason for the absence of the probability interpretation of $f(s)$ is in fact already indicated. It would be possible if the integration could be reordered after substitution of Equation (9.2) into (9.3), but this is impossible for the reasons already explained.

The expansion into the series of exponents for integrals with the source function $B(\omega, \Theta)$ does not differ in principle from other variants:

$$\Pi(z) = \frac{1}{\Delta\omega} \int_{\omega'}^{\omega''} d\omega\, B(\omega, \Theta)\, e^{-z\kappa(\omega)}. \tag{9.18}$$

In Equation (9.18) the Planck function $B(\omega, \Theta)$ is related to the spectrum of the external field or to the coefficient of the intrinsic emission of the medium. In order not to perform integration over g outside the interval $[0, 1]$, it is necessary to introduce the relation

$$U(\omega, \Theta) = \frac{B(\omega, \Theta)}{\Omega}, \qquad \Omega = \frac{1}{\Delta\omega} \int_{\omega'}^{\omega''} U(\omega, \Theta)\, d\omega.$$

Then, for Equation (9.18) the series (9.8) appears with the substitution of $s'(g)$ for $s(g)$, $s'(g)$ being the inverse of the function

$$g'(s) = \frac{1}{\Delta\omega} \int_{\kappa(\omega)\leq s} U(\omega, \Theta)\, d\omega. \tag{9.19}$$

The variant (9.19) differs from Equation (9.10) only in the positions of the points on axis g (owing to the change of variables $d\omega' = d\omega\, U(\omega, \Theta)$), and can readily be handled on a computer.

Application of the above relations suggests the preliminary computation of the absorption coefficient with all its intricate problems. It is quite clear however that many spectral details may turn out to be smoothed for the quantities integrated over the frequency spectrum, and so it becomes attractive to construct $s(g)$ immediately on the basis of experimental data on the transmission function. Direct calculation using

Equation (9.4) requires analytical prolongation of the function (9.2) in this case. This is mathematically cumbersome (Tvorogov, 2001), and so to date only statement of the problem has been formulated.

One more interesting problem for investigation is the relationship of the Dirichlet series theory with the characteristics of the fractal structure of spectra (Kistenev *et al.*, 2002, 2003). This issue requires a large introduction and is beyond the scope of this chapter.

9.3 THE SERIES OF EXPONENTS AND THE RADIATIVE TRANSFER EQUATION

9.3.1 Integration of the radiative transfer equation over the frequency spectrum (kinetic equation)

The equation of radiative transfer in an inhomogeneous aerosol–molecular medium

$$\begin{aligned} \mathbf{n}\,\mathrm{grad}\, I(\mathbf{r},\mathbf{n};\omega) &= -(\kappa(\mathbf{r},\omega)+\sigma(\mathbf{r},\omega))I(\mathbf{r},\mathbf{n};\omega) \\ &\quad + \int d\mathbf{n}'\,\varphi(\mathbf{n},\mathbf{n}';\omega)I(\mathbf{r},\mathbf{n}';\omega)+\eta(\mathbf{r},\omega) \end{aligned} \tag{9.20}$$

is written for the spectral intensity at frequency ω, $I(\mathbf{r},\mathbf{n};\omega)$, of a ray at point $\mathbf{r}$ traveling in the direction of the unit vector $\mathbf{n}$; $\kappa(\mathbf{r},\omega)$, $\sigma(\mathbf{r},\omega)$, $\eta(\mathbf{r},\omega)$, and $\phi(\mathbf{n},\mathbf{n}';\omega)$ are the coefficients of molecular absorption, aerosol extinction, emission, and the scattering phase function, respectively. Assume that we need a spectrally integrated quantity

$$A(\mathbf{r},\mathbf{n}) = \frac{1}{\Delta\omega}\int_{\omega'}^{\omega''} d\omega\, I(\mathbf{r},\mathbf{n};\omega), \qquad \Delta\omega = \omega''-\omega'. \tag{9.21}$$

Formally, this implies integration of the radiative transfer equation over frequency and the series of exponents is necessary to derive the corresponding kinetic equations (Tvorogov, 1999).

The practical problem which appears in Equations (9.20) and (9.21) is well known. The integral term in Equation (9.20) creates certain computational difficulties. This term describes scattering whose characteristics rather slightly depend on ω. At the same time, molecular absorption, essentially trivial in the problem of wave propagation, crucially increases the computation burden from the use of Equation (9.21) due to the huge number of spectral lines. Naturally, we would like to have an equation immediately for the quantity (9.21). At this point the series of exponents for the transmission function treated in Section 9.2 appear to be useful.

The idea of integration of the transfer equation over frequency has been put forward by Van de Hulst and Irvine (1963) as applied to a homogeneous medium without emission, $\eta = 0$ in Equation (9.20). The structure of this transformation is rather simple. Let us consider this case in order to extend it to the inhomogeneous medium with emission.

We introduce the function $J(\mathbf{r},\mathbf{n},l;\omega)$ which is the solution of the equation

$$\mathbf{n}\,\mathrm{grad}\,J + \frac{\partial J}{\partial l} = -(\kappa+\sigma)J + \int d\mathbf{n}'\,\varphi J(\mathbf{r},\mathbf{n}',l;\omega).$$

The boundary condition for function J is $J(l=0) = J(l=\infty) = 0$. Obviously, $I = \int_0^\infty dl\,J$.

The following substitution

$$J = E\,e^{-\kappa(\omega)l} \tag{9.22}$$

eliminates the selective summand $\kappa(\omega)$, and the equation for E

$$\mathbf{n}\,\mathrm{grad}\,E + \frac{\partial E}{\partial l} = -\sigma E + \int d\mathbf{n}'\,\varphi E(\mathbf{r},\mathbf{n}',l) \tag{9.23}$$

now contains only the ω-independent characteristics of scattering (in the given spectral range). By integrating Equation (9.21) over ω and l, it can be shown that the quantity (9.21) takes the form

$$A = \sum_\nu b_\nu A_\nu, \tag{9.24}$$

where A_ν is the solution of the following equation

$$\mathbf{n}\,\mathrm{grad}\,A_\nu(\mathbf{r},\mathbf{n}) = -(\sigma + s(g_\nu))A_\nu + \int d\mathbf{n}'\,\varphi A(\mathbf{r},\mathbf{n}').$$

As a result, we come back to the same radiative transfer equation (9.20). The difference is that in the new equation the function $\kappa(\omega)$, which varies widely with frequency ω, is replaced by the frequency-independent function $s(g_\nu)$. If the quadrature formula is chosen properly, Equation (9.24) involves only several terms and not thousands of terms, which is typical in the case of direct numerical integration of Equation (9.20). Convincing estimates in this respect can be found in the paper by Firsov *et al.* (2002).

The approach at hand is extended in the paper by Tvorogov (1999) to the case of an inhomogeneous medium (all aerosol and molecular characteristics are functions of $\mathbf{r}$), taking into account emission of the medium. In this instance $I = I^{(0)} + I'$, where $I^{(0)}$ is the general solution of the homogeneous equation (at $\eta = 0$) with the boundary conditions of the problem, and I' is a partial solution of the inhomogeneous equation. It is clear that A acquires a similar structure: $A = A^{(0)} + A'$.

Now, because of the dependence of κ on $\mathbf{r}$ when calculating $I^{(0)}$, substitution of Equation (9.22) into Equation (9.23) in E leads to:

$$-El\mathbf{n}\,\mathrm{grad}\,\kappa(\mathbf{r},\omega) + \mathbf{n}\,\mathrm{grad}\,E + \frac{\partial E}{\partial l} = -\sigma E + \int d\mathbf{n}'\,\varphi E(\mathbf{r},\mathbf{n}',l).$$

that is, it adds to the right-hand side of Equation (9.23) the term $(-El)\mathbf{n}\,\mathrm{grad}\,\kappa(\omega,\mathbf{r})$, where the frequency dependence of this term is determined by the absorption coefficient $\kappa(\mathbf{r},\omega)$. Approximation becomes inevitable, if there is a desire to get rid of this term. The condition of such a possibility follows from the form of the last equation:

$$|\text{grad}\,\kappa(\mathbf{r},\omega)|/\sigma \ll 1. \tag{9.25}$$

Equation (9.25) implies comparison of the "excess" term and the first term on the right-hand side of the equation in E and is almost unquestionably true in most applications of atmospheric optics. Formally, Equation (9.23) can be interpreted as a "time-dependent radiative transfer equation" (regarding $l \sim t$ as a certain "time") and for it, in any problem with scattering, $l = 0(1/\sigma)$ (Sobolev, 1972). Condition (9.25) returns us to the previous variant (9.23). The only difference is that $s(g_\nu) \to s(g_\nu;\mathbf{r})$.

The partial solution is equal to

$$I'(\mathbf{r}',\mathbf{n}';\omega) = \int d\mathbf{r}\,d\mathbf{n}\,\eta(\mathbf{r},\mathbf{n};\omega)G_\omega(\mathbf{rn}\,|\,\mathbf{r}'\mathbf{n}'),$$

where the Green function $G_\omega(\mathbf{rn}\,|\,\mathbf{r}'\mathbf{n}')$ is the solution of the problem

$$\begin{aligned}-\mathbf{n}\,\text{grad}\,G_\omega(\mathbf{r},\mathbf{n}\,|\,\mathbf{r}',\mathbf{n}') &= -(\kappa(\mathbf{r},\omega)+\sigma(\mathbf{r},\omega))G_\omega(\mathbf{r},\mathbf{n}\,|\,\mathbf{r}',\mathbf{n}') \\ &\quad + \int d\mathbf{n}''\,\varphi(\mathbf{r},\mathbf{n},\mathbf{n}'';\omega) + \delta(\mathbf{r}-\mathbf{r}')\,\delta(\mathbf{n}-\mathbf{n}')\end{aligned}$$

under zero boundary conditions. The last equation is transformed as pointed out above employing the series of exponents for the source function. Note that $A'_\nu = A_\nu^{(1)} + A_\nu^{(2)}$, where individual terms refer to the radiation of molecules and aerosol particles. Finally, we have the following relations:

$$A(\mathbf{r},\mathbf{n}) = \sum_\nu b_\nu A_\nu(\mathbf{r},\mathbf{n});$$

$$A_\nu = A_\nu^{(0)} + A_\nu^{(1)} + A_\nu^{(2)};$$

$$\mathbf{n}\,\text{grad}\,A_\nu^{(0)} = -(\sigma + s(g_\nu;\mathbf{r}))A_\nu^{(0)} + \int d\mathbf{n}'\,\varphi(\mathbf{r},\mathbf{n},\mathbf{n}')A_\nu^{(0)}(\mathbf{r},\mathbf{n}');$$

where $s(g_\nu;\mathbf{r})$ is the inverse function of

$$g(s,\mathbf{r}) = \frac{1}{\Delta\omega}\int_{\kappa(\omega,\mathbf{r})\le s;\omega\in[\omega',\omega'']} d\omega;$$

$$\mathbf{n}\,\text{grad}\,A_\nu^{(1)} = -(\sigma + \gamma(g_\nu;\mathbf{r}))A_\nu^{(1)} + \int d\mathbf{n}'\,\varphi(\mathbf{r},\mathbf{n},\mathbf{n}')A_\nu^{(1)}(\mathbf{r},\mathbf{n}') + \Omega(\mathbf{r})q;$$

$\gamma(g_\nu; \mathbf{r})$ is the inverse function of

$$g(\gamma, \mathbf{r}) = \frac{1}{\Delta\omega} \int_{\kappa(\omega,\mathbf{r}) \le \gamma; \omega \in [\omega', \omega'']} U(\omega)\, d\omega;$$

$$U = \frac{B}{\Omega};$$

$$\mathbf{n} \operatorname{grad} A_\nu^{(2)} = -(\sigma + \gamma'(g_\nu; \mathbf{r}))A_\nu^{(2)} + \int d\mathbf{n}'\, \varphi(\mathbf{r}, \mathbf{n}, \mathbf{n}') A_\nu^{(2)}(\mathbf{r}, \mathbf{n}') + \Omega'(\mathbf{r})\gamma'(g_\nu; \mathbf{r});$$

$$\Omega'(\mathbf{r}) = \frac{1}{\Delta\omega} \int_{\omega'}^{\omega''} B(\omega)\rho(\omega)\, d\omega;$$

and γ' is the inverse function of

$$g(\gamma', \mathbf{r}) = \frac{1}{\Delta\omega} \int_{\kappa(\omega,\mathbf{r}) \le \gamma'; \omega \in [\omega', \omega'']} U'(\omega)\, d\omega; \qquad U' = \frac{B\rho}{\Omega'}. \tag{9.26}$$

Thus, the above-written radiative transfer equations for the radiation characteristics integrated over the frequency spectrum incorporate the quantities that are non-selective with respect to ω and are valid for the inhomogeneous aerosol–molecular medium with emission. Note that condition (9.25) resulting in the existence of such an equation suggests the definite relation (9.25) between molecular absorption and aerosol scattering and has no bearing on CKD approximation.

9.3.2 Radiation fluxes in the aerosol–molecular medium

Some examples of computation of radiation fluxes in the Earth's atmosphere are given below in order to illustrate the use of the series of exponents.

Radiation codes applied in the Institute of Atmospheric Physics of the Siberian Branch of the Russian Academy of Aciences (IAP SB RAS) for solution of the problem of radiation fluxes in the aerosol–molecular medium use two algorithms for taking into account molecular absorption. One of them based on calculation of the transmission function (called Algorithm 1 below) rests upon the probability of the photon's "trajectory" (in terms accepted in the Monte Carlo method) in which molecular absorption is included phenomenologically. The commonly employed assumption concerning the rather slight spectral dependence of optical aerosol characteristics enables one to make good use of any information on the transmission function.

Another algorithm based on the use of photon single-scattering albedo (called Algorithm 2 below) invokes the series of exponents as outlined above. Expansion coefficients in the series of exponents were determined with the use of Equation (9.26), where the solar constant served as a source function. The molecular absorption coefficient was calculated by the line-by-line method.

9.3.2.1 Test calculations

Results of comparison of calculations of radiation characteristics taking the molecular absorption into account according to the two algorithms mentioned above and the benchmark line-by-line calculations are presented.

Table 9.1 presents the values of upward ($F^{\uparrow}_{clr}$) and downward ($F^{\downarrow}_{clr}$) fluxes of solar radiation calculated by two algorithms (1 and 2) in the case of an aerosol–molecular medium. Molecular absorption in Algorithm 1 was considered through the transmission function of atmospheric gases, and in Algorithm 2 through the probability of photon single-scattering albedo. Calculations were made in the region of the spectral band 10,000 cm^{-1}–10,500 cm^{-1} with $N_{\text{exp}} = 4$ (N_{exp} is the number of terms of the series of exponents) for the Maritime I aerosol model (WCP, 1986) and correspond to the 50th and 51st ICRCCM standard sets (Fouquart *et al.*, 1991). Spectral line parameters were taken from the database HITRAN-2000 (*http://www.hitran.com*), and the solar constant was taken from LOWTRAN-7 (Kneizys *et al.*, 1996). The difference between fluxes calculated by Methods 1 and 2 does not exceed 0.05%–0.1%, which is much smaller than the error due to the use of the short series of exponents.

Table 9.1 also includes the results of benchmark calculations from Fomin and Gershanov (1996), performed using the database HITRAN-92 by the line-by-line method. Comparison of the results in Table 9.2 shows that the present algorithms of taking molecular absorption into account are on the whole in good agreement with benchmark calculations. The existing discrepancy (tenths of watts per square meter) is caused by the use of different HITRAN databases and the different number N_{exp} of quadrature (Zhuravleva and Firsov, 2004).

9.3.2.2 *Comparison with the data of field experiments*

In this section we compare the calculated spectral fluxes of downward solar radiation and results of measurements obtained in the case of single-layer low-level overcast clouds during the 1997–1998 ARM campaign at the ARM SGP (Southern Great Plains) site in Oklahoma (Li *et al.*, 2000).

Calculations accounted for the spectral behavior of the surface albedo inferred from MFRSR measurements. The cloud extinction coefficient σ_{cl} was chosen so that the calculated and measured spectral fluxes coincided in the 500 nm–550 nm band. The effective radius of cloud droplets r_{ef} was varied in the range from 6 µm to 11 µm.

Calculations of spectral fluxes were performed for three different cloud situations (Table 9.2) using Algorithm 1. The effective molecular absorption coefficients were found taking into account the Gaussian filter function RSS radiometer (512 channels) (Zhuravleva and Firsov, 2004, 2005). The aerosol characteristics corresponded to the cont-I model of the continental aerosol (WCP, 1986). Scattering in clouds was simulated using the Henyey–Greenstein scattering phase function with the mean cosine $\langle \mu \rangle = 0.86$.

Figure 9.2 gives the spectral fluxes of the downward solar radiation $F^{\downarrow}(\lambda)$, measured at the surface with the RSS radiometer and calculated using our algorithm and MODTRAN-4 radiation code. The measured and numerical results agree quite well, except at the center of the H_2O band at 940 nm.

Table 9.1. Upward and downward radiation fluxes $F^{\uparrow}_{clr}(z)/F^{\downarrow}_{clr}(z)$ in the aerosol–molecular atmosphere in the 10,000 cm^{-1}–10,500 cm^{-1} spectral range calculated by different methods. The number of terms in the series of exponents $N_{\exp} = 4$, MLS meteorological model (Anderson *et al.*, 1986), Maritime I aerosol model (WCP, 1986), solar zenith angle $\xi_{\oplus} = 30°$.

z (km)	$A_s = 0$					$A_s = 0.8$				
	Fomin and Gershanov (1996)	*Our calculation*				*Fomin and Gershanov (1996)*	*Our calculation*			
	LBL	*Algorithm 1*	*Algorithm 2*	*Algorithm 1*	*Algorithm 2*	*LBL*	*Algorithm 1*	*Algorithm 2*	*Algorithm 1*	*Algorithm 2*
	HITRAN-92			*HITRAN-2000*		*HITRAN-92*			*HITRAN-2000*	
1	0 / 22.87	0 / 22.74	0 / 22.74	0 / 22.40	0 / 22.39	18.57 / 23.22	18.48 / 23.10	18.48 / 23.10	18.20 / 22.74	18.20 / 22.73
2	0.0592 / 24.96	0.061 / 24.68	0.0616 / 24.68	0.0599 / 24.40	0.0603 / 24.39	16.84 / 25.21	16.90 / 24.94	16.90 / 24.94	16.55 / 24.64	16.54 / 24.64
3	0.117 / 26.79	0.121 / 26.66	0.121 / 26.67	0.118 / 26.43	0.118 / 26.45	16.08 / 26.94	16.18 / 26.82	16.17 / 26.83	15.79 / 26.59	15.78 / 26.60
4	0.128 / 28.23	0.132 / 28.36	0.132 / 28.35	0.129 / 28.21	0.130 / 28.23	15.74 / 28.36	15.84 / 28.51	15.84 / 28.50	15.44 / 28.35	15.43 / 28.37
5	0.142 / 29.28	0.146 / 29.57	0.146 / 29.56	0.142 / 29.48	0.143 / 29.47	15.57 / 29.41	15.68 / 29.71	15.68 / 29.69	15.27 / 29.61	15.27 / 29.60
10	0.154 / 30.00	0.16 / 30.33	0.16 / 30.34	0.157 / 30.28	0.157 / 30.26	15.48 / 30.11	15.60 / 30.45	15.59 / 30.45	15.18 / 30.39	15.18 / 30.37
12	0.219 / 31.31	0.224 / 31.35	0.224 / 31.34	0.221 / 31.35	0.221 / 31.34	15.38 / 31.36	15.49 / 31.40	15.49 / 31.39	15.07 / 31.40	15.07 / 31.39

20	0.239 31.40	0.245 31.41	0.245 31.38	0.242 31.41	0.232 31.43	15.36 31.43	15.47 31.44	15.47 31.42	15.06 31.44	15.06 31.46
50	0.264 31.44	0.274 31.44	0.274 31.46	0.27 31.44	0.27 31.46	15.37 31.44	15.48 31.45	15.48 31.47	15.06 31.45	15.06 31.47
70	0.274 31.45	0.286 31.45	0.286 31.47	0.282 31.45	0.283 31.47	15.37 31.45	15.48 31.45	15.48 31.47	15.07 31.45	15.06 31.47
100	0.274 31.45	0.286 31.45	0.286 31.47	0.283 31.45	0.283 31.47	15.37 31.45	15.48 31.45	15.48 31.47	15.07 31.45	15.06 31.47
	0.274 31.45	0.286 31.45	0.286 31.45	0.283 31.45	0.283 31.45	15.37 31.45	15.48 31.45	15.48 31.45	15.07 31.45	15.06 31.45

Table 9.2. Atmospheric parameters used as input data in calculations of spectral solar radiation fluxes; experiments were performed at the Atmospheric Radiation Measurement Southern Great Plains site (U.S.A.).

Date	*Solar zenith angle* (°)	*Total content* (cm)			*Position of the cloud layer* (km)	*Cloud optical depth* (0.55 μm)	*Effective radius* (μm)
		Water	*Liquid*	*Ozone*			
October 19, 1997	47.15	1.6	0.008	0.34	0.58–0.85	16.5	7.2
April 3, 1998	31.17	1.4	0.034	0.38	1.0–1.5	55.1	9.3
August 5, 1998	24.39	4.1	0.019	0.33	1.49–1.88	25.9	9.1

9.3.3 Molecular atmosphere

In Section 9.3.1, application of the series of exponents was examined in the process of integration of the radiative transfer equation in its general form over the frequency spectrum. In the case of a purely molecular atmosphere the solution of the radiative transfer equation is well known, and the series of exponents can be used for immediate frequency integration of the solution, enabling one to obtain expressions for radiation fluxes in terms of series of exponents. This problem was treated in a paper by Nesmelova *et al.* (1999) for a horizontally homogeneous atmosphere under condition (9.15). Now frequency integration of the solution of the radiative transfer equation for a molecular atmosphere will be performed without any additional conditions.

Let us consider the radiative transfer equation

$$\cos\theta \frac{\partial I(\omega, z, \theta)}{\partial z} = -\kappa(\omega, z) I(\omega, z, \theta) + \eta(\omega, z), \tag{9.27}$$

where $I(\omega, z, \theta)$ is the spectral intensity I, at frequency ω, of radiation passing through the horizontally homogeneous atmosphere at height z at angle θ to the vertical; $\kappa(\omega, z)$ is the absorption coefficient; and $\kappa(\omega, z)$ refers to the mixture of gases in the given atmospheric model. The emission coefficient is equal to $\eta(\omega, z) = \kappa B(\omega, \Theta(z))$ under condition of local thermodynamic equilibrium, where $B(\omega, \Theta(z))$ is the Planck function, and $\Theta(z)$ is the temperature.

As usual, the problem of $I(\omega, z, \theta)$ is considered separately for downward ($\pi/2 \leq \theta \leq \pi$) radiation and upward ($0 \leq \theta \leq \pi/2$) radiation, with standard notations $I^{\downarrow}$ and $I^{\uparrow}$. The boundary conditions for downward radiation are taken at the top of the atmosphere, where $B = 0$ and $I(\omega, z, \theta) = 0$ (only infrared radiation is considered). The boundary conditions for upward radiation are taken at the surface, representing an external source of radiation with respect to the atmosphere. Assume

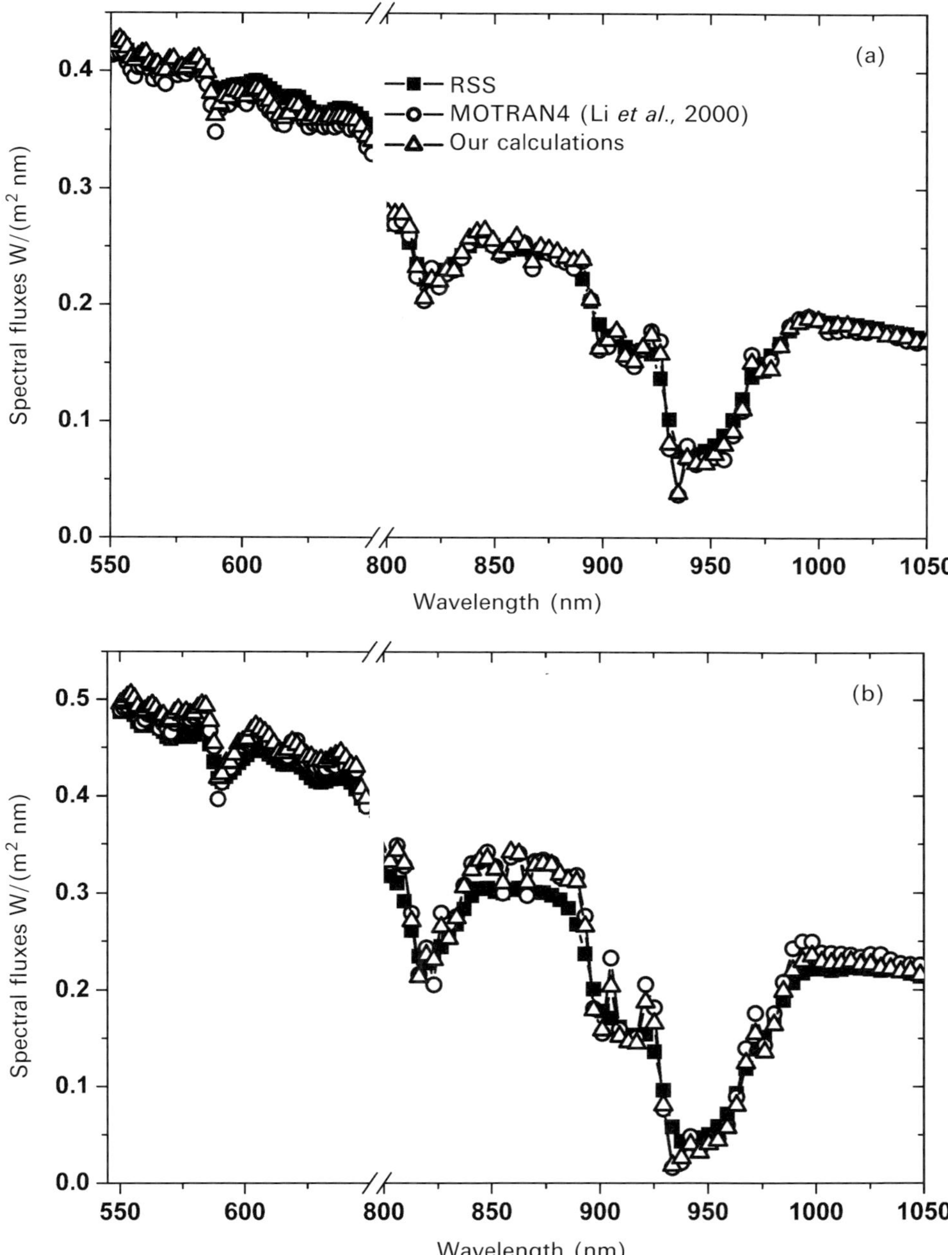

Figure 9.2. Downward solar fluxes at the surface level at the ARM SGP site and model calculations: (a) liquid water path LWP = 0.008 cm, ozone content is 340 DU; (b) liquid water path LWP = 0.019 cm, ozone content is 330 DU.

that the surface radiation is $B(\omega, \Theta)$ and $I(\omega, 0) = B(\omega, 0)$. Surface reflectance is also ignored; the appropriate extensions can be found in the paper by Nesmelova *et al.* (1999).

Let us introduce the function (for the sake of convenience of calculations)

$$D = I - B. \tag{9.28}$$

Then, the equation in D takes the form:

$$\cos\theta \frac{\partial D}{\partial z} = -\kappa D - \frac{\partial B}{\partial z}\cos\theta. \tag{9.29}$$

The boundary condition for D is $D = 0$ for upward and downward radiation. The solution of Equation (9.29) is the function

$$D = -\int_{z_0}^{z} dz' \frac{\partial B(z')}{\partial z'} e^{-\sec\theta \int_{z'}^{z} dz'' \kappa(z'')}, \tag{9.30}$$

where z_0 is the beginning of scale z in the solution of Equation (9.29).

For subsequent frequency integration of Equation (9.30) we use the series of exponents, and finally get

$$\tilde{D} = -\sum_{\nu} b_\nu \int_{z_0}^{z} dz' \frac{\partial A(z')}{\partial z'} e^{-s(g_\nu; z', z)\sec\theta}, \tag{9.31}$$

where g_ν and b_ν are the abscissas and ordinates of an appropriate quadrature formula. In Equation (9.31) the following notation is used. As usual, the quantity, $s(g_\nu; z', z)$ is the inverse function of

$$g(s; z', z) = \frac{1}{\Delta\omega} \int_{\tau(\omega) \le s, \omega \in [\omega', \omega'']} d\omega\, U(\omega, z'), \tag{9.32}$$

and

$$\tau(\omega; z', z) = \int_{z'}^{z} dz''\, \kappa(\omega, z''), \tag{9.33}$$

$$A(z') = \frac{1}{\Delta\omega} \int_{\omega'}^{\omega''} B(\omega, z')\, d\omega, \qquad \Omega(z) = \int_{\omega'}^{\omega''} B(\omega, z)\, d\omega, \tag{9.34}$$

$$U(\omega, z') = \frac{\dfrac{\partial B(\omega, z')}{\partial z'}}{\dfrac{\partial A(z')}{\partial z'}}. \tag{9.35}$$

In the case of $I^{\uparrow}$ the value $z_0 = 0$, $0 \le \theta \le \pi/2$, and $z' \le z$. Note that the corresponding optical depth is

$$\tau^{\uparrow}(\omega; z', z) = \int_{z'}^{z} dz''\, \kappa(\omega, z''), \tag{9.36}$$

and we find for it the value $s^{\uparrow}(g_\nu; z', z)$ using Equations (9.32) and (9.33). Then,

Equation (9.31) leads to

$$I^{\uparrow} = -\sum_{\nu} b_{\nu} \int_0^z dz' \frac{\partial \Omega(z')}{\partial z'} e^{-s^{\uparrow}(g_{\nu};z',z)\sec\theta} + \Omega(z'). \tag{9.37}$$

Now Equation (9.37) can be integrated by parts. The integral in Equation (9.37) is equal to

$$A(z')\, e^{-s^{\uparrow}(g_{\nu};z',z)\sec\theta}\Big|_{z'=0}^{z'=z} + \int_0^z dz'\, A(z') \sec\theta\, e^{-s^{\uparrow}(g_{\nu};z',z)\sec\theta} \frac{\partial s^{\uparrow}(g_{\nu};z',z)}{\partial z'}.$$

The free term at the low limit is

$$-A(0)\, e^{-s^{\uparrow}(g_{\nu};0,z)\sec\theta}.$$

The upper limit of the free term is equal to $\Omega(z)$, and after its substitution into Equation (9.37) it is canceled with the second term in the right-hand side of Equation (9.37) because $\sum_{\nu} b_{\nu} = 1$.

Some comment is required concerning the equality $s(g_{\nu}; z, z) = 0$. This limit can be found by setting $z' = z - \Delta z$ in the definition of the transmission function and subsequent passing to the limit $\Delta z \to 0$. Since $s^{\uparrow}(g)$ represents $\tau^{\uparrow}$ ordered in value, the quantity $s(g_{\nu}; z, z)$ can be expanded in a Taylor series with only first-term $O(\Delta z)$. The derivative of the transmission function with respect to the low limit after multiplication by $\Delta z \to 0$ will be positive and will define the quantities $s^{\uparrow}(g)$, which $\to 0$ when $|\Delta z| \to 0$.

Finally,

$$I^{\uparrow} = A(0) \sum_{\nu} b_{\nu}\, e^{-s^{\uparrow}(g_{\nu};0,z)\sec\theta} - \sum_{\nu} b_{\nu} \int_0^z dz'\, A(z')\, e^{-s^{\uparrow}(g_{\nu};z',z)\sec\theta} \sec\theta \frac{\partial s^{\uparrow}(g_{\nu};z',z)}{\partial z'}. \tag{9.38}$$

Apparently, the terms of Equation (9.38) correspond to the emission of source radiation and the emission of atmospheric radiation.

If the quantity $I^{\downarrow}$ is considered, we have $z_0 = \infty$, $\pi/2 \le \theta \le \pi$, and $z' \ge z$. After transformations similar to those above we obtain

$$I^{\downarrow} = -\sum_{\nu} b_{\nu} \int_z^{\infty} dz'\, A(z')\, e^{s^{\downarrow}(g_{\nu};z,z')\sec\theta} \sec\theta \frac{\partial s^{\downarrow}(g_{\nu};z,z')}{\partial z'}. \tag{9.39}$$

Further calculation of the fluxes and influxes is quite standard. Calculation of the fluxes integrated over angles enables one to write the expression for the radiation summand K in the equation $(\partial\Theta/\partial t \sim K)$ for the temperature distribution in the atmosphere:

$$K = 2\pi\Omega(0)\sum_{\nu} b_\nu E_2(s^{\uparrow}(g_\nu;0,z))\frac{\partial s^{\uparrow}(g_\nu;0,z)}{\partial z} + 2\pi\sum_{\nu} b_\nu \int_0^z dz'$$
$$\times\left\{-E_1(s^{\uparrow}(g_\nu;z',z))\frac{\partial s^{\uparrow}(g_\nu;z',z)}{\partial z}\frac{\partial s^{\uparrow}(g_\nu;z',z)}{\partial z'} + E_2(s^{\uparrow}(g_\nu;z',z))\frac{\partial^2 s^{\uparrow}(g_\nu;z',z)}{\partial z'\,\partial z}\right\}$$
$$+ 2\pi\sum_{\nu} b_\nu \int_z^\infty dz'\,\Omega(z')$$
$$\times\left\{-E_1(s^{\downarrow}(g_\nu;z,z'))\frac{\partial s^{\downarrow}(g_\nu;z,z')}{\partial z}\frac{\partial s^{\downarrow}(g_\nu;z,z')}{\partial z'} + E_2(s^{\downarrow}(g_\nu;z,z'))\frac{\partial^2 s^{\downarrow}(g_\nu;z,z')}{\partial z\,\partial z'}\right\} \tag{9.40}$$

where E_2 is the integral exponential function.

Thus, the rigorous mathematical theory of the series of exponents provides an opportunity to perform, in essence, exact calculation in the case of the molecular medium. Inclusion of realistic boundary conditions, horizontally inhomogeneous medium, refraction, specific features of surface reflection, and the spectrum of solar radiation, etc. will lead only to more lengthy formulas (Tvorogov, 1999).

9.4 THE SERIES OF EXPONENTS AS A MEANS FOR CALCULATION SIMPLIFICATIONS

The theory of the series of exponents provides an opportunity to solve some particular problems appearing in the process of calculating radiation characteristics. In this section some issues are discussed which have not yet quantitative illustrations but are worthy of note from both the theoretical and practical points of view due to their formulations themselves and suggested ways of solution. This permits us to consider overlapping bands with the help of introduction of an equivalent line: calculation formulas obtained using asymptotic analysis in the case of small pressures, derivation of the necessary condition of fulfillment of CKD approximation, application of one-parameter formulas for the expansion coefficients of the series of exponents.

9.4.1 Equivalent line and overlapping bands

Expression (9.17) given in Section 9.2 provides a rigorous solution to the problem but some of it is somewhat inconvenient for mass computations. Thus, every combination of concentrations in a mixture requires a new calculation by (9.17). It is desirable to have some approximate way of allowing one to use the separate functions $s_i(g)$ found for each individual gas, for the determination of function $s(g)$ of the mixture. This is not difficult to do, but the direct application of this variant leads

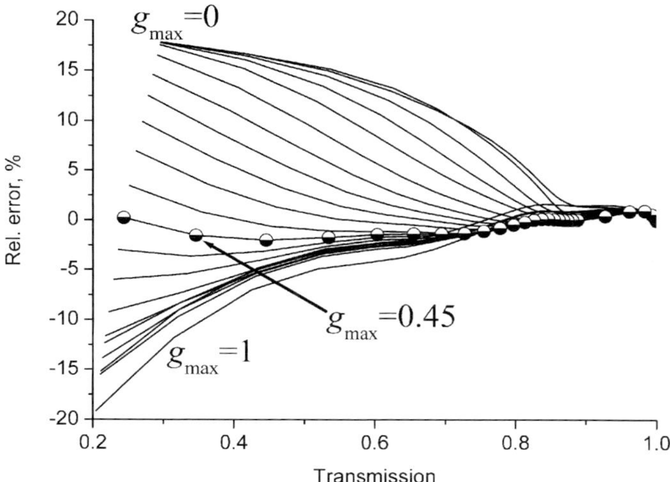

Figure 9.3. The error of accounting for overlap of H_2O and O_3 absorption bands in the spectral range 4,600 cm^{-1}–4,800 cm^{-1} at different positions of the maximum g_{max} of the equivalent line. Atmospheric transmission was calculated for the pathlength with the fixed upper limit 90 km, and with the lower limit varying from 90 km to 0 km.

to a significant increase in the terms of the total series. The idea to introduce an equivalent line to reduce the number of terms is approved in papers (Firsov and Chesnokova, 1998; Firsov *et al.*, 1998), where it appears as an approximation technique. In doing so the function $s(g)$ remains monotonic for one of the gases, and for another it is taken as a unimodal function with the varying position of the maximum. Computer experiments confirm the efficiency of this approach (see Figure 9.3).

Let us now generalize application of the equivalent line idea to obtain the sufficient condition of existence of a straightforward variant of calculation of the transmission function in the mixture of two gases through functions $s_i(g)$ pertaining to individual gases:

$$P = \int_0^1 e^{-s(g)z}\,dg \cong \int_0^1 e^{-s_1(g)u_1 - s_2(g)u_2}\,dg, \tag{9.41}$$

where u_1 and u_2 are the absorber amounts of corresponding gases.

In the 1960s, the idea of constructing an equivalent line for the spectral range $\Delta\omega$ was rather popular (Zuev, 1966). It should lead to the same value of the transmission function as the usual expression (9.2). The variant of the rigorous solution of a similar problem making good use of the technique producing the series of exponents is treated below. The function $s(g)$ constructed on the interval $[0, 1]$ for calculation of the transmission function P is monotonic (see Section 9.2). At the same time in this interval there may exist some other functions whose integral is equal to P. For instance, the function $f(g)$ shown in Figure 9.3 is symmetrical in the range $[0, 1]$, and its maximum and minimum are equal to s_{max} and s_{min} values of the function $s(g)$ (we will call these values invariants of $f(g)$).

It is apparent that the construction of the series of exponents for the spectrum $f(g)$ will lead to $P(z)$ if the transmission function is written as

$$P(z) = \int_0^1 dg\, e^{-zf(g)} = \int_0^1 dg\, e^{-zs(g)}. \tag{9.42}$$

Of course, the interrelation between $g(s)$ and $P(z)$ is unique, as follows from Equation (9.4). However, a unique relation between $P(z)$ and $\kappa(\omega)$ will exist only in the case of a monotonic absorption coefficient $\kappa(\omega)$. It is also clear that $f(g)$ should be non-monotonic (such as the line in Figure 9.3) because otherwise $f(g)$ simply coincides with $s(g)$.

Now let us return to the case of overlapping bands.

To estimate the approximation (9.41) the following function is introduced

$$\Lambda(g) = -s(g)z + u_1 s_1(g) + u_2 s_2(g) \equiv -s(g)z + s_0(g). \tag{9.43}$$

The consequence of Equations (9.41) and (9.43) is

$$P = \int_0^1 e^{-s_0(g)+\Lambda(g)}\, dg. \tag{9.44}$$

Let us consider the cases of monotonic and non-monotonic $\Lambda(g)$ separately. Assume that $\Lambda(g)$ is the monotonic function. Then the application of the second mean value theorem to the integral (9.44) gives

$$P = e^{\Lambda(0)} \int_0^{\xi} e^{-s_0(g)}\, dg + e^{\Lambda(1)} \int_{\xi}^1 e^{-s_0(g)}\, dg, \quad 0 < \xi < 1.$$

We have

$$P = P_0 \equiv \int_0^1 e^{-s_0(g)}\, dg$$

under conditions

$$|\Lambda(0)| \ll 1, \qquad |\Lambda(1)| \ll 1; \tag{9.45}$$

which is, in fact, equivalent to approximation (9.41).

Conditions (9.45) can be expressed through the invariants of the function $f(g)$ (see Figure 9.4):

$$|-s_{max(min)} + u_1(s_1)_{max(min)} + u_2(s_2)_{max(min)}| \ll 1. \tag{9.46}$$

Eventually, the fact that function $s(g)$ is the ordered spectral absorption coefficient and Equation (9.41) allow one to write Equation (9.43) in the equivalent form:

$$|-(k_1 u_1 + k_2 u_2)_{max(min)} + u_1(k_1)_{max(min)} + u_2(k_2)_{max(min)}| \ll 1. \tag{9.47}$$

The monotonicity of Equation (9.43) may hardly be guaranteed even in the case of the monotonicity of all summands in Equation (9.43). The consequences of the absence of monotonicity for the approximation (9.41) are clarified by using the function depicted in Figure 9.5.

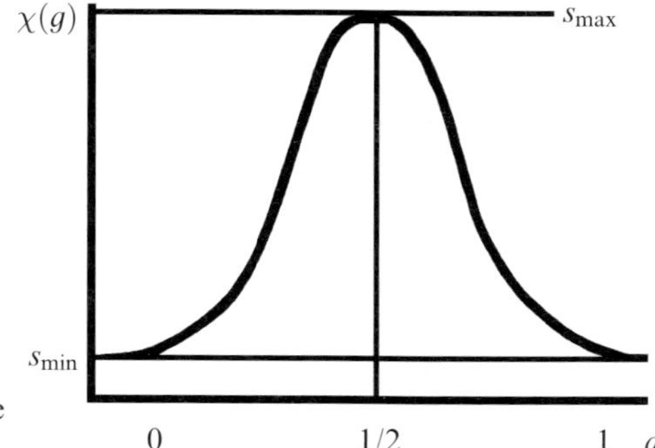

Figure 9.4. Equivalent line $f(g)$ constructed on the base of $s(g)$ and resulting in the same value of the transmission function $P(z)$.

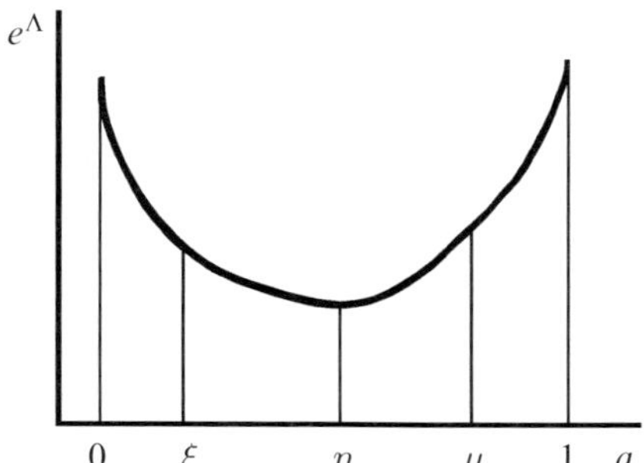

Figure 9.5. Behavior of the quantity e^{Λ} characterizing the error of approximation (9.36) in the case of non-monotonic e^{Λ}.

Let us consider the quantity (9.44) in this case. We split the interval [0, 1] at the point η, so that $P = \int_0^{\eta} + \int_{\eta}^1$, and apply to each integral the second mean value theorem. Then we have

$$P = e^{\Lambda}(0) \int_0^{\xi} e^{-s_0(g)}\, dg + e^{\Lambda}(\eta) \int_{\xi}^{\eta} e^{-s_0(g)}\, dg + e^{\Lambda}(\eta) \int_{\eta}^{\mu} e^{-s_0(g)}\, dg + e^{\Lambda}(1) \int_{\mu}^{1} e^{-s_0(g)}\, dg.$$

Condition (9.45) looks now like $e^{\Lambda}(0) = e^{\Lambda}(1) \cong 1$, and

$$P = P_0 + (e^{\Lambda}(\eta) - 1) \int_{\xi}^{\mu} e^{-s_0(g)}\, dg.$$

Thus, in the relation between quantities P and P_0 an additional summand appears:

$$(e^{\Lambda(\eta)} - 1) \int_{\xi}^{\mu} e^{-s_0(g)}\, dg.$$

The situation will be analogous if function e^{Λ} has a maximum. In the presence of several maxima and minima an appropriate sum of additional summands appears. We shall see that this difficulty can be avoided by passing to the equivalent line.

Denote equivalent lines replacing $s, u_1 s_1, u_2 s_2$ through f, f_1, f_2, constructed as in Figure 9.4. It follows from the definition of the equivalent line that

$$P = \int_0^1 dg\, e^{-f(g)}. \tag{9.48}$$

Let us introduce the function $\lambda(g)$ by analogy with Equation (9.43)

$$\lambda(g) = -f + f_1 + f_2 = -f + f_0, \tag{9.49}$$

and then transform Equation (9.48) in the same way as Equation (9.41). If function (9.49) is assumed to be monotonic in the interval $[0, 1/2]$ and $[1/2, 1]$ then application of the second mean value theorem and conditions (9.47) yields:

$$P \cong \int_0^1 e^{-f_1 - f_2}\, dg. \tag{9.50}$$

Again, some comment is required concerning the condition of monotonicity of $\Lambda(g)$ in the indicated intervals. Arbitrariness in construction of the equivalent lines of individual gases should be used to reach monotonicity. Thus, an n-parametric or asymmetric curve may be chosen and equations for parameters may include the monotonicity conditions of $\Lambda(g)$. It is clear that the condition of applicability of Equation (9.50) does not change in this procedure since it relates to the invariants of equivalent lines. Moreover, the real n-parametric procedure is not needed, and it is sufficient that it is in principle possible.

Therefore, the approximation (9.50) is applicable if the equivalent lines for individual gases are available. From the computational viewpoint, Equation (9.50) is equivalent to Equation (9.41) because s_1, s_2 and f_1, f_2 are calculated through the same spectrum $\kappa(\omega)$.

9.4.2 Small pressures

It is well known that functions $s(g)$ at small pressure show some specific features, creating computational problems (Chou *et al.*, 1995), which are usually solved by means of a substantial increase in the number of terms in the series of exponents. Curve 1 in Figure 9.6 demonstrates the typical behavior of $s(g)$ pertinent to the pressure characteristic for the upper layers of the atmosphere. For comparison, Curve 2 is for the case when the pressure is assumed to be sufficiently large. This situation is practically obvious: spectral lines are narrowing with decreasing pressure, and the values $\kappa(\omega)$ in the line center increase. Thus, it becomes clear that in the case of small pressures the behavior of $s(g)$ near the point $g = 1$ is defined by the peaks of the most intensive lines in the range $\Delta\omega$. (In this subsection $\sum_j$ denotes the sum over these individual strong lines, allowing further analytical calculation).

Application of the properties of the series of exponents enables one to use the specific features of $s(g)$ behavior at small pressures for the derivation of formulas providing more exact calculation. Thus, there appears the possibility of asymptotic estimate of the integral over g in the region of rapid decay of the curve of the type of curve 1 in Figure 9.6.

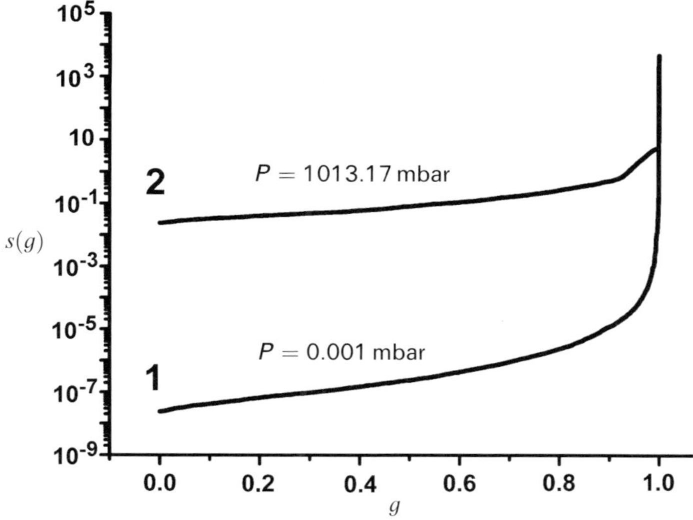

Figure 9.6. Functions $s(g)$ at different pressures. CO_2, $T = 296$ K, Voigt line shape to $10\,\text{cm}^{-1}$, grid step $= 0.001\,\text{cm}^{-1}$, $780\,\text{cm}^{-1}$–$800\,\text{cm}^{-1}$.

Let us give a derivation for the Lorentzian line (the cases of Voigt and Doppler lines can be considered in a similar manner).

The function $g(s)$ for lines with a Lorentzian line shape can be derived analytically and is equal to

$$g(s) = 1 - \sum_j \frac{2\alpha_j}{\Delta\omega} \sqrt{\frac{Q_j}{s\pi\alpha_j} - 1}, \tag{9.51}$$

where Q_j and α_j are the Lorentzian line intensity and half-width, and of course $(Q_j/s\pi\alpha_j) > 1$. Specific features of $s(g)$ at the point $g = 1$ can be most clearly seen by the example of an isolated Lorentzian line. In this case

$$g(s) = \frac{Q}{\pi\alpha}\left(\frac{2\alpha}{\Delta\omega}\right)^2 \frac{1}{\left(\dfrac{2\alpha}{\Delta\omega}\right)^2 + (1-g)^2}. \tag{9.52}$$

From Equation (9.52) it at once follows that

$$s'(1) = 0, \qquad s''(1) < 0. \tag{9.53}$$

Such a behavior of $s(g)$ and conditions (9.53) almost automatically pose the question of the asymptotic estimate of the integral in the transmission function for small pressures by the method of saddle point (Evgrafov, 1968). To comply with the rules of the method of saddle point, let us treat expression A related to the required transmission function:

$$A = \int_0^1 dg\, e^{\varphi(g)}, \qquad \varphi(g) = \ln(1 - e^{-us(g)}). \tag{9.54}$$

This replacement rests on the physical conditions of the problem, accounting for small pressures in the upper layers of the atmosphere, thus assuming the smallness of u. Next, the standard procedure of the method of saddle point and the relation $P = 1 - A$, following from Equation (9.54), give

$$P = 1 - \sqrt{\frac{\pi}{2}}(1 - e^{-us(1)})^{3/2} \frac{e^{\frac{1}{2}us(1)}}{\sqrt{u|s''(1)|}} < 1. \tag{9.55}$$

When $u \to 0$, Equation (9.55) leads to $P = 1$. However, in the formal passage $u \to \infty$, Equation (9.55) does not coincide with $\lim_{u\to\infty} P = 0$, as would be the case for the transmission function. The explanation straightforwardly follows from the physical conditions of the problem, which determine the region of rapid changes of $s(g)$ producing the predominant contribution to the value of the integral P. This region is called the affected zone Δg (in terms of asymptotical analysis). It is estimated in the standard way:

$$\Delta g = 0\left(1 \Big/ \sqrt{\frac{2}{\varphi''(g)}}\right), \qquad \varphi''(1) = \frac{e^{-us(g)}}{1 - e^{-us(g)}} u|s''(1)|.$$

It is worthy of note that at sufficiently small u the value of Δg is independent of u, because in this case $\varphi'(g) = |s''(g)|/s(g)$.

From this the restriction on u follows, which defines the applicability limits of Equation (9.55):

$$\sqrt{\frac{\pi}{2}} \frac{e^{\frac{1}{2}us(g)}}{\sqrt{u|s''(g)|}} < 1. \tag{9.56}$$

It may appear inconvenient that Equation (9.55) violates the structure of the series of exponents.

The variant when the range $[0, 1]$ is replaced with the affected zone looks like some compromise eliminating this inconvenience. It is reached through the introduction of the function $f(y) \equiv \psi(y\,\Delta g + 1 - \Delta g)$, $y \in [0, 1]$ instead of the integrand of P with subsequent application to the integral $\int_0^1 dy\, f(y)$ of the corresponding quadrature formula (its coefficients are automatically adjusted to fit the new interval on the formal change of variables $g \leftrightarrow y$). A formula of this kind returns to the initial form when Δg exceeds unity with pressure increase.

9.4.3 Inhomogeneous media

Use of the series of exponents enables one to obtain the necessary condition of the applicability of approximation (9.15) in inhomogeneous media.

As already mentioned in Section 9.2, practically always in the case of inhomogeneous media in the literature the technique referred to as the CKD approximation is applied instead of the series (9.13) with rigorous values (9.14) for expansion coefficients. For each l in Equation (9.12) its own function $g(s;l), s(g;l) = g^{-1}(s;l)$ is calculated by Equation (9.10), and then $\tilde{s}(g)$ is replaced with $\int_l dl\, s(g;l)$, see Equation (9.15). The problem of such replacement is discussed rather actively and is usually related with the problem of correlation of the absorption coefficients under different thermodynamic conditions (Goody *et al.*, 1989; Lacis and Oinas, 1991; West *et al.*, 1990; and a number of papers analyzing numerically various sides of these approximations). Examination of rigorous relations gives a possibility to view the mathematical part of the problem (Tvorogov *et al.*, 2000, 2005).

The transmission function in the inhomogeneous medium, Equation (9.11), can be written as $\int_0^1 dg\, I$, in terms explained in Section 9.3, when the radiative transfer equation was expressed through the series of exponents. I is the solution of equation

$$[\mathbf{n}\,\text{grad}\, I] = -(\mathbf{n}\,\text{grad}\, s(g;\mathbf{r},\mathbf{n}))I, \tag{9.57}$$

and $s(g;\mathbf{r},\mathbf{n})$ is now the inverse of the function

$$g(s;\mathbf{r},\mathbf{n}) = \frac{1}{2\pi i}\int_{c-i\infty}^{c+i\infty} \frac{dz}{z} e^{sz} \frac{1}{\Delta\omega}\int_{\omega'}^{\omega''} e^{-z\tau}\, d\omega = \frac{1}{\Delta\omega}\int_{\tau(\omega;\mathbf{r},\mathbf{n})\le s,\omega\in(\omega',\omega'')} d\omega. \tag{9.58}$$

The approximation (9.15) under discussion corresponds to the replacement of I for $\tilde{I}$, the solution of the equation

$$[\mathbf{n}\,\text{grad}\, \tilde{I}] = -\tilde{s}(\tilde{g};\mathbf{r})\tilde{I}, \tag{9.59}$$

and, as follows from the previous notation, $\tilde{s}(g;\mathbf{r})$ is the inverse of the function

$$\tilde{g}(\tilde{s},\mathbf{r}) = \frac{1}{2\pi i}\int_{c-i\infty}^{c+i\infty} \frac{dz}{z} e^{sz} \frac{1}{\Delta\omega}\int_{\omega'}^{\omega''} d\omega\, e^{-z\kappa(\omega,\mathbf{r})} = \frac{1}{\Delta\omega}\int_{\kappa(\omega,\mathbf{r})\le s;\omega\in[\omega',\omega'']} d\omega. \tag{9.60}$$

Comparison of the quantities

$$\mathbf{n}\,\text{grad}\, s(g;\mathbf{r},\mathbf{n}) \quad \text{and} \quad \tilde{s}(\tilde{g},\mathbf{r}) \tag{9.61}$$

enables one to elucidate the condition of nearness of solutions of Equations (9.57) and (9.59); in other words, the necessary condition of approximation (9.15). We recall that the values s from Equation (9.58) are ordered values $\tau(\omega)$, and $\tilde{s}$ from Equation (9.60) are ordered values $\kappa(\omega)$ at the point $\mathbf{r}$.

It follows from the definition of s that

$$g(s(g_\nu;\mathbf{r},\mathbf{n})) = g_\nu = \text{const}.$$

The last relation and Equation (9.58) yield:

$$\mathbf{n}\,\text{grad}\, s(g;\mathbf{r},\mathbf{n}) = \frac{\dfrac{1}{2\pi i}\displaystyle\int_{c-i\infty}^{c+i\infty} dz\, e^{sz} \frac{1}{\Delta\omega}\int_{\omega'}^{\omega''} d\omega\, \kappa(\omega,\mathbf{r})\, e^{-z\tau}}{\dfrac{1}{2\pi i}\displaystyle\int_{c-i\infty}^{c+i\infty} dz\, e^{sz} \frac{1}{\Delta\omega}\int_{\omega'}^{\omega''} d\omega\, e^{-z\tau}}. \tag{9.62}$$

We use the following technique to transform the integrals on the right-hand side of Equation (9.62). The integral over ω in the denominator of Equation (9.62) is written as an integral sum

$$\int_{\omega'}^{\omega''} d\omega\, e^{-z\tau(\omega)} = \sum_k (\delta\omega)_k\, e^{-z\tau(\omega_k)}.$$

In the last sum we regroup terms by combining those $(\delta\omega)_k$ belonging to intervals where $\tau(\omega_k)$ are identical. Then we arrange these values $\tau(\omega_k)$ in increasing order and denote the sequence obtained through τ_j. The appropriate sequence of intervals $\Delta\omega_j$ divided by $\Delta\omega$ becomes

$$\Delta\omega_j = \sum_{k<j} (\delta\omega)_k / \Delta\omega.$$

Further calculation of the integral over z transforms the denominator of Equation (9.62) to the form

$$\sum_j \Delta\omega_j\, \delta(s - \tau_j). \tag{9.63}$$

Change of order of summation and integration is possible here because the sum is finite and the sequence τ_j is monotonic. In Equation (9.63) only one term is retained, because of the presence of the δ-function, with $\tau_j = s$.

The integral in the numerator of Equation (9.62) is transformed in a similar manner, and finally we have

$$\sum_j \Lambda_j\, \delta(s - \tau_j), \qquad \Lambda_j = \frac{1}{\Delta\omega} \sum_{k<j} (\delta\omega)_k \kappa(\omega_k). \tag{9.64}$$

Now the relation (9.62) can be rewritten as

$$\mathbf{n}\,\mathrm{grad}\, s(g_\nu; \mathbf{r}, \mathbf{n}) = \tilde{\kappa}_j \equiv \frac{\sum_{k<j} (\delta\omega)_k \kappa(\omega_k)}{\sum_{k<j} (\delta\omega)_k}, \tag{9.65}$$

where subscript j corresponds to the value $\tau_j = s$.

As follows from the above, the guarantee of nearness of the quantities (9.61) is the nearness of values g_ν and g'_ν corresponding to sequences τ_j and $\tilde{\kappa}_j$; or, in other words, the roots of the following equation should coincide:

$$\tilde{s}(g'_\nu) = \tilde{\kappa}_j, \qquad s(g_\nu) = \tau_j. \tag{9.66}$$

This necessary condition seems to be quite rigid. It is not surprising because Equations (9.57) and (9.59) can be considered identical in this case. It is also understandable that the discussion of Equation (9.61) allows one to formulate some sufficient conditions.

A quite obvious consequence of Equation (9.65) is, for example: if the sequences τ_j and $\tilde{\kappa}_j$ are synchronous the values g_ν and g'_ν appear to be equal to each other (the case of CKD approximation). However, this does not mean that CKD approximation is the indispensable condition of the efficiency of Equation (9.59).

Therefore, strictly speaking, there is no need to synchronize artificially the absorption coefficients at different heights, because the necessary condition does not require the identity of spectral behaviors of $\kappa(\omega)$.

There are serious grounds to say that the approximation (9.57) is almost universal, at least in atmospheric problems. It is confirmed by the results of a number of test calculations (Kistenev *et al.*, 2002; Tvorogov *et al.*, 2005), when only exotic situations force the use of the rigorous variant (9.58).

9.4.4 One-parametric approximation formulas

The availability of the rigorous formulas for the expansion of radiation characteristics into series of exponents (Tvorogov *et al.*, 2000) enables one to address the approximation of their thermodynamic dependences.

Before the advent of mass line-by-line calculations, of wide use were the models of absorption bands, and different ways of reducing absorption along inhomogeneous paths to that along equivalent homogeneous paths were popular.

At the present time the usual practice is tabulation of the absorption coefficients at different temperatures and pressures and subsequent calculation of functions $s(g)$ for individual atmospheric layers (CKD approximation). One more way of accounting for the temperature and pressure dependences of expansion coefficients in the series of exponents is to obtain approximation formulas for them, which actually signifies the return to models of absorption along equivalent paths. This variant is realized in the paper by Chou *et al.* (1993), but for a number of reasons it has not acquired the versatility desired in a climate model.

Application of functions $s(g)$ implies transition to the smoothed values of absorption instead of very irregular absorption coefficients. The fact of similar "averaging" and the smooth height behavior of $s(g)$ enables one to hope to separate their regular thermodynamic dependences. It is noteworthy that the approximate expressions for absorber amounts, reducing them to some average temperature and pressure, are extended to the expansion into series of exponents:

$$s(g, \varepsilon) = f(\varepsilon, \varepsilon_0) s(g, \varepsilon_0),$$

where $s(g, \varepsilon_0)$ are the expansion coefficients at the chosen temperature and pressure, and $s(g, \varepsilon)$ are the expansion coefficients at arbitrary temperature and pressure.

Preliminary calculations (Bogdanova and Rodimova, 2005; Rodimova and Bogdanova, 2006) show that approximations of this kind may lead to reasonable results. There is an example below (Figure 9.7) of calculation of the O_3 cooling rate under MLS conditions in the spectral range $980\,\mathrm{cm}^{-1}$–$1{,}100\,\mathrm{cm}^{-1}$.

It is noteworthy that this approach can easily adapt to specific features of the behavior of the line shape in cases when it is different from the Lorentzian line shape.

9.5 CONCLUSION

The mathematically rigorous approach to the description of the series of exponents outlined in this chapter not only leads to a more fundamental understanding of

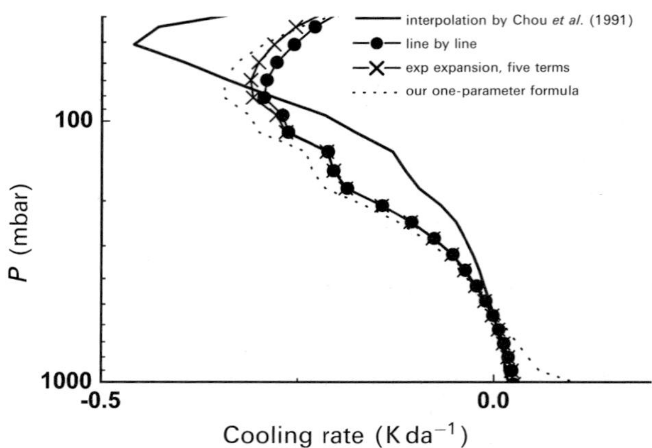

Figure 9.7. The O_3 cooling rate at MLS conditions in the spectral range 980 cm^{-1}–1,100 cm^{-1} calculated by various approximation formulas (Chou and Kouvaris, 1991).

calculation problems arising in the realization of this method but can serve as a base of application of present-day computer technologies.

It is useless to say that even in such problems as climatic ones, whose solution is impossible without application of modern computers, theoretical comprehension of different approximations may provide not only new understanding but also redirect computational effort towards new goals. In this regard, the probability interpretation of the series expansion of gas mixture absorption and the nature of the CKD approximation deserve note.

Acknowledgments. The experimental data on spectral fluxes and atmospheric parameters as well as the results of calculations based on the radiative transfer code MODTRAN-4 were kindly provided by Z. Li, A. Trishchenko and M. Cribb, Canada Center for Remote Sensing, Ottawa, Canada.

This work is supported by the Russian Foundation of Basic Investigations, grant Nos. 05-05-64256 and 06-05-64484.

9.6 REFERENCES

Ambarzumyan V.A. (1968). Present natural sciences and philosophy. *Achievements of Physical Sciences*, **96**(1), 3–19 [in Russian].

Anderson G.P., Clough S.A., Kneizys F.X., Chetwynd J.H., and Shettle E.P. (1986). AFGL Atmospheric Constituent Profiles (0–120 km), AFGL-TR-86-0110. *Environmental Research Papers*, No. 954.

Bogdanova Yu.V. and Rodimova O.B. (2005). One-parameter approximation for the CO_2 transmission functions in the 15 mm region. *Computer Technologies*, **10**(1), 87–93 [in Russian].

Chou M.-D. and Kouvaris L. (1991). Calculations of transmission functions in the infrared CO_2 and O_3 bands. *J. Geophys. Res.*, **96**(D5), 9003–9012.

Chou M.-D., Ridgway W.L., and Yan M.M.-H. (1993). One-parameter scaling and exponential-sum fitting for water vapor and CO_2 infrared transmission functions. *J. Atmos. Sci.*, **50**, 2294–2303.

Chou M.-D., Ridgway W.L., and Yan M.M.-H. (1995). Parameterizations for water vapor IR radiative transfer in both the middle and lower atmospheres. *J. Atmos. Sci.*, **52**, 1159–1167

Evgrafov M.A. (1968). *Analytical Functions*. Science, Moscow, 472 pp. [in Russian].

Firsov K.M. and Chesnokova T.Yu. (1998). A new method of treating overlapping absorption bands of atmospheric gases in radiative transfer parameterization. *Atmospheric and Oceanic Optics*, **11**, 356–360.

Firsov K.M., Mitsel A.A., Ponomarev Yu.N., and Ptashnik I.V. (1998). Parameterization of transmittance for application in atmospheric optics. *J. Quantitative Spectroscopy and Radiative Transfer*, **59**(3/5), 203–213.

Firsov K.M., Chesnokova T.Yu., Belov V.V., Serebrennikov A.B., and Ponomarev Yu.N. (2002). Series of exponents in computations of radiation transfer by Monte Carlo method in spatially non-homogeneous aerosol-gaseous media. *Computer Technologies*, **7**(5), 77–87 [in Russian].

Fomin B.A. and Gershanov Yu.V. (1996). *Tables of the Benchmark Calculations of Atmospheric Fluxes for ICRCCM Test Cases, Part II: Shortwave Results*. Russian Research Center, Kurchatov Institute, Moscow, IAE 5990/1, 42 pp.

Fouquart Y., Bonnel B., and Ramaswamy V. (1991). Intercomparing shortwave radiation codes for climate studies. *J. Geophys. Res.*, **96**, 8955–8968.

Goody R., West R., Chen L., and Crisp D. (1989). The correlated-*k* method for radiation calculations in nonhomogeneous atmospheres. *JQSRT*, **42**(6), 539–550.

Kistenev Yu.V., Ponomarev Yu.N., and Firsov K.M. (2002). Analysis of temperature dependence in cumulative spectra of rotational–vibrational absorption bands of atmospheric gases. *Atmospheric and Oceanic Optics*, **15**, 689–691.

Kistenev Yu.V., Ponomarev Yu.N., Firsov K.M., and Gerasimov D.A. (2003). Use of lacunarity parameter in analysis of the errors in atmospheric transmittance inhomogeneities calculated using exponential series. *Atmospheric and Oceanic Optics*, **16**, 247–250.

Kneizys F.X., Robertson D.S., Abreu L.W., Acharya P., Anderson G.P., Rothman L.S., Chetwynd J.H., Selby J.E.A., Shetle E.P., Gallery W.O., Berk A., Clough S.A., and Bernstein L.S. (1996). *The MODTRAN 2/3 Report and LOWTRAN 7 Model*. Phillips Laboratory, Geophysics Directorate, Hanscom Air Force Base, MA, 260 pp.

Kondratyev K.Ya. (1950). *Long-wave Radiation Transfer in the Atmosphere*. Gostechizdat, Leningrad, 278 pp. [in Russian].

Lacis A.A. and Hansen J.E. (1974). A parameterization for the absorption of solar radiation in the Earth's atmosphere. *J. Atmos. Sci.*, **31**, 118–133

Lacis A. and Oinas V. (1991). A description of the correlated k-distribution method for modeling nongray gaseous absorption, thermal emission, and multiple scattering in vertically inhomogeneous atmospheres. *J. Geophys. Res.*, **96**, 9027–9063.

Leont'ev A.F. (1976). *Series of Exponents*. Science, Moscow, 536 pp. [in Russian].

Leont'ev A.F. (1980). *Sequence of Polynomials of Exponents*. Science, Moscow, 384 pp. [in Russian].

Leont'ev A.F. (1983). Integral Functions. Series of Exponents. Science Publ. Moscow, 175 pp. [in Russian].

Li Z., Trishchenko A., and Cribb, M. (2000). Analysis of cloud spectral radiance/irradiance at the surface and top of the atmosphere from modeling and observations. *Proceedings of the*

Tenth Atmospheric Radiation Measurement (ARM) Science Team Meeting. Available at *http://www.arm.gov/publications/proceedings/conf10/abstracts/li-z.pdf*

Liou K.N. and Sasamori T. (1975). On the transfer of solar radiation in aerosol atmosphere. *J. Atmos. Sci.*, **32**, 2166–2177.

Nesmelova L.I. and Tvorogov S.D. (1996). Some applications of the exponential series for calculating the absorption function. *Atmospheric and Oceanic Optics*, **9**, 727–729.

Nesmelova L.I., Rodimova O.B., and Tvorogov S.D. (1997). Calculation of transmission functions in near infrared region using series of exponents. *Atmospheric and Oceanic Optics*, **10**, 923–927.

Nesmelova L.I., Rodimova O.B., and Tvorogov S.D. (1999). Application of exponential series to calculation of radiative fluxes in the molecular atmosphere. *Atmospheric and Oceanic Optics*, **12**, 735–739.

Rodimova O.B. and Bogdanova Yu.V. (2006). Calculation of radiation fluxes due to CO_2 in the IR spectral region. *Computer Technologies*, **11**, 44–51 [in Russian].

Sobolev B.V. (1972). *Light Scattering in the Planetary Atmospheres*. Science, Moscow, 335 pp. [in Russian].

Tarasova T.A. and Fomin B.A. (2000). Solar radiation absorption due to water vapor: Advanced broadband parameterizations. *J. Appl. Meteorol.*, **39**, 1947–1951.

Tvorogov S.D. (1994). Some aspects of the problem of representation of the absorption function by a series of exponents. *Atmospheric and Oceanic Optics*, **7**, 165–171.

Tvorogov S.D. (1997). Use of Dirichlet series in atmospheric optics. *Atmospheric and Oceanic Optics*, **10**, 249–254.

Tvorogov S.D. (1999). Application of exponential series to frequency integration of the radiative transfer equation. *Atmospheric and Oceanic Optics*, **12**, 730–734.

Tvorogov S.D. (2001). Construction of exponential series directly from information on the transmission function. *Atmospheric and Oceanic Optics*, **14**, 670–673.

Tvorogov S.D., Nesmelova L.I., and Rodimova O.B. (1996). Representation of the transmission function by the series of exponents. *Atmospheric and Oceanic Optics*, **9**, 239–242.

Tvorogov S.D., Nesmelova L.I., and Rodimova O.B. (2000). k-distribution of transmission function and theory of Dirichlet series. *J. Quantitative Spectroscopy and Radiative Transfer*, **66**, 243–262.

Tvorogov S.D., Rodimova O.B., and Nesmelova L.I. (2005). On the correlated k-distribution approximation in atmospheric calculations. *Optical Engineering*, **44**(7), 071202/1–071202/10.

Van de Hulst M.C. and Irvine W.M. (1963). Scattering in model planetary atmospheres. *Meteorol. Soc. Roy. Sci. Liège*, **5-7**(1), 78–86.

WCP (1986). *A Preliminary Cloudless Standard Atmosphere for Radiation Computation*. World Climate Research Program, Technical Report WCP-112, WMO/TD No. 24. World Meteorological Organization, Geneva, 60 pp.

West R., Crisp D., and Chen L. (1990). Mapping transformations for broadband atmospheric radiation calculations. *J. Quantitative Spectroscopy and Radiative Transfer*, **43**, 191–199.

Zhuravleva T.B. and Firsov K.M. (2004). Algorithms for calculation of sunlight fluxes in the cloudy and cloudless atmosphere. *Atmospheric and Oceanic Optics*, **17**, 799–806.

Zhuravleva T.B. and Firsov K.M. (2005). On variability of the radiative characteristics in the 940-nm band at variations of water vapor in the atmosphere: Numerically simulated results. *Atmospheric and Oceanic Optics*, **18**, 696–702.

Zuev V.V. (1966). *Atmospheric Transparency for Visible and Infrared Light*. Soviet Radio, Moscow, 318 pp. [in Russian].

10

Forecast of biosphere dynamics using small-scale models

Andrey G. Degermendzhi, Sergey I. Bartsev, Vladimir G. Gubanov, Dmitry V. Erokhin, Anatoly P. Shevirnogov

10.1 INTRODUCTION

The biosphere–climate system, inside which we are all living, is subjected to different influences (influx of anthropogenic CO_2 and pollutants, deforestation, harmful land management, etc.). There are reasons to consider the well-known change in global parameters (atmospheric CO_2 concentration, temperature, distribution of precipitation, soil erosion, etc.) as the result of these influences. It will be recalled that in Chapter 2 we discussed the fact that Kirill Kondratyev was critical of the fact that, while computer models of the climate could predict future climatic conditions on the basis of gradual change, they were most unlikely to be able to predict sudden catastrophic changes in climate. Therefore, the vital question arises: "Can changes in the various global parameters lead to irreversible negative changes in the climate–biosphere system or a global ecological catastrophe?" The possibility of irreversible changes may be not very high, but one cannot ignore it. So, the main aim of this chapter is to take irreversible changes into consideration. Kondratyev also argued that the global carbon cycle was not well enough understood or studied in enough detail. As well as addressing these issues, we shall discuss some experimental results on microecosystems as well in this chapter.

Projected into the future, the currently observed trends of global environmental parameters suggest the possibility of significant changes in the parameters of the biosphere and Earth's climate (IPCC, 2007). It seems evident that assessment of the rates, scales, and reversibility/irreversibility of these changes is not just the task of academic science only but is one of the currently most important practical issues. Humanity as a whole can be adversely affected by significant changes in the biosphere and climate. Knowledge of the key factors involved in global change can however provide a way to put an end to the negative trends in the development of the biosphere, which are now becoming more and more pronounced.

Whether this task is accepted as a practical one depends on an investigator's attitude to the data on changes in climate and the biosphere. There is an opinion that the observed change in climate and the biosphere are natural variations that have occurred many times throughout Earth's history, and thus nothing special should be done about them. This position seems to be inadequate for two reasons. First, having recognized that changes in the biosphere and climate are quite natural, one does not have to passively await possible cataclysms. The present scientific and technological power of humanity allows us to solve the issues of global security, such as the issue of anti-asteroid defense, which is now being discussed. Second, complex models of climate and the biosphere contain a great number of parameters whose values are known only to a limited accuracy. Thus, predicted trajectories can differ significantly due to variations of the parameters within their confidence intervals. There is a very illustrative example of a divergence between possible climate change scenarios within the framework of a single basic model (Stainforth *et al.*, 2005). Hence, a perfectly accurate answer to the question: "Will serious climate and biosphere changes occur?" can be obtained only when these changes become pronounced (i.e., when it may be too late to act). So, it would be irrational to wait for confirmation of possible cataclysms and then take action. One could relax and forget about cataclysms only when it had been confidently proven that no cataclysms would ever occur.

Let us determine the required confidence level, which depends on the value attached to the issue. For instance, at the present time, the probability of an aviation accident is about 10^{-5} per flight, and this is considered acceptable. Let us assume that in the case of an aviation accident, about 100 passengers die. Global human-induced changes in the biosphere–climate system and the ensuing processes may cause the death of all of humankind (due to global crop failure, starvation, epidemics, social conflicts, weather calamities, etc.). The total would be about 10^{10} human lives. Then, based on "aviation" statistics, the acceptable probability of irreversible global changes should be 10^{-13}. To compare, the probability of very serious climate change calculated using a set of climate models is about 10^{-2} (Stainforth *et al.*, 2005). Thus, if the life of future generations and sustainable development of civilization are really of high value to decision-makers (and people electing them), practical steps to prevent possible global cataclysms should be taken as early as possible.

The Rio Declaration (the United Nations Framework Convention on Climate Change, UNFCCC) was based on the following: working towards international agreements which respect the interests of all and protect the integrity of the global environmental and developmental system; and recognizing the integral and interdependent nature of the Earth, our home. Principle 15 of the Declaration proclaimed:

> "In order to protect the environment, the precautionary approach shall be widely applied by States according to their capabilities. Where there are threats of serious or irreversible damage, *lack of full scientific certainty shall not be used as a reason for postponing cost-effective measures to prevent environmental degradation.*"

Thus, the decision is evident but it has not been made. The question why these adverse, or even deadly, probable consequences of present-day inaction are ignored is a question of psychology and sociology rather than natural sciences. A scientific investigation of the biosphere–climate system can only provide more information on possible scenarios of its development. Let us consider the scenarios that are of the greatest interest to supporters having an active attitude to possible global change (supporters of the alternative position are not interested in anything other than favorable scenarios).

As a rule, developers of climate and biosphere models aim at predicting the most probable scenario. Thus, they have to take into account the maximum possible number of various, frequently mutually compensating, interactions of the components of these systems. However, assessment of the contribution of any climatic or biospheric process has finite accuracy and is represented by a confidence interval. As the consequences of global warming can be very significant, involving possible extinction of highly organized species (Karnaukhov, 2001), the significance level must be unprecedented, amounting to several hundredths and thousandths of a percent. Thus, the confidence intervals are very wide and researchers obtain a wide range of scenarios.

Hence, even confident knowledge of the most probable scenarios will not suffice for making practical decisions, because confidence intervals for the scenario are very wide. Wide confidence intervals imply that the probability of the scenario developing in the way that is the most unfavorable to us does not equal zero. Practically speaking, these possible developments towards the worst case scenario are the most important.

So, in the approach we have adopted, we make use of the principle of the worst scenario. This approach can be used for decision-making. We have to specify what is meant by the worst scenario. Most mathematical climate models demonstrate smooth or gradual changes in atmospheric CO_2 and near-surface temperature with an increase in total carbon due to fossil fuel combustion (IPCC, 2007; Semyonov, 2003; Stainforth *et al.*, 2005; Tarko, 2005). However, gradual global change is just one possible scenario for the future and a rather optimistic one at that because in this case it is theoretically possible to slow down negative changes by reducing the human impact. We have to take into account catastrophic, threshold variants of global change, in which the magnitude of changes exceeds a certain threshold of stability of the biosphere–climate system, giving rise to avalanche-like and irreversible changes in global parameters (see also Sections 2.5 and 18.4). Therefore, in this study the term "catastrophic" corresponds to its meaning in catastrophe theory—rather than to its everyday usage, when any serious natural cataclysm is perceived as a catastrophe.

Thus, the worst case scenario will be one in which the time left before the threshold of stability of the biosphere–climate system is exceeded is the shortest. It seems obvious that for practical reasons we should primarily investigate the worst but also possible variants, in which the contribution of compensatory and alleviating mechanisms is the smallest. It is also very important in practice to estimate the time of the onset of irreversible changes triggered by a certain mechanism. Based on the above, we can formulate the principle of the worst scenario, which will largely determine what follows in this study and the direction of the authors' research.

The worst case scenario principle is to isolate and investigate only those processes that can most quickly lead to negative changes in the studied system, deliberately ignoring possible compensatory mechanisms.

The worst case scenario principle applied to the interaction pattern in the biosphere–climate system is a means of finding the potentially quickest positive feedback loops, and thus use a zero-dimensional (minimal) mathematical model to describe the biosphere. Zero-dimensional mathematical models, such as that of Svirezhev and von Bloh (1997) appear to be an ideal tool for constructing scenarios of the development of the biosphere–climate system in accordance with the worst case scenario principle. Moreover, the worst case scenario principle naturally leads one to use minimal (i.e., extremely simple) models.

It is important that, due to the small number of parameters in the model, their arbitrary and independent fitting to the observed dynamics is ruled out. Thus, the minimal model can be used as an instrument to test a set of various data on global biospheric processes for their compatibility or agreement. Moreover, if all the available data are in agreement, the minimal model can serve as an instrument (though a rough one) for fitting global monitoring data to biota parameters. The aim of constructing this minimal model is to illustrate the probability that the catastrophic variant of the development of the biosphere–climate system can take place and to determine one of the mechanisms that can bring about early formation of the catastrophic regime.

The worst case scenario principle is consistent with the traditional methodology of scientific investigation: maximal reduction of the studied system in the early stages of investigation. Ecological biophysics usually works with a maximally simplified model which, nevertheless, reproduces the essential features of the studied system. This approach is used in the work with both mathematical and experimental models of ecosystems and the biosphere as a whole. It seems important and even necessary to study experimental models of the biosphere in order to verify the basic principles used in constructing mathematical models of the biosphere. Verification of biospheric models using the available time series (see Section 10.2.) is obviously insufficient to predict the transition to catastrophic regimes: to do this we need experimental validation of catastrophic processes. As experiments with the biosphere are out of the question, the only remaining possibility to verify biospheric models under critical conditions is to use experimental models of the biosphere. The advantages and difficulties of using experimental models of the biosphere are discussed in Section 10.4.

We have chosen atmospheric carbon dioxide as a simulation object for a number of reasons. First, available estimates suggest (Monin and Shishkov, 2000) that increasing atmospheric carbon dioxide concentration is the major aggregate contributor to the rise in global temperature due to the greenhouse effect (at least now and in the nearest future). Second, carbon dioxide is a major component of the carbon cycle, used in plant photosynthesis and soil respiration. Some estimates suggest that biota influences climate mainly through changing atmospheric CO_2 concentration (Semyonov, 2003). Third, anthropogenic emissions due to combustion of carbon-containing fossil fuels are considered to be the main cause of temperature elevation. This assertion is based on results obtained in a series of computational experiments with a number of climate models (IPCC, 2007).

10.2 THE WORST CASE SCENARIO PRINCIPLE AND MINIMAL MODELS OF THE BIOSPHERE

10.2.1 Initial minimal model of the biosphere

Even preliminary examination of the biosphere–climate system yields the following, rather complex, picture of interactions between the major processes and events (Figure 10.1). As is apparent from system representation even at this level, to model this complex system is not a trivial task.

Following the worst case scenario principle and the above plan of constructing a minimal model, we construct a simplest zero-approximation model based on the reduced scheme of interactions in the system (Figure 10.2). We have chosen them because these are high-rate processes that cause changes in the environment within an extremely short time. One of the key mechanisms of the biosphere–climate interaction is the temperature–carbon dioxide concentration positive feedback. It is assumed that the positive feedback is effected through two loops of interactions. Elevation of atmospheric CO_2 concentration causes, due to the greenhouse effect, a near-surface temperature rise, which, in turn, leads to an increase in soil microflora respiration and release of more CO_2. The second loop involves a decrease in the photosynthesis

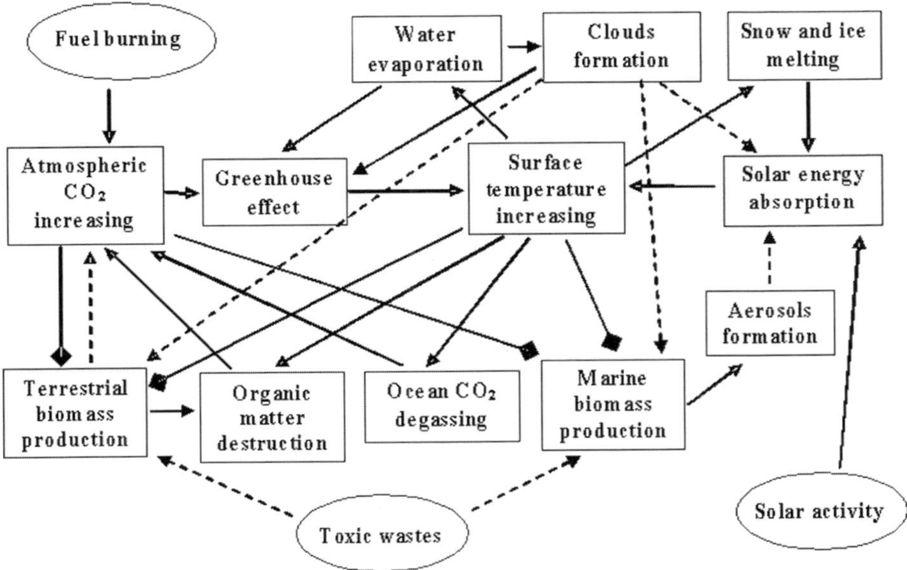

Figure 10.1. The scheme of the most obvious interactions between processes in the biosphere. ⟵ positive effect; ← – – negative effect; ◆— uncertain effect.

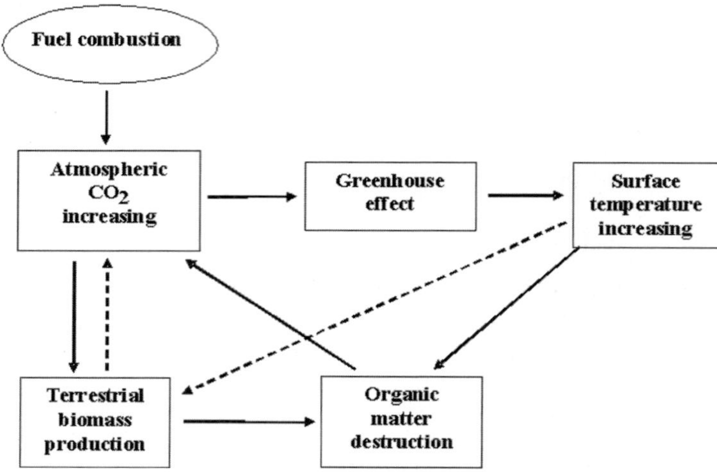

Figure 10.2. The simplified scheme of the most obvious interactions between processes in the biosphere. ⟵ positive effect; ←−− negative effect.

rate of land plants due to temperature departing from the optimum and, as a consequence, a decrease in the uptake of atmospheric CO_2, which amounts to the second positive feedback.

The global model of long-term carbon dioxide dynamics is based on a closed-loop carbon cycle. This consists of three compartments between which carbon dioxide is exchanged: the atmosphere, land plants, and respective dead organic residues. These compartments are interrelated through three processes: growth, death, and decomposition of biomass. To model fuel combustion, the model also contains the anthropogenic carbon source, which upsets the carbon balance of the system. The set of equations constituting the model has the following form (Bartsev *et al.*, 2005):

$$\left.\begin{aligned} \frac{dC}{dt} &= \text{fuel}(t) \\ \frac{dx}{dt} &= P(x, A, T(A)) - D(x) \\ \frac{dy}{dt} &= D(x) - S(y, T(A)) \\ A &= C - x - y \end{aligned}\right\} \qquad (10.1)$$

where C is the total amount of carbon involved in the biospheric cycles of matter turnover; x is the amount of carbon in the biomass of the plant compartment; y is the amount of carbon in dead biomass; A is the amount of carbon in the atmosphere; fuel(t) is an empirical function describing anthropogenic carbon emission due to fossil fuel combustion; function $P(x, A, T(A))$ describes plant biomass growth rate; $T(A)$ is a function describing the dependence of Earth's mean annual global near-surface temperature on the amount of carbon in the atmosphere; function $D(x)$ describes the biomass death rate; and function $S(y, T(A))$ describes the soil respiration rate (decomposition of dead organics). The functions are described below.

It is important to note that all variables are measured in GtC, thus accepting the volume of the Earth's atmosphere as invariable we use mass units in the equations as representing concentrations.

The first equation describes a change in total carbon in the Earth's global cycle, the second describes a change in the amount of carbon in the biomass of living plants, the third describes carbon dynamics in various organic residues, and the fourth equation describes the amount of carbon in the atmosphere and this is the law of conservation of carbon mass.

As the model is zero-dimensional, it would be natural to accept assumptions that (1) a mean annual global near-surface temperature of 15°C is optimal for biota. Any global temperature change leads to similar changes in local temperatures: if, for example, the mean annual global temperature increases by 5°C, mean annual local temperatures will rise likewise (this is, of course, an oversimplification), and (2) the time needed for equalizing concentrations of atmospheric gases is negligibly small compared with the characteristic times of changes in model variables (months, years). Moreover, as photosynthesis and soil respiration are influenced by the temperature of the atmosphere and the relatively thin layer of the Earth's surface, the thermal lag of the system is not taken into account in the model.

Exclusion of the precipitation factor may be regarded as oversimplification of the model. This simplification, however, is done in accordance with the worst case scenario principle, too. Indeed, it is difficult to predict how a change in the precipitation regime will influence plants and soil microflora (and flora), and as this influence may be ambivalent an unfavorable variant is accepted: as the temperature rises, soil respiration increases (i.e., water is not a limiting factor of the process).

The rates of the chosen biospheric processes are described by the following functions.

Plant biomass growth rate depends on the amount of biomass, temperature, and CO_2 concentration (GtC/yer) as a multiplicative function:

$$P(x, A, T) = V_p x(x_{\max} - x)V(A)f_p(T(A)), \tag{10.2}$$

where V_p is the scale factor (1/(GtC × year)); and $x_{\max}$ is the highest possible plant cover density (GtC).

Function $V(A)$ describes biomass growth vs. atmospheric CO_2 concentration. It was obtained based on the well-known Monod function:

$$V(A) = \frac{A}{K_A + A}. \tag{10.3}$$

The Monod equation usually contains concentrations, but as the volume of the reaction space (atmosphere) remains unchanged, the model uses total atmospheric carbon as a unit, to simplify data adjustment. Parameter K_A has been chosen to equal 900 GtC based on experimental data (Morgan *et al.*, 2001; Pritchard *et al.*, 2001).

The empirical dependence of plant biomass growth rate on temperature T and maximal growth temperature T_{max} has the following form:

$$f(T) = T^d (T_{max} - T), \quad 0 \le T \le T_{max}. \tag{10.4}$$

The empirical dependence of the increase in mean annual global near-surface temperature on CO_2 concentration has been taken from published data (Gifford, 1993):

$$T(A) = T_0 + T_{del} \log_2 \left(\frac{A}{A_0} \right), \tag{10.5}$$

where A_0 is the amount of atmospheric carbon at the moment of measuring mean near-surface temperature T_0, which is equal to 15°C at the present time; and T_{del} is the temperature increase as a result of doubling CO_2 concentration.

The biomass death rate (GtC/yr) is written in a simple form:

$$D(x) = V_d x, \tag{10.6}$$

where V_d is the scale factor.

The soil respiration rate (decomposition of dead organics) (GtC/year) and CO_2 release into the atmosphere is described by the following function:

$$S(y, T) = V_s y f_M(T), \tag{10.7}$$

where V_S is the scale factor; and $f_M(T)$ is a function of type (10.4) expressing temperature dependence of soil respiration.

Anthropogenic carbon emission due to fuel combustion until the present time is expressed by an empirical function which is in good agreement with available data (Gifford, 1993). The type of the function of anthropogenic emission in the future depends on the expected scenarios of fossil fuel consumption. For certainty, the type of this function was chosen to correspond to Scenario B2 (IPCC, 2001). According to this scenario the main forces are applied to solution of local problems of economical, social, and ecological stability. In terms of CO_2 emission rates, it occupies an intermediate position among the proposed scenarios.

Some aspects of the model should be explained. Introduction of the factor describing the maximum amount of carbon in biomass into formula (10.2) seems to be essentially important. First, the carbon taken up by wood is not removed from the atmosphere forever, but is only temporarily sequestered in the sink. Second, the ability of plants to consume carbon dioxide is generally limited either by a deficiency of nutrients (nitrogen, phosphorus, etc.) or by restrictions on available area.

The largest possible amount of carbon in biomass x_{max} is given in the model as $x_0 G$, where x_0 is the amount of land plant biomass in the late 1950s, and G is the coefficient characterizing the ability of plants to increase the amount of the biomass.

The issue of the temperature dependence of soil respiration is rather complicated. It has been reported in some studies that a soil temperature increase does not cause any increase in the emission of CO_2 from the soil (Luo *et al.*, 2001). Those were results of observations of the soils in the steppe (Fitter *et al.*, 1999), Arctic tundra (Johnson *et al.*, 2000), and boreal forests (Liski *et al.*, 1999). On the other hand, there are some studies that demonstrate the temperature dependence of soil respiration (Risk *et al.*, 2002; Rochette *et al.*, 1999). Thus, there is no unanimous opinion on the role of soil microorganisms in carbon dynamics. Some researchers believe that soil can be a safe carbon sink even under temperature rise. Others have the opposite opinion: namely, that microorganisms will make the soil a rich source of carbon dioxide, which will enhance the greenhouse effect. In accordance with the worst case scenario principle, we have chosen the pessimistic variant.

Different estimates of carbon pools in the atmosphere and land compartments and flows between them have been reported in the literature. This may be accounted for either by differences in the methods used or by an actual increase in these parameters (Brovkin *et al.*, 2002, 2004; IPCC, 2001; Schimel *et al.*, 1996; Watson, 2000): carbon in the atmosphere 600 GtC–760 GtC; in biomass 500 GtC–850 GtC; in the soil 1,080 GtC–2,000 GtC; gross primary production (GPP) 110 GtC–120 GtC/yr; photosynthesis net primary production (NPP) 55 GtC–60 GtC/yr; and soil respiration 55 GtC–60 GtC/yr.

In accordance with the worst case scenario principle, we have chosen the following values for the parameters of our model: the initial amount of carbon in plant biomass as 850 GtC and in organic residues 2,000 GtC. We should mention separately that numerical experiments have also been performed using the data reported by Krapivin and Potapov (2002), with the amount of carbon in plant biomass estimated as 466 GtC and carbon in organic residues as 2,011 GtC.

The choice of the scale factors in functions (10.2), (10.6), and (10.7) was aimed at attaining the following: the biomass growth rate should be equal to an earlier published estimate of 55 GtC/year and, in the absence of anthropogenic CO_2 flux, the model should be in a steady state, with parameters corresponding to the actual global values of the late 1950s.

The final choice of parameters was aimed at making the model describe the dynamics of the mean annual atmospheric CO_2 concentration between 1958, when Mauna Loa observations (Keeling and Whorf, 2001) were started, and 2000 (Figure

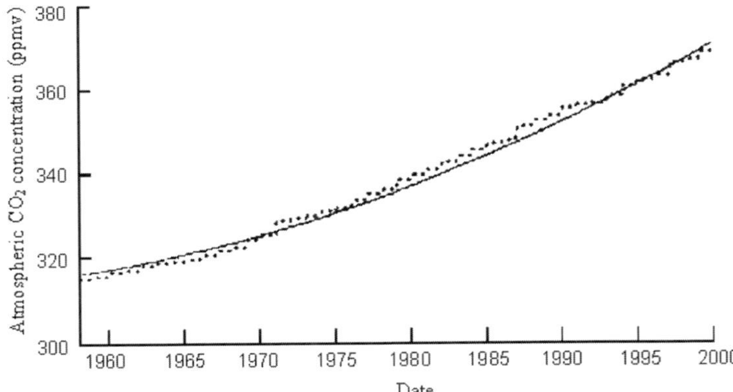

Figure 10.3. Comparison of the model data with measurements made at the Mauna Loa Observatory, Hawaii.

10.3). Figure 10.3 shows that, in spite of slight divergence, relative deviation of the model curve from the observation data does not exceed 2%.

To assess the realizability of the proposed mechanisms of catastrophic development and to estimate the approximate dates of the catastrophes, we have considered the most unfavorable values of parameters from the available confidence intervals and assumed that compensatory mechanisms were inactive. For instance, the maximum sensitivity of the Earth's climate to CO_2 doubling is estimated as 4.5°C (IPCC, 2001). As the model describes the land biosphere, the data that expect the near-land temperature to be 40% higher than the average global temperature yield $T_{del} \approx 6°C$.

10.2.2 Results of modeling

Computational experiments on modeling the future dynamics of carbon pools showed that even at moderate rates of fossil fuel combustion (Scenario B2), the model predicts the development of catastrophic processes, leading to irreversible changes in the biosphere. An example of catastrophic dynamics at $G = 1.5$ is shown in Figure 10.4. Different curves correspond to different dates when fuel combustion is

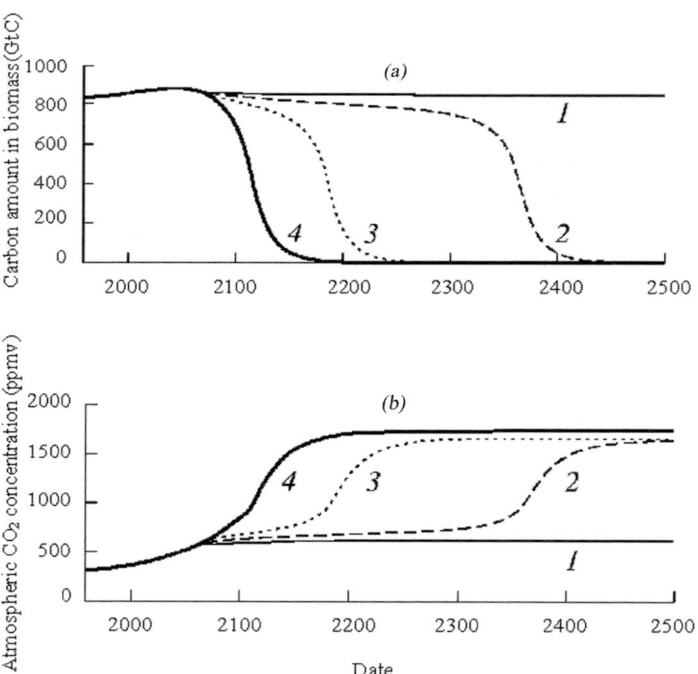

Figure 10.4. Variants of the dynamics of CO_2 in biomass (a) and atmospheric CO_2 concentration (b) at different dates of completely stopping the emission: (1) 2059; (2) 2064; (3) 2070; (4) 2090.

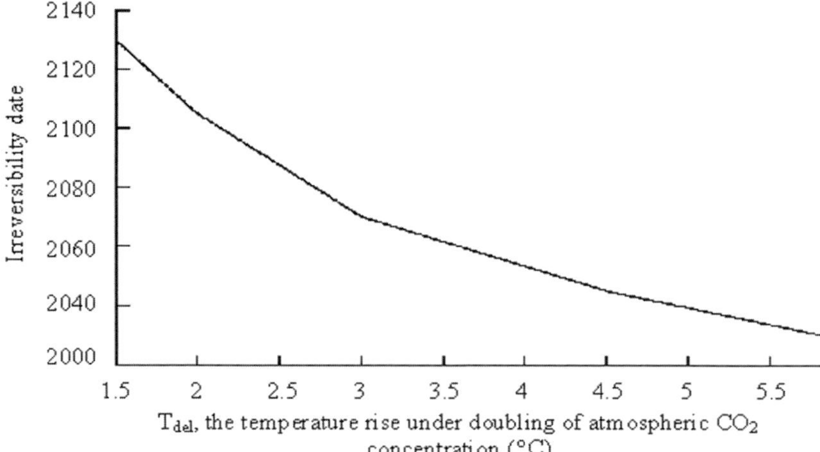

Figure 10.5. The theoretical curve "irreversibility date" vs. increase in T_{del}, characterizing temperature rise under doubling of atmospheric CO_2 concentration

stopped, thus digressing from Scenario B2. This unrealistic element is introduced in order to demonstrate that, even if fuel combustion is stopped completely, there still will be an irreversibility date after which the catastrophic process in the system becomes irreversible.

Varying the important greenhouse effect parameter T_{del}, which is the temperature rise under doubling of atmospheric CO_2 between 1.5°C and 6°C (IPCC, 2001), we do not alter the curves and we can construct the T_{del} dependence of the irreversibility date (i.e., test the sensitivity of the model, Figure 10.5). The implication of the curve is perfectly clear: as the greenhouse effect is reduced, the irreversibility date becomes more remote. It is also important to note that, prior to the irreversibility date and for some time after it, biosphere and climate parameters change in a gradual and unalarming manner.

The sensitivity of the model to variations in its key parameters was further tested in a series of computational experiments. The values of the model key parameters were varied within the ranges of their possible variations (taken from different sources). By varying combinations of different values we obtained both very optimistic and very pessimistic scenarios. The latter predicted irreversible changes in the biosphere around 2035.

Eventually, the zero-approximation model allows us is to conclude that the most significant parameters for triggering irreversible processes are the temperature dependence of the decay of dead organics, the ability of plants to consume excess carbon (buffer capacity of biota), and temperature rise under doubling of CO_2 concentration. The results of computational experiments are listed in Table 10.1.

We should note that the so-called irreversibility dates are certainly not exact dates, and changes within the framework of a given scenario can begin both before and after these dates. However, high accuracy is not the aim of computational

Table 10.1. Critical points in different scenarios.

Value of parameter T_{del} (°C)	T_{max} *value of soil respiration function* (°C)						*Value of parameter* *G*
	45		*40*	*35*		*30*	
6	2060	1990*	2045 2080 2320	2040	2050*	2035	1.5 1.3 1.1
4	2110	2140*	2060	2055	2150*	2058	1.5
3				2665			1.1
2	2120	2300*	2107	2095	2170*	2085	1.5

* Irreversibility dates obtained by using initial data by Krapivin and Potapov (2002).

experiments with a model. The constructed model puts emphasis on one of the quickest possible mechanisms of CO_2 release, which works on the principle of positive feedback. To develop the model and make it more accurate, it is essential to have reliable experimental data on the temperature dependences of the growth of plants and soil microorganisms and on the maximum possible increase in land biomass with an increase in atmospheric CO_2 concentration.

The main result of constructing the basic model was demonstration of the probability that catastrophic changes in the development of the biosphere–climate system can take place and determination of their characteristic time parameters, the most important of which is the irreversibility date.

The sphere of application of the basic minimal model of the biosphere is limited. First, for short-range forecasts, it does not give a sufficiently detailed description of the dynamics of atmospheric carbon dioxide concentration, which demonstrates seasonal changes against the background of the long-term trend. Second, for long-range forecasts, one cannot ignore the oceans' contribution to global carbon dynamics. The worst case scenario principle does justify ignoring probable compensatory mechanisms but it does not provide grounds for ignoring the mechanisms whose compensatory effect is doubtless, at least at the contemporary level of knowledge. Moreover, it would not be quite correct to extend the forecast range to over 50 years because the verification period for the parameters is 50 years.

Hence, there may be two directions for developing the model: (1) to model seasonal dynamics of atmospheric carbon dioxide concentration (see Section 10.3), and (2) to extend the forecast range by extending the verification period of the model. These two directions have different timescales and are aimed at attaining different objectives.

10.2.3 Integrated minimal model of long-term carbon dioxide dynamics in the biosphere

To extend the range and enhance the validity of the basic model, it is necessary to extend the duration of the time period for its verification. To do this, the oceans' contribution to global carbon dynamics should be taken into account; this is mainly the physical/chemical processes of CO_2 interactions with the oceans, the contribution of marine biota being of secondary significance. The worst case scenario principle is not violated by taking into account the compensatory contribution of the oceans, as the choice of the worst variant must not come into conflict with reliable data. In this case, it is well known that carbon dioxide is intensely exchanged between the oceans and the atmosphere (Brovkin *et al.*, 2002, 2004; Falkowski *et al.*, 1998; Kondratyev and Krapivin, 2004; Semyonov, 2004). The main matter fluxes considered in the modes are presented in Figure 10.6.

Most of the limitations and assumptions used in the initial model remain unchanged in the integrated model. The integrated long-range model describes the dynamics of the following carbon sinks: the atmosphere, living plants, and dead organic matter in the soil, the photic and deep layers of the ocean, marine autotrophs and heterotrophs.

The equations describing the growth dynamics of autotrophs and heterotrophs and decomposition of dead organics in the ocean are based on the same principle as the equations of land processes, though corrected for specific features of biological processes in the ocean.

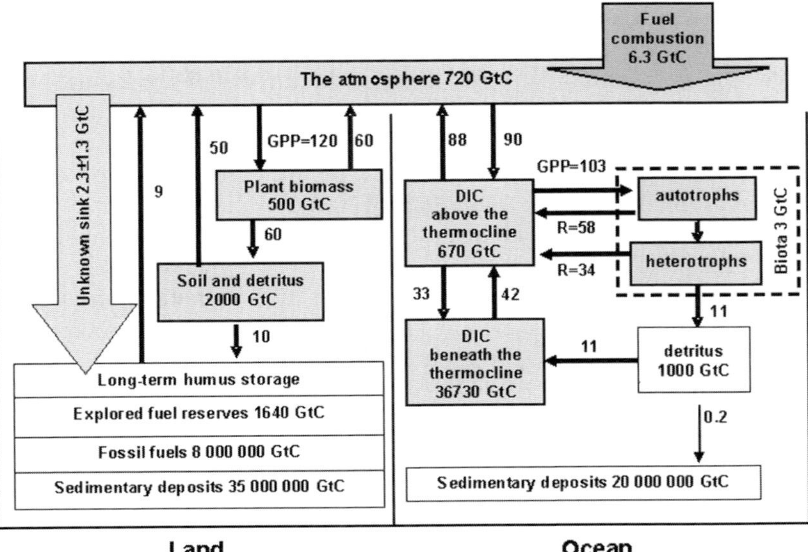

Figure 10.6. A carbon cycle scheme used to construct the integrated biota–atmosphere–ocean long-term minimal model. Gray parts of the scheme show the structure of the minimal model. DIC stands for dissolved inorganic carbon.

The model is represented by the following system of equations:

$$\frac{dA}{dt} = S(y, T(A)) + C_{a\_up}BM_{out}(B) - P(x, A, T(A)) - C_{a\_down}AM_{in}(A) + \text{fuel}(t) + (1 - H)F(t) + X(t) \quad (10.8)$$

$$\frac{dx}{dt} = P(x, A, T(A)) - D(x) - F(t) \quad (10.9)$$

$$\frac{dy}{dt} = D(x) - S(y, T(A)) + HF(t) \quad (10.10)$$

$$\frac{dB}{dt} = [RE(z) + C_{a\_down}AM_{in}(A) + C_{d\_up}U] - [C_{f\_down}B + C_{a\_up}BM_{out}(B) + N(m, A, B, t)] \quad (10.11)$$

$$\frac{dm}{dt} = N(m, A, B, t) - V(m, z) \quad (10.12)$$

$$\frac{dz}{dt} = V(m, z) - E(z) \quad (10.13)$$

$$\frac{dU}{dt} = C_{f\_down}B - C_{d\_up}U + (1 - R)E(z). \quad (10.14)$$

In this system, the first equation (10.8) describes the dynamics of atmospheric carbon concentration determined by the following processes: photosynthesis (P); respiration of soil microflora (S); sorption ($C_{a\_down}AM_{in}(A)$), where C_{a_down} is the flux rate, M_{in} is the factor determining the flux level, which will be described in greater detail later; and desorption ($C_{a\_up}BM_{out}(B)$), where B is CO_2 concentration in the given ocean layer, C_{a_up} is the flux rate, and M_{out} is the factor determining the flux level, which will be described in greater detail later; CO_2 in the surface layer of the ocean; deforestation ($F(t)$), where H is the fraction of forest biomass subjected to oxidation; and Houghton's flux (Houghton, 2003) ($X(t)$).

Equation (10.9) describes variations in the amount of carbon in living plant biomass as a result of photosynthesis (P) and death of biomass ($D(x)$).

Equation (10.10) describes carbon dynamics in dead organic matter as a balance between the processes of biomass dying ($D(x)$) and the decay of dead matter in the soil due to the vital activity of soil microflora ($S(y, T(A))$).

Equation (10.11) describes the balance between CO_2 fluxes in the photic layer of the oceans resulting from plankton photosynthesis ($N(m, A, B, t)$); respiration of heterotrophs (RE), where R is a coefficient denoting the amount of biomass consumed by heterotrophs that is used for respiration; CO_2 absorption and desorption between the atmosphere and the surface of the ocean, and the physical transfer of CO_2 between the photic layer of the ocean and its deep layers ($C_{d\_up}U$ is upwelling; $C_{f\_down}B$ is downwelling, U and B are the CO_2 concentrations in the deep and surface layers, respectively, and C_i is the rate of the respective fluxes); and m is the amount of carbon in the biomass of the plant compartment (GtC).

Equation (10.12) describes carbon dynamics in the biomass of oceanic autotrophs due to plankton photosynthesis ($N(m, B)$) and its consumption by heterotrophs ($V(m, z)$); and z is the amount of carbon in the biomass of animals.

Equation (10.13) describes carbon input ($V(m, z)$) and output ($E(z)$) in the biomass of heterotrophs.

Equation (10.14) is a description of carbon mass transfer in the deep ocean layer, taking into account the fluxes of detritus ($(1 - R)E$), where $1 - R$ is the biomass of heterotrophs transferred to the mortmass, and inorganic CO_2 ($C_{d\_up}U$ and $C_{f\_down}B$).

Equations (10.9) and (10.10) use the same functions as the initial model equations: (10.2), (10.6), and (10.7). However, the temperature dependence of soil respiration in the function $S(y, T(A))$ is of a different form.

At the present time, there is no unambiguous answer to the question about the type of this dependence. Different opinions on this subject were discussed above. However, knowing that the soil is inhabited by several groups of microorganisms that have different temperature preferences and succeed each other as the temperature rises, the last in succession being thermophilic organisms (Zavarzin and Kolotilova, 2001), we can assume that the level of respiration of soil microflora will remain as high as possible throughout the temperature range. Therefore, we can introduce an enveloping, or integrated, curve describing the temperature dependence of respiration of generalized soil microflora with a temperature maximum of soil respiration shifted towards high temperatures relative to the temperature maximum of the photosynthesis rate (Figure 10.7).

Phytoplankton increase is described by the function:

$$N(m, A, B, t) = V_M m(m_{\lim} - z - m)K_m W(B) f_m(T(A)), \tag{10.15}$$

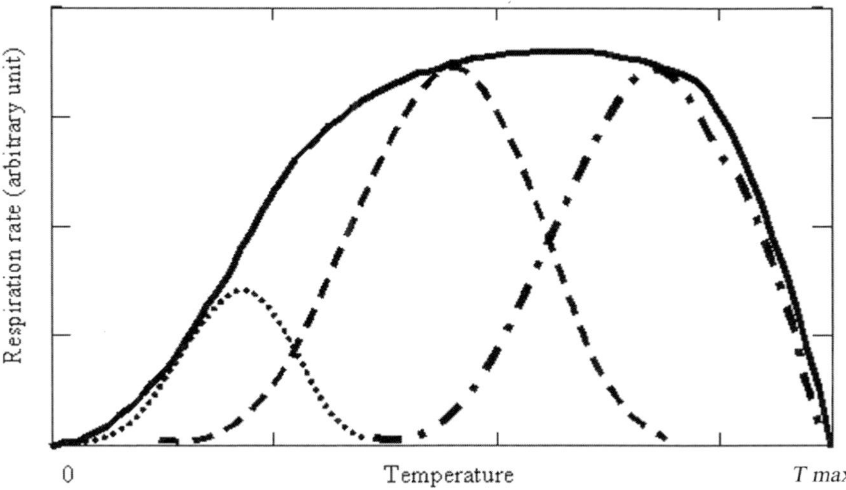

Figure 10.7. Integrated curve (——) describing temperature dependence of respiration rate of soil microflora, based on the principle of succession of microbial communities with temperature rise: psychrophiles (– – –), mesophiles (– · –·), and thermophiles (· · · · ·).

where m is the amount of carbon in the biomass of the plant compartment (GtC); V_M is the scale factor; $m_{\lim}$ is the maximum possible density of the total autotroph and heterotroph biomass, limited by nutrient (phosphorus, nitrogen, and iron) deficiency; K_m is the plankton biomass doubling time; $W(B)$ is a function of type (10.3); $f_m(T(A))$ is the temperature dependence of photosynthesis in the form of formula (10.4).

Equation (10.16) describes the rate of plankton biomass consumption by marine living organisms:

$$V(m,z) = V_{Dm} m K_Z z, \tag{10.16}$$

where m and z are the amounts of carbon in the biomasses of phytoplankton and heterotrophs; V_{Dm} is the scale factor; and K_Z is the animal biomass doubling time.

The average death rate of heterotrophs is described by the formula:

$$E(z) = V_{\text{out}} K_Z z, \tag{10.17}$$

where V_{out} is the scale factor.

The rates of absorption and desorption of CO_2 by the ocean surface are empirical approximations of the data on CO_2 concentration in water at different temperatures (Kondratyev and Krapivin, 2004):

$$M_i(x) = \frac{0.94e^{-0.03T(x)}}{0.94e^{-0.03T_0}}. \tag{10.18}$$

10.2.4 Model verification results

Adjustment of the model parameters was performed based on the core data on atmospheric carbon dioxide concentration in the past, from 1700 to the present time, the data on fossil fuel combustion, and the dynamics of the global temperature since 1860 (IPCC, 2001, 2007; Keeling *et al.*, 2001). Additionally, we used the data on atmospheric carbon dioxide concentration obtained at the Mauna Loa Observatory and the data on deforestation rates and changes in the types and methods of land management (Houghton, 2003).

The data on variations in carbon dioxide concentration and global temperature in the past can be used as a basis to calculate climate sensitivity immediately; this amounts to about 2°C under doubling of carbon dioxide concentration. This sensitivity value was used in the integrated model. To attain the observed fit (Figure 10.8), however, we had to assume that in 1700 carbon exchange between the surface and deep layers occurred at a rate that was half the present-day rate. This may be a price that we have to pay for using a minimal model, but, on the other hand, those carbon fluxes might have really been less intense.

10.2.5 Forecasts of the future dynamics of the biosphere

To calculate possible scenarios of the development of the biosphere, we used the variant of fossil fuel combustion labeled as A2 in the IPCC classification (IPCC, 2001). The purpose of computational experiments was to estimate the conditions

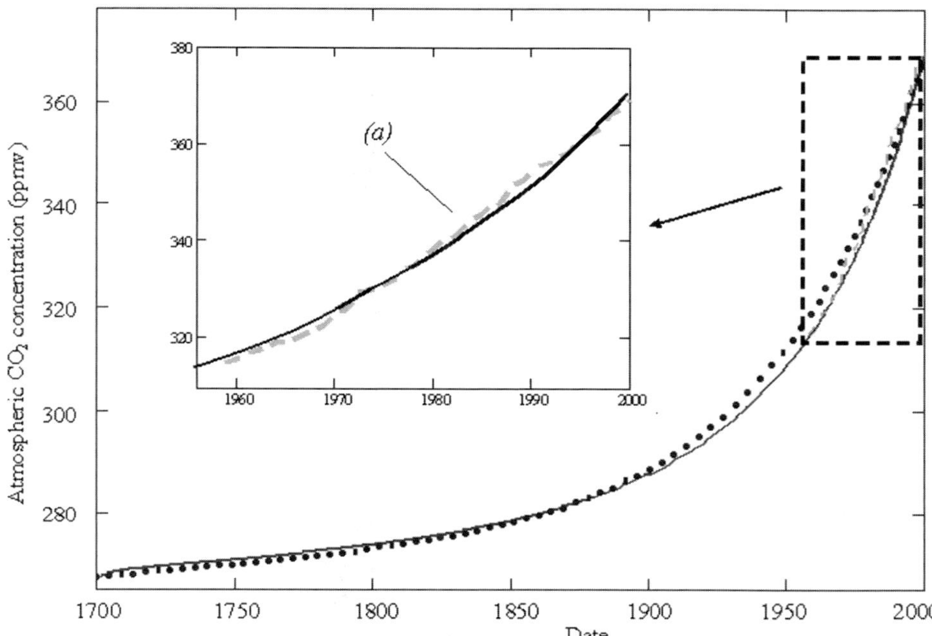

Figure 10.8. Comparison of the observation data and computational experiments on carbon dioxide dynamics; (a) is a fragment of the comparison with Mauna Loa data.

under which a catastrophic scenario is realized and determine the irreversibility dates corresponding to these conditions (i.e., the moments after which the system collapses, even if human-induced carbon emissions have been stopped completely). Scenarios were constructed under the assumption that climate sensitivity of CO_2 increase is not a fixed value, but can vary depending on conditions of Earth and, theoretically, can reach rather high values (Stainforth *et al.*, 2005).

Experiments showed (Figure 10.9) that under climate sensitivity to doubling of CO_2 less than 4.5°C, no avalanche-like processes occur in the biosphere–climate system, but some changes that are quite catastrophic in the ordinary sense can take place. However, at higher sensitivity values, irreversible changes do occur in the system; these changes are catastrophic in scientific terms and irreversibility dates may come very soon.

The difference between these results and the results obtained using the initial model is accounted for by the significant compensatory role of the ocean. However, as the results of modeling show, the oceans are not always able to maintain the biosphere–climate system within the favorable range for human beings.

In our opinion, the main result is that our model has shown and substantiated the probability of future irreversible negative changes caused by anthropogenic emission of carbon dioxide in the biosphere–climate system, which also includes the oceans (i.e., the conclusions suggested by the initial minimal model were confirmed).

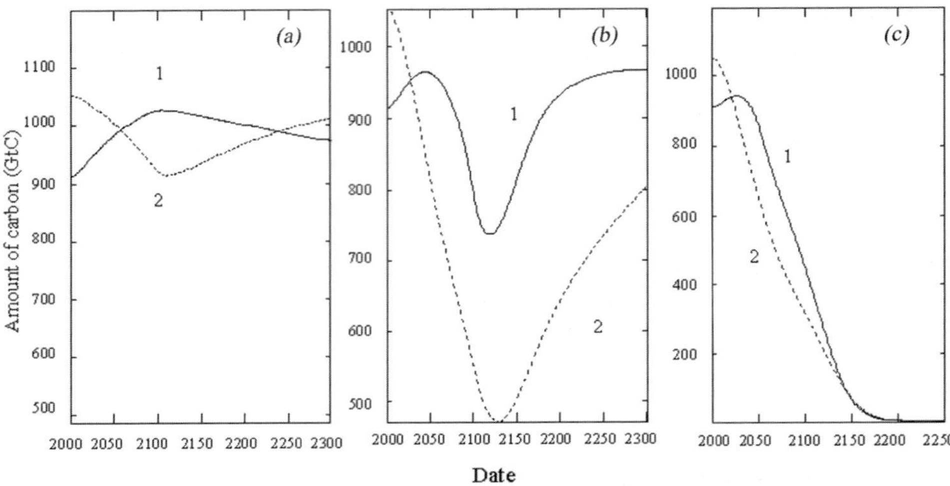

Figure 10.9. Variants of scenarios for the development of the biosphere under different values of climate sensitivity parameter T_{del}: (a) 2°C; (b) 4.5°C; (c) 6°C. In case (c) the "irreversibility date" is 2054. In the graphs, 1 is biomass and 2 is dead organic matter.

We should note that the integrated model yields more realistic dynamics of the biosphere's behavior when combustion is stopped before the irreversibility point is reached, as it predicts that the major parameters can acquire values close to their initial ones.

The main purpose of what is expounded in this chapter is to draw the attention of readers to the unfavorable, and thus most important practically, variants of future development instead of describing the most probable and milder scenarios. The results that we have obtained by isolating the critical mechanisms of a quick positive feedback actually pose a task for experimenters: to test and specify the parameters of these mechanisms (see Sections 10.4.2 and 10.4.3). These experiments are as important as biosphere–climate studies, whose significance is doubtless.

10.3 THE CARBON CYCLE; THE STUDY OF CHLOROPHYLL GLOBAL DYNAMICS AND NET PRIMARY PRODUCTION (NPP) BY SATELLITE METHODS

10.3.1 Introduction

To preserve the biosphere and to use it efficiently, it is necessary to gain a deep insight into the dynamics of the primary production process on our planet. Net primary production (NPP) determines the amount of net carbon fixed by plants. In fact, this is the beginning of the carbon biogeochemical cycle. NPP is the main indicator of the health of an ecosystem, resource recycling, and biospheric carbon flows. NPP variations in space and time are very important for environmental studies, control and monitoring of natural resources, and studying the global carbon cycle, which is connected, in its turn, with climate change. NPP is used to verify model data obtained in the process of modeling CO_2 accumulation in the atmosphere.

Terrestrial ecosystems are responsible for a large part of the seasonal and inter-annual variation in CO_2 concentration in the atmosphere (Prentice *et al.*, 2001). Atmospheric changes and inverse modeling presume that net terrestrial carbon sink uptake increased mainly from the 1980s to the 1990s (Battle *et al.*, 2000; Bousquet *et al.*, 2000), but the reasons for this increase are still not clear (Schimel *et al.*, 2001). The joint study of NPP and model data help to reveal the processes determining the carbon cycle. Optical radiation reflected from plants and registered by satellites can be used to calculate NPP and such radiation is connected with the above-ground biomass (Prince and Goward, 1995). Thus, we plan to estimate the biomass of green plants.

10.3.2 Trends in the global photosynthetic activity of land vegetation

We have measured NPP variability with a spatial resolution of 8 km and a periodicity of 10 days from 1981 to 2000. The basis was NOAA/AVHRR satellite data. These data were averaged by latitude with steps of 5° and 30°. Results were obtained using GLO-PEM (Global Production Efficiency Model). Almost all variables used in the model are satellite data. The result shows that there is a rising trend for the NPP of terrestrial ecosystems. The trend is observed both for the northern and southern hemispheres. This result for such a short period is matched to global increasing of biomass calculated by using an original low-dimensional model (Section 10.2.). To provide the model with data, a time series of satellite data of NOAA and SeaWiFS from 1981 to 2005 was analyzed.

Earlier many investigators had studied multiannual NDVI and NPP trends, but we were faced with the task of monitoring the dynamics of NDVI and NPP at different latitudes in accordance with the requirements of the global small-parametric model for corresponding latitudes and comparison of the data received by satellite remote sensing and by modeling (Gao *et al.*, 2004; Running *et al.*, 2004; Shabanov *et al.*, 2002).

10.3.2.1 Methods and materials

The most significant parameter for studying vegetation on a global and regional scale is NPP. The NDVI (normalized difference vegetation index) is connected with parameters such as the leaf area index and chlorophyll-bearing biomass, which reflect the instantaneous content of phytopigments in the area being measured, and are not directly connected with the value of primary production which, in its turn, depends on many other parameters (light, humidity, temperature, etc.). In various climatic zones these factors may differ. For studying the functions of land plant communities, investigation of their spacetime dynamics, and understanding of the functional links in plant communities, it is necessary to measure both the NDVI and NPP. At the same time it is important to take into account the fact that the NDVI is connected with plant biomass in a complex way and depends on vegetative types, as well as on other parameters (humidity, stratification, state of plants, etc.).

In terms of measurement technology, the NDVI is a simpler and more precise parameter. The NDVI is calculated on the basis of radiation received by spaceborne sensors in the red and near-infrared spectral ranges. A simplified estimation of NPP is possible, despite the complexity of the necessary calculations. At present there are several models for calculation of NPP with the help of satellite data. After a study of the literature, we have chosen the GLO-PEM model. The main argument for this model was that it is based on physiological principles; in particular, the amount of carbon fixed per canopy light absorption unit is modeled rather than fitted using field observations (Prince and Goward, 1995).

GLO-PEM allows the making of global maps of NPP. For modeling the ground NPP on a global scale, GLO-PEM uses spectral backscattered radiation and the temperature of the ground surface received by remote sensing.

The following data are used for calculations of NPP:

- AVHRR data.
- The amount of incident photosynthetically active radiation (PAR) obtained with the help of TOMS (Total Ozone Mapping Spectrometer) data. Irradiation of the Earth surface in the wavelength range of PAR (0.4 μm–0.7 μm) is estimated by calculating the difference between maximum possible irradiance (clear sky conditions) and PAR backscattered by clouds and aerosols (Eck and Dye, 1991).
- Backscattered radiation in the visible red and near-infrared spectral ranges is used for the calculation of spectral vegetation indices, the values of which are connected with the PAR fraction in a linear fashion, absorbed by ground vegetation. The spectral vegetation index in combination with PAR results in the measurement of canopy light absorption.
- Surface radiometric temperature and atmospheric column precipitable water vapor amount that were obtained by measuring the temperature in various spectral ranges.
- NPP = GPP (i.e., respiration).

Thus, GLO-PEM based on TOMS and AVHRR satellite data can be used to obtain the global distribution of NPP, which, in its turn, allows using the results of GLO-PEM (Goetz *et al.*, 2000; Goward and Dye, 1997; Prince and Goward, 1995) to verify our low-dimensional model.

10.3.2.2 Results on the perennial dynamics of global NDVI and NPP

The NDVI time series that we have used, determined from AVHRR data, began in 1981 and ended in 2000. To continue this time series we used SeaWiFS data from 2000 to 2005. The multisatellite data used overlaps for the year 2000. To superpose NOAA and SeaWiFS data we worked out an empirical equation (derived by

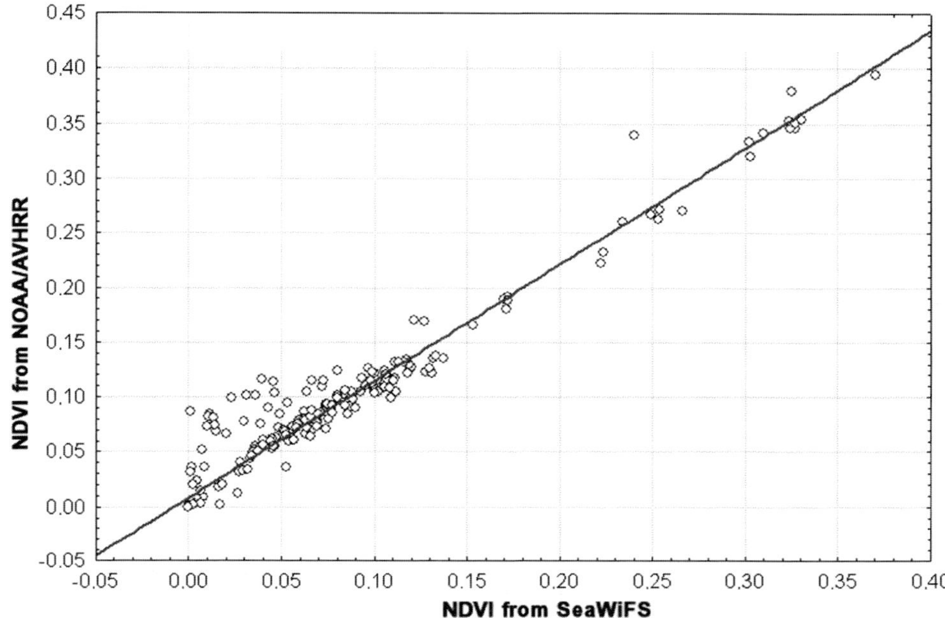

Figure 10.10. Regression analysis of common NOAA and SeaWiFS data. $r = 0.9771$, NOAA_AVHRR $= 0.0081 + 1.0648 \times$ SeaWiFS.

regression analysis) (Figure 10.10):

$$\text{NOAA\_AVHRR} = 0.0081 + 1.0648 \times \text{SeaWiFS},$$

where NOAA_AVHRR is the NDVI determined from AVHRR data; and SeaWiFS is the NDVI determined from SeaWiFS data.

During regression analysis the linear regression coefficient was 0.97. This means that it is possible to use the data of both sensors together.

For the investigation of NDVI time series at different latitudes, we averaged the NDVI values in the zones having the width of 30° (Figure 10.11a). We also averaged the global data (Figure 10.11b).

Analyzing NPP data, as in the NDVI case, we averaged the data in latitudinal zones with a width of 30° (Figure 10.12a) and globally (Figure 10.12b). The model data show a rather small increase in NPP values, which indicates that there is only qualitative correspondence between the model and satellite data on NPP so far.

The joint use of satellite data on global NPP dynamics along with the traditional method of model verification by atmospheric CO_2 improves our understanding of processes in the climate–biosphere system. This has a positive effect on the validity of the model results. The results of model verification with regard to the satellite data on global NPP dynamics are shown later (see Section 10.3.5).

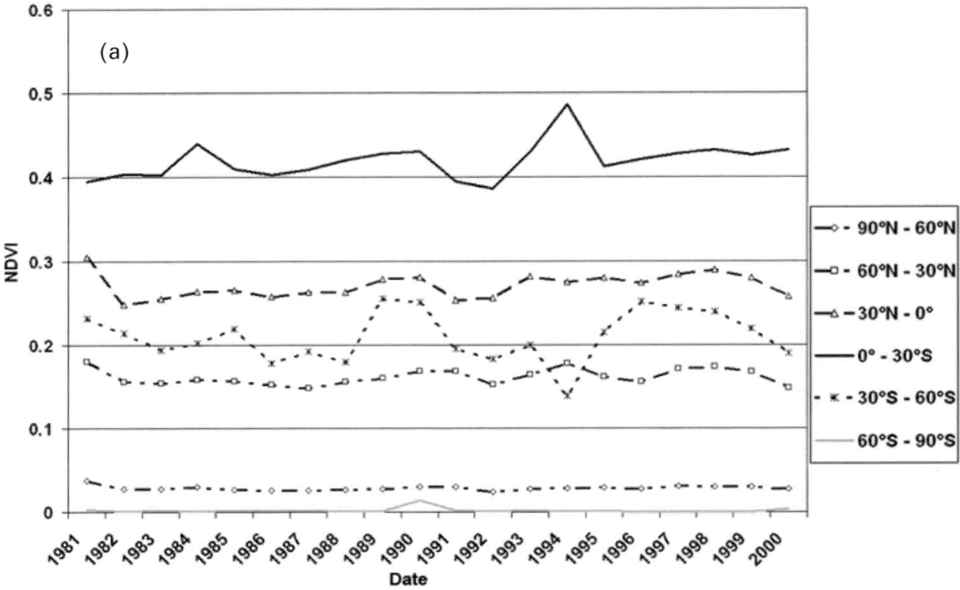

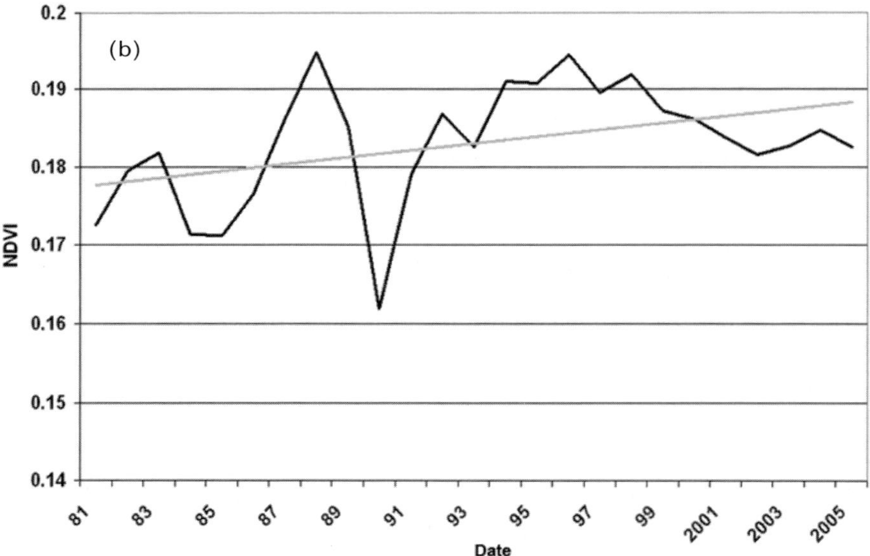

Figure 10.11. (a) Perennial dynamics of global NDVI at latitude zones. (b) Perennial dynamics of summary global NDVI.

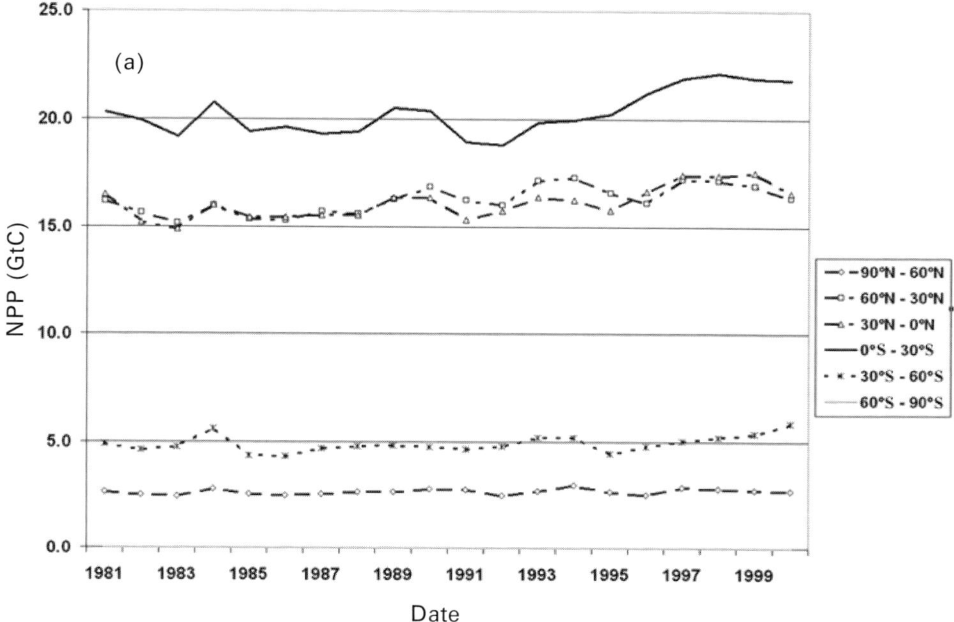

Figure 10.12. (a) Perennial dynamics of global NPP at latitude zones. (b) Perennial dynamics of summary global NPP.

To summarize, the following facts were discovered as a result of analyzing NPP data averaging:

— Maximum NPP values are observed in the tropical zone of the southern hemisphere from 0° to 30°S (this coincides with data for the NDVI).
— In comparison with NDVI, the trends from 60°N to 30°S are more distinct.
— The global NPP data trend is non-linear, has a near-exponential form, and the values increase by 10% within the time series considered.

The following facts were discovered as a result of the analyzing NDVI data averaging:

— Maximum NDVI values are observed in the tropical zone of the southern hemisphere from 0 to 30°S (tropical forests of South America).
— In 1991–1992 the consequences of the Mount Pinatubo volcanic eruption were observed (Gao *et al.*, 2004; Shabanov *et al.*, 2002). A decrease can be seen in the zones from 60°N to 60°S.
— In the 30°N–0° and 30°S–60°S latitude zones there is a small trend in averaged NDVI data.
— The global data show a dip as a result of the Mount Pinatubo volcanic eruption in 1991 and a small trend in global averaged NDVI data.

Comparison of the model and satellite data showed a sufficiently high degree of correspondence. It proves the hypothesis about the potential possibility of using remote-sensing data for the creation of not only distributed—but also minimal—models. Over time, after the increase in the number of parameters measured by remote sensing and of the accuracy and time of measurement, the connection that we have presented between theoretical and practical investigations may be studied in a more detailed way.

Thus:

(1) A technology for determining the global dynamics of NDVI and NPP according to latitudinal zones was worked out.
(2) The considerable difference between perennial changes in NPP compared with NDVI was demonstrated.
(3) The correspondence between global NPP dynamics from satellite and model data was shown.

10.3.3 Long-term dynamics of chlorophyll concentration in the ocean surface layer (from space data)

Information about atmospheric warming imparts particular significance to the task of determining the real-life dynamics of the biosphere. The actual contributions of land and ocean biotas have not been accurately determined, although there is a great body of literature on the subject.

The extensive scientific discussion of global warming causes a natural wish to relate this process to possible changes in the amount and dynamics of terrestrial and oceanic vegetation. The question arises as to whether this process influences variations in the amount and diversity of plants and whether it influences the pattern of their seasonal and long-term variations. It would seem that the continuing increase in the concentration of CO_2 and increase in the mean global temperature must cause permanent long-term changes in the amounts of phytopigments in the biosphere. But, this raises a couple of questions: "Is this is really so" and "How can we plot the direction of these changes?"

Thus, the initial task was to reveal long-term trends of phytopigment concentrations in the ocean. This task could be fulfilled based on daily satellite measurements conducted over a period of many years.

Of particular importance is the problem of determining the dynamics of primary production and phytopigment concentration in the World's oceans (Bode and Varela, 1998). The changes are of different scales in space and time (Mete Uz and Yoder, 2004; Shevyrnogov *et al.*, 2004). To determine the relationship between global changes in climate and the biosphere, it is particularly important to trace the long-term variability of phytopigment concentrations at different latitudes and in different biogeographical conditions (Bidigare and Ondrusek, 1996; Denman and Abbott, 1988; Dickey *et al.*, 1991).

This section, using SeaWiFS data, describes the long-term dynamics of seasonal variations in phytopigment concentrations in the global scale under different biogeographical conditions.

10.3.3.1 Methods

We have used 79 chlorophyll Level-3 standard mapped 9 km SeaWiFS images from September 1997 to March 2004, which were made available by NASA after the fourth reprocessing. These data were calculated using 8-day SeaWiFS composites with the weighted mean method. Because speckling of the imagery is possible, median filtration was done with 3×3 squares. For each year, we made maps of calculated average chlorophyll concentrations. Then, for every map we calculated the average chlorophyll concentration from the obtained dataset, taking into account latitude dependence of pixel area. We used only those pixels that had data for all the 6 years (coverage is 85%). Although the area of the zones with high chlorophyll concentration is relatively small, their contribution to the total concentration is large, so the curve in Figure 10.13 does not represent the processes occurring in the greater part of the oceans.

10.3.3.2 Long-term dynamics of chlorophyll concentration in the oceans

To illustrate the results of investigations of the long-term dynamics of chlorophyll concentration, three types of information are presented.

Figure 10.13 shows variations in the average chlorophyll concentration in the oceans from 1998 to 2003. The latitude dependence of the data has been taken into account; the origin of this dependence is the different area of pixels at different

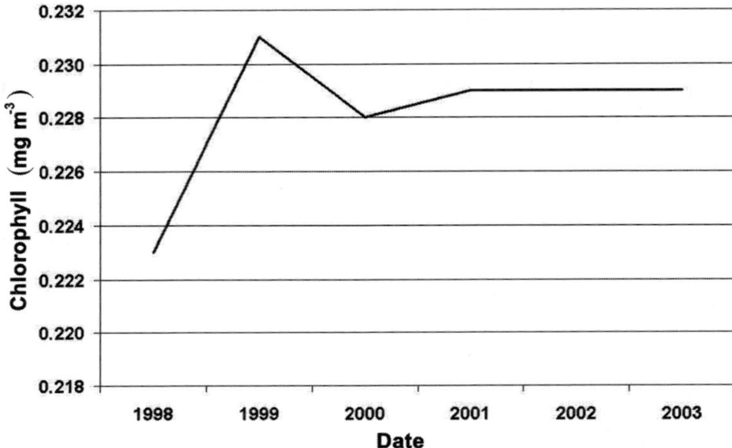

Figure 10.13. Dynamics of average chlorophyll concentrations.

latitudes. The graph indicates that in the period between 1998 and 2003 the average chlorophyll concentration in the oceans reached its minimum in 1998. The maximum chlorophyll concentration was registered in 1999.

Based on the graph of variations in the areas of significant minimal chlorophyll concentrations (Figure 10.14), it can be assumed that the process of their occurrence is periodic. So, winter 1997–1998 and 2003 were taken as the years of its most intense manifestations.

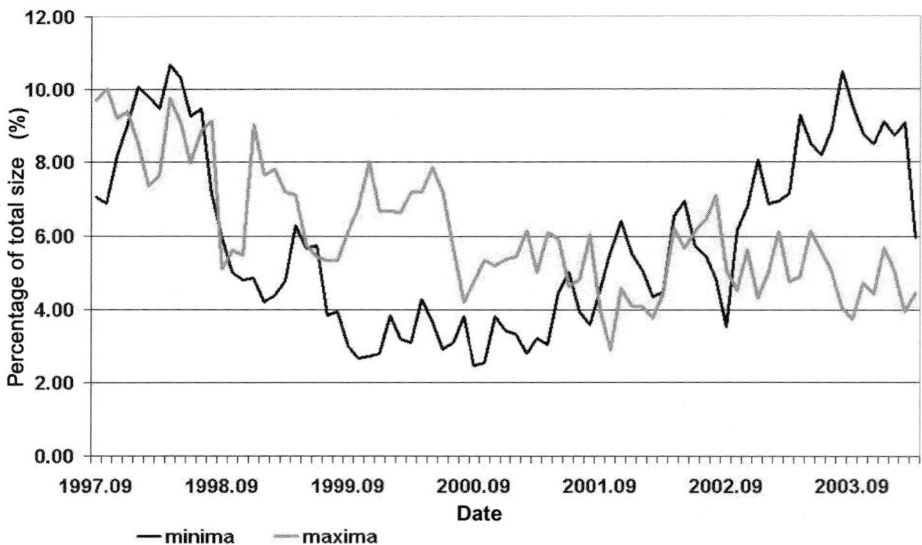

Figure 10.14. Areas occupied by minima and maxima.

Thus, we can conclude that in that period of time (6 years) long-term antiphase processes occurred in vast areas of the oceans, leading to phytopigment decrease in some regions and an increase in others. The question arises as to whether these changes are periodic or not. It is not unlikely that the process we registered on a global scale in this period of time, between winter 1997–1998 and 2003, could manifest itself like that quite by accident. That is, if the data for another time period had been analyzed, the result might have been processes with minima and maxima in the middle of the time period analyzed rather than unidirectional short-term trends. This pattern can be typical of the oscillatory process. The question is how long the period is. The answer can only be obtained by analyzing long time series, of at least 15–20 years.

Over the 7 years of the SeaWiFS operation, the proportions of the ocean areas occupied by minimal and maximal chlorophyll concentrations have been changing. Figure 10.14 clearly demonstrates this change. It shows changes in the areas occupied by significant minimal and maximal chlorophyll concentrations between 1997 and 2004. At the start of the observation time, from September 1997 to June 1998, the areas of both minimal and maximal concentrations grow. Thus, the distribution of areas with different chlorophyll concentrations in the oceans becomes more variable. At the end of the observation time, from February 2003 to March 2004, the graphs of maximal and minimal chlorophyll concentrations indicate that the areas occupied by minimal chlorophyll concentrations grow and the areas occupied by maximal chlorophyll concentrations become smaller. It could be supposed that total chlorophyll concentration in the ocean should decrease, but this is not so (see Figure 10.13). An opposite trend is observed at the beginning of 2002 when an increase in the total chlorophyll concentration could be expected, but again direct calculations do not confirm this (Figure 10.13). Thus, we can assume that (1) variations in the proportions of areas occupied by minimal and maximal chlorophyll concentrations are determined by different biogeochemical conditions and (2) the long-term dynamics of total phytopigment concentration is smoothed by averaging the values over the area of the oceans, in spite of acute local processes, which are often oppositely directed.

10.3.4 Seasonal variations in oceanic phytopigment values in the northern and southern hemispheres averaged by three climatic zones (northern hemisphere starting from 30°N, southern hemisphere starting from 30°S, and the tropical zone)

To compare the results of satellite measurements of chlorophyll concentration, we undertook sampling of satellite data and processed the data for 1997–2006. Chlorophyll concentration dynamics was determined on the basis of averaged values for latitude belts of 60°. Figure 10.15 shows changes in chlorophyll concentration from 1997 to 2006. It is clear how the chlorophyll concentration in the ocean in polar and subpolar belts differs in amplitude from that in tropical and subtropical belts. At the same time the role of the northern belt (30°N–90°N) is almost twice as high. Strong seasonal dynamics can be observed in polar regions, while in tropical regions

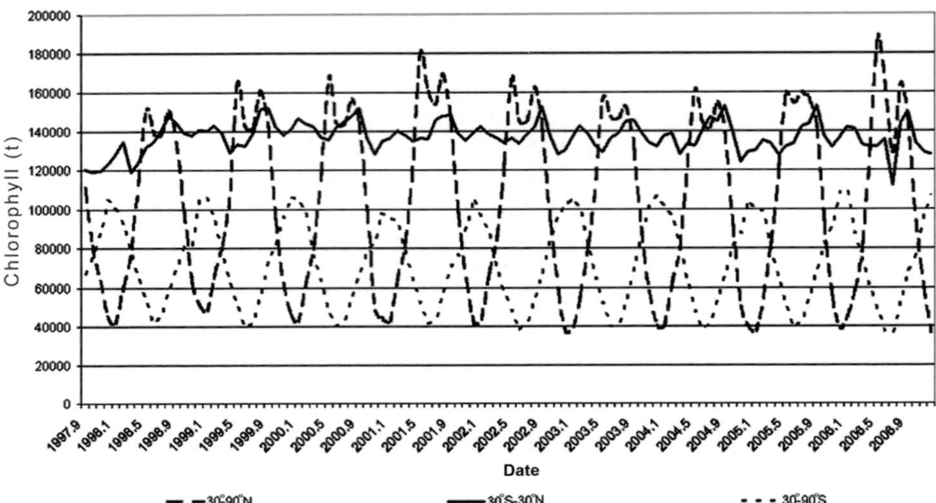

Figure 10.15. Dynamics of chlorophyll amount in the ocean surface layer (1 m).

it is feebly marked. Perennial trends spatially averaged by latitude belts are not apparent. Approximately 5% of oscillatory changes occur during 3–4 years. Nevertheless, the data obtained can be effectively used for comparison with models to discover the adequacy of ocean phytopigment global seasonal dynamics modeling.

Analyzing the perennial global dynamics of chlorophyll concentration in the oceans, one can observe a strong oscillation in the polar and subpolar regions of the southern hemispheres, oscillations in the northern polar region being a little smaller. Such dynamics do not have unidirectional trends during the time period being studied, except for the zone at 45°N–50°N, where during the years 2002–2003 a considerable increase in chlorophyll concentration was observed (Figure 10.16).

10.3.5 Minimal model of carbon dioxide seasonal dynamics

The global model of carbon dioxide seasonal dynamics is based on the model of multiannual dynamics, within the frames of the same concept and with the use of the same generalizations and suppositions. However, there are crucial supplements: all soil living biomass and organic substances were divided into three compartments: northern, tropical, and southern.

The model is aimed at searching for a minimal mechanism describing the seasonal dynamics of CO_2 and additional verification of the main locations of the model on the basis of additional data on system characteristic time, which shows in periodical dynamics. Moreover, in terms of model minimization, estimation of the marine biota contribution to seasonal oscillations of carbon dioxide global concentration is certainly of interest.

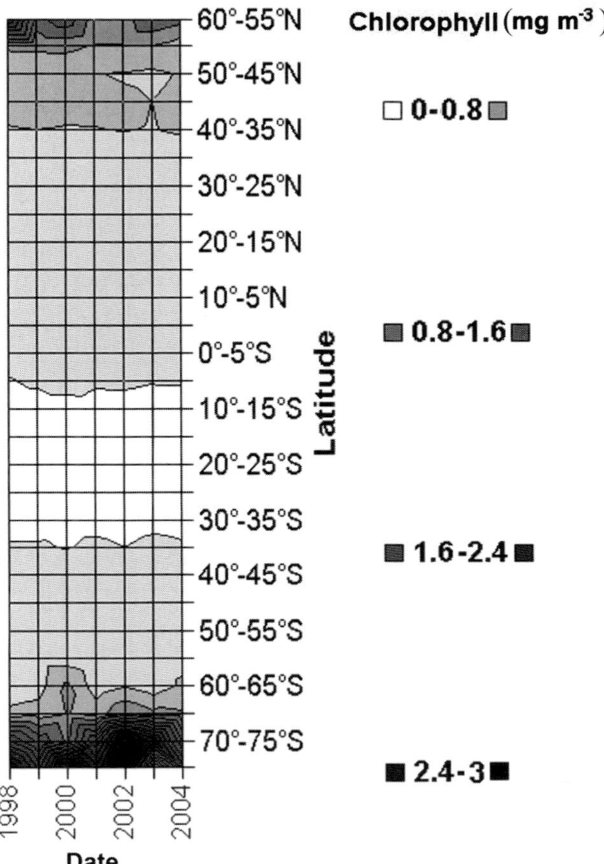

Figure 10.16. Spacetime diagram of chlorophyll concentration distribution in the ocean.

The model of multiannual dynamics was the simplest variant of carbon cycle representation, a kind of loop where carbon was consumed by plants, deposited in humus and returned to the atmosphere. The seasonal model consists of a whole set of such loops, where each of them describes the carbon cycle in one of the latitude compartments. A flowchart illustrating this approach is presented in Figure 10.17. Along with the terrestrial part, it includes the marine part of the biosphere.

Biotic reservoirs are divided into latitude compartments: tropical (from 30°S to 30°N), southern (to the south of 30°S), and northern (to the north of 30°N). They consist of two parts: plant biomass and dead organic substance.

The latitudinal division described is based on the seasonal variability of the activity of plants located to the north and south of the tropics. This phenomenon is especially intense in the northern vegetative compartment (Figure 10.18; Field *et al.*, 1998). The border of maximal seasonal differences in NPP of the northern hemisphere is located at 30°N, which was the reason for choosing this allocation of latitude belts.

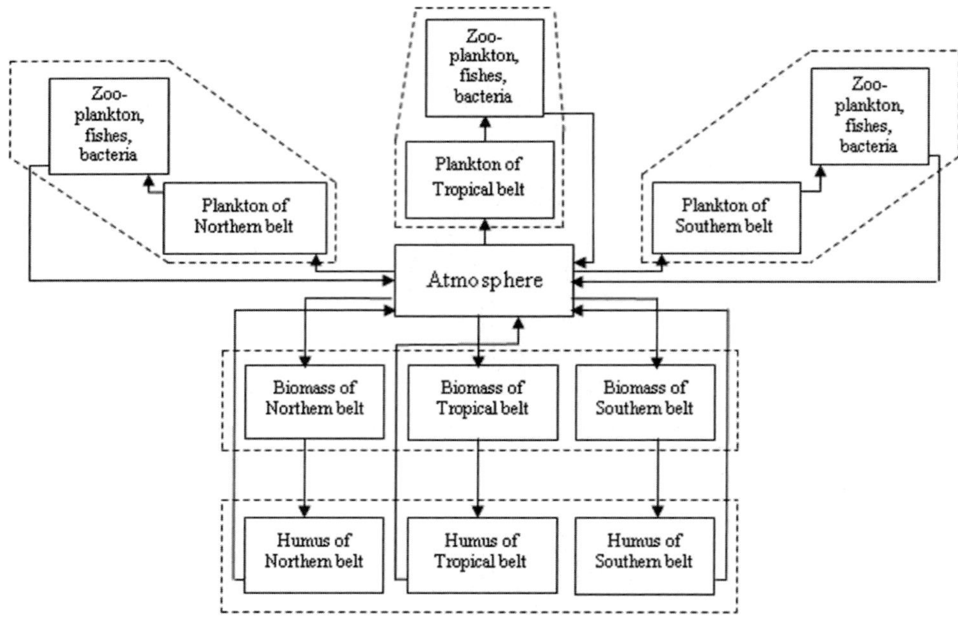

Figure 10.17. Flowchart of global seasonal model compartments.

The system of four equations, see equation (10.1), describing the dynamics of the terrestrial biota part becomes larger and assumes the following form:

$$\left.\begin{aligned} \frac{dC}{dt} &= \text{fuel}(t) \\ \frac{dx_i}{dt} &= P(x_i, A, T(A), t) - D(x_i, t) \\ \frac{dy_i}{dt} &= D(x_i, t) - S(y_i, T(A), t) \\ A &= C - \sum_i x_i - \sum_i y_i \end{aligned}\right\} \tag{10.19}$$

where i corresponds to n northern, t tropical, and s southern compartments. Several of the variables $(P(x, A, T(A), t), D(x, t), S(y, T(A), t))$ that we have used before are now also functions of time, t (in months), as well as the arguments we included before.

Seasonal variations in growth rates for each of the latitude compartments are specified with the help of phenomenological functions of time

$$V\_L_i(t) = B_i + A_i \sin^4\left(\pi\frac{t - t_w}{12} - \frac{\pi b_i}{2}\right), \tag{10.20}$$

where B_i is a constant specifying minimal rate level; A_i is the amplitude of oscillations

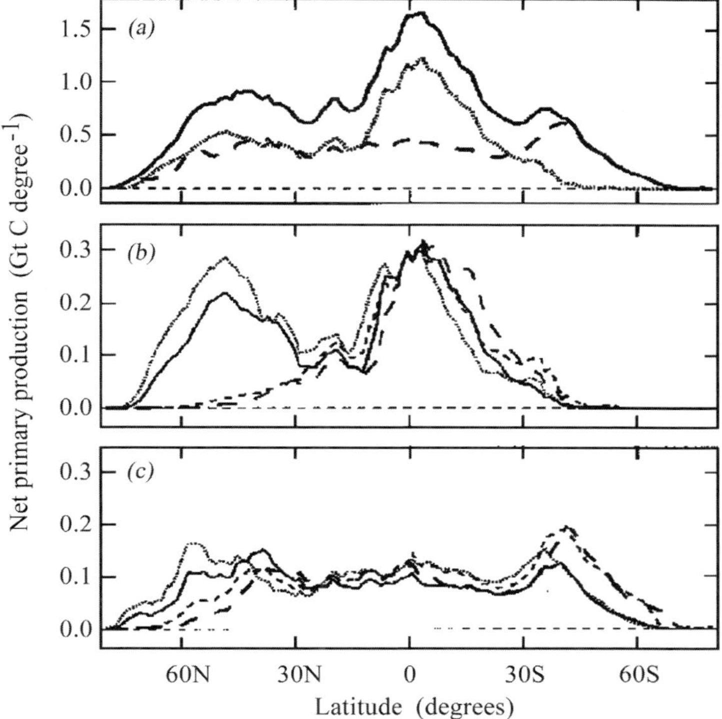

Figure 10.18. Latitudional distribution of global NPP. (a) NPP global value (solid line), NPP terrestrial value (dotted line), NPP value in the ocean (dashed line). (b) Dynamics of terrestrial NPP: from April to June (solid line), from July to September (dotted line), from October to December (short dashed line), from January to March (long dashed line). (c) NPP seasonal dynamics in the ocean, the same designation as in (b).

of seasonal factors; and t_w and b_i are parameters providing phase lag of the function for adjustment to the observation data (Figure 10.19). The sine function was raised to the fourth power to provide empirical correspondence between the functions of new production in the model and NPP values calculated according to the satellite monitoring data for each latitude belt (Figure 10.19).

The seasonal variation in biomass decomposition rates for each latitude belt was described with the help of the following empirical function

$$V\_R_i(t) = B_i + A_i \sin\left(2\pi\frac{t - t_w}{12} - \frac{\pi b_i}{2}\right). \tag{10.21}$$

In this case a simple sine function, rather than the fourth power, was used because, taking into account the general considerations and modeling results (Raich and Schlesinger, 1992), this function seems to be adequate for description of soil respiration averaged by latitude belts.

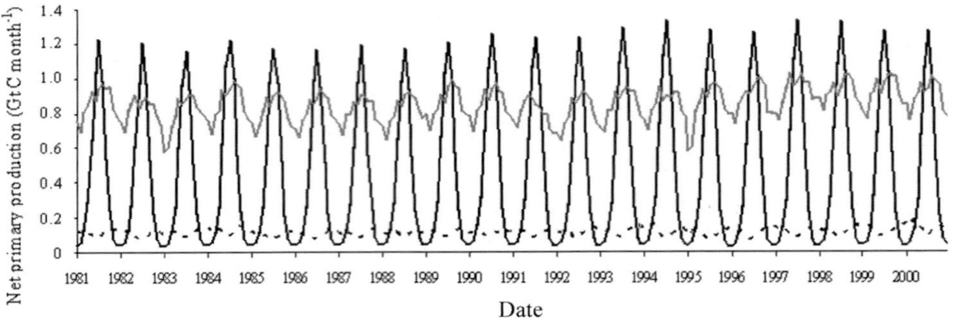

Figure 10.19. Seasonal dynamics of total NPP for selected latitude belts. —— 90°N–30°N; – · –· 30°N–30°S; · · · · · 30°S–90°S.

Functions $P(x, A, T(A), t)$ and $S(y, T(A), t)$ assumed the following form after adding a new factor:

$$P_i(x_i, A, T, t) = x_i(x^i_{\max} - x_i)V_p V(A) f_p(T(A))V\_L_i(t) \tag{10.22}$$

$$S_i(y_i, T, t) = y_i V_s f_M(T(A))V\_R_i(t). \tag{10.23}$$

The initial values of carbon concentration in the plant biomass and dead organic substance of the soils were calculated by division of the global values in proportion to the areas occupied by the given compartments. Thus, total carbon in the living biomass was: northern compartment 280 GtC, tropical 488 GtC, and southern 45 GtC. The initial values of carbon concentration in dead organic substances were: northern compartment 594 GtC, tropical 289 GtC, and southern 171 GtC (Gifford, 1993; WRI, 1998). It has to be noted that the model of seasonal dynamics is not strongly based on the principle of the worst scenario. The main goal of the model is to describe seasonal dynamics via the description of minimal complexity. Therefore, we have used available data on the distribution of carbon among compartments, and the total amount of carbon does not correspond to the principle of the worst scenario.

To study the contribution of marine biota to seasonal oscillations, the model of the terrestrial carbon cycle was expanded with equations describing the oceanic part of the carbon cycle. These equations describe the dynamics of growth of autotrophs and heterotrophs, as well as the decomposition of dead organic substance in the ocean. As a result, the model equation system was extended by six more equations:

$$\left.\begin{aligned} \frac{dm_i}{dt} &= N(m_i, A, B, t) - V(m_i, z_i) \\ \frac{dz_i}{dt} &= V(m_i, z_i) - E(z_i), \end{aligned}\right\} \tag{10.24}$$

where i corresponds to n northern, t tropical, and s southern compartments, and the first three equations describe the dynamics of phytoplankton (m) and the last three the dynamics of total heterotroph biomass (z).

The increase in phytoplankton quantity is described by function (10.16) multiplied by a seasonal factor:

$$N(m, A, B, t) = V_M m(m_{\text{lim}} - z - m) K_m W(B) f_m(T(A)) V\_Gm(t), \qquad (10.25)$$

where the function of seasonal variations of phytoplankton biomass growth $V\_Gm(t)$ has the same form as the analogous function (10.21) which was chosen out of considerations of the larger heat retention of the oceans, which provides smoothness of seasonal variations.

The rate of plankton biomass consumption by marine heterotrophs ($V(m, z)$) and the averaged death rate of heterotrophs ($E(z)$) are the same as formulae (10.16) and (10.17).

The following data were used in the model of the carbon cycle oceanic part. Based on the analysis of literature data (Brovkin *et al.*, 2002; Falkowski *et al.*, 1998; Kondratyev and Krapivin, 2004; Semyonov, 2004), it was assumed that marine biota assimilates about 45 GtC per year, its own biomass being only 3 GtC. This is explained by the high rates of growth, consumption and oxidation of phytoplankton remains (days, months), and by the inverse correlation of the biomass of autotrophs and heterotrophs in regard to the similar parameter over land. As a result, carbon does not accumulate at the stage of CO_2 consumption by phytoplankton, and practically the same amount of carbon (90%) returns to the atmosphere. The remaining 10% descends to ocean bottom layers, mainly in the form of organism remains and fecal pellets. At the same time organic carbon is oxidized by dissolved oxygen or as a result of bacterial activity, turning it into carbon dioxide and entering into reactions of hydrogen sulfide formation. As a result, only 1%–3% of organic substance reaches the bottom. Such carbon loss was ignored in the present variant of the model as insignificant in terms of the timescale under study. The remaining part of the dead organic substance returns to the annual carbon cycle.

Adjustment of the parameters of the seasonal dynamics model was done in the following way. First, the scale multipliers of the functions describing the processes in latitude belts were chosen so that the model variables would remain in a steady state, carbon anthropogenic emission and the functions of seasonal variations being switched off. Then the functions of seasonal variations and carbon fuel combustion were turned on, and the model parameters were chosen so that CO_2 dynamics would correspond, in the best way possible, to Mauna Loa measurements (Figure 10.20) and NPP dynamics calculated on the basis of satellite data (Figure 10.21). First of all, we adjusted the parameters of soil respiration as they were not supported by external data. The most important indicator of soil respiration that provided agreement with the calculated carbon dioxide dynamics to Mauna Loa data (while there was an evident qualitative difference between the form of registered carbon dioxide concentration dynamics and the form of the NPP curve) was a phase lag against the NPP curve. In other words, the seasonal maximum of soil respiration is behind the NPP maximum. As is seen in the graph, the agreement of the model data with experimental measurements is rather high. This proves the hypothesis about the potential possibility of using remote-sensing data for additional verification of minimal models. On the other hand, a minimal model allows the satellite monitoring and

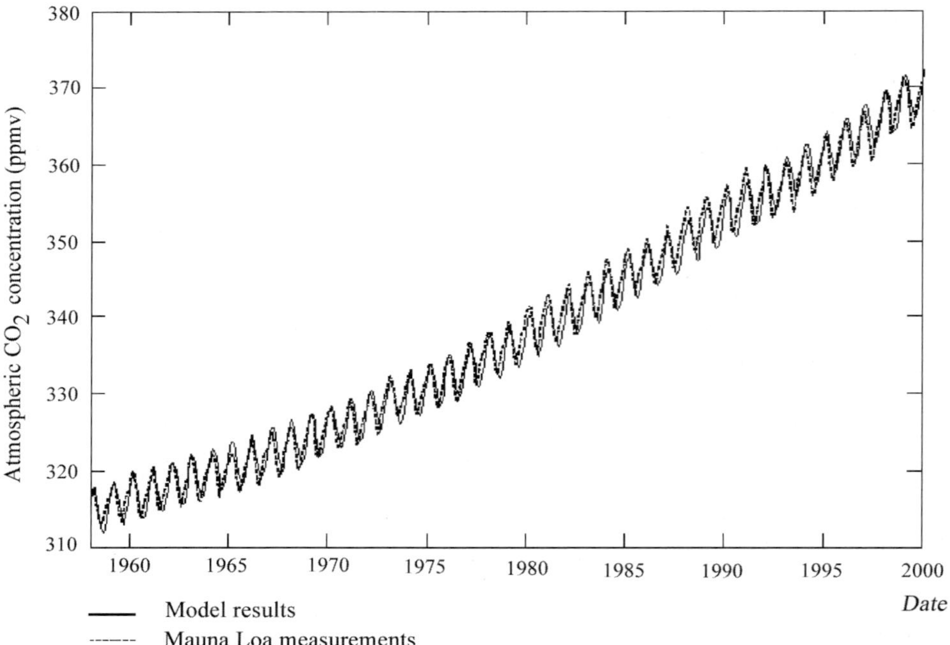

Figure 10.20. Comparison of model results with Mauna Loa data.

ground data to be linked, due to the close connection of global CO_2 dynamics, ground data on biomass distribution, and photosynthesis activity, including plant physiology data and satellite monitoring data.

To estimate the effect of marine biota on the seasonal dynamics of CO_2 global concentration we compared two models of seasonal dynamics, one of which describes

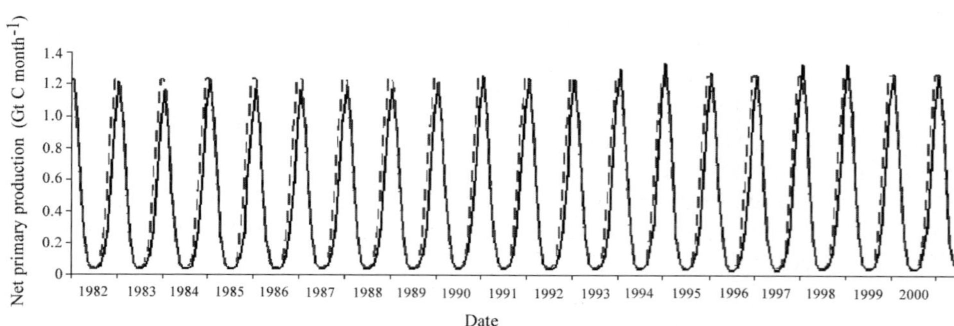

Figure 10.21. Comparison of the model results of calculated plant NPP with data received on the basis of satellite measurements in the northern geographic compartment (de-trended) from 30°N to 90°N. Solid line is the model simulation, dashed line NPP calculated by GLO-PEM.

both terrestrial and oceanic biota, and the other includes only the terrestrial part of the carbon cycle. The processes of adjustment of the model parameters did not differ from one another. The results of verification of the two models of seasonal dynamics indicate that the contribution of marine biota to the seasonal variations of CO_2 concentration, as compared to the contribution of terrestrial biota, is insignificant. So, marine biota can be excluded from consideration while modeling the seasonal dynamics of CO_2 concentration in the atmosphere.

10.4 UNICELLULAR ORGANISM BASED EXPERIMENTAL CLOSED MICROECOSYSTEMS AS MODELS OF BIOSYSTEMS SIMILAR TO THE BIOSPHERE

Artificial ecosystems with a high degree of closure of the material biotic cycle can be an efficient instrument for experimental modeling of biospheric processes, in particular for the investigation of their resistance to anthropogenic factors.

It is well known that systems based on the material biological cycle can be used as life support systems functioning over long periods of time. A high degree of closure of artificial ecosystems can be reached only if a thorough scientifically based choice of separate biological components is made to unite them into systems, and the necessary balance rates of substance transformation and congruence of chemical elements cycle are provided. Neglecting the closure mechanisms responsible for maintenance of a long-term and stable material cycle can lead to negative consequences for the whole system.

Closed ecosystems can be divided into two types. The first type (Type 1) includes ecosystems in which the intensity of material cycling is determined by the mutual balance of production–destruction processes, while the species composition and biomass values of all closed ecosystems species are changing. These are self-maintaining, self-regulating systems. Examples of such ecosystems are lake systems and the biosphere itself (at least, it has belonged to this type until recently). The experimental analog of such an ecosystem will be described in this section.

The second type (Type 2) unites all biosystems in which a cycle is dominated by one biological link (population), and the remaining components of the cycle adjust the characteristics of their kinetic processes in accordance with the requirements of this dominant component. Usually, the dominant component means mankind (the vehicle) in a specially created closed life support system, the activity of this component and its requirements to the material cycle (closure, intensity) remaining relatively unchanged over a long period of time. It is supposed that there is no dynamics of vehicle quantity.

One can suppose that in the future, with the growth of human population, the material cycle in the biosphere will be transformed from Type 1 to Type 2, if we take into account all kinds of human needs—not only physiological and metabolic ones. Both types of closed ecosystem are of profound importance in terms of their use for experimental modeling of the dynamic properties of biosystems similar to the biosphere and belonging to two extreme classes: one with variable intensity of the material cycle and the other with a preset permanent intensity value. The laws of stability and control of such systems are of exceptional interest for providing the explanation and understanding of the structure and dynamics of the biosphere and its development trends. Task-oriented experiments aimed at destabilization of such experimental systems (Type 1 and Type 2) will be useful when analyzing and interpreting the actual anthropogenic dynamics of the biosphere or similar biosystems.

In this part of our work we describe the evolution of experimental microecosystems based on unicellular organisms (Type 1). The functioning of natural ecosystems and the biosphere as a whole involves the continuous transport of substances cycling over trophic components by energy received from outside. The intensity of the material cycling, distribution of the matter over the trophic levels, stability of biotic cycling, and species composition of the ecosystem seem to be determined by the quantity of substances limiting biosynthesis and the intensity of energy supply. The qualitative analysis of dependencies between the said factors leading to this or that type of process limitation in an ecosystem is a problem of prime importance for theoretical biology and especially for biospherics (i.e., the science investigating the mechanisms of biosphere dynamics and evolution) and particularly needing new theoretical and experimental approaches. Methodologically, it is convenient to start solving this problem with consideration of the minimum possible communities capable of functioning autonomously for a long time with their characteristics under study amenable to control and monitoring. Artificial closed microcosms (MESs) were the first to be identified with objects of this kind.

10.4.1 A microecosystem (MES) mathematical model

10.4.1.1 Description of the model

Consider a homogeneous closed system consisting of one producer species (X, biomass of autotrophs) and one reducer species (R, biomass of heterotrophs). The mnemonic block scheme of the model is presented in Figure 10.22. The major processes in the system are the increment of organism biomass (photosynthesis, reproduction), respiration, death of the organisms, and mineralization of non-living organic matter.

In the process of vital activity the producers consume light energy (E), CO_2 (W), H_2O (V), and biogenic elements (S_j); build up their biomass; and emit O_2 (Q). We define the specific producer biomass growth rate (μ_X) according to the L-system concept (Abrosov *et al.*, 1982; Poletayev, 1966) by the expression:

$$\mu_X = \min\left\{\mu'_X, \beta_{Xi}S_j, \beta_{XW}W, \frac{GE(1-e^{-uX})}{X}, \beta_{XV}V\right\}, \tag{10.26}$$

where μ'_X is the maximum possible specific growth rate of producer ($\mu'_X = \text{const.}$); β_{Xi} is the coefficient of organism adaptation to Type i substance (i are S_j, W, V) equal to the product of the specific consumption rate for this substance C_{Xi} by the economic coefficient of its assimilation (economic yield coefficient) Y_{Xi} (i.e., $\beta_{Xi} = C_{Xi}Y_{Xi}$ ($C_{Xi}, Y_{Xi} = \text{const.}$), $j = 1, 2, \ldots, m$, m is the number of biogenic elements taken into

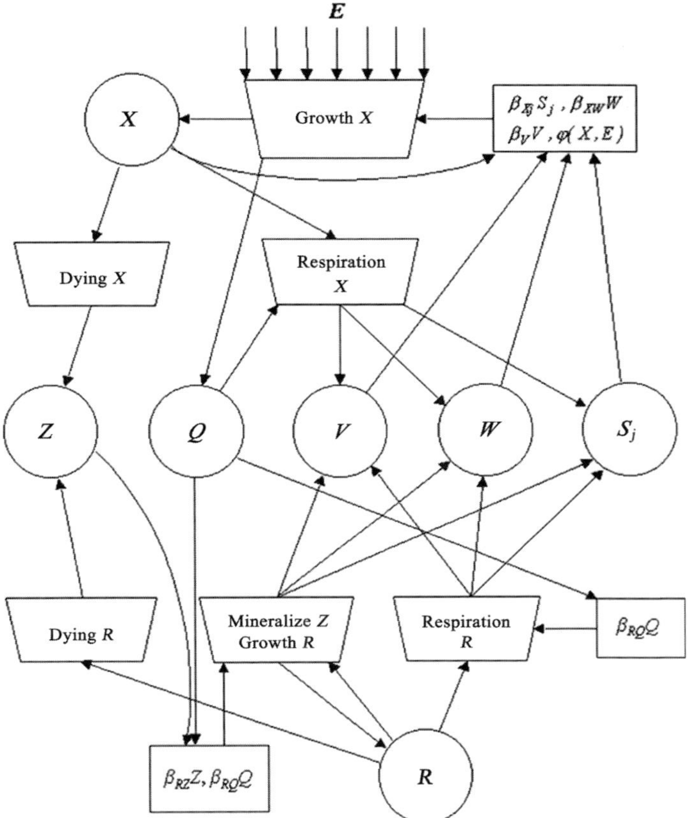

Figure 10.22. Mnemonical block scheme of the homogeneous closed microecosystem model. The following material components are shown with circles: X producer biomass, R reducer biomass, Z concentration of dead organic substance, S_j background concentration of j biogenous element, Q background concentration of oxygen (O_2), W background concentration of carbon dioxide (CO_2), V water (H_2O); the directions of their flows and light energy (E) are shown with arrows; processes with triangles; formation of state functions that are potential limiting factors with rectangles; $\phi(X, E) = G(1 - e^{-uX})/X$.

account); G specifies the producer's specific rate per unit energy consumed ($G = \text{const.}$); u is the coefficient of light absorption by unit autotroph biomass equal to the product of relative chlorophyll content in the cells and the coefficient of light absorption by chlorophyll ($u = \text{const.}$). The specific rate of producer respiration is assumed to be constant ($\gamma_X = \text{const.}$).

When mineralizing the dead biomass of organisms (Z) the reducers consume O_2 and the biogenic components CO_2 and H_2O are formed. Dissolved O_2 is also used for the respiration of microorganisms. The formation of CO_2 and H_2O and emission of biogenic elements are also taken into account. We define the specific growth rate of reducers as follows (μ_R):

$$\mu_R = \min\{\mu'_R, \beta_{RZ}Z, \beta_{RQ}Q\}, \tag{10.27}$$

where μ'_R is the maximum possible growth rate of reducers ($\mu'_R = \text{const.}$); β_{RZ}, β_{RQ} are the coefficients of reducer adaptation to non-living organic matter and oxygen ($\beta_{RZ}, \beta_{RQ} = \text{const.}$) and $\beta_{RZ} = C_{RZ}Y_{RZ}$, respectively.

We assume also that respiration of the reducers is proportional to the concentration of dissolved oxygen at its low values or maximum possible at high O_2 concentrations. The specific rate of heterotroph respiration (γ_R) is defined by the expression:

$$\gamma_R = \min\{\gamma'_R, \beta_Q Q\}, \tag{10.28}$$

where γ'_R is the maximum possible specific rate of reducer respiration ($\gamma'_R = \text{const.}$); and β_Q is the specific rate of reducer respiration at unit O_2 concentration (adaptation coefficient of reducers to oxygen in respiration process) ($\beta_Q = \text{const.}$, taking, for simplicity, $\beta_Q = \beta_{RQ}$).

The MES component concentration dynamics is described by the system of equations:

$$\left.\begin{aligned}
X'' &= (\mu_X - r_X - \gamma_X)X, \\
R'' &= (\mu_R - r_R - \gamma_R)R, \\
Z'' &= r_X X + r_R R - \frac{\mu_R}{Y_{RZ}}R, \\
S''_j &= -\frac{\mu_X X}{Y_{Xj}} + \alpha_j\left[\gamma_X X + \gamma_R R + \mu_R\left(\frac{1}{Y_{RZ}} - 1\right)R\right], \\
W'' &= \frac{\mu_X X}{Y_{XW}} + D_X\alpha_X\gamma_X X + D_R\alpha_R\gamma_R R + D_{RZ}\alpha_{RZ}\mu_R\left(\frac{1}{Y_{RZ}} - 1\right)R, \\
Q'' &= \frac{\mu_X X}{Y_{XW}B} - \alpha_X\gamma_X X - \alpha_R\gamma_R R - \alpha_{RZ}\mu_R\left(\frac{1}{Y_{RZ}} - 1\right)R, \\
V'' &= \delta_X\gamma_X X + \delta_R\gamma_R R + \delta_{RZ}\mu_R\left(\frac{1}{Y_{RZ}} - 1\right)R - \varepsilon_X\mu_X X,
\end{aligned}\right\} \tag{10.29}$$

where r_X and r_R are the specific death rates of organisms ($r_X = \text{const.}$, $r_R = \text{const.}$); α_j is the relative content of the jth biogenic element in organisms ($\alpha_j = \text{const.}$); B is the assimilation (photosynthetic) coefficient of the producers indicating the amount of CO_2 emitted by the autotrophs in respiration per unit O_2 emitted in (g of CO_2)/

(g of O_2) ($B = \text{const.}$); D_X is the producer respiration coefficient indicating the amount of CO_2 emitted by the autotrophs in respiration per unit O_2 consumed in (g of CO_2)/(g of O_2) ($D_X = \text{const.}$); D_R is the reducer respiration coefficient in burning their own biomass (no food, $Z = 0$) in (g of CO_2)/(g of O_2) ($D_R = \text{const.}$); D_{RZ} is the reducers respiration coefficient during their growth when consuming the non-living biomass of organisms ($Z \neq 0$) in (g of CO_2)/(g of O_2) ($D_{RZ} = \text{const.}$) (as the respiration coefficient depends on the composition of food in the general case $D_R \neq D_{RZ}$); α_X and α_R are the coefficients indicating the amount of O_2 consumed per unit biomass spent in the respiration process by producers and reducers, respectively ($\alpha_X = \text{const.}$, $\alpha_R = \text{const.}$); α_{RZ} is the amount of O_2 required to mineralize a unit biomass of the non-living matter of microorganisms ($\alpha_{RZ} = \text{const.}$); δ_X and δ_R are the amounts of H_2O that is formed per unit spent biomass in respiration of autotrophs and heterotrophs, respectively ($\delta_X = \text{const.}$, $\delta_R = \text{const.}$); δ_{RZ} is the amount of H_2O formed by reducers mineralizing per unit of non-living organic matter biomass ($\delta_{RZ} = \text{const.}$); and ε_X is the amount of water used by producers to photosynthesize a unit biomass ($\varepsilon_X = \text{const.}$).

10.4.1.2 Material cycling conditions

As the system is assumed to be closed, the total concentration M of each component in the ecosystem present in it both in free (background concentrations of biogenic components S_j), and in bound forms (biomass of organisms, CO_2, O_2, H_2O), remains constant; that is,

$$\left.\begin{aligned}
M_j &= \alpha_j(X + R + Z) + S_j = \text{const.} \\
M_C &= \alpha_C(X + R + Z) + \alpha_{CW} W = \text{const.} \\
M_H &= \alpha_H(X + R + Z) + \alpha_{HV} V = \text{const.} \\
M_O &= \alpha_O(X + R + Z) + \alpha_{OW} W + \alpha_{OV} V + 1 \cdot Q = \text{const.}
\end{aligned}\right\} \tag{10.30}$$

where $\alpha_C, \alpha_H, \alpha_O$ are the relative amounts of carbon, hydrogen, and oxygen, respectively, in the biomass of organisms ($\alpha_C, \alpha_H, \alpha_O = \text{const.}$). As the content of a certain element in every organism species is, on the average, similar for different species we assume, for simplicity, $\alpha_j, \alpha_C, \alpha_H, \alpha_O$ to be similar for all species; and α_{CW}, α_{OW} to be the relative contents of C and O, respectively, in carbon dioxide ($\alpha_{CW}, \alpha_{OW} = \text{const.}$) with $\alpha_{CW} + \alpha_{OW} = \alpha_{OV} + \alpha_{HV} = 1$.

It is assumed that $\sum \alpha + \alpha_O + \alpha_C + \alpha_H = 1$ (i.e., each organism species involves all the components considered without any others).

A number of the coefficients can be expressed as ratios following from the law of conservation of matter and the requirement of cycling closure:

$$\left.\begin{aligned}
&\text{(a)}\ \alpha_j = \frac{1}{Y_{Xj}},\\
&\text{(b)}\ \alpha_C = \frac{\alpha_{CW}}{Y_{XW}},\\
&\text{(c)}\ \delta_X = \varepsilon_X = \frac{\alpha_H}{\alpha_{HV}},\\
&\text{(d)}\ \delta_R = \delta_{RZ} = \frac{\alpha_H}{\alpha_{HV}},\\
&\text{(e)}\ \alpha_O = \frac{\alpha_{OW}}{Y_{XW}} + \alpha_{OV}\frac{\alpha_H}{\alpha_{HV}} - \frac{1}{Y_{XW}B},\\
&\text{(f)}\ D_X a_X = B a_X = \frac{1}{Y_{XW}},\\
&\text{(g)}\ D_R a_R = D_{RZ} a_{RZ} = \frac{1}{Y_{XW}},\\
&\text{(h)}\ D_X = D_R = D_{RZ}.
\end{aligned}\right\} \tag{10.31}$$

Conditions (10.31) are necessary to support prolonged cycling in system (10.29) (i.e., the conditions of complete closure in the biological sense). We should consider these correlations, or closure conditions, in more detail.

Condition (e) specifies the distribution of oxygen in the process of photosynthesis. The content of oxygen in CO_2 and H_2O supplied for biosynthesis is equal to the content of oxygen in the biomass of the organisms and molecular oxygen (O_2), formed in the process of photosynthesis. This seems to be valid for every autotroph species.

Condition (a) means that the relative content of a given element in the biomass of producers (α_j) is equal to the amount of this element required for the producers (with yield coefficient Y_{Xj}), to increase their biomass by a unit. For example, the specific nitrogen content in the biomass of the alga *Chlorella* $\alpha_N = 0.08$; consequently, the yield coefficient for nitrogen is $Y_{XN} = 1/\alpha_N = 10.5$ (i.e., for each gram of increment *Chlorella* must consume 0.08 g of nitrogen).

Similarly for condition (b): for producers to increase their biomass by a unit the amount of CO_2 to be consumed is $1/Y_{XW}$ (where Y_{XW} is the yield coefficient of producers for CO_2). But the consumed CO_2 contains α_{CW}/Y_{XW} of carbon (α_{CW} is the specific content of C in CO_2) (i.e., one gram of biomass of producers comprises $\alpha_{CW}/Y_{XW} = \alpha_C$ grams of carbon). So, assuming the organisms consist of carbon to the extent of 50% (i.e., $\alpha_C = 0.5$ and $\alpha_{CW} = 12/44 = 3/11$), the yield coefficient for CO_2 $Y_{XW} = 6/11 = 0.545$ (see Gitelson *et al.*, 1975, p. 104, where $Y_{XW} = 0.51$).

We can explain condition (f) $Ba_X = 1/Y_{XW}$, where the left-hand side is the amount of CO_2 formed in oxidation of a unit biomass of producers in respiration and the right-hand side is the amount of CO_2 necessary to increase the biomass of the producer by a unit (growth, photosynthesis). As growth and respiration is one (reciprocal) reaction

$$CO_2 + 2H_2O \overset{E}{\longleftrightarrow} (CH_2O) + H_2O + 2O,$$

these amounts of CO_2 should be equal, as reflected in (f); that is, condition (f) in Equation (10.31).

Equality of the respiratory coefficient (D_X) and the photosynthetic coefficient (B) (i.e., $B = D_X$), and the equality of coefficients δ_X and ε_X (c) is substantiated in analogy. As ε_X is the amount of H_2O used in the process of photosynthesis, the amount of hydrogen consumed to synthesize a unit biomass of producers is equal to $\varepsilon_X \alpha_{HV}$ (α_{HV} is the relative content of hydrogen in H_2O) (i.e., a biomass unit of the producer contains $\varepsilon_X \alpha_{HV} = \alpha_H$ of hydrogen) (c).

As the respiratory coefficient depends on the composition of food and the latter can be specified by the relative content of elements α_j in it (e.g., the content of C, O, H in carbohydrates differs from the content of these elements in proteins and fats), and in this case (by construction) α_j is identical for all organism species in the dead unit Z, it is apparent that the respiratory coefficient of reducers in the absence of food for them D_R ($Z = 0$) and the respiratory coefficient in the presence of food D_{RZ} ($Z \neq 0$) must be equal (i.e., $D_R = D_{RZ}$) (h). Therefore, $\alpha_R = \alpha_{RZ}$ (d) and $\alpha_R = \alpha_{RZ}$ (d). As α_R is the amount of water formed by respiration of reducers (absence of food, $Z = 0$), the amount of hydrogen liberated is $\alpha_X \alpha_{HV} = \alpha_H$ (d) and the amount of carbon dioxide formed is $D_R \alpha_R = 1/Y_{XW}$ (g).

From the assumption about the identical and invariable element composition of different species it follows that in the course of respiration different species of organisms consume identical amounts of oxygen; that is, (f) and (g) hold. This is also true for the equality of respiratory coefficients of producers (D_X) and reducers (D_R) (i.e., $D_X = D_R$) (h). The latter equality is one of the most necessary conditions to sustain prolonged material cycling in a biotic system from the biological standpoint (Gitelson *et al.*, 1975) and seems to possess the property of generality.

Taking into account conditions (10.31), model (10.29) will have a simpler form:

$$\left.\begin{aligned}
X'' &= (\mu_X - r_X - \gamma_X)X, \\
R'' &= (\mu_R - r_R - \gamma_R)R, \\
Z'' &= r_X X + r_R R - \frac{\mu_R}{Y_{RZ}} R, \\
S_j'' &= \alpha_j A, \\
W'' &= \frac{A}{Y_{XW}}, \\
Q'' &= -\frac{A}{Y_{XW} B}, \\
V'' &= \frac{\alpha_H}{\alpha_{HV}} A,
\end{aligned}\right\} \tag{10.32}$$

where

$$A = -\mu_X X + \gamma_X X + \gamma_R R + \mu_R\left(\frac{1}{Y_{RZ}} - 1\right)R.$$

10.4.1.3 Analysis of the model; elements of closed ecosystem steady-state control theory

For the system of differential equations (10.32) the phase space is divided by expressions (10.26)–(10.28) into $(3 + m) \times 3 \times 2$ regions of MES operations different in their limitation to biological processes, where m is the number of biogenic elements taken into account. Water, being the habitat of the organisms, is assumed not to limit the functioning of the latter. But stable (i.e., asymptotically stable, as can be demonstrated by Lyapunov's method in the first approximation; Merkin, 1996) steady-state conditions are possible in $(3 + m)$ regions only where

$$\left.\begin{array}{lll}
\text{(a)}\ \mu_X = \beta_{Xj}\overline{S_j}, & \mu_R = \beta_{RZ}\overline{Z}, & \gamma_R = \gamma'_R, \\[2ex]
\text{(b)}\ \mu_X = \beta_{XW}\overline{W}, & \mu_R = \beta_{RZ}\overline{Z}, & \gamma_R = \gamma'_R, \\[2ex]
\text{(c)}\ \mu_X = \dfrac{GE(1 - e^{-u\overline{X}})}{\overline{X}}, & \mu_R = \beta_{RZ}\overline{Z}, & \gamma_R = \gamma'_R, \\[2ex]
\text{(d)}\ \mu_X = \dfrac{GE(1 - e^{-u\overline{X}})}{\overline{X}}, & \mu_R = \beta_{RQ}\overline{Q}, & \gamma_R = \gamma'_R,
\end{array}\right\} \tag{10.33}$$

The bar above the symbols $\overline{S_j}, \overline{X}, \overline{Z}$, etc. indicates stationary values of the corresponding components of the system.

Steady states (in which the limitation of autotrophs by biogenic elements or CO_2, the growth of reducers, by oxygen and their respiration, is maximal) require a strict correlation between the coefficients, which is realized on the borders of steady-state regions and thus are objectively unlikely.

Steady states in which the functioning of the second trophic level is limited by oxygen ($\mu_R = \beta_{RQ}\bar{Q}, \gamma_R = \beta_{RQ}\bar{Q}$) are not realized because in this case a contradictory correlation $\beta_{RQ}\bar{Q} = r_R + \beta_{RQ}\bar{Q}$ would have to be true.

Thus, if the functioning of the first trophic level is limited by some biogenic element (say, the first element, i.e., $j = 1$), then the region of a stable steady state is realized (10.33(a)). From the point of view of biospheric modeling, this area is very important as the model of dynamics of biosphere components and their final (stationary) states where primary production is limited by some biogenic element (e.g., mineral phosphate). In this case the growth of reducers is limited by food (Z) and their maximum possible respiration (γ'_R). This region is realized under the following conditions:

$$
\left.
\begin{aligned}
&\text{(a)}\ M_1 > \frac{r_X + \gamma_X}{\beta_{X1}} + \alpha_1 \frac{r_R + \gamma'_R}{\beta_{RZ}} = M_{1X},\\
&\text{(b)}\ E > \frac{r_X + \gamma_X}{G\left(1 - e^{-u\frac{M_1 - M_{1X}}{\alpha_1 T}}\right)} \cdot \frac{M_1 - M_{1X}}{\alpha_1 T},\\
&\qquad T = 1 + \frac{r_X Y_{RZ}}{r_R(1 - Y_{RZ}) + \gamma'_R},\\
&\text{(c)}\ M_j > \frac{\alpha_j}{\alpha_1}\left(M_1 - \frac{r_X + \gamma_X}{\beta_{X1}}\right) + \frac{r_X + \gamma_X}{\beta_{Xj}}, \qquad j = \overline{2, m},\\
&\text{(d)}\ M_C > \frac{\alpha_C}{\alpha_1}\left[M_1 + \frac{r_X + \gamma_X}{\beta_{X1}}\left(\frac{C_{X1}}{C_{XW}} - 1\right)\right],\\
&\text{(e)}\ M_H \gg \frac{\alpha_H}{\alpha_1} M_1 + (r_X + \gamma_X)\left(\frac{\alpha_{HV}}{\beta_{XV}} - \frac{\alpha_H}{C_{X1}}\right),\\
&\text{(f)}\ M_O > \frac{\alpha_{OW}}{\alpha_{CW}} M_C + \frac{\alpha_{OV}}{\alpha_{HV}} M_H + \frac{r_R \gamma'_R}{\beta_{RQ}} - \frac{M_1 - \dfrac{r_X + \gamma_X}{\beta_{X1}}}{\alpha_1 Y_{XW} B} = M_{O1}.
\end{aligned}
\right\}
\qquad (10.34)
$$

Condition (10.34(a)) means that the concentration of the biogenic element limiting the growth of producers must be larger than a certain critical value (M_{1X}), below which a long-term material cycle in the system is impossible, and the MES dies.

It is possible to examine the other conditions (10.34(b)–10.34(f)) and discuss their significance in terms of the states of the ecosystem.

Below are some examples of system component values (10.32) in the stationary state for this region of functioning (i.e., such regions in which the first biogenic element is the factor limiting producer growth):

$$
\left.
\begin{aligned}
\overline{X} &= \frac{M_1 - M_{1X}}{\alpha_{1T}}, \\
\overline{R} &= \frac{M_1 - M_{1X}}{\alpha_1 T/(T-1)}, \\
\overline{Z} &= \frac{r_R + \gamma'_R}{\beta_{RZ}}, \\
\overline{S}_1 &= \frac{r_X + \gamma_X}{\beta_{X1}}, \\
\overline{S}_j &= M_j - \alpha_j(\overline{X} + \overline{R} + \overline{Z}) = \overline{S}_j(M_1, M_j), \qquad j = \overline{2,m}, \\
\overline{W} &= \frac{M_C}{\alpha_{CW}} - \frac{M_1}{\alpha_1 Y_{CW}} + \frac{r_X + \gamma_X}{\beta_{X1}} = \overline{W}(M_1, M_C), \\
\overline{Q} &= M_O - \frac{\alpha_{OW}}{\alpha_{CW}} M_C - \frac{\alpha_{OV}}{\alpha_{HV}} M_H + \left(\frac{M_1}{\alpha_1} - \frac{r_X + \gamma_X}{C_{X1}}\right) \frac{1}{BY_{XW}} \\
&= \overline{Q}(M_1, M_C, M_O, M_H), \\
\overline{V} &= \frac{M_H}{\alpha_{HV}} - \frac{M_1 \alpha_H}{\alpha_1 \alpha_{HV}} + \frac{r_X + \gamma_X}{C_{X1}\alpha_{HV}} \alpha_H = \overline{V}(M_1, M_H), \\
\overline{X} + \overline{R} = \overline{LB} &= \frac{M_1 - M_{1X}}{\alpha_1}, \\
\overline{X} + \overline{R} + \overline{Z} = \overline{TB} &= \frac{M_1 - M_{1X}}{\alpha_1} + \frac{r_R + \gamma'_R}{\beta_{RZ}}.
\end{aligned}
\right\} \tag{10.35}
$$

Further, the component values in other stationary states for the corresponding regions of functioning are not given, but all of them can easily be deduced analytically in an explicit form.

Transfer to another region is possible by changing the summary concentrations of the components in the system or the illumination (i.e., by changing the dependencies (10.34)). Adding CO_2, O_2, to the system, increasing summarized concentrations of biogenic (not limiting) elements within the limits of conditions (10.34) results only in the increase in concentrations of the corresponding components in the environment. Increase in the total concentration of carbon in the system leads to augmentation of the CO_2 background and reduction of the O_2 background, see (10.35). In the limits of the region concerning the stationary values of living components $(\overline{X}, \overline{R})$, living biomass $(\overline{LB} = \overline{X} + \overline{R})$ and total biomass $(\overline{TB} = \overline{Z} + \overline{LB})$ increase in proportion to the increase in the system of the total concentration of the biogenic element limiting the functioning of producers (M_1).

The pattern for other biogenic elements $(j > 1)$ is similar.

For the region of the stable steady state (10.33(b)) in which the growth of autotrophs is limited by CO_2 concentration in the environment (total concentration of carbon, M_C), the stationary values of living components $\overline{LB}$ and $\overline{TB}$ change in proportion to the value of the factor limiting the functioning of the first trophic level.

Addition of CO_2 to the system is analogous to the increase in total carbon concentration. This area is of especial interest in terms of experimental modeling of the consequences of carbon cycle change in the biosphere due to the increase in carbon amount by way of human-induced and natural inflow to the atmosphere, the issue of consequences having been described in the preceding Sections 10.2. and 10.3. An increase in summary concentrations of biogenic elements or O_2 results in the increase in the corresponding background values of the system. This region is realized under the following conditions:

$$\left.\begin{array}{ll}
\text{(a)}\ M_C > \alpha_{CW}\dfrac{r_X+\gamma_X}{\beta_{XW}} + \alpha_C\dfrac{r_R+\gamma'_R}{\beta_{RZ}} = M_{CX}, & \\[2ex]
\text{(b)}\ E > \dfrac{r_X+\gamma_X}{G\left(1-e^{-u\frac{M_C-M_{CX}}{\alpha_C T}}\right)}\cdot\dfrac{M_C-M_{CX}}{\alpha_C T}, & \\[2ex]
\text{(c)}\ M_j \gg \dfrac{\alpha_H}{\alpha_C}M_C + \dfrac{r_X+\gamma_X}{\beta_{Xj}}\left(1-\dfrac{C_{Xj}}{C_{XW}}\right), \quad j=\overline{1,m}, & \\[2ex]
\text{(d)}\ M_H \gg \dfrac{\alpha_H}{\alpha_C}M_C + (r_X+\gamma_X)\left(\dfrac{\alpha_{HV}}{\beta_{XV}}-\dfrac{\alpha_H}{C_{XW}}\right), & \\[2ex]
\text{(e)}\ M_O > \dfrac{M_C}{\alpha_{CW}}\left(\alpha_{OW}-\dfrac{1}{B}\right) + \dfrac{\alpha_{OV}}{\alpha_{HV}}M_H + \dfrac{r_R+\gamma'_R}{\beta_{RQ}} + \dfrac{r_X+\gamma_X}{B\beta_{XW}} = M_{OC}. &
\end{array}\right\} \qquad (10.36)$$

The total carbon concentration in the closed ecological system must exceed a certain critical value below which a long-term material cycle in the system is impossible and so the system dies (10.36(a)). Within the limits of the region concerned the stationary values of living components $(\overline{X}, \overline{R})$, living biomass $(\overline{LB} = \overline{X} + \overline{R})$, and total biomass $(\overline{TB} = \overline{Z} + \overline{LB})$ increase in proportion to the increase in the system of the total carbon concentration limiting the functioning of producers (M_C).

In the case when the growth of autotrophs is limited by light energy (a possible model of the biosphere in Arctic regions limited by light during certain seasons), two regions of stable steady states (10.33(c) and 10.33(d)) can be realized:

$$\left.\begin{array}{ll}
\text{(a)}\ E > \dfrac{r_X+\gamma_X}{uG} = E_X, & \\[2ex]
\text{(b)}\ M_j > M_{jX} + \alpha_j Tf(E), \qquad f(E) = \overline{X}, & \\[1ex]
\text{(c)}\ M_C > M_{CX} + \alpha_C Tf(E), & \\[1ex]
\text{(d)}\ M_H \gg \alpha_H\left[\dfrac{r_R+\gamma'_R}{\beta_{RZ}} + Tf(E)\right] + \alpha_{HV}\dfrac{r_X+\gamma_X}{\beta_{XV}}, & \\[2ex]
\text{(e)}\ M_O > \dfrac{\alpha_{OW}}{\alpha_{CW}}M_C + \dfrac{\alpha_{OV}}{\alpha_{HV}}M_H + (r_R+\gamma'_R)\left(\dfrac{1}{\beta_{RQ}} - \dfrac{1}{Y_{XW}B\beta_{RZ}}\right) + \dfrac{Tf(E)}{Y_{XW}B} = M_{OE}. &
\end{array}\right\}$$

(10.37)

In this case the value of illumination of the ecosystem must be larger than a certain critical value, making up for the minimal necessary power consumption of organisms and a long-term stable material cycle in the closed ecological system (Svirezhev, 1978). The dependence of $\overline{LB}$ and $\overline{TB}$ on the value of illumination takes the following form:

$$\frac{\omega}{1 - e^{-u\omega}} = F_\omega E + F_z, \tag{10.38}$$

where ω is $\overline{X}$, $\overline{R}$, $\overline{LB}$, or $\overline{TB}$.

The second region of the steady state is realized under the following conditions:

$$\left.\begin{aligned}
&\text{(a)}\ E > E_X,\\
&\text{(b)}\ M_j > \frac{r_X + \gamma_X}{\beta_{Xj}} + \alpha_j Y_{XW} B\left(\frac{\alpha_{OW}}{\alpha_{CW}} M_C + \frac{\alpha_{OV}}{\alpha_{HV}} M_H - M_O + \frac{r_R + \gamma'_R}{\beta_{RQ}}\right),\\
&\text{(c)}\ M_C > \frac{\alpha_{CW}}{1 - B\alpha_{OW}}\left[\frac{r_X + \gamma_X}{\beta_{XW}} + B\left(\frac{\alpha_{OV}}{\alpha_{HV}} M_H - M_O + \frac{r_R + \gamma'_R}{\beta_{RQ}}\right]\right) = M_{CQ},\\
&\text{(d)}\ M_H \gg \frac{\alpha_{HV}}{1 - Y_{XW} B \dfrac{\alpha_{OV}}{\alpha_{HV}} \alpha_H}\\
&\qquad \times \left[\frac{r_X + \gamma_X}{\beta_{XV}} + Y_{XW} B \frac{\alpha_H}{\alpha_{HV}}\left(\frac{\alpha_{OW}}{\alpha_{CW}} M_C - M_O + \frac{r_R + \gamma'_R}{\beta_{RQ}}\right)\right],\\
&\text{(e)}\ \max\{M_{OC}, M_{Oj}\} < M_O < M_{OE}.
\end{aligned}\right\} \tag{10.39}$$

Within the limits of this region the stationary concentrations of living components and $\overline{LB}$ change in proportion to illumination (10.38), while $\overline{TB}$ does not change, as the concentration of dead organic matter in the system is successively reduced. This is one of the characteristic features of the steady-state region (10.33(d)).

$\overline{LB}$ retains its constant value in the steady state, while $\overline{TB}$ decreases in proportion to the summary concentration of oxygen in the system. The addition of CO_2 to the system or an increase in the summary concentration of any biogenic element only increases their concentrations in the environment.

The third characteristic feature in this case is the linear growth of $\overline{TB}$ caused by the increase of carbon summary concentration in the ecosystem. In this situation $\overline{LB}$ retains its stationary value.

Thus, as model analysis shows, in a homogeneous producer–reducer closed ecological system stable steady states are realized in $3 + m$ of the $6 \times (3 + m)$ possible regions of functioning, differing by the nature of the limiting process. The realization of one or another stationary state (and the regime of functioning in general) depends on the correlation between closed ecological system characteristics: total quantity of matter suitable for biosynthesis (M_E, M_O, M_H, Mj) and energy supply (E), which are ecosystem control parameters. For instance, low illumination ($E < E_X$) or small summary concentrations of elements in the system ($Mj < Mj_X$ or $M_C < M_{CX}$ and so on) cause the system's death.

Stationary values of concentrations of living components ($\overline{X}, \overline{R}$) and living biomass ($\overline{LB}$) are proportionate, in linear mode or in correlation (10.38) within the limits of the region of the corresponding stationary state, to the change in the total value of the factor limiting the functioning of the first trophic level. Changes in the stationary concentration of total biomass ($\overline{TB}$) are similar.

The background concentrations of limiting factors (biogenic elements, CO_2, O_2, Z) in the corresponding stationary states do not depend on the general mass of the same matter circulating in the closed ecological system (autostabilization effect; Degermendzhi *et al.*, 1979), as the specific rate of organism elimination remains unchanged ($r + \gamma$). This paradoxical phenomenon will be discussed later in greater detail (see pp. 295–296).

10.4.2 Experimental technique

For the test probe we have chosen an experimental system with one species of the *Chlorella 21901* unicellular alga and two species of reducers, a *Pseudomonas* sp. and *Mycobacterium rubrum*, which we functionally considered as one unit with certain integral characteristics (Fishtein, 1981; Gubanov *et al.*, 1984; Kovrov and Fishtein, 1978, 1980; Kovrov *et al.*, 1976).

The aim of the experiment was (a) to realize the stationary MES conditions that limit the functioning of the biocenosis by carbon or nitrogen in the system, (b) to investigate the MES condition (species composition, population size of each species, mass characteristics, etc.) depending on each limiting factor value, and (c) to realize death regions by deficient carbon or nitrogen (population of organisms of all species or population of one of the trophic levels equals zero).

The coefficients to calculate the carbon-limited or nitrogen-limited regions of MES functioning and the stationary states possible herewith were taken from the literature and from analyses of the stationary states of experimental MESs to reach a better agreement between the mathematical model presented above (in the stationary state) and steady-state experimental systems (the chosen MES, in particular).

The microsystems had volumes 40 mL–50 mL, including 5 mL–10 mL of the liquid phase inhabited by microorganisms. Microorganisms were sealed in glass test-tubes, their lower part shaped as a rectangular cell with the liquid 16 mm thick (Figure 10.23). This allowed measuring the content of CO_2, chlorophyll, and undissolved organic matter in an MES in the course of the experiment without opening the test-tube. The test-tubes were placed in a luminostat with round-the-clock lighting intensity of 3 klux–4 klux and temperature of $28^\circ \pm 3^\circ$C.

The medium used in the MES was calculated by formulas (10.34) and (10.36) for the content of P, S, and other elements in the MES to be in ecological maximum with respect to the content of C and N.

The amount of suspension was determined by the ability to completely oxidize the introduced organic matter in the MES by available oxygen. In such a situation the oxygen (in the stationary condition, at least) was not a limiting factor.

Regions of MES death due to insufficient carbon or nitrogen, the regions where the functioning of the system in stationary condition would be limited by carbon, and

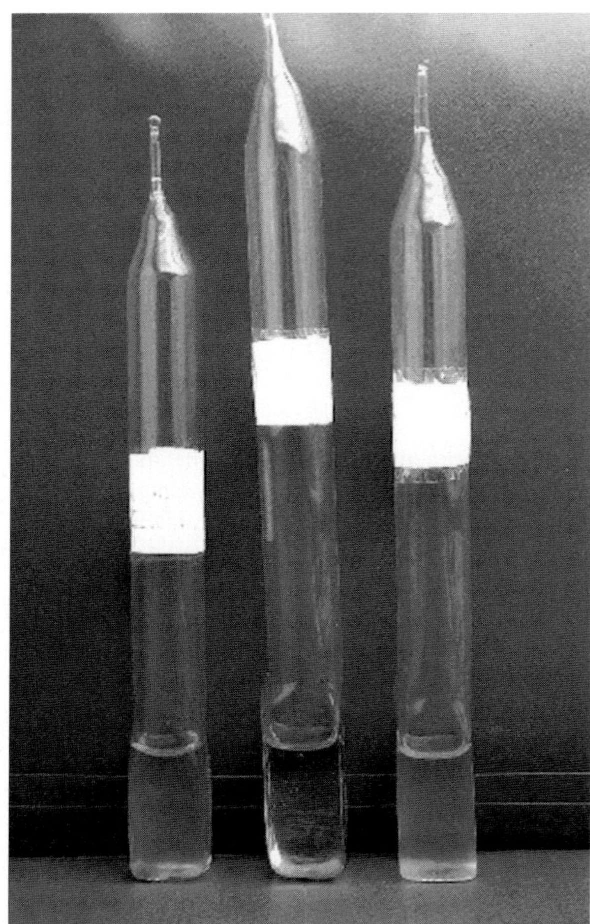

Figure 10.23. Unicellular organism based experimental closed microecosystems.

the nitrogen-limited region (i.e., the region where carbon is not the limiting factor) have already been calculated. The calculations showed that $M_{CX} = 0.325\,\mathrm{mg\,mL^{-1}}$, and $M_{NX} = 0.079\,\mathrm{mg\,mL^{-1}}$. We should remember that M_{CX} (M_{NX}) is the total concentration of carbon (nitrogen) in the MES, below it the cycling in the system cannot last long and the system dies ($X = 0, R = 0$). So, with $M_C < M_{CX} = 0.325\,\mathrm{mg\,mL^{-1}}$ or $M_N < M_{NX} = 0.078\,\mathrm{mg\,mL^{-1}}$, the development of the MES theoretically results in death. With the appropriate content of nitrogen ($M_N = 0.15\,\mathrm{mg\,mL^{-1}}$) for $0.325 < M_C < 0.625\,\mathrm{mg\,mL^{-1}}$ the functioning of the MES biocenosis in a stationary condition is limited by carbon, and at $M_C > 0.625\,\mathrm{mg\,mL^{-1}}$ by nitrogen. The mode of nitrogen limitation of MES biocenosis with appropriate carbon content ($M_C = 0.625\,\mathrm{mg\,mL^{-1}}$) should also occur with $0.078 < M_N < 0.15\,\mathrm{mg\,mL^{-1}}$, and with $M_N > 0.15\,\mathrm{mg\,mL^{-1}}$ carbon limitation starts.

10.4.3 Experimental results

The examples of undissolved organic matter dynamics for MESs C1, C4, C6 and N1, N4, N8 are given, respectively, in Figures 10.24 and 10.25 (C1, C4, C6, etc. are MESs in which total carbon content varied, while N1, N4, N8, etc. are MESs in which total nitrogen content varied). The dependence of undissolved organic matter quantity in the MES on the quantity of carbon introduced into the system is given in Figure 10.26. Here we also show the correlation of this experimental dependence with the theoretically (estimated) preassigned quantities of the organic matter in MESs (here and below experimental organic matter corresponds to theoretical $\overline{TB}$). The difference between dissolved organic matter and undissolved organic matter is probably the content of dissolved organic matter in the MES, while undissolved organic matter yields the sum of the biomass ($\overline{LB}$) and detritus.

The dependence of the quantity of undissolved organic matter in an MES on the total concentration of nitrogen is illustrated in Figure 10.27. It is seen that in the first phase of the undissolved organic matter distribution curve (up to system N7) undissolved organic matter concentration in the experimental MES is more than the theoretically estimated value of $\overline{TB}$ (organic matter). Excessive growth of organic matter contrary to that expected is probably related to the synthesis of a large amount of fats and carbohydrates on the nitrogen-rich medium with insufficient content of nitrogen. Furthermore, (after N8) the undissolved organic matter concentration is less than the corresponding organic matter value. The difference between organic matter and undissolved organic matter values here and above can be interpreted as dissolved organic matter.

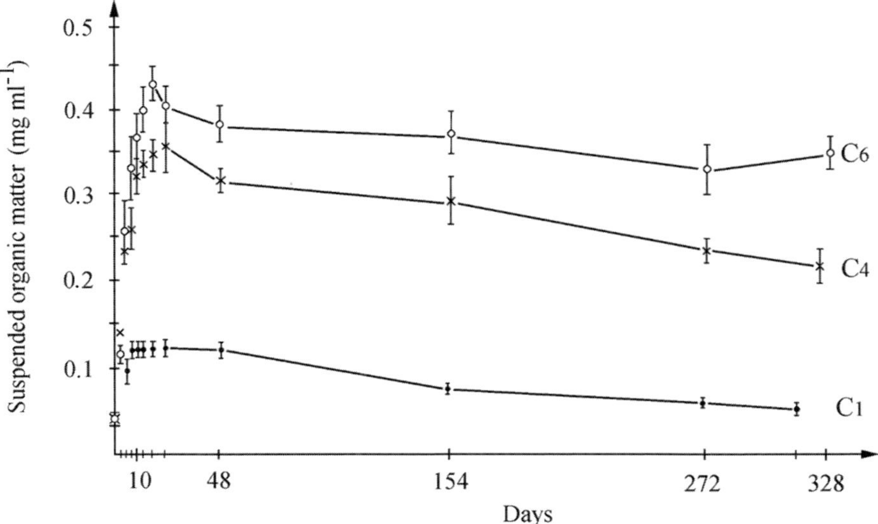

Figure 10.24. Dynamics of suspended organic matter for MES C1, C4, C6 with different carbon contents (M_C).

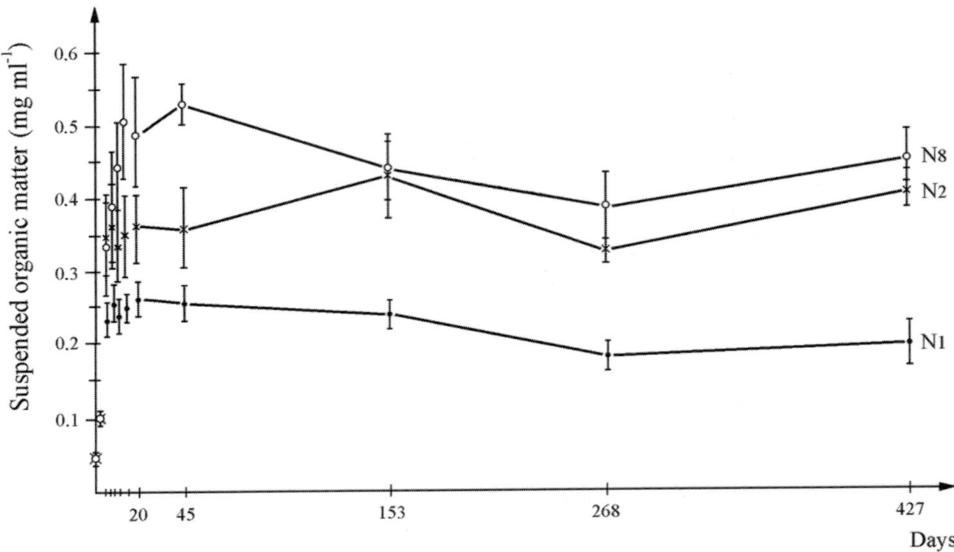

Figure 10.25. Dynamics of suspended organic matter for MES N1, N4, N8 with different nitrogen contents (M_N).

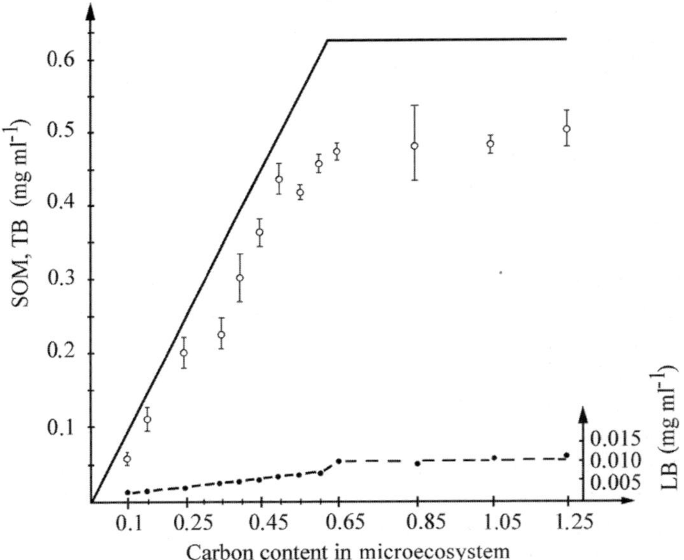

Figure 10.26. Theoretically calculated distribution of total biomass ($\overline{TB}$) (solid line); and experimentally derived distribution of suspended organic matter (SOM) (dashed line) estimates in systems with different carbon content (M_C) (steady state). — · — estimated living biomass ($\overline{LB}$) of cenosis of unicellular organisms.

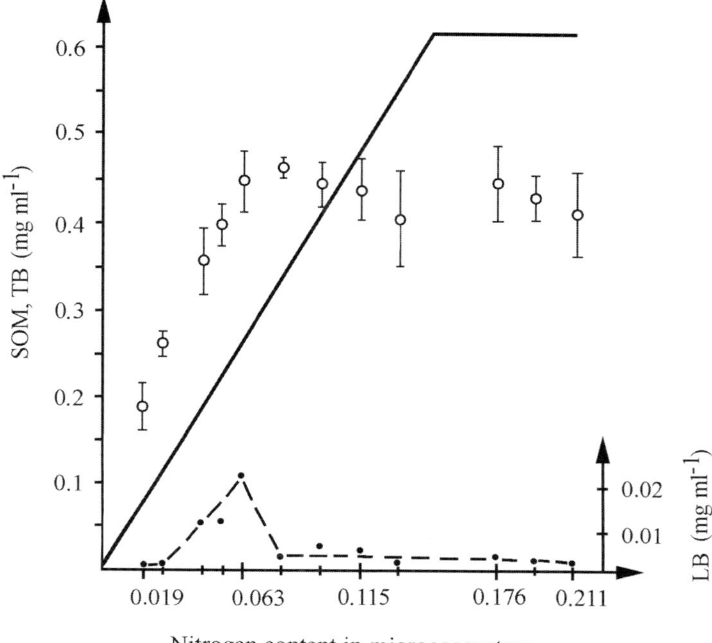

Figure 10.27. Theoretically calculated distribution of total biomass ($\overline{TB}$) (solid line); and experimentally derived distribution of suspended organic matter (SOM) (dashed line) estimates in systems with different nitrogen content (M_N) (steady state). — · — estimated living biomass ($\overline{LB}$) of cenosis of unicellular organisms.

It should especially be noted that in accordance with calculations, carbon as a limiting factor should be changed for nitrogen between Systems C9 and C10 at $M_N = 0.15\,\mathrm{mg\,mL^{-1}}$ and $M_C = 0.625\,\mathrm{mg\,mL^{-1}}$, and in the experiment this change occurred as calculated (Figure 10.26). The calculation of nitrogen to carbon change (indication of the change point) should be considered less successful. In accordance with calculations, the change should have occurred between Systems N9 and N10 at $M_C = 0.625\,\mathrm{mg\,mL^{-1}}$ and $M_N = 0.15\,\mathrm{mg\,mL^{-1}}$, but in the experiment it occurred in System N5 (Figure 10.27). However, the calculation did determine the region of the change from one limiting factor to the other. In the experiment, as distinct from calculation, one could find oneself in the region solely limited by nitrogen and then the experimental series would be a continuously increasing distribution at a certain angle to the X-line. Or one could find oneself in the region limited by carbon and then the experimental system distribution would be parallel (within statistical error) to the X-line. Since this did not happen and the change from nitrogen to carbon limitation did occur—moreover, it occurred in the region that was far from disastrous; e.g., not in System N2; Figure 10.27)—this result can be considered quite successful.

The MESs proposed can assist in solving certain problems of both general (development strategy, issues of closure and intensity of material cycling, similarity and scaling, etc.) and microbial ecology (contribution of individual microorganism species to formation and sustenance of material cycling, problems of interaction, etc.).

10.5 DISCUSSION AND CONCLUSION

First of all, we should again address the two types of attitudes, or approaches, to the mechanisms accounting for the climate trends observed over the past 10 years. In fact, our approach, the principle of the worst case scenario within the confidence interval which we described in brief in Section 10.1, does not need any additional explanation. However, the controversy between protagonists supporting either of the two approaches, anthropogenic and natural mechanisms of global climate change, is so heated that we have to present other solid arguments in favor of the approach that takes into account human-induced factors.

The main target attacked by our opponents is an increase in the greenhouse effect, which can lead to warming (re-emission of the visible light into the infrared region at +15°C, the Earth's mean surface temperature). In terms of physics, this is a proven fact: the existence of greenhouse components (CO_2, H_2O, and CH_4) can cause a rise in surface and atmospheric temperatures (Monin and Shishkov, 2000), and thus if there were no greenhouse effect the Earth's mean surface temperature would now be −20°C rather than the actual +15°C. There are similar records of the greenhouse effect on Venus; there it is a much more strongly pronounced effect because the atmospheric CO_2 concentration on Venus is very high. However, the greenhouse effect on Venus may be caused not only by the presence of greenhouse gases but also by a very thick atmosphere; so, both factors should be taken into account. Then again, the pattern of climate formation (Figure 10.1) involves important components that are difficult to calculate theoretically (aerosols, gas exchange with the ocean, albedo, dynamics of the planet's biota, cloudiness structure, etc.). They can have both positive and negative feedbacks with the temperature, and thus can be interpreted to prove the insignificance of the greenhouse effect.

The physical fact of the existence of the greenhouse effect does not necessarily imply that it is the major contributor to climate change. So, evaluation of the contribution of the greenhouse effect to climate, taking into account real components of the biosphere and (what is particularly important) their spatio-temporal dynamics is very difficult but necessary.

The second argument of anti-anthropogenic effect proponents is that in the distant past there were rises and falls of CO_2 concentration too, at the time when humanity did not burn any fuel (Figure 10.28). More precisely, temperature changes did not always occur in phase with CO_2 changes, which is interpreted as an argument against the existence of the greenhouse effect. However, there are models that account for the delay in the temperature response to CO_2 changes in the past via the mechanism of the thermal lag of biosphere (ocean) components, thus confirming the functioning of the greenhouse system (Karnaukhov, 2001). At the same time, we still have no reliable data as to the reasons for the former elevations of CO_2 con-

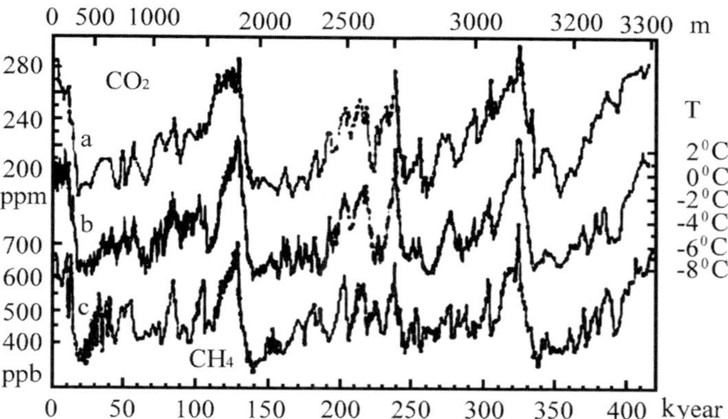

Figure 10.28. Paleoenvironmental data from the "Vostok" station. It is possible to directly observe the dynamics of CO_2 (a) and CH_4 (c) concentrations and temperature T (b) during the past 400,000 years. The top scale refers to the drilling depth of Antarctic ice. The bottom left-hand corner refers to the present

centration, intensity of incident solar radiation at that time, etc., though anyway contemporary rates of CO_2 increase are several orders of magnitude higher than past rates, and this may be an argument in favor of anthropogenism.

Thus, asserting that the greenhouse effect makes a significant contribution to contemporary global warming, let us discuss in greater detail some important aspects of the model results. We should note here that additional CO_2 emission due to soil heating was taken into account in another, similarly structured model (Lenton, 2000). However, in our opinion, some features of our model make a significant contribution to the emergence of "irreversibility dates", which we did not find in Lenton's model. First of all, this is the worst case scenario principle, the basis for choosing the least favorable possible values of the model parameters. Second, our model takes into account the natural limitation of biomass increase, which does not allow biota to take up arbitrary amounts of carbon dioxide from the atmosphere.

The hopes for compensatory removal of CO_2 from the atmosphere due to photosynthesis on land and in the ocean are almost groundless. We have shown theoretically that with the growth of atmospheric CO_2 concentration, even if we assume that the entire land-based photosynthesis is limited by carbon dioxide. the total biomass that takes up some of the carbon increases slightly and the plant mass increment is restricted physically by the surface area on which they are located and by possible antagonistic allelopathic interactions, which are formally described by the coefficient x_{max}. Respiration and fires gradually balance out photosynthesis, in the limit leading to zero carbon accumulation. The situation becomes even more serious when the temperature increases and photosynthesis is inhibited.

We should note that the model we have presented is limited by the fact that we do not consider the known positive feedback loops that can bring nearer the stability threshold date, such as temperature elevation and albedo decrease due to a shift in the upper border of the snow cover, and temperature elevation and methane emissions due to permafrost melt. These and other aggravating mechanisms together will be taken into account in our future studies.

An important part of modeling is testing and verification of the model. In fact, the global data on the dynamics of all variables included in the model are partly testing parameters and partly verifying ones, of different accuracy and reliability. For instance, the data on CO_2 dynamics of ice cores and the data on atmospheric CO_2 concentration are sufficiently detailed and accurate, particularly those obtained at the Mauna Loa Observatory in the Pacific. Verification based on the greenhouse hypothesis showed a good agreement with global atmospheric CO_2 dynamics. The degree of uncertainty of other variables (the amount of soil organics, land and ocean biomass quantity, the amount of burnt fuels, etc.) is very high. Future models could involve integrated satellite data on radiation and surface temperatures to further formalize the greenhouse hypothesis, taking into account the major climate feedbacks (Figure 10.1), within the framework of the worst case scenario minimal model.

Even worse is the situation with the data about the first process derivative (i.e., about the speed of the processes and flows that can be used as verifying parameters, too). Here it is necessary to note that there is also a theory and procedure of verification using the second derivative of biomass with respect to time (growth acceleration) that, theoretically, has a deep meaning of calculation of the combined effect of all limiting factors, but it was verified only for water ecosystems and microorganism continuous cultivation systems (Degermendzhy *et al.*, 1989, 1993). That is why there was a great temptation to use satellite monitoring data for which the maximum length of time series is about 20 years. The most difficult, if not completely impossible thing, is to state exactly the total amount and dynamics of the living terrestrial biomass, while the models based on the 20-year period of calculations show a minor increase in the latter. That is why it was reasonable to compare photosynthesis production calculated theoretically and satellite measurements over a period of 20 years for terrestrial and oceanic compartments.

The methods and problems of using satellite data for comparison with models should be discussed critically (see Section 10.3). The parameters calculated on the basis of spectral radiation scattered by above-ground vegetation and registered by a spacecraft can be used in the process of complex studying of natural vegetative communities using full-scale space measurements and environmental models. Such parameters are the NDVI (normalized difference vegetation index) and NPP (net primary production).

Though the fraction of absorbed photosynthetically active radiation can be represented by a function of the NDVI, the increase in biomass will depend upon a number of factors (vegetation type, environmental conditions, air temperature, soil humidity, pressure, etc). That is why the direct use of NDVI for estimation of global productivity is not very suitable (Golubyatnikov and Denisenko, 2006).

Taking into account the above-mentioned, we can state at the present stage: it is NPP determined with the help of GLO-PEM (the global production efficiency model) that is a more effective characteristic, being suitable for use in environmental models (Goetz *et al.*, 1999, 2000; Prince and Goward, 1995). To obtain a large amount of statistical material it is important that only parameters registered by means of remote sensing of the Earth's surface are used for NPP calculation.

The second method of verification is not directly connected with verification of the structural adequacy of the model proposed by us, but has to do with the question of how accurate is the key mechanism (effect) of additional CO_2 emission from the soil, as well as perhaps other hypotheses and mechanisms that can emerge in the future as a consequence of the development of the theory. The question is mainly about closed material cycle experimental ecosystems that model the principle of the biospheric components cycle (Section 10.4). Section 10.4 describes the microcosms the full cycle of which is realized by means of completely balanced processes of photosynthesis and decomposition (respiration) and which models the most important mechanism of biosphere stability: material cycle. If different amounts of carbon are introduced to such microcosms at the initial instant, this may be interpreted as the model of additional carbon inflow into the biosphere as a result of combustion. The experiments showed that there is a limit to system biomass growth, and further introduction of the limiting factor (carbon) does not lead to an increase in total biomass. However, a remark should be made that the experiment deals with a series of community stationary states, while the biosphere at present is not in a stationary state. In fact, the particular model created on the basis of these experiments yielded the diagrams of control of the composition and total biomass of community, in which the role of control lever is played by the variation of the total amount of some limiting biogenic component.

It should be noted that the autostabilization effect of limiting factors was demonstrated for closed ecosystems for the first time. Earlier this effect had been revealed in open flow systems and consisted in the following: the background concentration of a limiting substance in the environment paradoxically does not depend (!) on the inflow of the same substance into the system (the mechanisms of this autostabilization effect, theory, observations, and experiments are presented in detail by Degermendzhy *et al.*, 1979, 1989, 2002). Theoretically, for ecosystems with a full cycle (microcosms with an incomplete cycle always die), the autostabilization effect is formulated as follows: the background concentration of a limiting substance in a microcosm does not depend on the total amount of this substance that was initially introduced into the microcosm. Experiments conducted and theory proved this statement to be true. This conclusion allows predicting variations in community composition (correlation of species biomasses) in biosphere-like systems, varying the total amount of the cycling component and the components whose concentrations will be autostabilized.

In conclusion, we should mention the obvious insufficiency of the biological mechanisms of terrestrial and oceanic CO_2 withdrawal from Earth's atmosphere, as stated in this chapter, which does not make up for continuation of mineral fuel combustion, thus dismissing the hope that CO_2 concentration in the atmosphere will stabilize spontaneously. It is supposed that the key process in the CO_2–temperature chain is the efficiency of the greenhouse effect. That is why revelation of the actual role of the greenhouse effect and atmosphere density, creation of a strict quantitative model of variations of the average planetary temperature of the Earth taken into account, along with the growth of CO_2 concentration in the atmosphere, aerosol pollution of its upper layers, as well as other factors (Figure 10.1), such as the response rate of the oceans' temperature regime variations, all these issues present a very complicated but important task that can be solved if the specialists of various spheres of science unite their efforts.

At the same time, the development of mathematical models connecting the dynamics and biospheric carbon cycle, including biological constituents, with climate changes, is necessary in any case, because it is not expensive, but provides international experts with various cause–effect mechanisms of changes in the global climate.

The "inreversibility dates" obtained in the process of this chapter are new results in the analysis of the biosphere's biological role in climate formation in terms of formulation of a list of irreversible mechanisms. The task-oriented use of satellite methods and development of experimental closed ecosystem methods demonstrated in the chapter will be very helpful in the process of increasing the validity of mathematical models. The model hierarchy, from minimal to more detailed ones (e.g., including marsh and tundra compartments), are supposed to be the main line of development of the biosphere and global climate stability theory.

Acknowledgments. The work was supported by an "Origin and Evolution of Biosphere" grant from the RAS Presidium, SB RAS, and Project N2004 0.47.011.2004.030 (the Russian Foundation for Basic Research and the Netherlands Organization for Scientific Research).

10.6 REFERENCES

Abrosov N.S., Kovrov B.G., and Cherepanov O.A. (1982). *Environmental Mechanisms of Coexistence and Species Regulation.* Nauka, Novosibirsk, 302 pp. [in Russian].

Bartsev S.I., Degermendzhi A.G., and Erokhin D.V. (2005). Global minimal model of long-term carbon dynamics in the biosphere. *Doklady Earth Sciences*, 401(2), 326–329.

Battle M., Bender M., Tans P.P., White J.W.C., Ellis J.T., Conway T., and Francey R.J. (2000). Global carbon sinks and their variability, inferred from atmospheric O_2 and C. *Science*, **287**(24), 67–70.

Bidigare R.R. and Ondrusek M.E. (1996). Spatial and temporal variability of phytoplankton pigment distributions in the Central Equatorial Pacific Ocean. *Deep Sea Res., Part II: Topical Stud. Oceanogr.*, **43**, 809–833.

Bode A. and Varela M. (1998). Primary production and phytoplankton in three Galician Rias Altas (NW Spain): Seasonal and spatial variability. *Scientia Marina*, **62**, 319–330.

Bousquet P., Peylin P., Ciasis P., Le Quéré C., Friedlingstein P., and Tans PP. (2000). Regional changes in carbon dioxide fluxes of land and oceans since 1980. *Science*, **290**(1342), 6.

Brovkin V., Bendsen J., Claussen M., Ganapolski A., Kubatzki C., Petoukhov V., and Andreev A. (2002). Carbon cycle, vegetation, and climate dynamics in Holocene: Experiments with the CLIMBER-2 model. *Glob. Biogeochem. Cycles*, **16**(4), 11–39.

Brovkin V., Sitch S., von Bloh W., Claussen,M., Bauer E., and Cramer W. (2004). Role of land cover changes for atmospheric CO_2 increase and climate change during the last 150 years. *Glob. Change Biol.*, **10**, 1253–1266.

Degermendzhy A.G., Pechurkin N.S., and Shkidchenko A.N. (1979). *Autostabilization of Growth-controlling Factors in Biological Systems*. Nauka, Novosibirsk, 139 pp. [in Russian].

Degermendzhy A.G., Adamovich V.A., and Pozdiaev V.N. (1989). On the cybernetics of bacterial communities: Observations, experiments and theory. *J. of Cybernetics and Systems*, **20**(6), 501–541.

Degermendzhy A.G., Adamovich V.V., and Adamovich V.A. (1993). A new experimental approach to the search for chemical density factors in the regulation of monoculture growth. *J. of General Microbiology*, **139**, 2027–2031.

Degermendzhy A.G., Belolipetsky V.M., Zotina T.A., and Gulati Ramesh D. (2002). Formation of the vertical heterogeneity in the Shira Lake ecosystem: The biological mechanisms and mathematical model—"The structure and the functioning of the lake Shira ecosystem: An example of Siberian brackish water lakes". *Aquatic Ecology*, Special Issue, **2**, 271–297.

Denman K.L., and Abbott M.R. (1988). Time evolution of surface chlorophyll patterns from cross-spectrum analysis of satellite color images. *J. Geophys. Res.*, **93**, 6789–6798.

Dickey T., Marra J., and Granata T. (1991). Concurrent high-resolution biooptical and physical time series observations in the Sargasso Sea during the spring of 1987. *J. Geophys. Res.*, **96**, 8643–8663.

Eck T.F. and Dye D.G. (1991). Satellite estimation of incident photosynthetically active radiation using ultraviolet reflectance. *Remote Sens. Environ.*, **38**(2), 135–146.

Falkowski P.G., Barber R.T., and Smetacek V. (1998). Biogeochemical controls and feedbacks on ocean primary production. *Science*, **281**, 200–206.

Field C.B., Behrenfeld M.J., Randerson J. T., and Falkowski P. (1998). Primary production of the biosphere: Integrating terrestrial and oceanic components. *Science*, **281**, 237–240.

Fishtein G.N. (1981). Vidovaya struktura zamknutykh mikroekosistem (Species structure of closed micro-ecosystems). The manuscript registered at VINITI on 29 January 1981: No. 374-81, Moscow, 32 pp. [in Russian].

Fitter A.H., Self G.K., Brown T.K., Bogie D.S., Graves J.D., Benham D., and Ineson P. (1999). Root production and turnover in an upland grassland subjected to artificial soil warming respond to radiation flux and nutrients, not temperature. *Oecologia*, **120**, 575–581.

Gao M., Prince S.D., Small J., and Goetz S.J. (2004). Remotely sensed interannual variations and trends in terrestrial net primary productivity 1981–2000. *Ecosystems*, **7**, 233–242.

Gifford R.M. (1993). Implications of CO_2 effects on vegetation for the global carbon budget. In: M. Heimann (ed.), *The Global Carbon Cycle*. Springer-Verlag, Berlin, pp. 159–199.

Gitelson J.I., Kovrov B.G., Lisovsky G.M., Okladnikov Yu.N., Rerberg M.C., Sidko F.Ya., and Terskov I.A. (1975). Experimental ecological manned systems. In: A.A. Nichiporovich (ed.), *Problemy Kosmicheskoi Biologii (Problems of Space Biology)*. Nauka, Moscow, 312 pp. [in Russian].

Goetz S.J., Prince S.D., Goward S.N., Thawley M.M., and Small J. (1999). Satellite remote sensing of primary production: An improved production efficiency modeling approach. *Ecological Modelling*, **122**, 239–255.

Goetz S.J., Prince S.D., Small J., and Gleason A. (2000). Interannual variability of global terrestrial primary production: Results of a model driven with global satellite observations. *J. of Geophysical Research*, **105**(D15), 20077–20091.

Golubyatnikov L.L. and Denisenko E.A. (2006). Interrelation between the vegetation index and the climatic parameters and structural characteristics of vegetation cover. *Izvestiya of RAS: Physics of Atmosphere and Ocean*, **42**(4), 524–538.

Goward S.N. and Dye D. (1997). Global biospheric monitoring with remote sensing. In: H.L. Gholtz, K. Nakane, and H. Shimoda (eds.), *The Use of Remote Sensing in Modeling Forest Productivity*. Kluwer Academic, New York, pp. 241–272.

Gubanov V.G., Kovrov B.G., and Fishtein G.N. (1984). Closed microecosystems: A new test-object for biophysical and ecological research. In: I.A. Terskov (ed.), *Biophysical Methods of Ecosystem Research*. Nauka, Novosibirsk, pp. 34–44 [in Russian].

Houghton R.A. (2003). Revised estimates of the annual net flux of carbon to the atmosphere from changes in land use and land management 1850–2000. *Tellus*, **55B**, 378–390.

IPCC (2001). *Climate Change, 2001: Scientific Aspects*. United Nations Environment Programme, Zurich, Switzerland (UNEP), 881 pp. Available at *http://www.ipcc.ch/*

IPCC (2007). Contribution of Working Group I to the Fourth Assessment Report of the Intergovernmental Panel on Climate Change. In: S. Solomon, D. Qin, M. Manning, Z. Chen, M. Marquis, K.B. Avery, M. Tignor, and H.L. Miller (eds.), *Climate Change, 2007: The Physical Science Basis*. Cambridge University Press, Cambridge, U.K., 996 pp. Available at *http://www.ipcc.ch/*

Johnson L.C., Shaver G.R., Cades D.H., Rastetter E., Nadelhoffer K., Giblin A., Laundre J., and Stanley A. (2000). Plant carbon–nutrient interactions control CO_2 exchange in Alaskan wet sedge tundra ecosystems. *Ecology*, **81**, 453–469.

Karnaukhov A.V. (2001). Role of biosphere in the formation of the Earth's climate: The greenhouse catastrophe. *Biofizika*, **46**(6), 1078–1089 [in Russian].

Keeling C.D. and Whorf T.P. (2001). *Atmospheric Carbon Dioxide Record from Mauna Loa*. University of California, La Jolla, CA. Available at *http://cdiac.ornl.gov/trends/co2/sio-mlo.htm*

Kondratyev K.Ya. and Krapivin V.F. (2004). *Carbon Global Cycle Modeling*. Fizmatgiz, Moscow, 336 pp.

Kovrov B.G. and Fishtein G.N. (1978). Experimental closed microecosystems containing unicellular organisms. In: *Continuous Cultivation of Microorganisms: Seventh International Symposium, Prague*, p. 43.

Kovrov B.G. and Fishtein, G.N. (1980). Biomass distribution in synthetic closed microbiocenoses, depending upon their species structure. *Izv. SO AN SSSR, No. 1: Ser. Biol.*, 35–40 [in Russian].

Kovrov B.G., Mamavko G.A., and Fishtein G.N. (1976). Experimental models of closed ecosystems on unicellular microorganisms. In: *Proceedings of XI All-Union Workshop on Material Cycling in Closed Systems Based on Vital Activity of Lower Organisms*. Naukova Dumka, Kiev, pp. 61–63 [in Russian].

Krapivin V.F. and Potapov I.I. (2002). *Methods of Ecoinformatics*. VINITI RAN, Moscow, 496 pp. [in Russian].

Lenton T.M. (2000). Land and ocean carbon cycle feedback effects on global warming in a simple Earth system model. *Tellus*, **52B**(5), 1159–1188.

Liski J., Ilvesniemi H., Makela A., and Westman C.J. (1999). CO_2 emissions from soil in response to climatic warming are overestimated: The decomposition of old soil organic matter is tolerant of temperature. *Ambio*, **28**, 171–174.

Luo Y., Wan S., Hui D., and Wallace L.L. (2001). Acclimatization of soil respiration to warming in a tall grass prairie. *Nature*, **413**, 622–625.

Merkin D.R. (1996). *Introduction to the Theory of Stability*. Springer-Verlag, New York, 312 pp.

Mete Uz B. and Yoder J.A. (2004). High frequency and mesoscale variability in SeaWiFS chlorophyll imagery and its relation to other remotely sensed oceanographic variables. *Deep Sea Research, Part II: Topical Studies in Oceanography*, **51**(10/11), 1001–1017.

Monin A.S., and Shishkov Yu.A. (2000). Climate as a problem in physics. *Uspekhi Fizicheskikh Nauk*, **170**(4), 436–445.

Morgan J.A., LeCain D.R., Mosier A.R., and Milchunas D.G. (2001). Elevated CO_2 enhances water relations and productivity and affects gas exchange in C3 and C4 grasses of the Colorado shortgrass steppe. *Global Change Biol.*, **7**, 451–466.

Poletayev, I.A. (1966). On mathematical models of elementary processes in biogeocenoses. In: S.V. Yablonskiy (ed.), *Problems of Cybernetics*. Nauka, Moscow, pp. 171–190 [in Russian].

Prentice I.C., Farquhar G.D., Fasham M.J.R., Heimann M.L., Jaramillo V.J., and Kheshgi H.S. (2001). The carbon cycle and atmospheric carbon dioxide. *Climate Change, 2001: The Scientific Basis*. Cambridge University Press, Cambridge, U.K., pp. 183–237.

Prince S.D. and Goward S.J. (1995). Global primary production: A remote sensing approach. *J. of Biogeography*, **22**, 815–835.

Pritchard S.G., Davis M.A., Mitchell R.J., Prior A.S., Boykin D.L., Rogers H.H., and Runion G.B. (2001). Root dynamics in an artificially constructed regenerating longleaf pine ecosystem are affected by atmospheric CO_2 enrichment. *Environmental and Experimental Botany*, **46**, 35–69.

Raich J.W. and Schlesinger W.H. (1992). The global carbon dioxide flux in soil respiration and its relationship to vegetation and climate. *Tellus*, **44B**, 81–99.

Risk D., Kellman L., and Beltrami H. (2002). Carbon dioxide in soil profiles: Production and temperature dependence. *Geophysical Research Letters*, **29**(6), 111–114.

Rochette P., Angers D.A., and Flanagan L.B. (1999). Maize residue decomposition measurement using soil surface carbon dioxide fluxes and natural abundance of Carbon-13. *Soil Science Society of America Journal*, **63**, 1385–1396.

Running S.W., Nemani R.R., Heinsch F.A., Zhao M., Reeves M., and Hashimoto H. (2004). A continuous satellite-derived measure of global terrestrial primary production. *BioScience*, **54**(6), 547–560.

Schimel D., Alves D., Enting I., Heimann M., Joos F., Raynaud D., Wigley T., Prather M., Derwent R., Ehhalt D. *et al.* (1996). Radiative forcing of climate change. In: J.T. Houghton, L.G.M. Filho, B.A. Callander, N. Harris, A. Kattenberg, and K. Maskell (eds.), *Climate Change, 1995: The Science of Climate Change*. Cambridge University Press, Cambridge, pp. 65–131.

Schimel D.S., House J.I., Hibbard K.A., Bousquet P., and Ciasis C. (2001). Recent patterns and mechanisms of carbon exchange by terrestrial ecosystems. *Nature*, **414**, 169–172.

Semyonov D.A. (2003). Impact of biota on global climate. Ph.D. thesis, Krasnoyarsk, IBP SB RAS, 117 pp. [in Russian].

Semyonov S.M. (2004). *Greenhouse Gases and Modern Climate of the Earth*. Meteorology and Hydrology, Moscow, 176 pp. [in Russian].

Shabanov N.V., Zhou L., Knyazikhin Y., Myneni R.B., and Tucker C.J. (2002). Analysis of interannual changes in northern vegetation activity observed in AVHRR data from 1981 to 1994. *IEEE Transaction on Geoscience and Remote Sensing*, **40**(1).

Shevyrnogov A.P., Vysotskaya G.S., and Shevyrnogov E.A. (2004). A study of the stationary and the anomalous in the ocean surface chlorophyll distribution by satellite data. *International Journal of Remote Sensing*, **25**(7/8), 1383–1387.

Stainforth D.A., Aina T., Christensen C., Collins M., Faull N., Frame D.J., Kettleborough J.A., Knight S., Martin A., Murphy J.M., Piani C., Sexton D.L., Smith A.R., Spicer A.A., Thorpe J., and Allen M.R. (2005). Uncertainty in predictions of the climate response to rising levels of greenhouse gases. *Nature*, **433**, 403–406.

Svirezhev Yu.M. (1978). On the length of the trophic chain. *Zhurnal obshchey biologii*, **39**(3), 373–379 [in Russian].

Svirezhev Yu.M. and von Bloh W. (1997). Climate, vegetation, and global carbon cycle: The simplest zero-dimensional model. *Ecol. Mod.*, **101**, 79–95.

Tarko A.M. (2005). *Anthropogenic Changes of Global Biospheric Processes: Mathematical Modeling*. Fizmatgiz, Moscow, 232 pp.

Watson R.T., Noble I.R., Bolin, B. *et al.* (eds.) (2000). *Land Use, Land-use Change, and Forestry*, Special Report on the IPCC. Cambridge University Press, Cambridge, U.K., 377 pp.

WRI (1998). *World Resources: A Guide to the Global Environment 1998–99*. World Resources Institute, Washington, D.C.

Zavarzin G.A. and Kolotilova N.N. (2001). *Introduction to Naturalistic Microbiology*. University Books, Moscow, 256 pp. [in Russian].

11

Air temperature changes at White Sea shores and islands in the 19th and 20th centuries

Olga A. Shilovtseva and Feodor A. Romanenko

11.1 INTRODUCTION

The scientific interests of Kirill Kondratyev, an outstanding Soviet and Russian scientist–geophysicist, academician of the Academy of Science of the USSR and the Russian Academy of Science, and honorary member of many authoritative international scientific institutes and organizations (see Chapter 1), were very wide and various. His research included work on satellite meteorology, atmospheric optics, actinometry, and problems of climate change and global ecology. During the last two decades of his life he concentrated on global problems of the the environment (global change) and on the interactions in the system nature–society (Demirchian *et al.*, 2006; Kondratyev, 1998, 2001, 2003a; Kondratyev and Cracknell, 1999; Kondratyev and Galindo, 1997; Kondratyev and Varotsos, 2000). We have seen in Chapter 2 that Kondratyev was very concerned about the need for good observational data when discussing climate change in recent decades and, especially, when attempting to make predictions about future trends in our climate. He wrote (Kondratyev, 2003b):

> "... measurements data (for the present they are inadequate from the point of view of their completeness and reliability) do not contain the exact existence of anthropogenic caused confirmation of 'global warming' at all (especially it concerns the ground-based observations in the USA, in Arctic regions) ..."

Among the observational data that are important is the near-surface air temperature (Kondratyev, 2004). He questioned the claims of modelers and of the IPCC that the strengthening of climate warming in high latitudes of the northern hemisphere was a characteristic attribute of anthropogenically caused global warming. He claimed that one can conclude, from the analysis of measurements of ground-based temperature of the air at Arctic stations

> "for 30 years and dendroclimatic indirect data for the last 2–3 centuries described in the work by Adamenko and Kondratyev (1999), ... that the ... homogeneous strengthening of warming was not observed, and climate changes both of the last century and the last decade were characterized by a strong spatiallytemporal heterogeneity: in (the) Arctic there were simultaneously forming the regions of climate warming as well as the regions with cold snaps of a climate ..."

In other words global models represent a simplification and the response to human activities is likely to show local or regional variations. What is of interest to many people is how their own local climate will be altered as a result of climate change, whether as a result of natural causes or as a result of human activities. There is therefore a need for regional and local climate models, and there is a need to study local variations in climate-related parameters such as near-surface air temperature. In the interests of following up Kondratyev's work on near-surface air temperature in the Arctic we have analyzed data from a large number of stations around the White Sea.

Towards the end of the 20th century, interest in global climate change has extended from scientific publications and climatologists, who have dealt with these subjects for a long time (the first publications on this theme were in the 19th century, e.g., Veselovsky, 1857 and Vrangel, 1891), to the general public. Experts of various disciplines, from mathematicians to economists and politicians, are now interested in climate change problems and articles about global warming appear in a number of different publications, which sometimes are very far from climatology (Kasimov and Klige, 2006a, b; Izrael, 2004). At the same time the main points of questions of modern climate change processes (such as whether it varies, why does it vary if it does, with what intensity, and to what extent are these climate changes steady in time) often disappears under emotional discussions. The basic method of modern climate change analysis (i.e., studying the results of long-term ground-based observations) also remains somewhat in the shadows.

Air temperature is arguably the most important parameter for indicating the tendency of climate change for a territory (Demirchian *et al.*, 2006). Therefore, its analysis over the longest possible period provides valuable evidence of recent climatic change. Following the invention of the thermometer at the beginning of the 17th century and the subsequent establishment of a network of meteorological stations, we now have a valuable resource in the form of a long-term record of air temperature with a good geographical distribution. As a contribution to the study of climate change this chapter is concerned with the estimation of long-term changes of air temperature at the coast and on the islands of the White Sea, which is the most southerly and the smallest of the Arctic seas and is almost entirely surrounded by land. This investigation continues a series of studies of the climate dynamics of Arctic seas (Filatov *et al.*, 2005; Shilovtseva and Romanenko, 2005). This work was supported by the Russian Fund for Basic Researches (Project No. 05-05-64872).

11.2 MATERIALS AND METHODS

The White Sea is virtually a bay of the Northern Polar Ocean deeply cut into the continent. Its area of water stretches approximately 500 km from north to south and approximately 550 km from west to east. There are four large bays: Kandalaksha Bay in the northwest, Dvina Bay and Onega Bay in the south, and Mezen Bay in the east. The White Sea shores have their own geographical names: Tersky, Kandalaksha, and Karelian (Karelsky) in the northwest and west; Pomorsky, Onegsky (Lyamitsky), and Letny in the south; and Zimny, Abramovsky, Konushisky, and Kaninsky in the southeast and east (see Figure 11.1).

The coast of the White Sea is rather well provided with meteorological information (Glukhovsky *et al.*,1989; Kondrasheva, 1954; Poznitskiy, 1966; Soboleva, 1956; NAHS, 1970; and IC, 1975). For this investigation the stations which were chosen

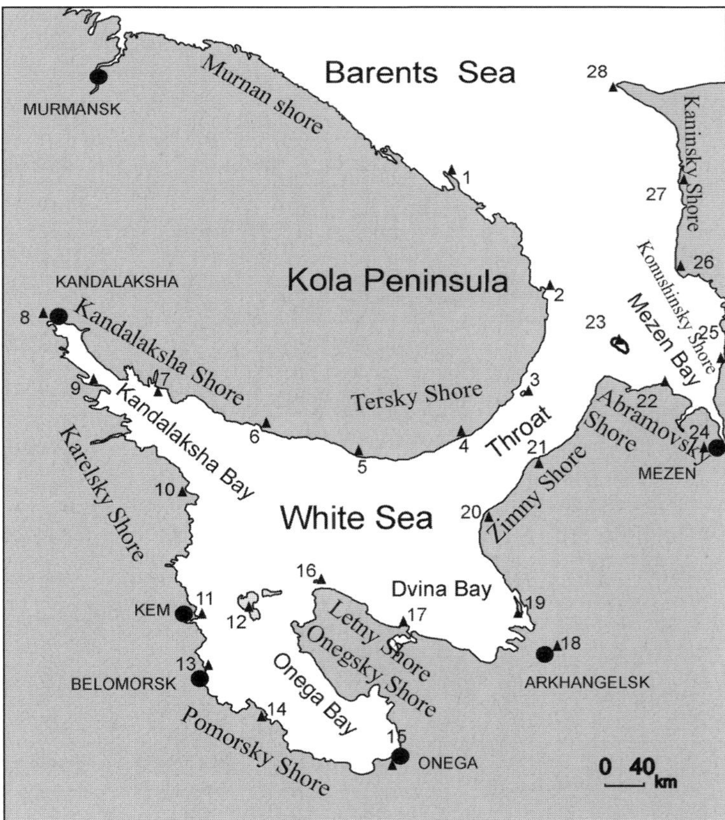

Figure 11.1. Meteorological stations whose data were used in the chapter (the list of stations and corresponding numbers are presented in Table 11.1). Filled triangles are meteorological stations; filled circles are cities and towns.

Table 11.1. List of meteorological stations whose data were used.

No. on Figure 11.1	*Meteorological station*	*Observation period*
1	Cape Svyatoy Nos	1896–2004
2	Tersko-Orlovsky Lighthouse	1896–1996
3	Sosnovets Island	1897–2004
4	Pyalitsa	1916–2004
5	Chavan'ga	1946 ... 1960–2004
6	Kashkarantsy	1946 ... 1961–2004
7	Umba	1933–2004
8	Kandalaksha	1913–2004
9	Kovda	1913–2004
10	Gridino	1918–2004
11	Kem	1966–2004
12	Solovki Archipelago	1888–2004
13	Raz-Navolok	1919–2004
14	Kolezhma	1938–2004
15	Onega	1987–2004
16	Zhizhgin Island	1896–2004
17	Unsky Lighthouse	1930–2004
18	Arkhangelsk	1813–2004
19	Mud'yug Island	1915–2004
20	Zimnegorsky Lighthouse	1896–2004
21	Intsy	1930–1997
22	Abramovsky Lighthouse	1930–2004
23	Morzhovets Island	1896–2004
24	Mezen	1884–2004
25	Nes'	1951–2004
26	Cape Konushin	1940–1996
27	Shoyna	1933–2004
28	Cape Kanin Nos	1916–2004

were those with records for no fewer than 50 years and which are still working at the present time (except for the Tersko-Orlovsky Lighthouse, Intsy, and Cape Konushin stations which stopped working in the late 1990s) (see Figure 11.1 and Table 11.1). From the 28 meteorological stations examined 11 were established in the 19th century, and in Arkhangelsk the observations were recorded practically continuously since October 1813. The station with the second longest records is Kem, which has continuous records since 1866. Seven stations were established near the beginning of the 20th century (1913–1918), another seven in the 1930s and the station in Nes' Village was established in 1951.

Practically all the series of data are lacking some observation periods. Especially, there were many gaps in the beginning of the observation period and, unfortunately, in the mid-1990s. The gaps in the records of average monthly air temperature have been filled by interpolation (Naumova, 1983). For example, the data for Kem Port were reconstructed for the period 1866–1916 from synchronous observations in Kem Town (1866–1944) (see Figure 11.2).

In order to study the trend in air temperature over a given period, the time series of air temperature was approximated by a linear regression equation:

$$t_a = AY + B, \tag{11.1}$$

where t_a is the air temperature (monthly, or annual average) in a particular year Y ($Y = 1936, 1937, \ldots$); and A and B are the coefficients calculated by the technique of least squares. Coefficient A determines a regression slope and describes mean temperature variation from year to year. When it is positive this means that t_a increases over the period considered. Negative values of A indicate that there is a decrease of annual (monthly) average temperatures over the period considered. The greater the

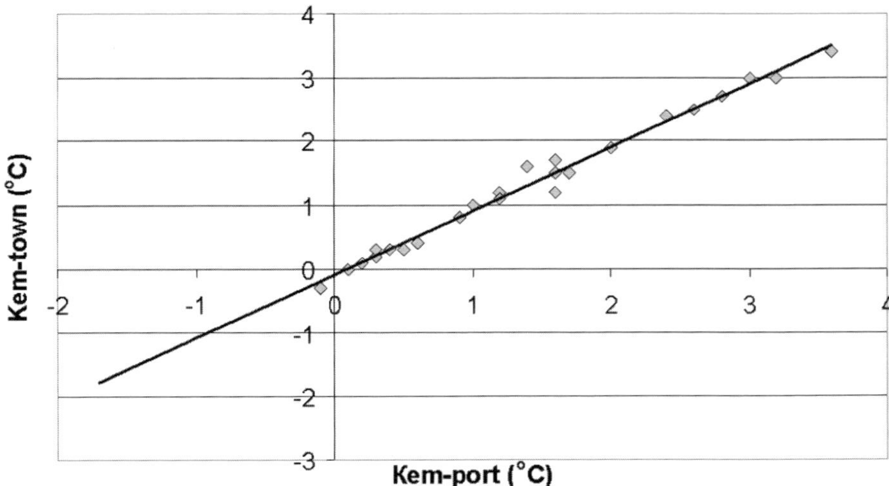

Figure 11.2. Correlation between yearly air temperature in Kem Town (Karelia) and Kem Port (1917–1944).

regression slope, the larger the temperature variations. Coefficient B is a certain initial value of air temperature, relative to which temporal variation occurs. If there is no tendency for an increase or decrease in temperature (i.e., the coefficient A tends to zero), then coefficient B is just equal to the average long-term value of t_a.

The statistical confidence of the trends was calculated with the Student test, using the value of the confidence level of squared correlation coefficient R^2 between t_a and Y. The calculated tendency of temperature change was considered to be statistically significant, when its confidence level R^2 (squared correlation coefficient) was equal to, or exceeded, 95% ($p > 0.95$) (Polyak, 1975).

11.3 THE REGIME OF AIR TEMPERATURE

The first aim of our examination was to analyze the new values of the air temperature regime at White Sea shores and islands.

The average values of yearly air temperature on the coasts of the White Sea Funnel (Voronka) and Throat (Gorlo) are negative and close to −1°C. Cape Svyatoy Nos is the exception to the rule: here the temperature is close to 0°C because of the influence of the Gulf Stream. On the Kandalaksha Bay coast the coldest place is in the environs of Kandalaksha Town, and to the south the average temperature gradually increases, reaching a maximum (+1.5°C) in the environs of Onega Town. On the coast of Mezen Bay and especially on the Konushinsky shore the average annual temperature is lowest (−1.5°C).

Analysis of the monthly average air temperature throughout the year is presented in Tables 11.2 and 11.3 and Figure 11.3.

Winter is prolonged (lasting almost half the year), with negative average temperatures observed at all stations from November until April. Only in Onega is the average monthly air temperature in April positive (+0.3°C).

During the coldest month of the year the air temperature varies from −14.6°C in Mezen to −8.5°C at Cape Svyatoy Nos. The influence of the Gulf Stream is evident. For more than half of the stations the minimum value of t_a occurs in February. This is common to Tersky, Karelsky, Zimny, and Kaninsky shores and is observed at most of the island stations. It is connected mainly with the process of continuous formation of sea ice cover, which usually finishes by the end of January (Glukhovsky *et al.*, 1989). The minimum of t_a in January is observed mainly at those stations that are situated farther from the coast (e.g., Arkhangelsk, Mezen, and Onega).

It is possible to consider April, May, and the first half of June as spring months. For example, the average air temperature in May at Cape Kaninsky Nos is negative (down to −0.8°C). The transition of the average air temperature through 0°C is observed in Onega approximately in the middle of April; at Kandalaksha, Karelian (Karelsky), Pomorsky, and the Onega coasts in the last ten days of April; and at the coast of the Throat of the White Sea in the middle of May (see Tables 11.2 and 11.3 and Figure 11.3).

In the summer the average temperature rises above 10°C at approximately half of the stations (mainly on the southern coasts of the White Sea), and in the northern

Table 11.2. Average and extreme air temperature of White Sea shores (°C).

Station		Jan	Feb	Mar	Apr	May	Jun	Jul	Aug	Sep	Oct	Nov	Dec	Year
Tersky Shore														
Cape Svyatoy Nos	average	−7.6	−8.5	−6.9	−3.0	0.9	5.4	8.9	9.0	6.4	1.7	−2.2	−5.2	−0.1
	max	−2.8	−1.8	−2	0.8	7.6	11.1	16.2	12.9	9.8	6	1.6	−0.7	1.9
	year of max	1930	1990	1967	1990	1897	1989	1960	1940	1938	2000	1967	1953	1938
	min	−13.8	−18.9	−15.4	−8.1	−3	0.8	4.9	5.2	3.6	−3.3	−5.8	−12.1	−3.3
	year of min	1907	1966	1966	1909	1899	1902	1918	1918	1902	1902	1902	1901	1966
Tersko–Orlovsky Lighthouse	average	−9.7	−10.5	−8.2	−3.5	1.0	5.7	9.0	9.3	6.4	1.0	−3.5	−6.8	−0.8
	max	−4.2	−2.9	−1.9	1	8.6	12.3	15.2	13.4	9.8	5.5	0.9	−1.4	2.2
	year of max	1930	1990	1989	1921	1897	1989	1960	1972	1974	1987	1967	1974	1989
	min	−18	−20	−15.5	−9.1	−3.4	0.5	4.5	4.3	3.8	−4.9	−11.8	−15.6	−4.1
	year of min	1968	1966	1902	1929	1918	1969	1918	1918	1968	1992	1992	1915	1902
Pyalitsa	average	−10.2	−11.0	−8.4	−3.6	1.5	6.9	10.2	9.8	6.5	1.4	−3.1	−6.6	−0.5
	max	−4.2	−3.1	−2.5	0.9	5.1	11	14.4	13.3	9.1	6.1	1.6	−1.1	1.7
	year of max	1930	1990	1989	1921	1989	1989	1960	1967	1992	2000	1967	1974	1938
	min	−20.1	−21.8	−15.6	−8.8	−2.5	3	7.3	5.3	3.7	−4	−7.9	−14.5	−3.4
	year of min	1985	1966	1941	1929	1999	1969	1926	1918	1993	1992	1988	1978	1966
Chavanga	average	−10.5	−11.2	−7.6	−3.0	2.4	8.3	12.0	10.7	7.1	1.8	−3.1	−6.5	0.1
	max	−4	−2.7	−2.2	−0.2	5.2	11.8	16.2	14.7	9.6	6.4	1.8	−0.5	2.0
	year of max	2001	1990	1989	1962	1989	1989	1960	1967	1992	1961	1967	1974	1989
	min	−21.1	−21.7	−15.4	−7.2	−0.9	5.2	9	8.1	3.9	−3.9	−7.7	−14.1	−2.8
	year of min	1985	1966	1966	1979	1999	1969	1949	1978	1993	1992	1988	1978	1966
Kashka–rantsy	average	−10.2	−11.0	−7.3	−2.6	2.6	8.8	12.8	11.4	7.4	2.0	−3.0	−6.6	0.3
	max	−3.6	−2.7	−2	0.3	5.5	12	16.1	15.4	9.6	6.6	1.6	−0.6	2.3
	year of max	2001	1990	1975	1967	1984	1989	1972	1967	1992	1961	1967	1974	1989
	min	−20.7	−21.3	−15.3	−6.5	−2.6	5	9.8	9.3	4.1	−3.8	−7.2	−14.1	−2.6
	year of min	1985	1966	1966	1979	1986	1966	1949	1992	1993	1992	1988	1978	1966

(continued)

Table 11.2 (*cont.*)

Station		*Jan*	*Feb*	*Mar*	*Apr*	*May*	*Jun*	*Jul*	*Aug*	*Sep*	*Oct*	*Nov*	*Dec*	*Year*
Kandalaksha Shore														
Umba	average	−11.2	−11.5	−7.4	−1.6	4.1	10.6	14.3	12.4	7.5	1.5	−4.0	−8.1	0.5
	max	−4.4	−1.9	−1.6	1.6	8.8	14.4	19.0	15.9	10.2	6.8	0.7	−1.5	3.2
	year of max	2001	1990	1967	1950	1963	1953	1960	1967	1974	1961	1967	1972, 1974	1938
	min	−23.3	−21.8	−15.8	−6.5	0.6	6.7	10.7	9.6	4.1	−4.3	−9.8	−17.2	−2.2
	year of min	1985	1966	1966	1958	1999	1982	1968	1987	1993	1992	2002	1955	1966
Kanda–laksha	average	−12.1	−12.0	−7.7	−1.5	4.3	10.9	14.7	12.5	7.1	1.0	−4.8	−9.1	0.3
	max	−4.9	−2.2	−1.2	3.1	9.2	14.7	18.3	15.5	10.2	6.2	0.2	−0.8	3.3
	year of max	1925	1990	1989	1921	1963	1953	1938	1937	1920	1961	1967	1970	1938
	min	−23.2	−21.4	−15.9	−5.7	1.2	4.6	10.9	9.5	3.3	−5.7	−10.9	−18.6	−2.6
	year of min	1985	1966	1966	1956	1916	1924	1968	1987	1993	1992	2002	1915	1966
Kovda	average	−11.4	−11.6	−7.4	−1.5	4.0	10.5	14.5	12.6	7.6	1.5	−3.9	−8.3	0.6
	max	−4.2	−1.7	−1.3	2.6	9.1	14.5	18.5	16.5	10.5	6.5	−0.3	−1.6	3.5
	year of max	1930	1990	1989	1921	1963	1953	1938	1951	1934	1961	1936	1972	1938
	min	−22.6	−21.0	−15.7	−6.6	−0.4	5.7	11.2	9.1	4.2	−3.5	−9.1	−17.5	2.4
	year of min	1985	1966	1966	1929	1916	1924	1968	1918	1993	1992	1992	1956	1966
Karelian (Karelsky) Shore														
Gridino	average	−10.2	−10.5	−6.6	−1.3	3.9	10.1	14.0	12.8	8.1	2.3	−2.9	−6.8	1.1
	max	−3.4	−1.2	−0.2	3.2	8.9	14.2	18	16.1	10.8	6.9	1	−0.8	3.5
	year of max	1930	1990	1967	1921	1963	1937	1938	1951	1974	1961	1967	1972	1938
	min	−19.8	−20	−15.4	−6.5	−0.5	6.3	10.4	8.6	5.1	−4.9	−6.5	−14.4	−2
	year of min	1968	1966	1966	1979	1933	1925	1950	1918	1986	1968	1919	1941	1966

Kem	average	−10.7	−10.8	−6.8	−1.1	4.1	10.2	13.8	12.8	8.2	2.1	−3.6	−8.1	1.0
	max	−3	−0.9	−0.2	4	11.6	15.0	18.0	16.5	11.8	7.1	1.7	−0.7	3.5
	year of max	1930	1990	1967	1921	1897	1881	1938	1868	1866	1961	1877	1972	1989
	min	−20.9	−21.0	−14.0	−6.5	−2.0	4.8	9.8	8.5	4.6	−5.5	−10.5	−18.6	−1.9
	year of min	1985	1893	1966	1909	1867	1899	1879	1918	1894	1992	1867	1876	1902
Pomorsky Shore														
Raz–Navolok	average	−10.9	−10.6	−6.5	−0.9	4.9	11.0	14.4	13.2	8.6	2.3	−3.2	−7.6	1.2
	max	−3	−0.8	−0.4	3.7	9.4	15.5	18.5	16.2	11.9	7.2	1.5	−0.7	3.7
	year of max	1930	1990	1967	1921	1920, 1921	1989	1938	1967	1934	1961	1996	1972	1989
	min	−21.7	−20.6	−13.3	−6.6	0.4	7.1	10.7	9.5	4.9	−3.6	−8	−18.6	−1.8
	year of min	1969	1985	1966	1929	1933	1976	1950	1923	1973	1992	1980	1978	1941
Kolezhma	average	−11.2	−10.5	−6.4	−0.3	5.6	11.9	14.9	13.2	8.3	2.5	−3.3	−8.2	1.4
	max	−4.5	−0.3	0	3.6	10.6	16.2	19.3	16.3	11.8	6.9	1.8	−0.7	3.9
	year of max	1949	1990	1989	1950	1963	1989	1938	1967	1937	1961	1996	1972	1989
	min	−21.9	−20.9	−14.6	−4.7	1	7.2	11.2	10.6	4.7	−2.4	−8.4	−19.3	−1.8
	year of min	1985	1966	1963	1956	1969	1941	1956	1987	1973	1992	1987	1955	1941
Onega	average	−11.9	−11.4	−6.7	0.3	6.6	13.0	16.1	13.8	8.4	2.1	−3.9	−9.0	1.5
	max	−3.4	−0.9	0	5.7	14.5	18.1	21.4	17.9	12.1	6.8	1.6	−1.3	4.1
	year of max	1930	1990	1989	2001	1920	1989	1960	1967	1992	1961	1996	1936	1989
	min	−22.9	−21.7	−14.7	−6.9	1.2	8.6	11.8	9.6	4.6	−3.6	−10.2	−20.2	−1.6
	year of min	1985	1966	1963	1929	1918	1941	1968	1923	1993	1902	1956	1955	1902
Letny Shore														
Unsky Lighthouse	average	−10.4	−10.3	−6.5	−1.0	4.5	10.4	13.8	12.9	8.5	3.0	−2.4	−6.8	1.3
	max	−2.7	−1.2	−0.3	2.8	10.9	15.1	18.2	16.7	12	6.8	2	−0.9	3.775
	year of max	1930	1990	1989	1950	1972	1989	1938	1972	1974	1961	1967	1936	1989
	min	−20.6	−20.7	−15.1	−5.6	−0.1	5.4	9.4	9.7	5.5	−1.3	−7.3	−17	−1.92
	year of min	1968	1966	1963	1956	1941	1941	1950	1987	1993	1992	1987	1955	1941

(continued)

Table 11.2 (*cont.*)

Station		*Jan*	*Feb*	*Mar*	*Apr*	*May*	*Jun*	*Jul*	*Aug*	*Sep*	*Oct*	*Nov*	*Dec*	*Year*
Zimny Shore														
Arkhangelsk	average	−13.2	−12.3	−7.3	−0.7	5.6	12.5	15.9	13.7	8.2	1.5	−5.1	−10.4	0.7
	max	−3.9	−1.6	0.6	5.2	14	18.7	21.3	17.9	14.3	6.3	1.3	−2	3.3
	year of max	1930	1990	1822	1921	1897	1823	1938	1847	1847	1961	1877	1936	1989
	min	−25.3	−24.8	−15.9	−7.8	−1.2	8.1	10.4	8.9	4.2	−5	−13.8	−23.7	−2.5
	year of min	1985	1871	1963	1929	1867	1982	1837	1918	1993	1902	1864	1864	1902
Zimnegorsky Lighthouse	average	−10.4	−10.7	−7.0	−1.4	3.9	9.9	13.3	11.9	7.4	1.7	−3.3	−7.4	0.7
	max	−3.3	−2.2	−1.2	3.6	13.2	15.4	19.5	16.8	11.4	6.1	1.5	−1.4	3.1
	year of max	1930	1990	1967	1921	1897	1989	1960	1967	1974	1961	1967	1936	1989
	min	−19.4	−21.1	−13.9	−7.9	−1.9	5.1	8.2	6.5	4.3	−4.6	−8	−17.2	−2.4
	year of min	1985	1966	1963	1929	1918	1899 1902	1926	1918	1986	1902	1956	1955	1902
Intsy	average	−10.9	−11.0	−7.9	−2.8	2.3	7.4	10.6	10.7	7.2	2.1	−3.1	−7.6	−0.3
	max	−3.2	−2.5	−1.4	1.9	6.9	12	15.2	14.6	10.7	6.3	1.5	−1.2	2.2
	year of max	1930	1990	1967	1995	1989	1989	1954	1932	1974	1961	1967	1936	1989
	min	−21.1	−21.6	−16	−7.4	−1.8	2.4	7.1	7.9	4.4	−3.1	−9.2	−18.2	−3.1
	year of min	1985	1966	1963	1956	1969	1969	1956	1969	1986	1992	1956	1955	1941
Abramovsky Shore														
Abramovsky Lighthouse	average	−12.2	−12.0	−8.5	−3.3	1.8	7.5	11.6	11.2	7.1	1.3	−4.5	−9.0	−0.7
	max	−4.7	−3.7	−2.3	1.2	6.8	13	16.6	15.1	10	5.4	0.2	−2.8	1.5
	year of max	1930	1990	1989	1950, 1951	1989	1989	1954	1967	1974	1961	1967	1936	1989
	min	−22.4	−21.8	−16.2	−7.8	−2.9	1.2	7.6	7.9	4.5	−3.9	−9.5	−17.7	−3.8
	year of min	1968	1966	1941	1956	1999	1969	1968	1969	1986	1992	1987	1955	1941

Konushinsky Shore														
Mezen	average	−14.3	−13.4	−8.9	−2.5	3.5	10.3	13.9	11.9	6.7	0.1	−6.4	−11.4	−0.9
	max	−5.5	−2.9	−1.3	3.3	12.6	16.5	19.5	16.5	10.9	5.4	−0.1	−3.5	1.9
	year of max	1930	1990	1967	1921	1897	1989	1974	1967	1992	1961	1996	1936	1920
	min	−26.9	−25.5	−18.7	−9.1	−2.3	4.3	8.9	6.2	3.5	−7.6	−14.2	−23	−4.3
	year of min	1985	1966	1963	1929	1918	1969	1926	1918	1953	1902	1885	1978	1902
Nes'	average	−14.3	−13.7	−9.3	−4.3	2.0	8.8	13.1	11.0	6.3	0.1	−6.9	−10.9	−1.5
	max	−7.2	−3.6	−2	1.2	6.7	14.5	18.2	15.1	10	5.1	−0.7	−3.6	0.6
	year of max	1996	1990	1967	1951	1989	1989	1974	1967	1992	1961	1996	1974	1989
	min	−25.8	−25.6	−19.6	−10	−3	2.9	8.4	7.9	3.3	−6.4	−12.5	−22.7	−4.3
	year of min	1985	1966	1963	1992	1999	1969	1968	1969	1952	1992	1956	1978	1966
Cape Konushin	average	−11.0	−11.2	−8.7	−4.6	0.6	6.7	11.0	10.3	6.7	1.2	−3.9	−7.9	−0.9
	max	−5.4	−3.4	−2.2	−0.2	51	1.7	16	13.9	9.7	5.6	1.1	−2.2	1.3
	year of max	1944	1990	1967	1951	1989	1989	1974	1967	1974	1961	1967	1953	1943
	min	−20.7	−22.5	−16.9	−9.7	−3.4	2	7.5	7.3	3.8	−3	−9	−16.3	−3.7
	year of min	1985	1966	1963	1979	1969	1969	1950	1969	1993	1992	1956	1978	1966
Kaninsky Shore														
Shoyna	average	−11.4	−12.3	−9.9	−5.4	−0.2	6.1	10.7	9.9	6.4	1.3	−3.8	−8.0	−1.4
	max	−3.6	−4.7	−3.1	−0.8	3.7	11.3	15.5	13.1	9.4	5.1	1	−1.6	0.9
	year of max	1937	1990	1967	1951	1989	1989	2004	1967	1938	1944	1936	1953	1943
	min	−21.1	−23.8	−18.3	−11.4	−4.2	0.7	6.8	6.5	3.5	−3.4	−9.5	−17.2	−4.3
	year of min	1985	1966	1966	1979	1969	1968	1947	1969	1939	1992	1998	1978	1966
Cape Kanin Nos	average	−8.5	−9.7	−8.3	−4.8	−0.8	4.4	8.6	8.5	5.9	1.7	−2.0	−5.6	−0.9
	max	−1.6	−3.2	−1.8	−0.9	3.2	10	14.2	11.6	8.7	5.1	1.7	−0.4	1.3
	year of max	1995	1995	1995	1921	1989	1993	1960	1931	1938	1944, 1961	1967	1953	1954
	min	−15.9	−20.5	−16.8	−10.3	−4.6	−0.3	4	4.9	3.4	−1.7	−7	−14.9	−3.8
	year of min	1986	1966	1966	1929	1999	1969	1918	1969	1939	1992	1998	1915	1966

Table 11.3. Average and extreme air temperature of White Sea islands (°C).

Station		*Jan*	*Feb*	*Mar*	*Apr*	*May*	*Jun*	*Jul*	*Aug*	*Sep*	*Oct*	*Nov*	*Dec*	*Year*
Sosnovets Island	average	−9.5	−10.5	−8.2	−3.6	0.6	4.9	8.3	8.8	6.6	1.7	−2.8	−6.3	−0.8
	max	−3.6	−3.2	−2.4	0.6	5	9.1	11.5	11.7	9.2	6.2	1.9	−1.2	1.5
	year of max	1995	1995	1967	1921	1897	1989	1954	1967	1938	1961	1967	1974	1989
	min	−18	−20.7	−15	−9.6	−3	1.1	5.6	4.7	4.2	−3.7	−6.4	−15.8	−4.2
	year of min	1985	1966	1941	1929	1918	1899	1899	1918	1902	1902	1988	1915	1902
Solovetsky Arkhipelago	average	−9.4	−9.9	−6.6	−1.3	3.7	9.5	13.1	12.1	8.1	2.9	−2.0	−6.1	1.2
	max	−2.3	−1.3	−1	3	10.3	14.6	19.1	15	11.5	7.3	2.3	−0.7	3.7
	year of max	1930	1990	1967	1921	1897	1989	1938	1967	1938	1961	1967	1972	1989
	min	−18	−19.6	−14	−6.3	−0.7	4.5	9.7	7.8	4.9	−2	−5.2	−14.7	−2.2
	year of min	1985	1966	1899	1909	1899		1904	1918	1894	1902	1902	1915	1902
Zhizhgin Island	average	−9.3	−10.2	−7.0	−1.9	2.8	8.1	11.9	11.5	7.9	2.7	−1.8	−5.6	0.8
	max	−2.3	−1.4	−0.7	2.4	9.8	12.8	17.9	14.6	11.1	6.8	2.4	−0.7	3.3
	year of max	1930	1990	1967	1921	1897	1989	1993	1932	1974	1961	1967	1936	1989
	min	−18.1	−19.6	−14.4	−7.5	−1.1	3.8	8.5	7.5	3.1	−1.8	−7.3	−14.9	−2.4
	year of min	1862	1966	1899	1845	1918	1899	1950	1918	1996	1912	1864	1915	1902
Mud'yug Island	average	−12.3	−12.1	−8.3	−2.0	4.6	12.1	15.5	13.5	8.4	2.1	−3.7	−8.7	0.8
	max	−3.7	−2.2	−1.7	2.5	10.3	17.1	20.4	17.8	11.7	6.4	0.9	−1.5	3.1
	year of max	1930	1990	1989	1921	1920	1989	1938	1967	1938	1961	1996	1936	1989
	min	−23.8	−23.9	−18.2	−9.4	−1.8	7.4	10.8	8.4	5.1	−2.8	−9.7	−21.1	−2.4
	year of min	1985	1966	1963	1929	1918	1969	1926	1918	1993	1992	1956	1955	1941
Morzhovets Island	average	−9.7	−10.0	−7.7	−3.5	0.6	5.9	9.9	9.8	7.0	2.0	−2.7	−6.7	−0.4
	max	−3.4	−1.2	−0.3	0.3	6.5	11.8	14.6	13.9	9.8	6.1	1.6	−1.4	2.0
	year of max	1995	1899	1905	1921	1897	2000	1974	1967	1967	1938	1961	1953	1989
	min	−17.9	−19.7	−14.4	−8.8	−3.7	0.9	5.4	4.9	4.6	−3.5	−7.2	−15.8	−3.7
	year of min	1985	1966	1941	1929	1918	1969	1918	1918	1902	1902	1902	1915	1902

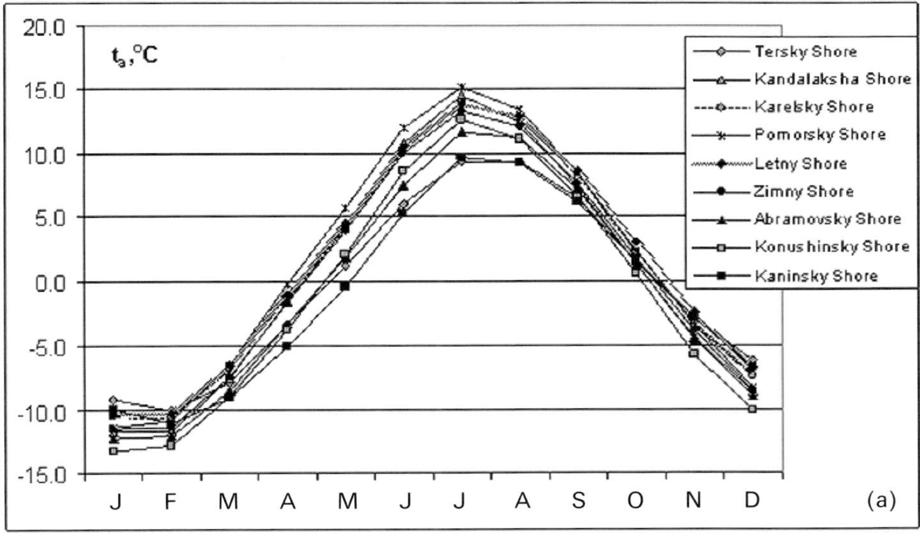

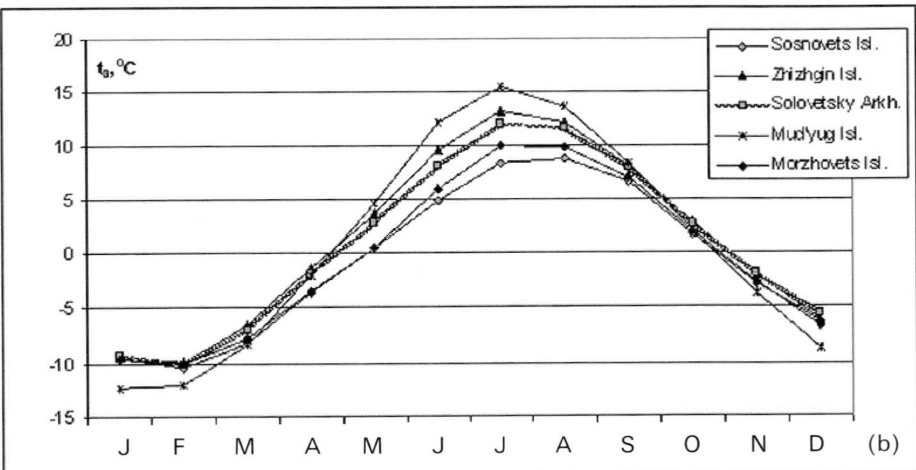

Figure 11.3. The annual course of average air temperature (°C) on the White Sea shore (a) and islands (b).

part of Tersky and Kaninsky shores the average air temperatures of the warmest month do not rise above 9°C. At four stations (Cape Svyatoy Nos, Tersko-Orlovsky Lighthouse, Sosnovets Island, and Intsy) the maximum of t_a occurs in August.

Autumn is warmer than spring; this is connected with the warming influence of the White Sea. So, the difference of temperatures between September and May is positive everywhere (i.e., September is warmer than May by 2°C–7°C), and towards the north the difference between monthly temperatures increases. On the northern part of the Tersky and Kaninsky shores, September is warmer than June and November is warmer than April (see Tables 11.2 and 11.3 and Figure 11.3).

The difference between t_a on the shores and islands of the White Sea is maximal in June (9°C), and minimal in September–October (3°C) (see Figure 11.3) (i.e., the temperature differences in the White Sea region during the periods of reorganization of atmospheric circulation are smaller than in the winter and summer).

The annual amplitudes of monthly average values, t_a, at the White Sea coasts and islands vary from 17.5°C–20°C in the northern part of the region up to 25°C–29°C in the southern part of the region. The increase in climate continentality is typical for both directions from the north to the south, and from the west to the east.

Analysis of the variability of t_a has shown that the average standard deviation (σ) of mid-year air temperature at the White Sea coast and islands varies by 1.1°C–1.3°C (Table 11.4). For monthly average temperatures, a precisely expressed annual course of σ with a minimum in September ($\sigma = 1.2$°C–1.6°C) and a maximum in February ($\sigma = 3.4$°C–4.3°C) is observed. The coefficients of variation V of the monthly average values of t_a in winter are 30%–40% and in summer 10%–20%. A significant increase in V is observed during the months when the temperature passes through zero (Figure 11.4). So, in the spring this occurs in April on the Kandalaksha, Karelian (Karelsky), Pomorsky, Letny, and Zimny coasts and in May on the Tersky, Abramovsky, Konushinsky, and Kaninsky coasts. In autumn the temperature passes through zero in most of the territory in October.

Histograms of the frequency of average monthly t_a for each station allow presentation of the typical mesoclimatic features of each place. So, by comparing the three stations located approximately along latitude 66°N (Gridino, Intsy, Mezen), it is evident that the amplitude of the limits in variation of t_a increase in direction from west to east (Figure 11.5). For example, in Mezen the amplitude of monthly temperature (A_t) in January is the greatest and reaches nearly 21°C. In the Throat of the White Sea (Intsy) it is equal to 17°C, and in Gridino to 16°C (Figure 11.5a). The shift in monthly t_a frequency is largest in the area of lower temperatures in Mezen (interval of maximum frequency is $\delta t_{\max} = -16$°C to -14°C with relative probability $P = 21\%$), which also testifies to the more severe climate of the Mezen Gulf in comparison with the area of the White Sea Throat region (Intsy: $\delta t_{\max} = -12$°C to -11°C, $P = 15\%$) and with western areas of the White Sea (Gridino: $\delta t_{\max} = -11$°C to -10°C, $P = 14\%$). The same is evident from analysis of the curves of repeatability of mid-year temperatures (Figure 11.5c).

By contrast, in Mezen and Gridino the maximum repeatability of average July temperatures is in the range of 12°C–14°C ($p = 36\%$ and 43%, respectively), and in Intsy it is shifted to a lower temperature interval of 10°C–11°C ($p = 22\%$) (Figure 11.5b).

Temperature variability in the direction from north to south becomes evident from comparing the t_a frequency curves for the three stations Cape Svyatoy Nos, Zimnegorsky Lighthouse, and Mud'ug Island, which are located close to longitude 40°E (Figures 11.5d–f). In January on Cape Svyatoy Nos the maximal repeatability of t_a is shifted to the interval of higher values of t_a than the other two stations (Figure

Table 11.4. The average standard square deviation of air temperature (°C).

Station	*Jan*	*Feb*	*Mar*	*Apr*	*May*	*Jun*	*Jul*	*Aug*	*Sep*	*Oct*	*Nov*	*Dec*	*Year*
The White Sea Shores													
Tersky	3.2	3.5	3.0	2.0	1.7	1.7	1.7	1.4	1.3	1.9	2.1	2.8	1.1
Kandalakshsky	3.6	3.9	3.2	1.9	1.7	1.9	1.7	1.4	1.4	2.0	2.7	3.8	1.2
Karelsky	3.4	3.7	3.0	2.0	2.1	2.0	1.7	1.4	1.4	1.9	2.2	3.2	1.1
Pomorsky	4.2	4.1	3.2	2.1	2.2	2.1	1.9	1.5	1.6	2.0	2.6	3.8	1.2
Letny	3.6	3.7	3.1	2.0	2.1	2.0	1.9	1.4	1.3	1.7	2.0	3.0	1.2
Zimny	3.9	3.9	3.1	2.3	2.3	2.1	2.2	1.7	1.5	2.0	2.6	3.8	1.2
Abramovsky	3.5	3.6	3.1	2.2	1.9	2.0	2.1	1.6	1.3	1.9	2.3	3.2	1.3
Konushinsky	3.9	4.3	3.7	2.8	2.1	2.0	2.3	1.6	1.5	2.3	3.0	4.0	1.3
Kaninsky	3.5	3.8	3.4	2.4	1.6	1.9	2.2	1.4	1.2	1.7	2.3	3.0	1.3
The White Sea Islands													
Sosnovets Island	3.1	3.4	3	2	1.5	1.4	1.4	1.1	1.2	1.9	2	2.8	1.2
Solovetsky Arkhipelago	3.2	3.5	2.9	1.8	1.9	1.9	1.7	1.4	1.3	1.7	1.8	2.9	1.2
Zhizhgin Island	3.3	3.5	3	2	1.9	1.8	1.7	1.4	1.2	1.6	1.8	2.6	1.1
Mud'yug Island	4.1	4.1	3.3	2.1	2.3	2.1	2.3	1.7	1.5	1.9	2.5	4.0	1.2
Morzhovets Island	3.2	3.4	2.9	2	1.8	1.9	2	1.5	1.2	1.8	1.9	2.9	1.2

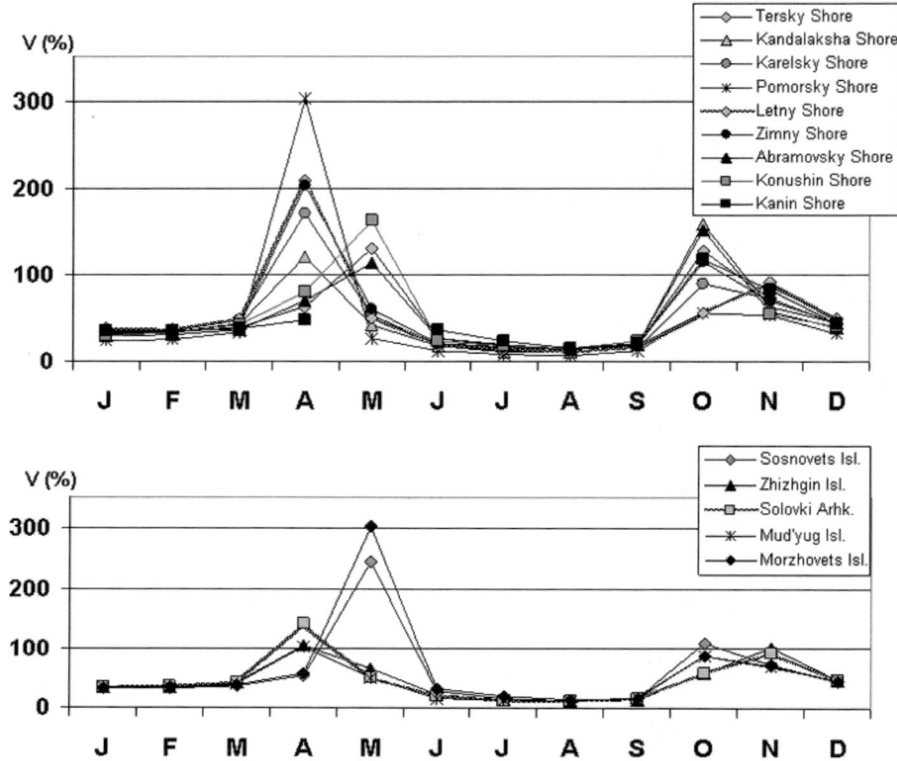

Figure 11.4. Variation coefficient (V, %) of monthly air temperatures on the White Sea shore (a) and islands (b).

11.5d). The opposite situation is observed in July (Figure 11.5e). Analysis of average yearly air temperature variability has shown the shift of the t_a repeatability maximum to higher temperatures in the direction from north to south (Figure 11.5f).

11.4 LONG-TERM CHANGES OF AIR TEMPERATURE

The question of long-term tendencies of air temperature values all over the world, especially during the last 10 years, is of very great interest to climatologists and to the general public. The Arctic region attracts special attention because, according to the modeling calculations, the most catastrophic warming is expected in the polar regions (ACIA, 2005; Katsov, 2006). For an examination of climate change peculiarities in the White Sea region, the linear trends of average temperatures for a year, a season, and a month, as well as their statistical significance were calculated for 26 stations (Tables 11.5 and 11.6). Observations in Kashkarantsy and Chavan'ga were not taken into account because of their short measurement period (only since the 1960s).

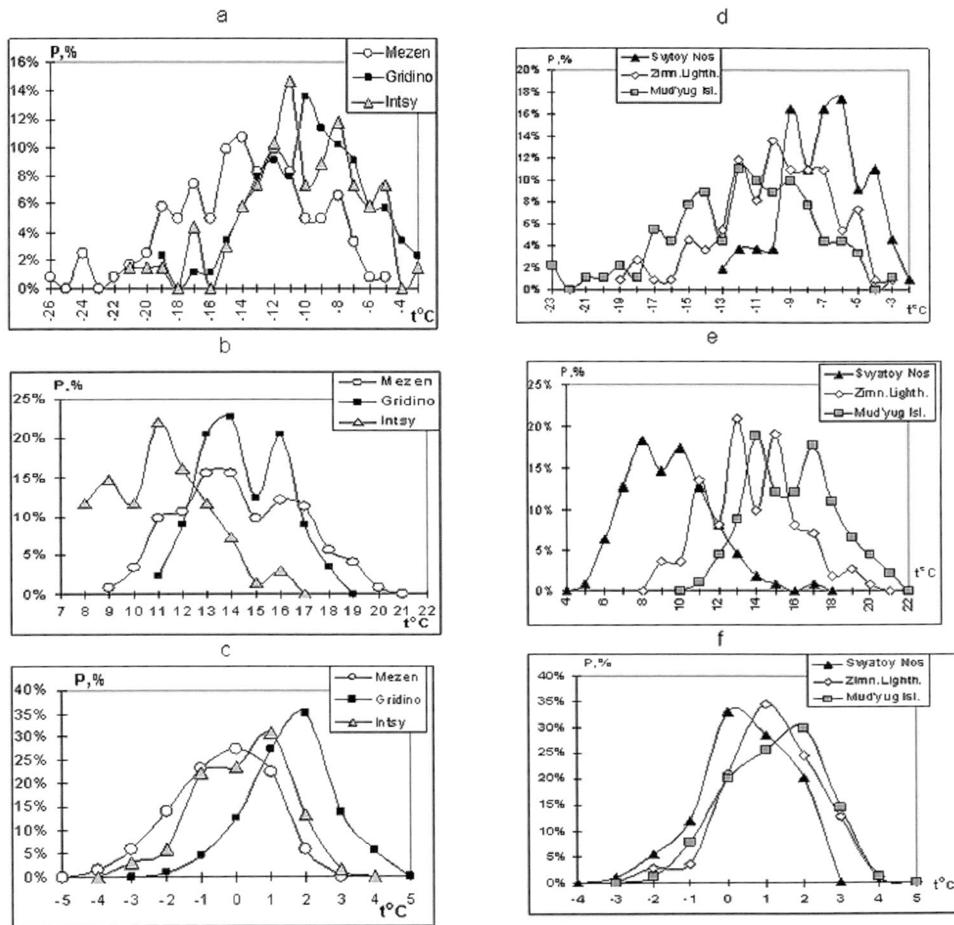

Figure 11.5. The frequency curves of air temperature in January (a, d), July (b, e), and year (c, f): latitude profile (a–c) and longitude profile (d–f).

The main difficulty encountered was the different lengths of the time series of t_a records for the various stations. This problem complicates carrying out the analysis. Comparison of the long-term records of average yearly air temperatures in Arkhangelsk, Kem, Mezen, Solovki, and Onega has shown good agreement among them. Practically all the trends of t_a variations from year to year and its anomaly (A_t) coincide very well (Figure 11.6b). This allows one to conclude that the temporal variability of the temperature regime in the area of the White Sea is determined mainly by the global processes of general atmosphere circulation. In turn this means that the conclusions about climate change obtained as a result of analyzing the long-term record of temperature in Arkhangelsk can be considered characteristic for all of the White Sea region as a whole.

Table 11.5. Estimation of linear trend of air temperature in the White Sea region: *year and season*.

Station	*Observation period*	*Linear trend of air temperature* Δt (°C per 100 years)					*Statistically significant level* (*p*)				
		Year	*Winter*	*Spring*	*Summer*	*Autumn*	*Year*	*Winter*	*Spring*	*Summer*	*Autumn*
Tersky Shore											
Svyatoy Nos	1896–2004	+1.5	+2.0	+1.2	+0.9	+1.5	0.999	0.999	0.98	0.97	0.999
Tersko-Orlovsky Lighthouse	1896–1996	+1.3	+1.70	+0.8	+1.2	+0.8	0.999	0.997	0.78	0.985	0.92
Pyalitsa	1916–2004	+0.2	−0.3	+0.7	+0.7	+0.1	0.29	0.32	0.74	0.85	0.22
Kandalaksha Shore											
Umba	1933–2004	−0.2	−0.8	+1.2	−0.1	−0.1	0.18	0.46	0.84	0.15	0.10
Kandalaksha	1913–2004	−0.5	−1.0	+0.3	−0.2	−0.7	0.77	0.79	0.42	0.49	0.77
Kovda	1913–2004	−0.1	−0.5	+1.2	+0.9	−0.5	0.25	0.45	0.96	0.95	0.63
Karelsky Shore											
Gridino	1918–2004	+0.1	−0.4	+0.6	+0.7	−0.1	0.19	0.36	0.62	0.87	0.15
Kem	1866–2004	+0.6	+1.1	+1.2	+0.8	+0.7	0.99	0.99	0.999	0.999	0.98
Pomorsky Shore											
Raz-Navolok	1919–2004	+0.1	−0.5	+0.8	+0.5	+0.2	0.12	0.43	0.73	0.68	0.23
Kolezhma	1938–2004	+0.7	+0.7	+1.6	+0.3	+0.2	0.61	0.42	0.86	0.35	0.18
Onega	1987–2004	+1.0	+1.0	+0.7	+1.1	+0.9	0.996	0.93	0.85	0.998	0.997

Unsky Lighthouse	1930–2004	+0.3	−0.1	+1.2	+0.4	+0.3	0.35	0.11	0.81	0.45	0.35
Zimny Shore											
Arkhangelsk	1814–2004	+0.5	+0.7	+0.8	+0.1	−0.01	0.996	0.99	0.998	0.36	0.04
Zimnegorsky Light.	1896–2004	+0.9	+0.9	+0.5	+1.1	+0.6	0.99	0.93	0.63	0.99	0.86
Intsy	1930–1997	−0.5	−0.5	+0.2	−0.7	−0.7	0.43	0.30	0.16	0.58	0.60
Abramovsky Shore											
Abramovsky Light.	1930–2004	+0.04	−0.2	+0.5	+0.02	+0.3	0.05	0.17	0.36	0.02	0.33
Konushinsky Shore											
Mezen	1884–2004	+0.9	+1.0	+0.7	+0.8	+1.1	0.993	0.92	0.78	0.98	0.993
Nes'	1951–2004	+0.6	+1.0	+0.6	−0.6	+1.4	0.39	0.40	0.26	0.36	0.72
Cape Konushin	1940–1996	+0.8	+1.8	+0.3	+0.6	−0.7	0.60	0.78	0.17	0.39	0.49
Kaninsky Shore											
Shoyna	1933–2004	−0.5	−1.6	−0.2	+0.8	−0.3	0.54	0.82	0.17	0.69	0.37
Cape Kanin Nos	1916–2004	+0.2	−0.1	+0.6	+0.7	0.0	0.35	0.13	0.61	0.75	0.1
White Sea Islands											
Sosnovets Island	1897–2004	+0.8	+1.0	+0.2	+0.9	+0.8	0.98	0.93	0.35	0.995	0.97
Solovetsky Lighthouse	1888–2004	+1.1	+1.0	+0.9	+1.5	+0.9	0.999	0.97	0.965	0.999	0.99
Zhizhgin Island	1896–2004	+0.8	+0.9	+0.5	+0.9	+0.5	0.98	0.93	0.74	0.99	0.82
Mud'yug Island	1915–2004	+0.5	+0.2	+1.4	+0.5	+0.5	0.74	0.23	0.94	0.68	0.60
Morzhovets Island	1896–2004	+0.8	+0.6	+0.4	+1.4	+0.7	0.98	0.77	0.63	0.999	0.96

Table 11.6. Estimation of the linear trend of air temperature in the White Sea region: month.

Station		*Jan*	*Feb*	*Mar*	*Apr*	*May*	*Jun*	*Jul*	*Aug*	*Sep*	*Oct*	*Nov*	*Dec*
Tersky Shore													
Cape Svyatoy Nos	Δt^*	+1.2	+2.5	+3.3	+0.8	+1.5	+0.9	+1.2	+0.5	+1.3	+1.7	+1.1	+1.7
	p	0.91	0.995	0.999	0.81	0.99	0.88	0.94	0.73	0.999	0.999	0.96	0.99
Tersko-Orlovsky Lighthouse	Δt^*	+0.5	+2.5	+3.9	+0.1	+1.4	+1.6	+1.2	+0.9	+0.5	+1.0	+0.2	+1.5
	p	0.38	0.98	0.999	0.13	0.94	0.97	0.89	0.87	0.77	0.88	0.27	0.86
Pyalitsa	Δt^*	−1.7	+0.1	+3.5	+0.1	+1.3	+0.6	+1.4	−0.1	+0.1	+0.2	−2.1	−1.5
	p	0.76	0.06	0.996	0.14	0.95	0.69	0.97	0.13	0.09	0.24	0.97	0.79
Kandalakshsky Shore													
Umba	Δt^*	−2.5	+1.2	+4.9	+1.2	+1.2	+0.6	+0.5	−1.5	−0.2	0.0	−4.1	−3.3
	p	0.77	0.38	0.99	0.72	0.78	0.44	0.38	0.94	0.17	0.0	0.993	0.86
Kandalaksha	Δt^*	−1.9	0.0	+2.6	−0.1	+0.8	+1.1	−0.5	−1.3	−1.1	−0.3	−3.0	−2.9
	p	0.81	0.0	0.97	0.13	0.79	0.85	0.58	0.98	0.94	0.26	0.993	0.93
Kovda	Δt^*	−1.4	+0.5	+3.2	+0.7	+1.8	+2.4	+0.6	−0.2	−0.8	−0.2	−2.4	−2.3
	p	0.69	0.26	0.99	0.64	0.99	0.999	0.66	0.32	0.83	0.21	0.99	0.90
Karelsky Shore													
Gridino	Δt^*	−1.4	+0.8	+2.6	+0.3	+0.9	+1.6	+1.1	−0.5	−0.3	+0.1	−1.7	−2.1
	p	0.66	0.40	0.94	0.29	0.72	0.95	0.87	0.58	0.47	0.13	0.94	0.91
Kem	Δt^*	−0.1	+1.1	+1.5	+0.7	+1.7	+1.4	+0.7	+0.4	+0.7	+0.7	+1.1	+1.8
	p	0.05	0.82	0.98	0.92	0.999	0.999	0.94	0.80	0.98	0.89	0.96	0.98

* In °C/100.

Pomorsky Shore													
Raz-Navolok	Δt^*	−1.7	+0.9	+2.7	+0.8	+0.8	+1.5	+1.1	−1.1	−0.4	+0.7	−2.2	−2.3
	p	0.67	0.42	0.96	0.67	0.61	0.91	0.86	0.94	0.41	0.59	0.95	0.85
Kolezhma	Δt^*	+0.7	+2.0	+6.0	+1.0	+2.1	+1.2	+1.8	−2.0	−0.5	+0.9	−3.5	−1.5
	p	0.23	0.55	0.995	0.55	0.89	0.63	0.87	0.987	0.38	0.56	0.97	0.47
Onega	Δt^*	−0.8	+1.5	+2.9	+0.8	+0.7	+1.5	+1.2	+0.5	+0.6	+1.1	+0.05	+1.3
	p	0.50	0.82	0.999	0.79	0.72	0.99	0.97	0.72	0.85	0.95	0.06	0.78
Letny Shore													
Unsky Lighthouse	Δt^*	−2.3	+0.1	+5.5	+1.0	+1.4	+1.1	+1.6	−1.4	−0.2	−0.4	−3.3	−2.0
	p	0.72	0.03	0.999	0.65	0.75	0.68	0.84	0.94	0.18	0.36	0.997	0.74
Zimny Shore													
Arkhangelsk	Δt^*	+0.4	+0.7	+0.7	+0.7	+1.0	+0.5	+0.2	−0.5	−0.25	+0.2	+0.5	+1.2
	p	0.52	0.80	0.90	0.97	0.997	0.92	0.56	0.96	0.74	0.60	0.80	0.97
Zimnegorsky Lighthouse	Δt^*	−0.5	+1.5	+2.9	−0.4	+1.4	+1.4	+1.6	+0.4	+0.3	+0.9	−0.2	+0.9
	p	0.35	0.86	0.999	0.38	0.94	0.97	0.97	0.49	0.46	0.88	0.22	0.64
Intsy	Δt^*	−4.1	+1.6	6.0	−0.6	+1.0	+0.8	−0.4	−2.4	−0.5	−0.8	−4.6	−1.2
	p	0.92	0.50	0.998	0.33	0.65	0.54	0.23	0.99	0.43	0.53	0.997	0.39
Abramovsky Shore													
Abramovsky Lighthouse	Δt^*	−2.5	+1.6	+5.7	+0.3	+0.6	+0.8	+1.2	−2.0	+0.3	+0.3	−4.5	−1.3
	p	0.80	0.57	0.999	0.17	0.44	0.52	0.69	0.98	0.36	0.20	0.999	0.50

* In °C/100.

(continued)

Table 11.6 (*cont.*)

Station		*Jan*	*Feb*	*Mar*	*Apr*	*May*	*Jun*	*Jul*	*Aug*	*Sep*	*Oct*	*Nov*	*Dec*
Konushinsky Shore													
Mezen	Δt^*	−0.4	+0.6	+2.4	+0.6	+0.8	+1.3	+0.8	+0.4	+0.8	+1.4	+0.6	+1.7
	p	0.28	0.41	0.992	0.57	0.77	0.975	0.82	0.59	0.95	0.97	0.51	0.86
Nes	Δt^*	−2.5	+3.4	+7.8	+0.3	+0.9	+0.3	+0.01	−2.2	+1.3	+1.6	−2.3	−1.6
	p	0.50	0.61	0.98	0.09	0.36	0.14	0.005	0.88	0.65	0.54	0.55	0.33
Cape Konushin	Δt^*	−1.4	+3.6	+9.8	−0.8	+1.4	+2.3	+1.4	−2.0	−0.9	−0.6	−2.8	+0.1
	p	0.40	0.77	0.999	0.30	0.70	0.90	0.59	0.92	0.56	0.27	0.85	0.02
Kaninsky Shore													
Shoyna	Δt^*	−4.8	−0.5	+4.5	−1.1	+0.6	+0.9	+2.2	−0.6	0.0	−0.7	−4.3	−2.8
	p	0.97	0.17	0.98	0.52	0.49	0.62	0.92	0.56	0.0	0.49	0.997	0.86
Cape Kanin Nos	Δt^*	−1.7	0.0	+3.2	+0.2	+0.9	+0.3	+1.9	−0.2	+0.1	0.0	−1.6	−0.5
	p	0.80	0.0	0.99	0.19	0.84	0.31	0.97	0.22	0.15	0.0	0.95	0.39
White Sea Islands													
Sosnovets Island	Δt^*	−0.4	+1.5	+3.0	−0.2	+0.6	+0.9	+1.2	+0.7	+0.6	+1.1	0.0	+0.8
	p	0.30	0.84	0.999	0.25	0.82	0.96	0.996	0.95	0.91	0.94	0.0	0.65
Solovetsky Archipelago	Δt^*	+0.2	+1.5	+2.2	+0.7	+1.2	+1.8	+1.7	+1.0	+0.9	+0.9	+0.5	+0.7
	p	0.22	0.88	0.995	0.81	0.97	0.999	0.999	0.99	0.99	0.95	0.63	0.65
Zhizhgin Island	Δt^*	−0.2	+1.6	+2.5	+0.2	+0.8	+1.3	+1.2	+0.4	+0.2	+0.7	−0.1	+0.8
	p	0.14	0.88	0.995	0.29	0.87	0.98	0.98	0.67	0.48	0.87	0.14	0.67
Mud'yug Island	Δt^*	−1.7	+0.8	+4.7	+1.1	+1.8	+0.8	+1.3	−0.5	+0.2	+0.7	−2.2	−0.4
	p	0.68	0.37	0.999	0.78	0.95	0.64	0.84	0.51	0.23	0.66	0.965	0.20
Morzhovets Island	Δt^*	−0.6	+0.3	+2.5	−0.1	+1.0	+1.6	+1.8	+0.8	+0.6	+0.9	−0.2	+0.9
	p	0.46	0.26	0.996	0.18	0.94	0.995	0.998	0.92	0.88	0.91	0.23	0.70

* In °C/100.

a

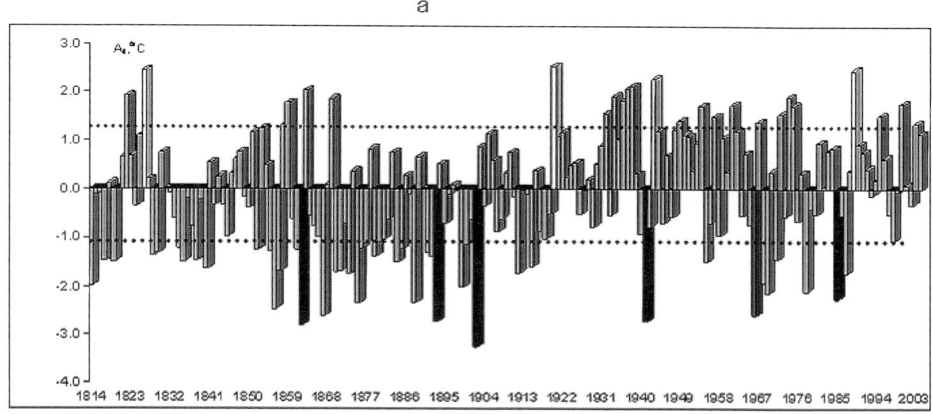

b

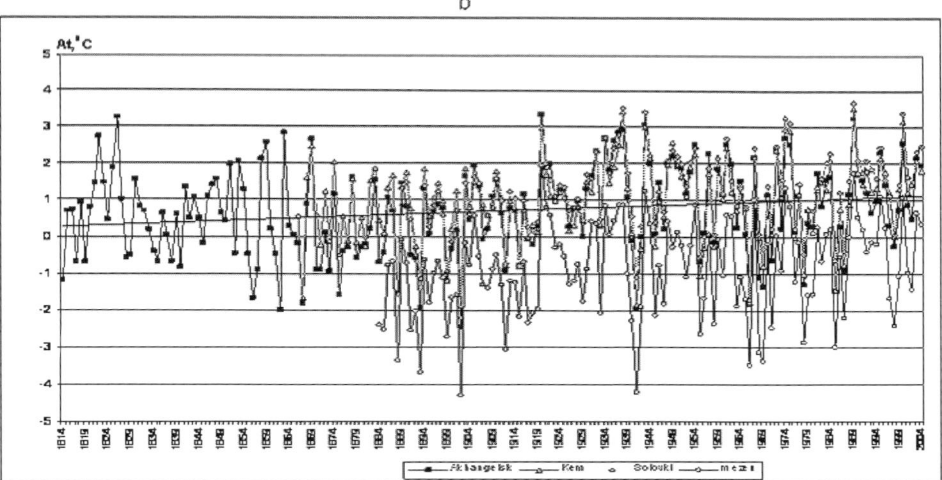

Figure 11.6. The long-term course of air temperature anomaly on stations with the longest observation period situated in the White Sea region: (a) yearly temperature anomaly in Arkhangelsk (white column maximal positive A_t, black column maximal negative A_t, pointed lines $\pm\sigma$ of A_t); (b) black line linear trend of A_t in Arkhangelsk.

A record of almost 200 years of measurements of t_a in Arkhangelsk allows us to have a clear view of the air temperature during the 19th and 20th centuries. It is characterized by cyclic variability. For every cycle, the value of the linear trend ($\Delta t/100$ years) according to formula (11.1) and its significance level (p) were calculated:

- Until approximately the mid-1820s an increase in temperature was observed.
- Then, approximately up to the 1880s the temperature was decreasing linearly with $\Delta t = -1.3°\text{C}/100$ years and $p = 0.77$.

- Subsequently an increase was observed until the mid-1940s with a linear trend with $\Delta t = +2.5°C/100$ years and $p = 0.998$.
- This was followed again with a small reduction in mean annual temperature during the next 35 years approximately up to 1979. During this period the trend was about $\Delta t = -3.6°C/100$ years with $p = 0.89$.
- Finally, from the beginning of the 1980s the temperature again increased, and for the period from 1980 to 2004 the trend value was $\Delta t = +2.8°C/100$ years with $p = 0.65$.

Overall, in Arkhangelsk a statistically significant increase in average yearly temperature was observed of approximately 0.5°C per 100 years (see Table 11.5 and Figure 11.6). Thus, for nearly 200 years the trend increase was almost 1°C, which is approximately half the estimate obtained for the whole Russian territory and for all territory of the northern polar area according to other time series analyses (ACIA, 2005).

Analysis of variation of the air temperature anomaly, A_t, over the whole period was calculated as recommended by the World Meteorological Organization for the basic period 1961–1990; that is, the average temperature calculated for the whole period of observations in Arkhangelsk only differs a little from t_a calculated for the basic period (namely, −0.7°C for 1814–2004 and −0.8°C for 1961–1990). This has shown that approximately 75% of A_t values do not exceed the value of its mean square deviation from the average anomaly ($\sigma_{A_t} = 1.25°C$) (Figure 11.7c). The maximum value of the positive anomaly was 2.6°C and was observed in 1920. Values close to maximal ($A_t = 2.5°C$) were noted both in the 19th century (in 1826) and in the 20th century (in 1989). Also 1943 was remarkable when A_t was 2.3°C (see Figure 11.7a).

The largest negative anomaly is greater than the positive one ($A_t = -3.3°C$), which was observed in 1902. That year was the coldest one for the whole two centuries. For all stations with a long record that year was characterized by the largest negative temperature anomaly. Its value varied from −2.4°C at Zimnegorsky Lighthouse and Morzhovets Island to −4°C at Sosnovets Island and Tersko-Orlovsky Lighthouse. A slightly smaller anomaly (from −2.5°C to −2.8°C) was also observed four times in the 19th century (1856, 1862, 1867, 1893) and only twice in the 20th century (1941, 1966). Thus, the cold snap, which was observed in the middle of the 19th century, was more significant than the cold snap in the second half of the 20th century (Figure 11.6a).

At 16 of the 28 stations for which data were analyzed, the year 1989 appeared to be the warmest; yearly air temperature anomalies varied from +2.2°C to +3.1°C. In Mezen ($A_t = 2.8°C$) and at Cape Svyatoy Nos the warmest year was 1920 as well as at Arkhangelsk. At the Umba, Kandalaksha, Kovda, Pyalitsa, and Gridino stations the warmest year was 1938 ($A_t = 2.5°C$–$3.5°C$), at the Shoyna station the warmest year was 1943 ($A_t = 2.6°C$), and at the Cape Kanin Nos station it was 1954 ($A_t = 2.4°C$) (see Tables 11.2 and 11.3).

If we compare the diagram in Figure 11.6a with the well-known figure of change in the average annual temperature anomaly in the northern hemisphere or for the

European part of Russia (ACIA, 2005; Gruza and Rankova, 2003, 2004) it becomes obvious that Arkhangelsk does not show that unambiguity which is characteristic of the anomaly for the northern hemisphere or for the European territory of Russia. For example, there is no negative anomaly since 1980 according to the graphs (ACIA, 2005; Gruza and Rankova, 2003, 2004), while in Arkhangelsk negative anomalies were observed in the years 1985–1987, 1997–1998, and there was also a small negative anomaly in 2002. It was also the case that the warming of the White Sea region was expressed more strongly in 1920–1930 than during the last 20 years. Such a situation is observed for the average temperatures in the Atlantic sector of the Arctic region as a whole (Alekseev, 2006).

To reveal the features of air temperature variations during a year long-term changes in average seasonal temperature were analysed. For the winter season the period from November to March was used, for the spring April and May, for the summer June to August, and for the autumn September and October. In Arkhangelsk there was a positive tendency for change in average seasonal temperature, with a high degree of significance, for winter and spring. In the summer and in the autumn there is no such tendency. This is confirmed by estimating the linear trend significance (Table 11.5).

Analysis of the tendencies of monthly average temperature change has shown that the winter season becomes warmer mainly due to December, for which the most significant positive and statistically significant trend of temperature ($\Delta t = +1.2^{\circ}\text{C}/100$, $p = 0.97$) from all winter months is found (see Figure 11.7c). In March the tendency to warming is also observed, although the significance of the trend is slightly lower ($\Delta t = +0.7^{\circ}\text{C}/100$, $p = 0.90$). In spring a positive significant trend is observed both in April ($\Delta t = +0.7^{\circ}\text{C}/100$, $p = 0.97$) and in May ($\Delta t = +1.0^{\circ}\text{C}/100$, $p = 0.997$) (Figure 11.7a). On the contrary, in the summer the long-term change of monthly air temperature in August is characterized by a negative trend with a high degree of reliability ($\Delta t = -0.5^{\circ}\text{C}/100$, $p = 0.96$) (Figure 11.7b).

The annual course of extreme anomalies of the monthly air temperature in Arkhangelsk has shown that positive A_t were observed in the 19th century during 6 months out of 12, and negative ones during 5 months (see Figure 11.8).

The value of the linear trend and its significance level were calculated for all 26 meteorological stations (Tables 11.5 and 11.6). They all have different lengths of observation period. The presented data of calculated Δt and p evidently show its dependence on trends of temperature, and that the sign of the trend of temperature and the degree of its reliability depend on the length of the observation period. Analysis of the stability of a trend sign against the background of the intracentury course of temperature for all the territory of the Russian Federation, obtained at the Main Geophysical Observatory named after A.I.Voeykov (in St. Petersburg), has shown that the seasonal nature of prevailing trends during the century was not constant. It was characterized by cyclic changes with a period of 25–30 years (Mirvis, 1999). This is apparent from analysis of our data. For example, temperature trend estimates and their statistically significant levels were calculated for different periods of time for Arkhangelsk (Figure 11.9a). The shortest analyzed period was for 15 years (from 1990 to 2004), and the longest was for 191 years (from 1814 to 2004). It is

a

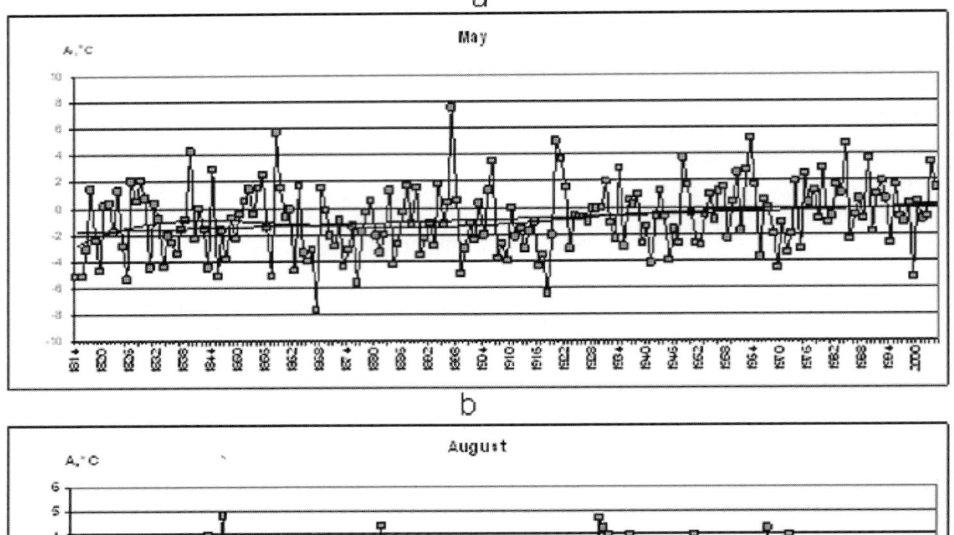

b

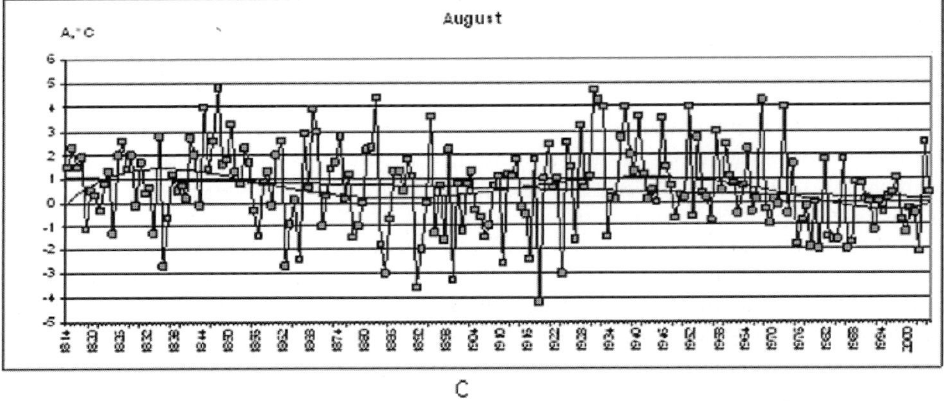

c

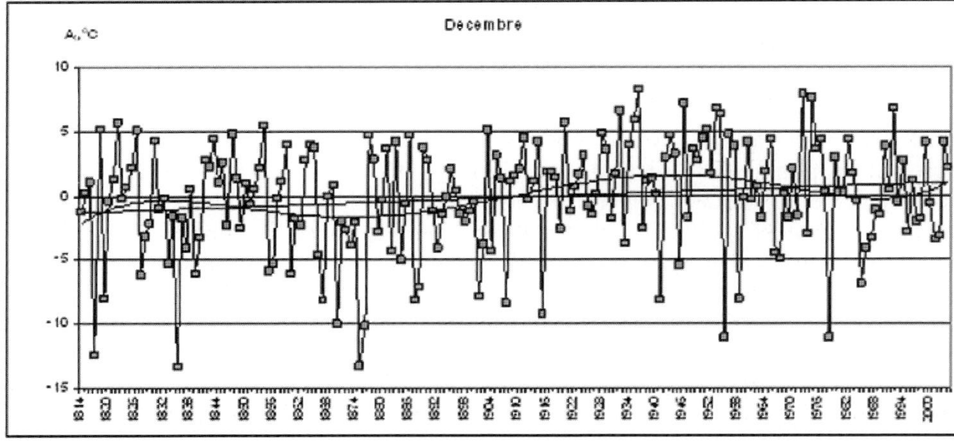

Figure 11.7. The long-term course of monthly air temperature in Arkhangelsk for months with a statistically significant trend level.

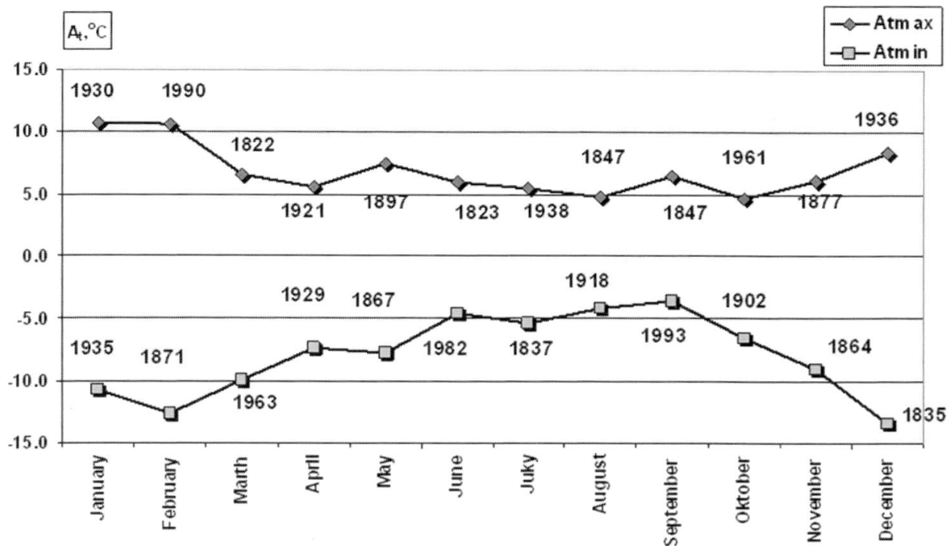

Figure 11.8. The annual course of extreme values in monthly air temperature anomalies in Arkhangelsk (1814–2004).

evident from Figure 11.9a that, due to the cyclicity of temperature change, the size and sign of a trend vary over a wide range if the period of observation is less than 100 years. At the same time all trend estimates for these periods are not statistically significant. Values of Δt become statistically significant ($p \geq 0.95$), when the measurement period exceeds 110 years. The dependence of Δt and p on the observation period becomes weaker with further increase in the period: for periods of more than 120 years the value of a trend gradually decreases from $\Delta t = 0.9°\text{C}/100$ years to $\Delta t = 0.5°\text{C}/100$ years for a period of more than 170 years. It is interesting to note that the values of trends are maximal for the shortest periods (1960–2004 and 1980–2004). Moreover, for some time series (1950–2004, 1910–2004) the values of temperature trends are negative (Figure 11.9a).

To examine these parameters (Δt and p) for monthly temperature, January and July were chosen as months with a non-significant tendency of t_a and December, May, and August as months with a significant tendency of t_a (Figures 11.9b–f). First of all it is necessary to note the appropriateness of the decrease in Δt variability with the increase of the period of observations, which is characteristic for all months. The greatest linear increase of monthly air temperature during the last 190 years was obtained for May ($\Delta t = +1°\text{C}/100$) (Figure 11.9b). The same situation is evident in December (Figure 11.9f). For other months the results are not so unambiguous. For example, in July the statistically significant increasing trend of t_a was obtained only for the shortest period (1990–2004), but for January there is no such period (Figures 11.9d, e). The variability of p from period to period in August clearly demonstrates the peculiarities of climate forming: the

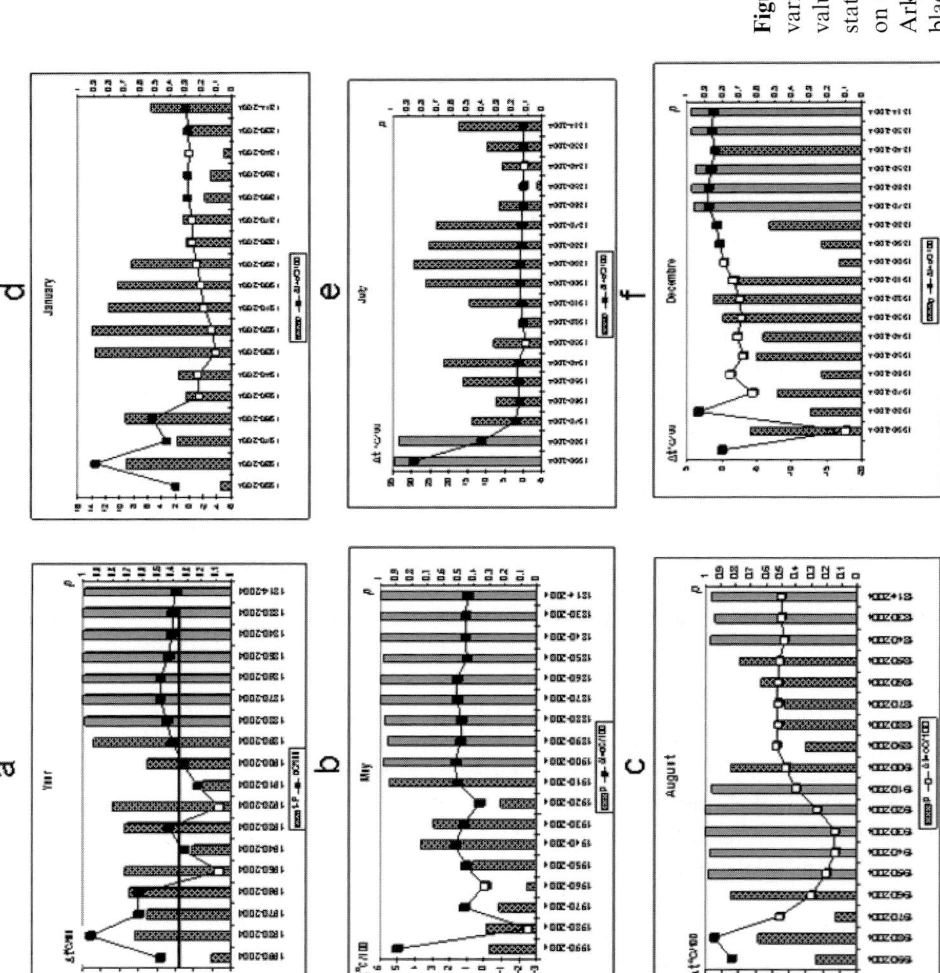

Figure 11.9. The dependence of variations of the air temperature trend value (Δt, °C/100, points) and its statistically significant level (p, columns) on observation period length in Arkhangelsk (gray columns $p \geq 0.95$; black points $\Delta t > 0$, white points $\Delta t < 0$).

most statistically significant level of p was obtained for a period of 75–85 years and for a period of more than 165 years. This is connected with two periods of air temperature increase in this month: during the 1840s and in 1920s–1930s. So, the very intensive warming during the 1920s–1930s is the main reason for the steady negative tendency in August, which has changed sign only during the last 25 years and is still not statistically significant.

This analysis of the trends of the mean monthly temperatures clearly demonstrates that the longer the period of observations that is analyzed the more representative are the conclusions about climate change.

For the group of stations with an observation period of more than 110 years a positive tendency in yearly temperature with a high level of significance is characteristic (Tables 11.5 and 11.6). The maximum value, $\Delta t = 1.5^{\circ}$C/100 years, is observed at Cape Svyatoy Nos, while for other stations the trend values vary from $\Delta t = 0.8^{\circ}$C/100 years to $\Delta t = 1.1^{\circ}$C/100 years. It is noticeable that for Kem, whose period of observations is 140 years, the trend value ($\Delta t = 0.6^{\circ}$C/100 years) is close to the same estimate obtained for Arkhangelsk (i.e. $\Delta t = 0.5^{\circ}$C/100 years).

Analysis of the tendencies of seasonal temperature variations has shown that in Kem, Solovky, and Cape Svyatoy Nos the statistically significant tendency of warming within the limits of $\Delta t = 0.7^{\circ}$C–2°C/100 years is observed for all seasons. It is especially remarkable for the summer and autumn seasons ($\Delta t = 0.8^{\circ}$C–1.5°C/100 years).

The trend of monthly t_a values is even larger. At all stations which started recording after 1880, in March the size of a linear trend changed from $\Delta t = 2.4^{\circ}$C/100 years up to $\Delta t = 3.9^{\circ}$C/100 years and this appears to be statistically significant. The same tendency is observed also in June ($\Delta t = 0.9^{\circ}$C–1.8°C/100 years) except for the Cape Svyatoy Nos station. The same value of Δt for Arkhangelsk for the period of 1880–2004 is significant with $p = 0.99$ and is equal to 0.75°C/100 years.

Stations that were established after 1910 form the second group. Analysis of the character of yearly air temperature changes allows us to conclude that for this group of stations there is no precisely expressed tendency. Temperature during the last 95 years changed mainly in a cyclic way. By analyzing seasonal changes at these stations, positive and statistically significant tendencies were revealed only at Kovda station in the spring and in the summer (see Table 11.5). Regarding the change of monthly temperature a positive and statistically significant trend of temperature in March is observed here, as well as at stations of the first group. At Kovda a significant increase of temperature is observed also in May and June, and in Pyalitsa in May and July. In November, on the contrary, at all stations of the second group there is a marked and statistically significant cold snap (see Table 11.6).

The non-significant tendency for a cold snap in Kandalaksha for the yearly means of t_a is a result of the fact that it is one of the few stations where the presence of a statistically significant negative tendency is observed for two months (August, November), and in September and December it is also quite significant with $p > 0.90$ (Table 11.6). It should be noted that for the examined period (1910 to 2004) in Arkhangelsk there is a similar weak tendency to a cold snap which is not statistically significant.

With the reduction of the period of observations to 75 years and less (the third group of stations) the variability of mid-annual values of temperature does not have any statistically significant tendency at any of these stations. It is interesting, that already at three stations belonging to this group (Umba, Shoyna, and Intsy) the trend values of yearly t_a are negative (see Table 11.6). There are no strongly expressed tendencies according to long-term changes in seasonal temperature. It is remarkable that again in March at all stations there is warming and the trend values have increased up to $\Delta t = 5°C–6°C/100$ years with a significance value of more than $p = 0.98$. In August and November there is a negative tendency of temperature change at practically all stations (see Table 11.6).

11.5 CONCLUSIONS

This analysis shows with confidence that

- in the White Sea region within the last nearly 200 years there has been an observed warming with a linear trend of nearly $\Delta t = 0.5°C/100$ years;
- most noticeably this tendency is observed in the northern part of the Tersky shore (Cape Svyatoy Nos, Tersko-Orlovsky Lighthouse), in Onega, Mezen, at Zimnegorsky Lighthouse, and on the Solovki Archipelago. At the same time, on the Kandalaksha and Karelian (Karelsky) shores, and also in Shoyna, there is the opposite tendency to a cold snap, but this is weak and not statistically significant;
- the average temperature in March increases practically everywhere at a statistically significant level, but in November the opposite occurs and there is a decrease;
- the warmest years during the period from 1814 to 2004 were 1826, 1920, 1938, and 1989 and the coldest were 1856, 1862, 1867, 1893, 1902, 1941, and 1966;
- the sign of the trend in temperature and the degree of its reliability depend on the length of the observation period. It is possible to consider the estimates of a linear trend of mid-annual air temperatures as being the most reliable, if the length of the period of observations is not less than 120 years. If the duration of measurements is shorter, estimates of the trend can change considerably, reflecting only temporary climatic fluctuations.

11.6 REFERENCES

ACIA (2005). *Arctic Climate Impact Assessment*. Cambridge University Press, Cambridge, U.K., 1,042 pp.

Adamenko V.N. and Kondratyev K.Ya. (1999). Global climate change and its empirical diagnostics. In: Yu.A. Izrael', G.V. Kalabin, and V.V. Nikonov (eds.), *Anthropogenic Impact on Northern Nature and Its Ecological Consequence*. Kola Scientific Center of the Russian Academy of Science, Apatity, pp. 17–37 [in Russian].

Alekseev G.V. (2006). Arctic climate change in the 20th century. In: Yu.A. Izrael' (ed.), *The Possibilities of Climate Change Prevention and Its Negative Consequences: Kyoto Protocol Problem*. Science, Moscow, pp. 391–400 [in Russian].

Demirchian K.S., Demirchian K.K., and Kondratyev K.Ya. (2006). The IPCC reports do not substantiate the need for realization of the Kyoto Protocol. In: Yu.A. Izrael' (ed.), *The Possibilities of Climate Change Prevention and Its Negative Consequences: Kyoto Protocol Problem*. Nauka, Moscow, pp. 183–226 [in Russian].

Filatov N., Pozdnyakov D., Johannessen O.M., Pettersson L.H., and Bobylev, L.P. (2005) *White Sea: Its Marine Environment and Ecosystem Dynamics Influenced by Global Change*. Springer/Praxis, Chichester, U.K.

Glukhovsky B.Kh., Lagutin B.L., and Rzheplinskiy G.V. (eds). (1989). *Hydrometerological Conditions of the Shelf Zone of SSSR Seas, 5: White Sea*. Hydrometeoizdat, Leningrad, 236 pp. [in Russian].

Gruza G.V. and Rankova E.Ya. (2003). Variations and changes of climate on the territory of Russia. *News of the Russian Academy of Science: Physics of Atmosphere and Ocean*, **39**(2), 166–185 [in Russian].

Gruza G.V. and Rankova E.Ya. (2004). Climate change determination: Present state, variability and extremity. In: Yu.A. Izrael' (ed.), *World Climate Change Conference*. Hydrometeoizdat, Moscow, pp. 101–110 [in Russian].

IC (1975). *USSR Climate Reference Book, Issue 3a: Karelskaya ASSR. Meteorological Data in Separate Years, Part 1: Air Temperature*. Informational Centre, Obninsk, 173 pp. [in Russian].

Izrael Yu.A. (ed.) (2004). *World Climate Change Conference*. Hydrometeoizdat, Moscow, 620 pp. [in Russian].

Kasimov N.S. and Klige R.K. (eds.) (2006a). *Recent Global Changes in the Natural Environment*, Vol. 1. Scientific World, Moscow, 696 pp. [in Russian].

Kasimov N.S. and Klige R.K. (eds.) (2006b). *Recent Global Changes in the Natural Environment*, Vol. 2. Scientific World, Moscow, 776 pp. [in Russian].

Katsov V.M. (2006). Arctic climate in the 21st century: ACIA experience. In: Yu.A. Izrael' (ed.), *The Possibilities of Climate Change Prevention and Its Negative Consequences: Kyoto Protocol Problem*. Science, Moscow, pp. 371–390 [in Russian].

Kondrasheva E.T. (ed.) (1954). *Climatological Reference Book of the USSR, Issue 2: Karelo-Finskaya SSR. Meteorological data in separate years, Part 1: Air Temperature*. Leningrad Administration of Hydrometeorological Service, Leningrad, 184 pp. [in Russian].

Kondratyev K.Ya. (1998). *Multidimensional Global Change*. Wiley/Praxis. Chichester, U.K., 761 pp.

Kondratyev K.Ya. (2001). Key issues of global change at the end of the second millennium. Our fragile world: Challenges and opportunities for sustainable development. *EOLSS Vorruner*, **1**, 147–165.

Kondratyev K.Ya. (2003a). High-latitude environmental dynamics in the context of global change. *Idojaras*, **107**(1), 1–29.

Kondratyev K.Ya. (2003b). Uncertainty of the observation data and climate modeling. Available at *http://cleanerproduction.ru/uncert.htm* [in Russian].

Kondratyev K.Ya. (2004). Key aspects of global climate change. *Energy and Environment*, **15**, 469–503.

Kondratyev K.Ya. and Cracknell A.P. (1999). *Observing Global Climate Change*. Taylor & Francis, London, 562 pp.

Kondratyev K.Ya. and Galindo I. (1997). *Volcanic Activity and Climate*. A. Deepak, Hampton, VA, 382 pp.

Kondratyev K.Ya. and Varotsos C.A. (2000). *Atmospheric Ozone Variability: Implications for Climate Change, Human Health, and Ecosystems*. Springer/Praxis, Chichester, U.K., 614 pp.

Mirvis V.M. (1999). Estimation of air temperature change over the territory of Russia in recent centuries. In: M.Ye. Berlyand and V.P. Melesko (eds.), *Contemporary Investigation at the Main Geophysical Observatory*, Vol. 1. Hydrometeoisdat, St. Petersburg, pp. 220–235 [in Russian].

NAHS (1970). *USSR Climate Reference Book, Issue 1: Arkhangelskaya and Vologodskaya Oblast' and Komi SSR. Meteorological Data in Separate Years, Part 1: Air Temperature.* Northern Administration of Hydrometeorological Service, Arkhangelsk, 460 pp. [in Russian].

NAHS (1975). *USSR Climate Reference Book, Issue 3a: Karelskaya ASSR. Meteorological Data in Separate Years. Part 1: Air Temperature.* Information Centre, Obninsk, Russia, 173 pp. [in Russian].

Naumova L.P. (1983). Estimation of the misses in observations on the value of climatic features. *Main Geophysical Observatory Works*, **475**, 20–25. [in Russian].

Polyak I.I. (1975). Estimation of the long-term meteorological data linear trend. *Main Geophysical Observatory Works*, **364**, 51–55 [in Russian].

Poznitskiy B.N. (ed.) (1966). *Climatological Reference Book of the USSR. Murmanskaya Oblast: History and physico-geographical description of hydrometeorological stations and posts.* Murmansk Administration of the Hydrometeorological Service, Murmansk, 100 pp. (in Russian).

Shilovtseva O.A. and Romanenko F.A. (2005). Air temperature long-term variations in the North-Western Taimyr and Lower Yenisei during the 20th century. *Meteorology and Hydrology*, **3**, 53–68 [in Russian].

Soboleva A.N. (ed.) (1956). *Climatological Reference Book of the USSR, Issue 1: Murmanskaya, Arkhangelskaya Oblast' and Komi SSR. Meteorological Data in Separate Years, Part 1: Air Temperature.* Leningrad Administration of the Hydrometeorological Service, Leningrad, 562 pp. [in Russian].

Veselovsky K.C. (1857). *About the Climate of Russia.* Imperial Academy of Science, St. Petersburg, 326 pp. [in Russian].

Vrangel F.F. (1891). *Climate Variations: Lectures.* R. Golike, St. Petersburg, 18 pp. [in Russian].

12

Climatic characteristics of temperature, humidity, and wind velocity in the atmospheric boundary layer over western Siberia

Valery S. Komarov and Nataly Ya. Lomakina

12.1 INTRODUCTION

It is well known that estimation of the vertical distributions of meteorological parameters (primarily temperature, humidity, and wind velocity) in the Earth's atmosphere is one of the main subjects of research for the analysis and modeling of climates (Kondratyev, 1987), solution of inverse problems of remote diagnostics of the environment from space (Kondratyev, 1988; Kondratyev and Timofeev, 1978), and the development of methods and means of laser sensing of atmospheric parameters (Ippolitov *et al.*, 1985; Zuev and Zuev, 1992). The application of statistics is very important in addressing these problems; this is because in actual practice the spacetime distribution of the required meteorological parameter in the atmosphere is highly specific to the time and place being studied and is often unknown. In addition, the quality of solutions to these problems depends substantially on the completeness and adequacy of the statistics employed.

This emphasizes the importance of obtaining adequate statistical information on the vertical profiles of temperature, humidity, and wind velocity fields in individual regions based on data of regular aerological observations. However, it should be noted that while the statistical structures of the vertical temperature, humidity, and wind velocity profiles in the free atmosphere have now been well investigated (Czelnai *et al.*, 1976; Zuev and Komarov, 1987), the same structures in the atmospheric boundary layer (ABL) are still poorly understood, especially above western Siberia, which is a large area that is insufficiently covered by observations. This resulted from the fact that data registered at standard and very widely spaced isobaric surface altitudes (without invoking information about singular points) were usually used, and hence it was impossible to study the structure of these fields in the atmospheric boundary layer.

Results of investigations into the vertical structure of meteorological fields in the atmospheric boundary layer above western Siberia have been presented in three papers (Komarov *et al.*, 1995; Nevzorova and Odintsov, 2005; Zuev *et al.*, 1997). However, the data presented in these papers are of limited significance, because they were obtained from data of long-term radiosonde temperature and wind velocity observation at only one station near Novosibirsk (Komarov *et al.*, 1995), from experimental short-term lidar (Zuev *et al.*, 1997), or from sodar measurements of wind velocity characteristics in the region of Tomsk (Nevzorova and Odintsov, 2005). In addition, only the vertical profiles of average wind velocity and its variance were considered by Zuev *et al.* (1997), and only the interlevel correlations of orthogonal wind velocity components measured at altitudes up to 300 m were considered (Nevzorova and Odintsov, 2005). Thus, there is a demand for statistical information on the vertical structure of the temperature, humidity, and wind velocity fields in the atmospheric boundary layer for Western Siberia. In this regard, intensive studies on the atmospheric boundary layer climatology for western Siberia have been carried out at the Institute of Atmospheric Optics of the Siberian Branch of the Russian Academy of Sciences based on physical–statistical analysis of the vertical profiles of temperature, water vapor mass fraction (commonly referred to as humidity), and zonal and meridional wind velocity components. This chapter presents the results of these studies.

12.2 DESCRIPTION OF INITIAL DATA AND SOME METHODOLOGICAL ASPECTS OF THEIR STATISTICAL PROCESSING

We have used twice-daily (at 00 : 00 h and 12 : 00 h GMT) radiosonde data for a 5-year period (2001–2005) from eight aerological stations: Salekhard (66°32′N, 66°40′E), Turukhansk (65°47′N, 87°56′E), Khanty-Mansijsk (61°01′N, 69°02′E), Aleksandrovskoe (60°26′N, 77°52′E), Verkhnee Dubrovo (56°44′N, 61°40′E), Omsk (54°56′N, 73°24′N), Novosibirsk (54°58′N, 82°57′E), and Emel'yanovo (56°11′N, 92°37′E). These data served as raw material for the study of the special features of the vertical statistical structure of the temperature, humidity, and wind velocity fields in the atmospheric boundary layer above western Siberia. The location of the study area is shown in Figure 12.1.

Since averaging was performed over a five-year period, questions arise as to whether this period is representative and whether the statistical characteristics averaged over it can be used as climatic norms. To answer these questions, we estimated the significance of deviation of average values and variances calculated for two independent samples included in a certain general set. According to Dlin (1975), to estimate the significance or randomness of deviation of average values and variances, the criterion for t_s of the form

$$|t_s| = \frac{|\bar{\xi}_1 - \bar{\xi}_2|}{\sqrt{\sigma_1^2/N_1 + \sigma_2^2/N_2}} \leq t_s(P,k) \tag{12.1}$$

was used, where $\bar{\xi}_1$ and $\bar{\xi}_2$ are average values of the meteorological parameter for the

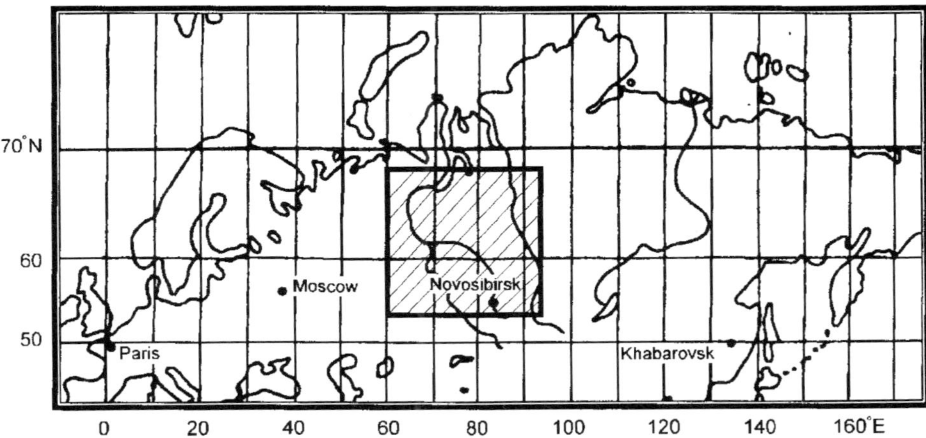

Figure 12.1. Location of the study area.

two samples being compared; σ_1 and σ_2 are the sample variances corresponding to them; N_1 and N_2 characterize the sample lengths; $t_s(P,k)$ is the threshold value of the significance criterion for the probability $P = 0.95$ and the number of degrees of freedom $k = (N_1 + N_2 - 2)$ as well as Fisher's criterion for T_H of the form

$$T_H = \frac{\sigma_1^2}{\sigma_2^2} \leq F_{1-P}(N_1, N_2), \tag{12.2}$$

where σ_1^2 and σ_2^2 are the variances calculated for the two samples, with a larger value placed in the numerator; and $F_{1-P}(N_1, N_2)$ is the threshold value of the criterion T_H for the significance level $q = 1 - P = 0.05$, determined from the special Fisher tables compiled for various combinations of the number of degrees of freedom N_1 and N_2. When conditions $|t_s| \leq t_s(P,k)$ and $T_H \leq F_{1-P}(N_1, N_2)$ are met, the difference between the average values $\bar{\xi}_1$ and $\bar{\xi}_2$ and the variances σ_1 and σ_2 are random and insignificant, and the samples themselves belong to the same general set.

By way of example, Table 12.1 gives the results of comparison of average values and variances for the temperature and orthogonal wind velocity components using the t_s and T_H criteria calculated for the Novosibirsk station from samples N_1 (1961–1970) and N_2 (2001–2005); average values and variances for sample N_1 were taken from Komarov (1972) (for temperature) and subsequently complemented by calculations for wind velocity. Sample lengths N_1 and N_2 were, respectively, 230 and 138 observations in January and 260 and 152 in July.

Analysis of Table 12.1 demonstrates that the t_s criterion is less than its threshold value $t_s(P,k) = 1.96$ calculated for probability $P = 0.95$ in all cases (i.e., irrespective of the meteorological parameter, month, and altitude level in the atmosphere), and the number of degrees of freedom is $k = 230 + 138 - 2 = 366$ for January and $k = 260 + 152 - 2 = 410$ for July. The criterion $T_H < F_{1-P}(N_1, N_2)$ is equal, respectively, to 1.31 and 1.28 for the significance level $q = 1 - P = 0.05$ and the same N_1

Table 12.1. Average values of temperature (t, °C), zonal (U, m s^{-1}), and meridional wind velocity components (V, m s^{-1}) and their variances (σ^2) calculated for the Novosibirsk station over the periods 1961–1970 and 2001–2005, and significance criteria t_s and T_H.

Altitude level (hPa)	*January*						*July*					
	1961–1970		*2001–2005*		t_s	T_H	*1961–1970*		*2001–2005*		t_s	T_H
	$\bar{\xi}_1$	σ_1	$\bar{\xi}_2$	σ_2			$\bar{\xi}_1$	σ_1	$\bar{\xi}_2$	σ_2		
Temperature												
Ground	−17.1	8.5	−16.5	7.7	0.70	1.22	18.8	4.9	18.4	5.5	0.74	1.26
925	−13.0	6.3	−11.8	5.7	1.87	1.22	15.0	4.2	15.6	4.3	1.37	1.05
850	−13.8	5.5	−12.8	5.3	1.74	1.08	9.8	3.8	10.2	4.0	0.99	1.11
Zonal wind velocity component												
Ground	0.5	2.2	0.9	2.1	1.74	1.10	0.0	1.6	0.1	1.8	0.57	1.26
925	5.4	7.4	6.3	6.9	1.18	1.15	0.4	4.4	0.6	4.6	0.43	1.09
850	6.5	8.4	7.3	7.7	0.93	1.19	1.0	4.8	1.1	4.9	0.20	1.04
Meridional wind velocity component												
Ground	1.1	1.6	0.8	1.8	1.63	1.26	−0.1	1.5	−0.6	1.6	0.61	1.14
925	2.5	6.0	1.4	5.5	1.76	1.19	−0.2	4.5	−0.7	4.7	1.14	1.20
850	1.4	6.3	0.3	5.7	1.72	1.13	−0.1	4.8	−0.4	5.0	0.59	1.08

and N_2 values. Hence, both the average values and variances of these meteorological parameters calculated for two independent samples vary randomly and do not differ significantly. Therefore, the sample we used is representative, and the statistical characteristics calculated for it can be considered climatic norms.

Let us now dwell briefly on some special features of forming the initial statistical sets used to calculate the following characteristics of vertical meteorological field profiles: average values $\bar{\xi}(h_j)$, standard deviations $\sigma_\xi(h_i)$, and autocorrelation functions $\mu_{\xi\xi}(h_i, h_j)$.

To form these sets, we used the following procedures:

- first, all the aerological data were interpolated (using the linear interpolation method) from the altitudes of standard isobaric surfaces at 1,000 hPa (or ground level), 925 hPa, 850 hPa, and 700 hPa and singular-point altitudes to geometrical altitudes of 0 m, 100 m, 200 m, 300 m, 400 m, 600 m, 800 m, 1,000 m, 1,200 m, and 1,600 m;

- second, aerological measurements at different stations were synchronized in time; as a result, the total number of synchronous measurements (for all stations) for each long-term month (January and July) was 138 and 152, respectively. This allowed us to obtain samples that were uniform over altitude and space and that in the first approximation were random and independent (i.e., they obey the laws of statistics from the viewpoint of obtaining statistically justified characteristics based on these samples);
- third, these samples were formed for the long-term month averaging period to exclude the non-stationarity of the meteorological data series typical of annual or seasonal averaging that can significantly distort the statistical characteristics being estimated;
- fourth, to form statistical datasets for humidity, we used the values of humidity (q, ‰) that cannot be measured directly; therefore, they were calculated from the formula

$$q = 622\frac{e}{p} = 622\frac{E_w(T_d)}{p}, \tag{12.3}$$

  where $e = E(T_d)$ is the partial water vapor pressure, in hPa, estimated with respect to water (here $T_d = (273.16 + t_d)$ is the dew point, in K, and t_d is the same dew point but in °C); and p is the atmospheric pressure, in hPa; and
- fifth, after formation of statistical datasets, each term of the set examined was climatically controlled using the expression (Zuev and Komarov, 1987):

$$|\xi_i - \bar{\xi}| \geq 3\sigma_\xi, \tag{12.4}$$

  where ξ_i and $\bar{\xi}$ are the controllable values of the meteorological parameter and its climatic norm at a given altitude level in the atmosphere; and σ_ξ is the standard deviation for the same altitude level. As a result, no more than 1%–3% of measurements were rejected. Since vertical profiles $\xi_i(p)$ with erroneous values at some altitude levels were excluded from further analysis, the statistical datasets so obtained were uniform functions of altitude. They were subsequently used to calculate all statistical characteristics.

Now we consider some methodological principles for calculating the statistical characteristics of the vertical distributions of temperature, humidity, and zonal and meridional wind velocity components. In this work, calculations were performed for long-term datasets that were registered at each aerological station, and all multidimensional observations for the given terms of the month in question were combined.

We use the term multidimensional observation to mean a certain k-dimensional vector (a vertical profile) whose components are discrete values of the meteorological parameter ξ at preset altitudes h_k $(k = 1, 2, \dots, K)$. This vector can be written as follows:

$$\xi = \|\xi(h_1), \xi(h_2), \dots, \xi(h_K)\|^T, \tag{12.5}$$

where T inidicates the transpose. After the rejection of erroneous data and formation of the refined statistical dataset, we calculated the following statistical (climatic) characteristics:

(1) the vector of average values $\mathbf{m}_\xi$ where:

$$\mathbf{m}_k = \left\| \begin{array}{c} m_\xi(h_1) = \dfrac{1}{N}\sum_{v=1}^{N} \xi_v(h_1) \\ \vdots \\ m_\xi(h_K) = \dfrac{1}{N}\sum_{v=1}^{N} \xi_v(h_K) \end{array} \right\|, \tag{12.6}$$

where $m_\xi(h_k)$ is the average value of the meteorological parameter ξ at the kth altitude level; $\xi_v(k_k)$ is the value of the same meteorological parameter measured at the kth altitude level; and N is the number of measurements;

(2) the vector of average standard deviations $\boldsymbol{\sigma}_\xi$ where:

$$\boldsymbol{\sigma}_\xi = \left\| \begin{array}{c} \sigma_\xi(h_1) = \sqrt{\dfrac{1}{N}\sum_{v=1}^{N} (\xi_v(h_1) - m_\xi(h_1))^2} \\ \vdots \\ \sigma_\xi(h_K) = \sqrt{\dfrac{1}{N}\sum_{v=1}^{N} (\xi_v(h_K) - m_\xi(h_K))^2} \end{array} \right\|; \tag{12.7}$$

(3) the correlation coefficients $r_{\xi\xi}(h_i, h_j)$ are given by:

$$r_{\xi\xi}(h_i, h_j) = \frac{1}{N} \frac{\sum_{v=1}^{n} (\xi_v^{(h_i)} - m_\xi^{(h_i)})(\xi_v^{(h_j)} - m_\xi^{(h_j)})}{\sigma_\xi(h_i) - \sigma_\xi(h_j)}, \tag{12.8}$$

where $\sigma_\xi(h_i)$ and $\sigma_\xi(h_j)$ are standard deviations of the meteorological parameter ξ at altitude levels h_i and h_j, respectively.

The correlation coefficients given by Equation (12.8) form the correlation matrix

$$\mathbf{r}_{\xi\xi}(h_i, h_j) = \left\| \begin{array}{cccc} r_{11} & r_{12} & \cdot & r_{1K} \\ r_{21} & r_{22} & \cdot & r_{2K} \\ \cdot & \cdot & \cdot & \cdot \\ r_{K1} & r_{K2} & \cdot & r_{KK} \end{array} \right\|. \tag{12.9}$$

The statistical characteristics listed above were used for climatic analysis of the vertical profiles of temperature, humidity, and orthogonal wind velocity components in the atmospheric boundary layer above western Siberia.

12.3 SOME SPECIAL FEATURES OF THE VERTICAL STRUCTURE OF AVERAGE TEMPERATURE, HUMIDITY, AND WIND VELOCITY FIELDS IN THE ATMOSPHERIC BOUNDARY LAYER

It is well known that the basic features of the vertical structure of a meteorological field are most clearly manifested when its background (average climatic) characteristics showing the general features of this field are analyzed. The present section is devoted to some results of this analysis carried out for western Siberia.

We immediately emphasize that we present results of the statistical analysis of the background characteristics separately for the temperature–humidity complex (Zuev and Komarov, 1987) (they are closely correlated) and for the wind velocity vector characterized by its zonal and meridional components. Application of these wind velocity components is caused by the fact that data on the wind speed and direction measured by a radiosonde cannot be used in our analysis; this is because sometimes the average wind direction makes no physical sense, because the sum of two oppositely directed vectors can be equal to zero (Czelnai *et al.*, 1976). For simplicity, we shall use the term *zonal* (or *meridional*) *wind*. In this case, positive values of the zonal wind correspond to western air mass transport, and negative values correspond to eastern air mass transport. At the same time, positive values of the meridional wind correspond to southern air mass transport, and negative values correspond to northern air mass transport.

12.3.1 Basic features of the vertical distribution of average temperature and humidity

We first dwell on the basic features of the vertical distribution of average temperature and humidity with special emphasis on the features revealed for background (average climatic) characteristics calculated for the atmospheric boundary layer with altitude resolution much better than by Zuev and Komarov (1987).

To estimate special features of the vertical distributions of average temperature and humidity, we take advantage of the data in Table 12.2, which gives vertical profiles of the average values of these meteorological parameters (denoted by $\bar{t}(h_k)$ and $\bar{q}(h_k)$, where h_k is the altitude of the kth level). They were recorded at eight aerological stations located in different parts of western Siberia (their names and geographical coordinates have already been given).

Analysis of the data in Table 12.2 demonstrates some basic features inherent in the vertical distributions of average temperature ($\bar{t}$) and humidity ($\bar{q}$).

In particular, in winter for most of the atmospheric boundary layer over western Siberia, average temperature and air humidity significantly increase with altitude up to 800 m–1,000 m for temperature and 600 m–1,000 m for humidity, rather than decrease which is typical of the majority of regions of the northern hemisphere (Zuev and Komarov, 1987). Thus, for example, in the polar regions of western Siberia (at the Salekhard station), the average temperature in the layer 0 m–1,000 m increases by 7.4°C (from −21.1°C at the ground level to −13.7°C at an altitude of 1,000 m), and humidity increases by 0.46‰ (from 0.70‰ to 1.16‰, respectively).

Table 12.2. Average values of temperature (t, °C) and humidity ($\bar{q}$, ‰) for the Salekhard (1), Turukhansk (2), Khanty-Mansijsk (3), Aleksandrovskoe (4), Verkhnee Dubrovo (5), Omsk (6), Novosibirsk (7), and Emel'yanovo (8) stations.

Altitude	$\bar{t}$								$\bar{q}$							
(m)	*1*	*2*	*3*	*4*	*5*	*6*	*7*	*8*	*1*	*2*	*3*	*4*	*5*	*6*	*7*	*8*
January																
0	−21.1	−23.2	−15.2	−17.1	−11.2	−15.6	−16.5	−17.4	0.70	0.56	1.09	1.00	1.41	1.08	1.05	0.95
100	−19.5	−22.1	−14.3	−16.3	−10.6	−14.5	−15.1	−15.7	0.77	0.60	1.13	1.02	1.48	1.13	1.10	0.99
200	−17.9	−21.1	−13.4	−15.4	−10.1	−13.5	−14.1	−14.3	0.87	0.64	1.19	1.06	1.52	1.20	1.14	1.05
300	−16.6	−20.3	−12.6	−14.5	−9.7	−12.2	−13.2	−13.3	0.96	0.68	1.25	1.12	1.56	1.30	1.19	1.10
400	−15.7	−19.6	−11.4	−13.7	−9.4	−11.2	−12.5	−12.6	1.03	0.72	1.31	1.18	1.58	1.36	1.22	1.13
600	−14.5	−18.2	−10.6	−12.7	−9.1	−10.2	−12.0	−12.5	1.12	0.81	1.38	1.24	1.58	1.41	1.28	1.14
800	−13.8	−17.5	−10.5	−12.5	−9.0	−10.0	−11.6	−12.4	1.16	0.87	1.39	1.24	1.57	1.40	1.27	1.11
1,000	−13.7	−17.2	−10.7	−12.6	−9.2	−10.1	−11.7	−13.0	1.16	0.91	1.38	1.23	1.52	1.36	1.26	1.08
1,200	−14.2	−17.3	−11.1	−12.8	−9.7	−10.2	−11.8	−13.6	1.14	0.90	1.35	1.20	1.47	1.30	1.23	1.06
1,600	−15.1	−18.3	−11.9	−13.7	−11.3	−11.1	−12.8	−14.9	1.07	0.83	1.20	1.11	1.30	1.22	1.12	0.97

July																
0	15.1	16.2	17.2	17.3	18.1	18.5	18.4	7.50	7.37	8.72	8.67	9.33	8.93	9.67	9.83	7.50
100	14.8	15.6	17.3	17.5	18.5	18.8	18.4	7.27	7.02	8.28	8.61	9.13	8.57	9.31	9.47	7.27
200	14.4	15.1	16.9	17.3	18.4	18.6	18.1	7.05	6.76	7.91	8.32	8.83	8.30	9.06	9.22	7.05
300	13.9	14.6	16.3	17.0	18.1	18.3	18.0	6.83	6.58	7.61	7.96	8.47	8.02	8.79	8.96	6.83
400	13.4	14.1	15.6	16.3	17.4	17.8	17.7	6.63	6.40	7.41	7.64	8.14	7.78	8.48	8.70	6.63
600	12.3	12.8	14.0	14.9	15.8	16.5	16.6	6.29	6.11	7.02	7.08	7.69	7.40	8.03	8.16	6.29
800	11.0	11.5	12.4	13.4	14.2	15.0	15.2	6.09	5.86	6.68	6.74	7.42	7.11	7.60	7.81	6.09
1,000	9.7	10.1	10.8	11.9	12.6	13.5	13.7	5.79	5.66	6.44	6.48	7.19	6.87	7.27	7.50	5.79
1,200	8.5	8.7	9.4	10.4	11.1	12.0	12.2	5.51	5.44	6.13	6.19	6.95	6.65	6.90	7.22	5.51
1,600	6.2	6.3	6.9	7.6	8.3	9.1	9.4	5.01	4.89	5.27	5.60	6.02	6.04	6.08	6.28	5.01

The inverse vertical distribution of temperature and humidity observed in most of the atmospheric boundary layer in winter above the region studied is caused by two factors. The main factor is strong radiative cooling, and as a consequence the drying of surface air above the cold underlying surface of western Siberia under conditions of the dominant anticyclonic regime of atmospheric circulation. The second factor that contributes to the formation of surface inversions is manifested to a greater extent in the polar regions of western Siberia (see the Salekhard station again) and is caused by warm air mass transport above the cold underlying surface under conditions of the cyclonic regime of atmospheric circulation. Thus, according to Drozdov *et al.* (1989), the moving cyclone recurrence (i.e., the long-term ratio of the number of days with moving cyclones to the number of observation days in the month in question, in %) is about 20% in the region of Salekhard, which is much greater than the moving anticyclone recurrence of about 5%. At the same time, in the south of western Siberia where the radiative cooling of surface air plays the main role in the formation of temperature and air humidity inversions, the moving cyclone recurrence, according to Drozdov *et al.* (1989), is only 4%–7%.

We note one more interesting fact. It can be observed from Table 12.2 that in winter in the entire atmospheric boundary layer above western Siberia, a significant reduction in average air temperature and humidity is observed in the northeastern direction. Indeed, whereas at the Verkhnee Dubrovo station, located in the southwest of western Siberia (near Ekaterinburg), surface temperature and humidity are −11.2°C and 1.41‰, and temperature and humidity, for example, at an altitude of 800 m are about −9.0°C and 1.57‰ at the Turukhansk station located in the northeast of the region studied their surface values are −23.2°C and 0.56‰, and at an altitude of 800 m they are −17.5°C and 0.87‰.

This special feature of the spatial behavior of average air temperature and humidity, which is characteristic of the entire atmospheric boundary layer, is caused by the fact that the northeastern part of western Siberia is very close to the Siberian cold pole (the region of the northern hemisphere where minimum surface air temperatures are observed) located in Yakutiya and formed under the influence of strong snow cover radiation and intensive cooling (and hence drying of air) under conditions of low cloud amount in the region of the extensive Asian anticyclone (Drozdov *et al.*, 1989).

In contrast to winter, in summer when the Eurasian continent warms up, strong temperature and humidity inversions are no longer observed in western Siberia. Therefore, the basic features of summer vertical distributions of the average temperature and humidity in the atmospheric boundary layer above western Siberia generally decrease to their minimum values at the upper surface of the atmospheric boundary layer. Only in the central and southern parts of the region studied is the tendency towards an air temperature increase or constancy with altitude observed in the lower 100 m layer (for humidity, this tendency is not traced). Thus, for example, in summer the air temperature increases from 18.5°C at the ground to 18.8°C at an altitude of 100 m in the region of Omsk.

Such an air temperature distribution in the lower 100 m layer is due to the night cooling of surface air from the underlying surface that is cooled, and temperature stratification becomes so stable that a surface inversion, though weak, starts to develop. Stable stratification in the lower atmospheric boundary layer is formed by subsidence inversions resulting from descending air motions and adiabatic air heating in anticyclones. Since the water vapor content in these inversions remains the same as before the subsidence (Khromov and Petrosyants, 2004), they do not affect vertical humidity distribution (Table 12.2).

Thus, the results of our analysis of the vertical distributions of average temperature and humidity in the atmospheric boundary layer above western Siberia not only confirm the previously established character of their altitude changes in this layer, but also allow some special features of these changes to be elucidated for separate parts of the region studied due to better altitude resolution.

12.3.2 Special features of the vertical distributions of average zonal and meridional wind

Alongside analysis of the vertical distributions of average air temperature and humidity, it is also of interest to study the basic features of altitude changes in the wind characteristics typical of the atmospheric boundary layer above western Siberia. In this case, to analyze the vertical structure of the wind field, we study zonal and meridional wind velocity components.

Let us now proceed directly to analysis of special features of the vertical distributions of orthogonal wind velocity components and consider first the special features of zonal wind. We consider the data in Table 12.3. From Table 12.3 it follows that the western zonal wind is observed in winter in the atmospheric boundary layer above the whole of western Siberia, and its speed increases everywhere with altitude (from 0.4 m s^{-1}–0.9 m s^{-1} at ground level up to 5.6 m s^{-1}–9.4 m s^{-1} at an altitude of 1,600m). A weak eastern wind rather than a western one is observed only in the extreme northeastern part of western Siberia and only in the lowest 200 m layer, and the eastern wind speed rapidly decreases with altitude. Thus, for example, in the region of Turukhansk, it decreases from −1.4 m s^{-1} at the ground to −0.3 m s^{-1} at an altitude of 200 m. At a level of 200 m and higher, the western wind increasing with a high rate dominates; however, the rate of increase above the given region is a minimum (compared with the whole of western Siberia), and the wind speed does not exceed 6 m s^{-1} even at an altitude of 1,600 m.

We mention one more interesting feature: the occurrence of maximum western wind speeds in the atmospheric boundary layer above the southeastern part of western Siberia (see data for the Emel'yanovo station). The maximum increase in western wind speed with altitude is also observed here. Thus, in the region of Omsk the western wind speed increases with altitude by 6.1 m s^{-1} (from 0.7 m s^{-1} at ground level to 6.8 m s^{-1} at an altitude of 1,600m), whereas in the region of Emel'yanovo (located near Krasnoyarsk) it increases by 8.1 m s^{-1} (from 1.3 m s^{-1} to 9.4 m s^{-1} at an altitude of 1,600 m).

Naturally, there is physical evidence for all these special features (Drozdov *et al.*, 1989; Zuev and Komarov, 1987).

Table 12.3. Average values of zonal ($\bar{u}$, m s^{-1}) and meridional wind velocity components ($\bar{v}$, m s^{-1}) for the Salekhard (1), Turukhansk (2), Khanty-Mansijsk (3), Aleksandrovskoe (4), Verkhnee Dubrovo (5), Omsk (6), Novosibirsk (7), and Emel'yanovo (8) stations.

Altitude	$\bar{u}$								$\bar{v}$							
(m)	*1*	*2*	*3*	*4*	*5*	*6*	*7*	*8*	*1*	*2*	*3*	*4*	*5*	*6*	*7*	*8*
January																
0	0.4	−1.4	0.7	0.6	0.8	0.7	0.9	1.3	0.7	1.5	0.8	0.9	0.6	0.9	0.8	0.5
100	1.6	−0.9	2.2	2.0	3.4	1.9	2.0	2.9	0.8	3.6	1.5	2.0	1.5	1.4	1.5	1.6
200	2.8	−0.4	3.5	3.6	4.8	2.8	3.2	4.3	0.8	4.8	1.6	2.3	1.6	1.5	1.9	2.1
300	4.0	0.6	4.3	4.9	6.1	3.7	4.1	5.5	0.5	4.8	1.2	2.1	1.4	1.2	1.8	2.3
400	4.9	1.5	5.0	6.0	6.9	4.3	4.9	6.6	0.4	4.7	0.9	2.0	1.2	0.8	1.6	2.2
600	6.0	2.9	6.0	7.2	7.7	5.1	6.0	8.4	0.3	4.0	0.5	1.8	0.6	0.6	1.5	2.0
800	6.4	3.7	6.6	7.7	7.9	5.5	6.5	9.0	0.4	3.0	0.3	1.5	0.3	0.4	1.3	1.4
1,000	6.9	4.3	7.0	7.9	8.0	5.9	6.7	9.1	0.5	2.3	0.3	1.2	0.3	0.3	0.9	1.0
1,200	7.1	4.8	7.5	8.1	8.0	6.4	7.0	9.2	0.4	1.8	0.3	1.1	0.2	0.2	0.4	0.6
1,600	7.2	5.6	8.4	8.5	8.0	6.8	7.4	9.4	0.1	0.9	0.1	0.7	0.2	0.2	0.2	0.2

July																
0	−0.1	0.4	0.1	0.1	0.1	0.1	0.1	0.4	−0.9	−1.1	−0.7	−0.7	−0.2	−1.2	−0.6	−0.3
100	0.1	0.5	−0.1	−0.1	0.4	−0.1	−0.6	0.9	−1.2	−1.1	−1.0	−1.1	−1.3	−1.6	−1.1	−0.3
200	0.3	0.7	−0.2	−0.1	0.4	−0.2	−0.4	1.1	−1.5	−1.1	−1.1	−1.2	−1.6	−2.0	−1.0	−0.3
300	0.5	0.8	−0.3	0.0	0.2	−0.3	−0.2	1.0	−1.7	−1.2	−1.2	−1.1	−1.7	−2.2	−0.9	−0.4
400	0.7	0.9	−0.1	0.0	0.2	−0.3	−0.1	0.9	−1.9	−1.3	−1.3	−1.2	−1.9	−2.3	−0.8	−0.5
600	1.0	1.0	0.4	0.1	0.3	−0.2	0.1	0.7	−2.2	−1.4	−1.5	−1.3	−2.0	−2.3	−0.8	−0.5
800	1.3	1.2	0.6	0.2	0.4	−0.1	0.4	0.8	−2.6	−1.3	−1.6	−1.3	−2.1	−2.3	−0.7	−0.3
1,000	1.5	1.4	0.8	0.3	0.6	0.1	0.7	0.9	−2.7	−1.2	−1.5	−1.3	−2.3	−2.4	−0.7	−0.1
1,200	1.9	1.6	1.0	0.4	0.8	0.4	0.9	1.1	−2.9	−1.2	−1.4	−1.2	−2.5	−2.5	−0.6	0.1
1,600	2.5	1.9	1.4	0.8	1.2	0.9	1.4	1.8	−3.3	−1.1	−1.5	−1.3	−2.7	−2.6	−0.3	0.5

In particular, the prevalence of the western zonal flow in winter in the atmospheric boundary layer above western Siberia is caused by a well-known mechanism: the development of strong western winds throughout the entire thickness of the troposphere observed above the Asian continent. At the same time, the occurrence of weak eastern winds in the northeastern part of western Siberia in the lower 200 m layer is due to the predominant eastern circulation along the Siberian coast. According to Guterman (1965), eastern circulation is formed here under the influence of the northern periphery of baric minima and the southern periphery of the region of relatively elevated pressure above central Arctic regions.

Finally, the occurrence of maximum western wind speeds above the southeastern part of western Siberia, traced throughout the entire thickness of the atmospheric boundary layer, is connected with the so-called angular effect (Khromov and Mamontova, 1974), according to which wind always strengthens when it flows round a hill or ridge from the side, leaving it to the right. In our case, the western wind strengthens when it flows round the Altai and Sayan mountains.

In contrast with winter, the intensity of tropospheric circulation in summer decreases significantly (Drozdov *et al.*, 1989; Guterman, 1965). This special feature is clearly manifested in the atmospheric boundary layer. Indeed, from Table 12.3 it follows that weak western winds are dominant in the entire atmospheric boundary layer above western Siberia; in the lower layer they even alternate with eastern winds. In this case, a certain increase in western wind speed with altitude is observed by analogy with winter; however, western wind speed does not exceed 2.0 m s^{-1}–2.5 m s^{-1} even at an altitude of 1,600 m.

It is well known that analysis of only the zonal wind velocity component does not describe the complete pattern of the wind velocity distribution in the atmospheric boundary layer above a given territory, because it is based only on the meridional component. Therefore, we now consider the character of the distribution of meridional wind in the atmospheric boundary layer above western Siberia. To this end, we take advantage of Table 12.3. It can be seen from Table 12.3 that weak southern meridional winds are observed in winter above western Siberia, and they prevail in the entire atmospheric boundary layer. In this case, southern wind speed increases with altitude in the lower 200 m–300 m layer; above this layer it decreases significantly reaching 0.1 m s^{-1}–0.9 m s^{-1} at an altitude of 1,600 m. An interesting feature of the distribution of meridional circulation above western Siberia is the occurrence of a region having increased southern wind speeds in the northeastern part. Indeed, southern wind speeds are maximum for the given part of western Siberia (at the Turukhansk station) irrespective of altitude level, reaching a maximum value (about 4.8 m s^{-1}) at altitudes of 300 m–400 m. Low meridional wind speeds are observed almost everywhere and the revealed special features of its behaviour are caused by the fact that meridional flows in the atmospheric boundary layer in winter, especially in its lower 200 m–300 m layer, are significantly influenced by friction as a dynamic factor, and their southern direction is connected with the region of southern winds observed (according to Guterman, 1965), in the lower troposphere above western Siberia. In addition, the decrease in meridional speed with altitude above 200 m–300 m, observed for the whole of western Siberia, is caused by the corresponding strengthening of western zonal circulation with altitude and its increased stability in the upper part of the boundary layer adjacent to the free atmosphere. As for the region of maximum southern wind speeds observed in the atmospheric boundary layer above the northeastern part of western Siberia, its occurrence here is caused by the same mechanism as the occurrence of eastern zonal winds (the reason for this has already been discussed).

In contrast to winter, northern meridional winds prevail in summer above western Siberia in the entire atmospheric boundary layer. This can be seen clearly from Table 12.3. Moreover, the strengthening of northern winds from east to west in the region studied is typically observed. Thus, whereas at the Emel'yanovo station located in the east of western Siberia the northern wind speed, for example, at an altitude of 1,600 m is only $0.5\,\mathrm{m\,s^{-1}}$, at the Verkhnee Dubrovo station located in the west of the region studied it already increases to $2.7\,\mathrm{m\,s^{-1}}$.

Such special features of the meridional circulation peculiar to the atmospheric boundary layer above western Siberia are caused by the formation of a trough of low pressure near the eastern boundary of the region. The axis of the trough is directed from the Taimyr Peninsula towards western Siberia and further towards India (Drozdov *et al.*, 1989; Guterman, 1965). In the back part of the trough where most of western Siberia is situated, northern winds prevail, and their speed is a minimum at the eastern boundaries of the study area (see Figure 12.1) located near the line dividing the northern and southern circulations prevailing in the front part of the same trough.

12.4 SPECIAL FEATURES OF THE VERTICAL DISTRIBUTIONS OF TEMPERATURE, HUMIDITY, AND WIND VELOCITY VARIABILITY ABOVE DIFFERENT PARTS OF WESTERN SIBERIA

Special features of the vertical statistical structure of the temperature, humidity, and wind fields in the atmospheric boundary layer above the territory studied can be inferred more fully if we consider their variability parameters in addition to the average climatic characteristics illustrating the basic features of the fields. This will allow us to estimate possible field variations caused by the spatio-temporal variability of atmospheric processes.

As already indicated above, we used the standard deviation σ_ξ of temperature, humidity, and wind speed to characterize their variability. However, this statistical parameter cannot always serve as a comparative characteristic of the variability of the humidity, especially for comparison of the σ_q values calculated for different seasons. Therefore, to estimate the spread of values of humidity q_i about the average value $\bar{q}$, we used the variation coefficient $\theta_q = \sigma_q/\bar{q}$, expressed as a percentage, which is accepted in the meteorological literature (Zuev and Komarov, 1987).

In addition, by analogy with average climatic values, we analyzed basic features of the vertical distributions of variability parameters separately for the temperature–humidity complex and both zonal and meridional wind velocity components.

12.4.1 Some special features of the vertical distributions of the variability of air temperature and humidity

Let us consider first basic features inherent in the vertical distribution of air temperature and humidity variability in the atmospheric boundary layer using the data in Table 12.4, where average standard deviations are given for temperature (σ_t, °C), and Table 12.5, where the standard deviations of humidity (σ_q, ‰) are given together with its variation coefficients (θ_q, %).

Table 12.4. Standard deviations of temperature (σ_t, °C) for the Salekhard (1), Turukhansk (2), Khanty-Mansijsk (3), Aleksandrovskoe (4), Verkhnee Dubrovo (5), Omsk (6), Novosibirsk (7), and Emel'yanovo (8) stations.

Altitude	σ_t							
(m)	*1*	*2*	*3*	*4*	*5*	*6*	*7*	*8*
				January				
0	7.9	9.4	7.8	7.9	6.7	7.2	7.7	9.0
100	7.2	8.0	6.7	6.8	5.9	6.2	7.0	7.6
200	6.7	6.9	6.2	6.1	5.4	5.6	6.3	6.7
300	6.4	6.3	5.9	5.7	5.1	5.2	5.9	6.2
400	6.2	5.8	5.7	5.6	5.0	5.0	5.7	6.1
600	6.1	5.3	5.2	5.4	5.0	4.8	5.7	6.1
800	6.0	5.2	5.1	5.3	5.0	4.7	5.7	5.9
1,000	5.9	5.1	5.0	5.1	4.9	4.6	5.6	5.8
1,200	5.7	5.0	5.0	4.9	4.9	4.6	5.3	5.6
1,600	5.6	4.9	4.9	4.8	4.8	4.5	5.2	5.4
				July				
0	6.1	5.0	5.7	5.7	6.5	5.8	5.5	5.4
100	5.8	4.9	5.4	5.1	5.7	5.0	5.1	5.2
200	5.7	4.9	5.4	5.0	5.4	4.8	4.7	4.7
300	5.7	4.8	5.4	4.9	5.3	4.7	4.6	4.5
400	5.7	4.8	5.4	4.8	5.2	4.6	4.5	4.3
600	5.7	4.7	5.4	4.7	5.0	4.5	4.4	4.3
800	5.7	4.6	5.3	4.6	4.9	4.4	4.3	4.1
1,000	5.6	4.5	5.2	4.6	4.7	4.3	4.2	4.1
1,200	5.5	4.4	5.0	4.5	4.6	4.3	4.1	4.0
1,600	5.3	4.0	4.6	4.3	4.3	4.2	4.0	3.9

Analysis of the data in Tables 12.4 and 12.5 demonstrates that maxima of temperature and humidity variability are clearly manifested near the surface (σ_t, σ_q, and θ_q values here are within the limits of 6.7°C–9.4°C, 0.46‰–0.75‰, and 49%–78%, respectively). Above the maxima, σ_t, σ_q, and θ_q values decrease with altitude. These special features of the vertical distributions of long-term temperature and humidity variability are caused by the following mechanisms. For example, the surface maximum of meteorological parameter variability recorded for the whole of western Siberia is formed under the joint influence of radiative and circulation factors. According to Guterman (1965) and Zuev and Komarov (1987), intensive latitudinal and zonal airmass exchange is observed above the Asian continent (including western Siberia), which causes significant surface temperature and humidity variability under conditions of intensive radiative cooling and drying of the surface air. The influence of the radiative factor weakens with increasing distance from the surface, and variability of the meteorological parameter in question decreases.

From winter to summer, air temperature and humidity variability characterized by the parameter θ_q significantly decreases in the entire atmospheric boundary layer. In addition, in summer (by contrast with winter) the surface variability maximum is observed only for temperature, and moreover,it is weakly pronounced. Above this maximum, temperature variability decreases with altitude everywhere (to the upper boundary of the atmospheric boundary layer). By contrast with temperature, humidity variability characterized by the variation coefficient θ_q increases with altitude and reaches a maximum near 1,600 m.

These special features of the vertical profiles of long-term temperature and humidity variability can be explained as follows. The total decrease in temperature and humidity variability observed from winter to summer is caused by the decay of the Asian anticyclone and the significant decrease in cyclonic activity in summer. At the same time, retention of the surface temperature variability maximum is connected with significant diurnal temperature variations near the underlying surface. As for humidity, each temperature rise from night to day is accompanied by an increase in the water content in the surface air, since actual evaporation is much less than its maximum possible value because of the low value for soil moisture.

These are the basic features peculiar to the vertical distribution of temperature and humidity variability parameters in the atmospheric boundary layer above western Siberia.

12.4.2 Special features of the vertical distributions of zonal and meridional wind variability

Rather extensive material on the distribution of the variability parameters of orthogonal wind velocity components has now been accumulated and generalized for the free atmosphere (Guterman, 1965; Oort, 1983; Rechetov, 1973); however, it does not cover the planetary boundary layer where the character of this variability is more complex than at higher altitudes.

Therefore, it is of interest to study long-term zonal and meridional wind variability in the atmospheric boundary layer above western Siberia. To analyze special

Table 12.5. Average values of standard deviations of humidity (σ_q, ‰) and variation coefficients (θ_q) for the Salekhard (1), Turukhansk (2), Khanty-Mansijsk (3), Aleksandrovskoe (4), Verkhnee Dubrovo (5), Omsk (6), Novosibirsk (7), and Emel'yanovo (8) stations.

Altitude	σ_q								θ_q							
(m)	*1*	*2*	*3*	*4*	*5*	*6*	*7*	*8*	*1*	*2*	*3*	*4*	*5*	*6*	*7*	*8*
								January								
0	0.49	0.46	0.66	0.69	0.69	0.74	0.75	0.68	68	78	61	70	49	68	71	71
100	0.48	0.44	0.63	0.64	0.66	0.70	0.69	0.64	64	73	56	63	44	62	63	65
200	0.49	0.42	0.62	0.60	0.65	0.66	0.64	0.60	56	66	52	57	43	55	57	58
300	0.50	0.40	0.60	0.57	0.65	0.63	0.60	0.56	52	59	48	51	42	48	50	51
400	0.52	0.39	0.61	0.55	0.66	0.62	0.58	0.55	50	54	46	47	42	46	48	49
600	0.54	0.39	0.62	0.56	0.68	0.63	0.58	0.53	48	48	45	44	43	45	46	47
800	0.54	0.40	0.60	0.54	0.70	0.64	0.59	0.53	46	46	45	45	44	46	47	48
1,000	0.53	0.41	0.59	0.53	0.66	0.62	0.58	0.49	46	46	46	43	43	45	46	47
1,200	0.52	0.41	0.58	0.53	0.65	0.60	0.57	0.48	46	46	47	44	44	46	46	47
1,600	0.51	0.38	0.57	0.52	0.60	0.56	0.53	0.47	48	46	48	47	46	46	47	48

July															
0	2.52	1.93	2.30	2.35	2.02	1.98	2.16	2.31	34	26	25	24	22	22	22	23
100	2.46	1.95	2.08	2.23	2.10	2.05	2.21	2.38	34	28	25	26	23	24	24	25
200	2.43	1.97	1.98	2.17	2.01	2.09	2.21	2.37	34	29	25	26	23	25	24	24
300	2.37	2.01	1.91	2.09	1.92	2.11	2.21	2.30	35	31	25	26	23	26	25	26
400	2.32	2.04	1.85	2.02	1.89	2.08	2.21	2.24	35	32	25	26	23	27	26	26
600	2.23	2.03	1.84	1.93	1.84	2.02	2.18	2.22	35	33	26	27	23	27	27	27
800	2.13	1.98	1.80	1.86	1.82	1.92	1.98	2.10	35	33	27	27	24	27	27	27
1,000	2.07	1.90	1.78	1.75	1.77	1.87	1.95	2.07	36	34	27	27	25	27	27	28
1,200	2.05	1.83	1.65	1.65	1.72	1.85	1.91	1.99	37	34	27	27	25	28	28	28
1,600	2.01	1.73	1.53	1.50	1.64	1.74	1.84	1.89	40	35	29	28	27	29	29	30

features of the vertical distribution of the variability of orthogonal wind velocity components, we use the same parameters σ_u and σ_v that characterize perturbations in zonal and meridional motions under the influence of cyclones, anticyclones, or smaller scale vortices. The values of these statistical parameters are given in Table 12.6.

Analysis of the data in Table 12.6 demonstrates that in both summer and winter the variability in zonal wind speed increases with altitude, especially in winter when the average standard deviations σ_u increase everywhere from 1.6 m s^{-1}–2.3 m s^{-1} at the ground to 6.1 m s^{-1}–8.2 m s^{-1} at an altitude of 1,600 m. In addition, zonal wind variability decreases in summer; this is true except at the surface level, where the average standard deviation σ_u is within the range 1.9 m s^{-1}–2.4 m s^{-1} in the whole of western Siberia (i.e., it differs only slightly from winter values).

The presence of minimum surface zonal wind variability is caused by the influence of a well-known dynamic factor: air friction with the surface. This factor causes a decrease in zonal flow velocity in the surface layer (Table 12.3). At the same time, the decrease in zonal wind variability observed from winter to summer is in good agreement with a significant decrease in zonal circulation intensity and wind in the atmosphere (especially above temperate latitudes) in summer (Drozdov *et al.*, 1989; Guterman, 1965; Rechetov, 1973).

We note one more circumstance connected with estimation of the degree of zonal wind stability. To this end, relative variability (or the variability coefficient) $K_v = 100(\sigma_v/\bar{v})$ is conventionally used in meteorological studies (Rechetov, 1973), where σ_v is the average standard deviation of wind speed, and $\bar{v}$ is its average value (in our case, they are σ_u and $\bar{u}$).

If we consider this statistical parameter (which is easily calculated from the data in Tables 12.3 and 12.6), we find that near the surface, where weak zonal flows are recorded, the maximum relative zonal wind variability, which decreases with altitude, is observed almost everywhere (this altitude dependence of K_u is especially vividly pronounced in winter). In this case, in winter the K_u value near the surface varies within the limits 177%–500% in almost the whole of western Siberia, and even at an altitude of 100 m it lies in the interval 110%–181%. In summer the relative variability of zonal wind, because of its very low speeds (Table 12.3), is much greater than in winter, and this peculiarity is observed for the entire thickness of the atmospheric boundary layer. All this testifies to the instability of weak zonal wind near the Earth's surface, especially in summer when weak winds are observed in the entire atmospheric boundary layer.

Let us now consider meridional wind variability using the data in the same table (Table 12.6). From Table 12.6 it can be seen that, by analogy with zonal wind, meridional wind speed variability everywhere has a minimum near the Earth's surface and increases with altitude for both seasons. Thus, in winter values of the parameter σ_v increase from 1.0 m s^{-1}–2.9 m s^{-1} at the surface to 4.5 m s^{-1}–7.5 m s^{-1} at an altitude of 1,600 m, and in summer their values increase from 0.8 m s^{-1}–2.5 m s^{-1} at the surface to 3.6 m s^{-1}–6.4 m s^{-1} at an altitude of 1,600 m. In addition, by analogy with zonal wind variability, a decrease in meridional wind variations from winter to summer is clearly traced for the entire atmospheric boundary layer in western Siberia.

We immediately emphasize that the presence of minimum meridional wind variability in the surface layer and its decrease from winter to summer are caused by the same factors that influence zonal wind variability: the friction of air with the Earth's surface and the decrease in atmospheric circulation at temperate latitudes from winter to summer.

Along with these special features, attention is drawn to one more special feature that follows from analysis of Table 12.6: namely, that for both seasons a region of the least meridional wind variability is clearly traced in the entire atmospheric boundary layer above the southeastern part of western Siberia (where the Emel'yanovo station is situated). Thus, for example, values of parameter σ_v are less than $5\,\mathrm{m\,s^{-1}}$ in winter and $4\,\mathrm{m\,s^{-1}}$ in summer even at an altitude of 1,600 m.

The presence of a region with the least meridional wind variability above the southeastern part of western Siberia for both seasons is caused by strengthening of the western zonal flow and its stability (which, in turn, also affects meridional wind stability) when it flows round the Altai and Sayan mountains.

These are basic features of the vertical distributions of the variability parameters for orthogonal wind velocity components typical of the atmospheric boundary layer above western Siberia.

12.5 BASIC LAWS AND SPECIAL FEATURES OF THE VERTICAL CORRELATION RELATIONS FOR TEMPERATURE, HUMIDITY, AND WIND VELOCITY

Climatic description of the vertical statistical structure of the temperature, humidity, and wind velocity fields in the atmospheric boundary layer will be incomplete if we do not consider, alongside the background (average) characteristics and variability parameters, special features of the interlevel correlation relations for these meteorological parameters. We shall now consider the basic laws and special features of these relations.

12.5.1 Interlevel correlation of temperature and humidity

Let us consider the temperature and humidity data obtained at Salekhard, Turukhansk, Khanty-Mansijsk (Figure 12.2), and Aleksandrovskoe stations, which are located to the north of the 60th parallel and for the Verkhnee Dubrovo, Omsk, Novosibirsk, and Emel'yanovo stations which are located to the south of the 60th parallel (Figure 12.3). Figures 12.2 and 12.3 depict the isopleths of the temperature correlation coefficients between temperature data at certain height levels and those at higher levels (up to an altitude of 1,600 m). The isopleths of the humidity correlation coefficients between the humidity data at certain height levels and those at higher levels are presented in the same figure. Evidently, the diagonal straight line (which corresponds to the correlation coefficients equal to unity) separates temperature isopleths (above the diagonal) from humidity isopleths (below the diagonal).

Table 12.6. Standard deviations of zonal (σ_u, m s^{-1}) and meridional wind velocity components (σ_v, m s^{-1}) for the Salekhard (1), Turukhansk (2), Khanty-Mansijsk (3), Aleksandrovskoe (4), Verkhnee Dubrovo (5), Omsk (6), Novosibirsk (7), and Emel'yanovo (8) stations.

Altitude	σ_u								σ_v							
(m)	*1*	*2*	*3*	*4*	*5*	*6*	*7*	*8*	*1*	*2*	*3*	*4*	*5*	*6*	*7*	*8*
January																
0	2.0	1.7	1.7	1.6	1.3	2.1	2.1	2.3	2.6	2.9	2.2	1.7	1.0	2.2	1.8	1.4
100	2.9	2.5	2.8	2.7	3.3	3.4	3.3	2.8	3.2	3.7	3.4	3.3	3.2	3.4	3.0	1.8
200	3.9	3.1	4.3	3.6	4.5	4.3	4.2	3.5	3.8	4.7	4.4	4.5	4.1	4.3	4.1	2.5
300	4.9	3.8	5.2	4.5	5.6	5.1	5.0	4.0	4.7	5.7	5.5	5.3	5.0	5.2	4.8	3.0
400	5.5	4.3	5.9	5.2	6.2	5.8	5.7	4.5	5.4	5.6	5.6	5.8	5.3	5.7	5.0	3.5
600	6.3	4.8	6.5	6.1	7.0	6.9	6.2	5.6	6.3	6.3	6.1	6.3	5.7	6.2	5.5	4.0
800	6.3	4.9	6.7	6.5	7.4	7.3	6.9	6.2	6.7	6.5	6.4	6.4	5.9	6.5	5.5	4.1
1,000	6.4	5.1	6.8	6.7	7.6	7.5	7.2	6.7	6.8	6.6	6.7	6.7	6.1	6.7	5.5	4.2
1,200	6.5	5.6	7.0	6.9	7.7	7.9	7.6	7.0	7.1	6.7	6.8	6.8	6.2	6.9	5.6	4.3
1,600	6.6	6.1	7.3	7.4	8.0	8.1	7.7	7.3	7.5	6.8	7.0	6.9	6.3	7.5	5.7	4.5

July																
0	2.4	2.2	2.2	2.1	2.0	1.9	1.9	1.9	2.5	2.5	2.3	1.8	0.8	2.4	1.6	1.2
100	3.2	2.8	3.0	2.7	3.0	2.9	3.0	2.7	3.1	3.6	3.3	3.1	3.0	3.5	2.9	1.6
200	4.0	3.4	3.4	3.3	3.7	3.6	3.7	3.5	3.7	3.9	3.6	3.5	3.4	4.2	3.4	2.0
300	4.7	3.5	3.8	3.8	4.2	4.1	4.1	3.9	4.2	4.4	3.8	3.8	3.7	4.9	3.9	2.2
400	5.2	3.6	4.2	4.1	4.4	4.3	4.3	4.2	4.6	4.5	4.2	4.1	3.8	5.3	4.1	2.5
600	5.7	3.7	4.7	4.6	4.5	4.6	4.6	4.5	5.0	4.8	4.4	4.3	3.9	5.7	4.5	2.6
800	5.8	3.9	4.8	4.7	4.5	4.7	4.7	4.6	5.1	4.9	4.8	4.6	4.0	5.9	4.6	2.7
1,000	5.9	4.1	4.8	4.8	4.6	4.8	4.8	4.7	5.2	5.0	4.9	4.7	4.1	6.0	4.7	3.0
1,200	6.0	4.4	4.9	4.9	4.7	5.0	4.9	4.8	5.3	5.1	5.1	5.0	4.2	6.1	4.9	3.3
1,600	6.0	5.1	5.0	5.4	4.8	5.7	5.0	4.9	5.4	5.4	5.2	5.1	4.3	6.4	5.1	3.6

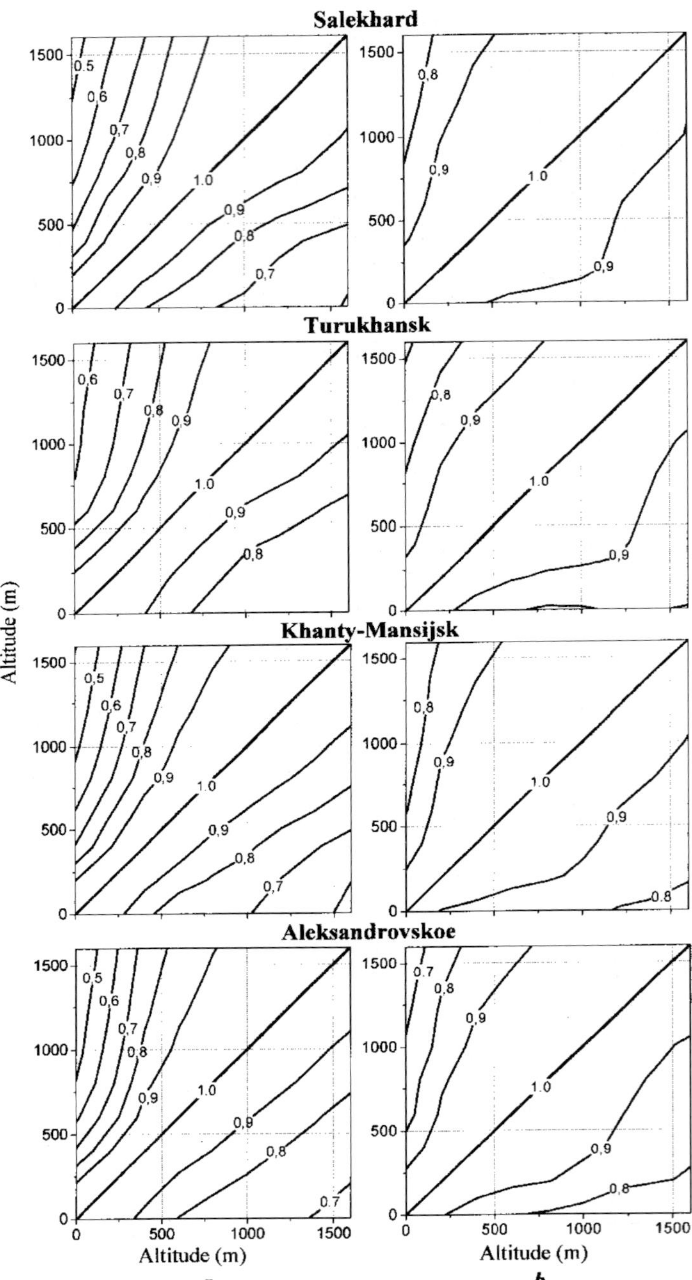

Figure 12.2. Plots of interlevel temperature (above the diagonal) and humidity correlations (below the diagonal) for typical stations of western Siberia located to the north of the 60th parallel in (a) January and (b) July.

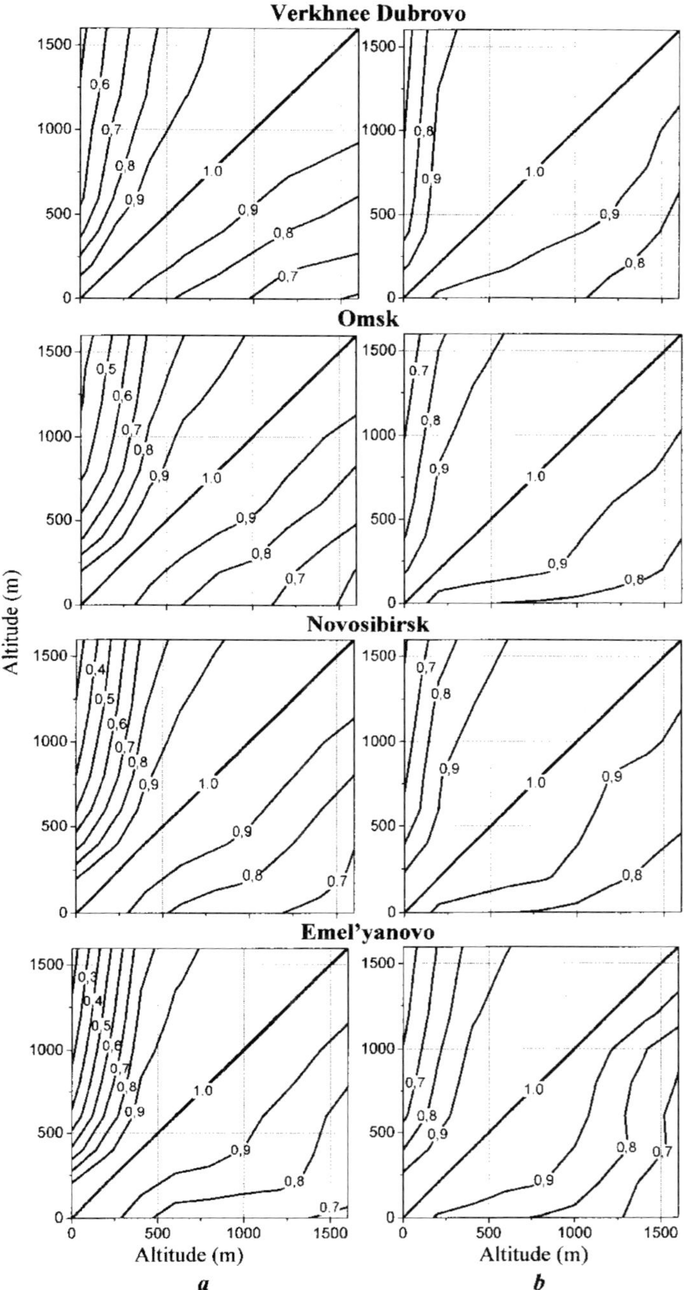

Figure 12.3. Plots of interlevel temperature (above the diagonal) and humidity correlations (below the diagonal) for typical stations of western Siberia located to the south of the 60th parallel in (a) January and (b) July.

Analysis of Figures 12.2 and 12.3 demonstrates that the correlations between temperature and air humidity variations at the ground and higher altitude levels in the atmospheric boundary layer are positive and decrease with increasing distance between the levels studied. This type of behavior of air temperature and humidity correlations between the initial ground level and all higher levels is observed everywhere for both summer and winter.

Along with this general behavior, some special features are also characteristic of the vertical correlation relations between temperature and humidity in the atmospheric boundary layer. In particular, in winter the rates of decrease of the correlation coefficients between temperature and humidity variations at the ground and higher levels are a maximum, as a rule, in the lower 600 m layer, where the interlevel correlations of these meteorological parameters are significantly reduced due to the occurrence of strong surface inversions, and the rates of decrease slow down at higher altitudes. Thus, for example, the temperature correlation coefficient $r_{tt}(h_0, h_j)$ at the Salekhard station decreased in this layer by 0.35 (from 1.00 at the ground to 0.65 at an altitude of 600 m), and in the layer 600 m–1,200 m (i.e., for the same spacing as the altitude levels), this correlation coefficient decreased only by 0.14 (from 0.65 to 0.50, respectively).

In addition, by contrast with winter, in summer without strong surface inversions the rate of decrease of the interlevel temperature and air humidity correlation coefficient slows down with increasing distance between the levels studied, and it behaves more smoothly. Indeed, in winter in the region of Salekhard the temperature correlation coefficient $r_{tt}(h_0$, 1,600 m) is equal to 0.46, whereas in summer it is much greater (about 0.70).

All these general and specific features are also characteristic of correlation coefficients $r_{tt}(h_i, h_j)$ and $r_{qq}(h_i, h_j)$ calculated between any initial (fixed) altitude level and all higher levels of the atmospheric boundary layer.

12.5.2 Interlevel correlation relations for wind velocity

Data on interlevel correlations of the wind velocity vector, in addition to their intrinsic interest, are necessary for solution of various applied problems connected, for example, with an increase in the efficiency of lidar sensing by means of reconstructing the vertical profiles of these meteorological parameters in a complex meteorological situation (fog, intensive precipitation, etc.) and with three-dimensional multielement optimal interpolation (one of the procedures of assimilation of four-dimensional information on the atmospheric state).

To analyze the basic laws of interlevel wind velocity correlation characterized by orthogonal wind components, we used Figures 12.4 and 12.5 illustrating, by way of example, plots of the distribution of interlevel correlation coefficients for zonal and meridional wind speeds vs. altitude in the atmospheric boundary layer drawn for the same two groups of stations (as in the case of temperature and humidity correlation).

These plots were drawn by analogy with plots of Figures 12.2 and 12.3, and each illustrates the distribution of zonal (above the diagonal) and meridional wind velocity components (below the diagonal).

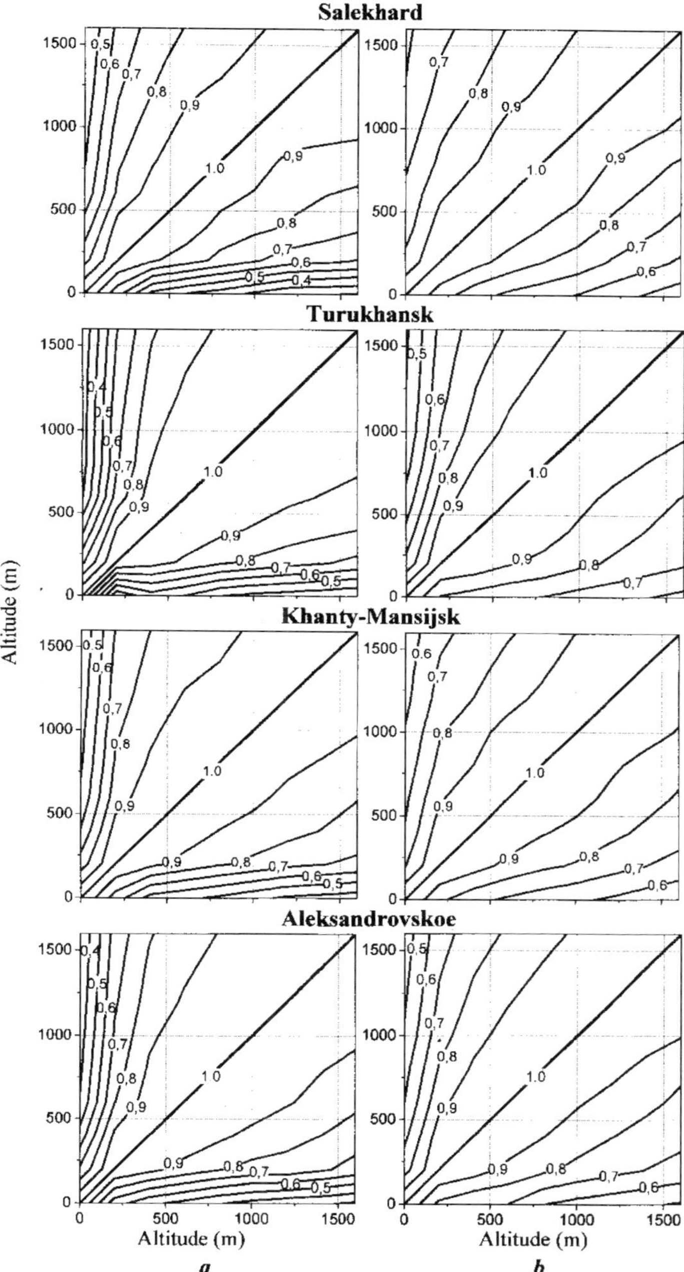

Figure 12.4. Plots of interlevel correlations of zonal (above the diagonal) and meridional wind velocity components (below the diagonal) for typical stations of western Siberia located to the north of the 60th parallel in (a) January and (b) July.

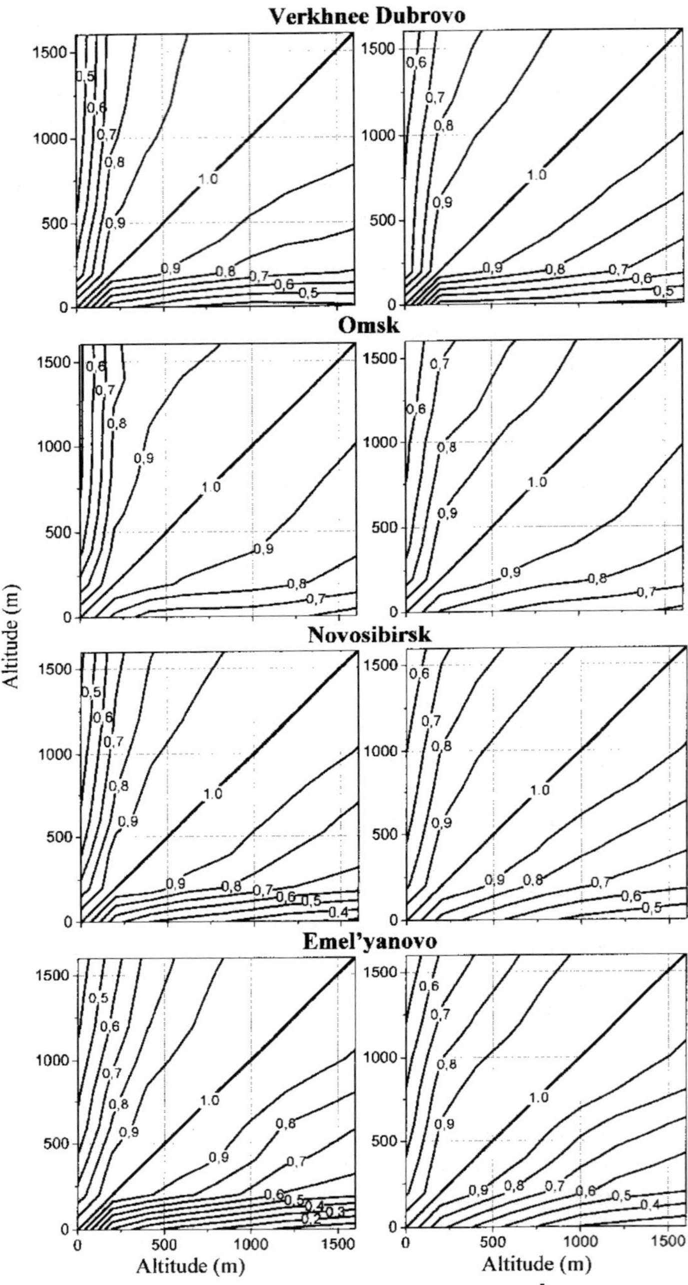

Figure 12.5. Plots of interlevel correlations of zonal (above the diagonal) and meridional wind velocity components (below the diagonal) for typical stations of western Siberia located to the south of the 60th parallel in (a) January and (b) July.

Analysis of Figures 12.4 and 12.5 demonstrates that zonal and meridional wind velocity components obey the same general laws as interlevel temperature and humidity correlation. The general feature is that the interlevel correlation of orthogonal wind velocity components is positive for the entire atmospheric boundary layer everywhere and irrespective of the season; it decreases with increasing distance between the altitude levels examined. In this case, in winter the interlevel correlations of zonal and meridional wind between the ground and higher levels is a minimum in the lower level 400 m–500 m layer (rather than in the 600 m layer as is the case for temperature and humidity correlation) (i.e., the rate of decrease of $r_{uu}(h_0, h_j)$ and $r_{vv}(h_0, h_j)$ is higher for wind velocity).

In addition, the different altitude dependences of interlevel correlation are characteristic of wind velocity (similarly as temperature and air humidity). In particular, in winter (in comparison with summer), the rate of decrease of the interlevel correlation between zonal and meridional wind velocity components with increasing distance between the levels studied is higher. This is caused by the fact that the interlevel wind velocity correlation in winter is significantly reduced due to the occurrence of surface inversions everywhere; these inversions are not observed in summer.

All the general features indicated above were also observed for the correlation coefficients $r_{uu}(h_i, h_j)$ and $r_{vv}(h_i, h_j)$ calculated between other initial (fixed) altitude levels and all higher altitude levels in the atmospheric boundary layer.

Finally, one interesting feature in the behavior of the interlevel correlation coefficient $\mu_{uu}(h_i, h_j)$ characteristic of the northeastern part of the region studied that manifests itself in winter should be mentioned: the minimum interlevel correlation of zonal wind speed calculated between the ground level or an altitude of 100 m and all higher levels is observed in winter. Thus, whereas the interlevel correlation coefficient, for example, $\mu_{uu}(h_0, 1{,}600) = 0.22$ in the region of Turukhansk in winter, in all other parts of western Siberia it was within the limits 0.30–0.42.

This behavior of the interlevel correlation of zonal wind speed above the northeastern part of western Siberia is caused by the fact that the eastern circulation observed there in the lower 200 m layer was changed by the western circulation that prevailed in the remaining part of the atmospheric boundary layer.

In conclusion, it should be noted that all data on the interlevel correlation of orthogonal wind velocity components (together with data on the interlevel temperature and humidity correlation) are not just of intrinsic (climatic) importance, but also can find wide application in solving various practical problems, and in particular problems of the optimal description of atmospheric perturbations in the atmospheric boundary layer using eigenvalues of the correlation matrices. However, the solution of this problem will be the subject of our future research.

12.6 REFERENCES

Borisov A.A. (1970). *Climatology of the Soviet Union*. Publishing House of Leningrad State University, Leningrad, 311 pp. [in Russian].

Czelnai R., Gandin L.S., and Zachariew W.I. (1976). *Statistische Struktur der Meteorologischen Felder*. Springer-Verlag, Vienna, 364 pp. [in German].

Dlin A.M. (ed.) (1975). *Mathematical Statistics*. Vyshaya Shkola Publishing House, Moscow, 398 pp. [in Russian].

Drozdov O.A., Vasiljev V.A., Kobysheva N.V., Raevskii A.N., Smekalova L.K., and Shkolnyi E.P. (1989). *Climatology*. Hydrometeoizdat, St. Petersburg, 567 pp. [in Russian].

Guterman I.G. (1965). *Distribution of Wind over the Northern Hemisphere*. Hydrometeoizdat, Leningrad, 251 pp. [in Russian].

Ippolitov I.I., Komarov V.S., and Mitzel A.A. (1985). Optical-meteorological model of the atmosphere for modeling lidar measurements and calculation of radiation propagation. In: V.E. Shuev (ed.), *Spectroscopic Methods of Atmospheric Sensing*. Nauka, Novosibirsk, pp. 4–44 [in Russian].

Khromov S.P. and Mamontova L.I. (1974). *Meteorological Dictionary*. Hydrometeoizdat, Leningrad, 568 pp. [in Russian].

Khromov S.P. and Petrosyants M.A. (2004). *Meteorology and Climatology*. Kolos Publishing House of Moscow State University, Moscow, 582 pp. [in Russian].

Komarov V.S. (ed.) (1972). *Handbook of the Statistical Characteristics of Temperature and Humidity in the Free Atmosphere over the USSR, Part I*. Hydrometeoizdat, Moscow, 160 pp. [in Russian].

Komarov V.S., Akselevich V.I., Kreminskii A.V., and Lomakina N.Ya. (1995). Regional climatic models for temperature and wind vertical distribution within boundary atmospheric layer. *Atmospheric and Oceanic Optics*, **8**(42), 1855–1865.

Kondratyev K.Ya. (1987). Global climate. *Meteorology and Climatology*, **17**, 1–316.

Kondratyev K.Ya. (1988). *Satellite Meteorology*. Hydrometeoizdat, Leningrad, 264 pp. [in Russian].

Kondratyev K.Ya. and Timofeev Yu.M. (1978). *Meteorological Sounding of the Atmosphere from Outer Space*. Hydrometeoizdat, Leningrad, 280 pp. [in Russian].

Nevzorova I.V. and Odintsov S.L. (2005). Correlation of wind velocity components in the atmospheric boundary layer. *Atmospheric and Oceanic Optics*, **18**(1/2), 124–129.

Oort V.P. (1983). *Global Atmospheric Circulation Statistics, 1958–1973*. NOAA Professional Paper No. 14, Government Printing Office, Washington, D.C., 180 pp.

Rechetov V.D. (1973). *Variability of the Meteorological Parameters in the Atmosphere*. Hydrometeoizdat, Leningrad, 215 pp. [in Russian].

Zuev V.E. and Komarov V.S. (1987). *Statistical Models of the Temperature and Gaseous Components of the Atmosphere*. D. Reidel, Dordrecht, The Netherlands, 306 pp.

Zuev V.E. and Zuev V.V. (1992). *Remote Optical Sensing of the Atmosphere*. Hydrometeoizdat, St. Petersburg, 232 pp. [in Russian].

Zuev V.E., Komarov V.S., and Kreminskii A.V. (1997). Application of correlation lidar data to modeling and prediction of wind components. *Applied Optics*, **36**(9), 1906–1914.

13

Ecological safety and the risks of hydrocarbon transportation in the Baltic Sea

Victor I. Binenko and A.V. Berkovits

13.1 INTRODUCTION

The book by Krapivin and Kondratyev (2002) stimulated our interest in the problem of environmental change in the Baltic Sea. The Baltic Sea in the 21st century is under the increasing influence of anthropogenic factors, and in particular those connected with the growth in the transport of hydrocarbons from newly constructed ports in the Gulf of Finland (each with a throughput capacity for oil of up to 60 Mt) and with the start of the construction of the land part of the North European Gas Pipeline (NEGP, also known as the "Nord Stream"), which then goes through the bottom of Portovaya Bay near Vyborg (Russia) to Greifswald (Germany) with the throughput capacity of 55×10^9 m^3 per hour and then on land to the Netherlands. The whole length of the NEGP is going to be almost 2,500 km, with the undersea part of the pipeline accounting for 1,200 km. It is being planned to extend one of its branches to the coast of England after 2010. Nowadays oil and natural gas meet more than 60% of the world's energy needs, but if they escape into the environment during extraction, transportation, processing, and storage, they have a negative impact on ecosystems. Thus, of the numerous anthropogenic factors that have bad effects on the environment oil takes the leading place, owing to the fact that it can escape into the environment during extraction, transportation, processing, storage, its actual use, and of course as a result of accidental oilspills. Oil consists of at least 1,000 (according to some sources of information more than 2,000) individual substances, most of which are poisonous for the overwhelming majority of animal organisms. Toxic oil components and products obtained from it, on entering an organism, can destroy its normal vital functions at the molecular, biochemical, physiological, and whole-organism levels. Among the oil components there are mutagens that can cause alterations in the genome (the set of chromosomes of an organism that deal with a cell's heredity) and carcinogens causing cancerous tumors. Also, oil can include substances influencing the biosynthesis of vital compounds and chemical compounds preventing cells from fission, embryogenesis, growth, breathing, reproduction, immune activity, and overall vital functions. What is more, almost 85 million people live and work along the Baltic coast. The entire range of ecological risks can occur, but an awareness of them together with data of an appropriate monitoring system can provide an acceptable level of ecological safety of the sea.

Therefore, the aim of this chapter is connected with

- Estimating the ecological risks connected with hydrocarbon transportation through the Baltic Sea using a module fitted into the GIS to estimate the ecological threat and economic loss caused by hydrocarbon spillage.
- Creating the geographically linked database of the Baltic Sea ecosystem within GIS MapInfo.
- Setting out the main objectives for ecological monitoring of hydrocarbon transportation routes.
- Applying GIF technology potentialities to optimize methods of working, thus eliminating the consequences of possible accidents during the extraction, transportation, and trans-shipment of hydrocarbons.

13.2 OBJECTS OF THE STUDY AND METHODS OF GENERALIZATION

The objects of our study are

(1) analysis of data about the Baltic Sea ecosystem, especially in the areas of intensive navigation and hydrocarbon transportation;
(2) analysis of information concerning the state of the underwater ecosystem at the places of oil extraction and where oil terminals are situated;
(3) making information on chemical and explosive weapons buried during and since World War II available; and
(4) gathering data on emergencies and accidents caused by hydrocarbon pollution of water.

The latter data were used to estimate the ecological risks connected with the possible pollution of the environment in the areas of extraction platforms and hydrocarbon transportation systems.

The creation of a database concerning the Baltic Sea ecosystem based on monitoring and the archiving of these data on a cartographic background using rated modules (which rate not only ecological risks but also possible pollution of water areas with oil products, the probability of these kinds of emergency, the impact of hydrocarbons in the case of an emergency, economic losses it can cause, etc.), based on geo-informational technologies (in our case based on GIS MapInfo, Version 7.5), the use of initial data received with the help of electronic maps, the attributive and rating parts compared with the available database, along with the prognostic GIS block: all this can facilitate making prompt decisions when eliminating the consequences connected with hydrocarbon exploitation.

The other important object of our research is estimation of the impact on the environment (EIE), connected with construction of the North European Gas Pipeline in the Baltic Sea, as well as securing the ecological support and ecological safety of this pipeline and the sea ecosystem. The arrangement of an environmental monitoring system at different levels of North European Gas Pipeline construction is also an important constituent part of the research, providing for elimination of ecological risk and possible damage to the underwater ecosystem and various benthos organisms along the path of the NEGP. Thus, the means of generalizing available data on the ecosystem and technosphere, connected with hydrocarbon transportation, should be implemented on the basis of GIS technologies which are designed to be used for practical purposes (Rastoskuev and Shalina, 2006).

13.3 ECOLOGICAL RISK

Analysis of ecological safety (according to Russian legislation) should be based on the concept of acceptable risk. Risk is the prognostic estimation of the probability of an emergency. The quantitative estimation of risk, R, is connected with the frequency of occurrence of emergencies; in other words, it is the correlation between the number of different negative consequences, n, of emergencies that have occurred and their possible number for a definite period of time. Thus, in 2001 of the 145 million people (N) living in Russia 2.058 million (n) died; consequently, the individual risk connected with residence in Russia gives

$$R = n/N = 2.058 \times 10^6/(1.45 \times 10^8) = 1.42 \times 10^{-2} \text{ per year.} \tag{13.1}$$

The individual risk of a person's death at the work place, or in a traffic accident, can be calculated in the same way using the same parameters R and N. The risk of human accidental death per year in Russia comes to $(1\text{–}1.7) \times 10^{-3}$, including murders 6×10^{-5}, suicides 1.9×10^{-4}, and traffic accidents 2.7×10^{-4}.

In the research carried out by Binenko *et al.* (2004) and Turkin (2004) the value of acceptable risk for staff is 10^{-5} and for regional residents 10^{-6} per year. The value of 10^{-6} per year is usually considered as the maximum acceptable level of risk of human death. Often the risk of human injury and damage to any object $R_{\text{dam}} = RP_d$ is calculated as the product of the frequency of some event, R, and the probability of damage, P_d, for which the risk is calculated. Thus, the probability of accidents in the technosphere can be divided into calculated and real ones. The theory of ecological risk is considered as the basic concept of ecological safety in the world. Ecological threat can be diminished but it cannot be entirely eliminated.

For biota, and for human beings in particular, ecological risk is determined by the possible failure of tendencies of natural–anthropogenic system development on its own. Because of this failure, changes in conditions will be negative for vital functions and can lead to different emergencies and even to ecological catastrophes. While natural–ecological risk is considered the natural condition of evolving geosystems, anthropogenic–ecological risk is the result of human activity, often the consequence of unpremeditated actions. These two constituent parts of ecological risk are essential for humankind, especially when their consequences coincide or provoke each other.

Quantitative estimation of economic loss R_e (per year), connected with ecological risk, can be defined with the following equation $R_e = RY$, where R is the value of ecological risk per year, and Y is the loss in euros.

At the same time, ecological damage to the underwater environment as a result of an oilspill can be calculated using the formula:

$$Y = \mu H_{\rm BW} K_{\rm EW} M_r, \tag{13.2}$$

where $\mu = 5$ is an increasing coefficient that calculates the excessive emission of poisonous substances; $H_{\rm BW}$ is the basic fine for an oilspill on the surface of the water ($H_{\rm BW} = 755$ euros per tonne); $K_{\rm EW}$ is a coefficient of the ecological situation and of the ecological importance of water objects ($K_{\rm EW} = 2.04$); M_r is the mass of oil that is considered to be polluting the water, taking into account that after the cleaning procedures the oil film is entirely removed; this is calculated with the formula:

$$M_r = 5.8 \times 10^{-3} M_s(C_s - C_b). \tag{13.3}$$

where M_s is the mass of oil spilled over the surface of the water (tonnes); C_s is the concentration of saturation of water with oil, $C_s = 26\,\mathrm{g\,m^{-3}}$; C_b is the background value of the concentration of dissolved and emulsion oil in the water before an accident ($C_b = 0.05\,\mathrm{g\,m^{-3}}$, water quality standard). According to equations mentioned above, the rated value of damage caused to an underwater ecosystem as a result of a 1-tonne oilspill represents a loss of approximately 1,000 euros.

Table 13.1 represents the data of ecological risks as the frequency of accidents per year on different objects connected with hydrocarbon pollution of the sea and coastal territories. On the basis of these data a conclusion can be reached that the ecological risk of gas pipelines is less than that of oil transportation and oil extraction.

The experience gained from running the deepwater (over 2,150 m) "Blue Stream" gas pipeline over two years (from the port of Dzhugba in Russia to Samsun in Turkey) which runs for 396 km along the bed of the Black Sea) and also the experience of a gas pipeline in the North Sea confirms the value of acceptable risk for gas pipelines as 10^{-5}–10^{-6}. The rated probability of serious accidents per year on gas pipelines with a length of 1,000 km is 10^{-4}, and the real probability is around 10^{-2} especially when the pipeline is operated for many years (Turkin, 2004; Binenko and Berkovits, 2006). The most serious leak occurred on the Kharyaga-Usinsk Pipeline in August 1994, when from 70,000 t to 100,000 t of oil were spilt, and in 1989 as a result of a gas pipeline rupture and fire near the railway line close to Ufa, where two passenger trains were passing: 575 people died and 118 people suffered various burns. In 2004–2005 over 20 terrorist acts were directed at pipeline transport, thus the urgency of strengthening antiterrorist activity along the entire length of the pipeline. In the last five years, 3,200 illegal inserts into pipelines have been revealed. These inserts lead to economic loss and ecological damage which have been estimated at tens of millions of euros (Khristenko, 2006; Binenko, 2006). Therefore, in order to minimize such damage as well as potential ecological catastrophes and human

Table 13.1. Ecological risks connected with some objects on or under the Baltic Sea.

Object name	*Ecological risk*
1. Extraction platform	1.9×10^{-3}
2. Technology platform	5.6×10^{-3}
3. Floating oil storage tank	1.0×10^{-3}–1.0×10^{-2}
4. Oil pipeline (coast)	2.8×10^{-3}
5. Underwater ecosystem of the Finnish Gulf/the Baltic Sea	2×10^{-4} 10^{-4}–2×10^{-5}
6. Gas pipeline	10^{-5}–10^{-6}

tragedies as a result it is necessary to follow technical, technological, and ecological requirements to comply with safety rules when servicing such potentially dangerous (flammable and highly explosive) objects as pipelines.

13.4 NORTH EUROPEAN GAS PIPELINE AND ECOLOGICAL SAFETY OF THE BALTIC SEA

The project of constructing the North European Gas Pipeline has been fully planned, but construction of its land part from Portovaya Bay near Vyborg to the town of Gryaznovec and farther to the south Russian oil–gas deposits (a total length of 920 km) was only started in 2006. The planned length of two pipelines as the underwater part of the gas pipeline on the bed of the Baltic Sea is 1,200 km and the pipeline length through Germany up to the main European connection point is planned to reach 400 km. For gas pipeline construction it is necessary to use steel pipes with a strength class of K60 and 36 mm thickness with an external three-layered anti-corrosion coating with a thickness of 6.0 mm, which will then be covered with a concrete layer 8 cm–10 cm thick. There are two possible variants for North European Gas Pipeline laying: direct laying or laying through an intermediate compressor station constructed on a metal platform on a sandbank near Gogland Island. Near Portovaya Bay, a compressor station with a power of 425 MW is going to be built, which will be able to pump over 55 billion m^3 of natural gas. Gas-pumping compressor stations should maintain high pressure (calculated by the Poiseuille formula) up to 22 MPa in the two pipelines of the NEGP. In order to cut off parts of the North European Gas Pipeline safely in case of an accident, ball pneumo-hydraulic cranes and remotely controlled linear cranes will be used as locking and regulating valves. In the event of an emergency the possibility of accident-free cessation of the pumping process using a remotely controlled system for gas transportation can be provided.

In order to increase the rate of ecological safety the North European Gas Pipeline should be deepened, and placed into trenches in the dangerous shallow places of the Baltic Sea. To provide the necessary stability for a gas pipeline (i.e., to prevent it from surfacing) it should be ballasted with concrete fillers.

Compared with land pipelines, underwater ones are notable for their safety from explosion and fire because of the fact that water lacks any great amount of oxygen. Nevertheless, the absence of ignition in the case of a leak in an underwater pipeline does not yet demonstrate the ecological safety of this object. For example, natural gas, leaking from a damaged pipeline, rises up and creates a poisonous cloud above the water surface, which is then spread by the wind. Surfacing of the gas occurs in the form of a two-phase stream, consisting of separate bubbles, which form some kind of a "boiling layer" with a diameter of 100 m on the water surface. In this offshore area, which is not deep, the leaking gas (as a result of a sudden pipe rupture) can form a gas–water fountain which can be 60 m high. Fountains do not form in depths of over 100 m.

When pipeline construction is destined to be set deep into the bottom, a trench is cut into the loose soil (several meters in width and depth) and a large amount of material suspension is formed. This is one of the main influences of pipeline laying on the bottom of the sea. Among other influences, the following can be mentioned (Binenko and Berkovits, 2006):

- a change in the morphology and the distribution of precipitation because of the physical presence of pipes and trench digging;
- a change in bottom-living biocenosis composition at the expense of biofouling, if the pipe lies on the surface;
- obstacles hindering movable benthos organism migration, if the pipe lies on the bottom; and
- sound, thermal, and electromagnetic influences.

Evidently, the most hazardous influence in the process of underwater pipeline laying is that on benthos forms, particularly in the spawning grounds of cod in the Baltic Sea (Smirnova and Smirnov, 2005). Figure 13.1 shows a map of the North European Gas Pipeline made using GIS MapInfo (Version 7.5) and the main spawning grounds (dark spots); arrows show the directions of cod-spawning migrations, and the places of shipwrecks with oilspill occurrence for 2005 according to Helsinki Commission (HELCOM) data (Fourman *et al.*, 2002). The average number of incidents connected with navigation for 1998–2005 amount 60 ± 3 (including 8 ± 2 for ship collisions). The largest number of shipwrecks happen in the coastal area, near ports, and the Kattegat Strait (over 2,000 large ships can be on the water at the same time). The statistical risk of such accidents will double by 2015, as a result of the increased number of vessels in the Baltic Sea and a doubling of the capacity to transport oil. Nevertheless, pollution in the Baltic Sea also depends on water from the 250 rivers that flow into it. These rivers carry the waste products of industrial and agricultural activity of more than 80 million people who live in the vicinity of the Baltic Sea.

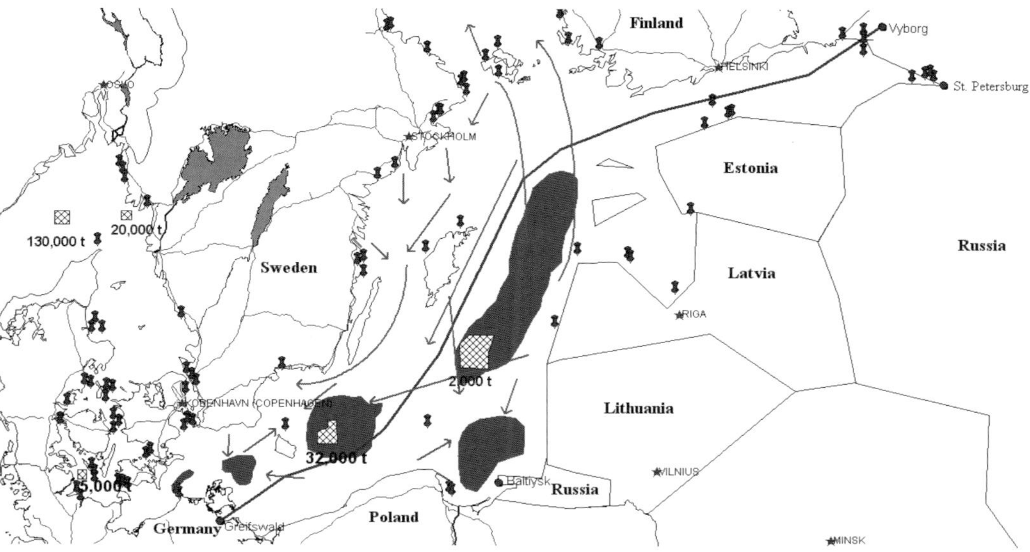

Figure 13.1. Thematic map of the NEGP route using the GIS MapInfo application. Main spawning areas (dark areas). Arrows show the directions of cod spawning migration, shipwreck and accident locations where there were oilspills in 2005. The locations of buried chemical weapons, poisonous substances, and explosives in the Baltic Sea are also shown.

Furman *et al.* (2002) represent the bathymetric characteristics of the Baltic Sea, its ice conditions, and data about the vertical and horizontal profiles of salinity, oxygen, and hydrogen sulfide content in the main parts of the Baltic Sea (Figure 13.2). The depth of the Baltic Sea reaches 459 m, but the average depth is 86 m. Data on ice forming in winter show that there are additional difficulties in ship transportion, especially in the Gulf of Finland. Data presented by Furman *et al.* (2002) indicate the relatively low salinity of Baltic Sea water, especially the surface water that is connected with the geographical location of the sea, which is surrounded by land. For this reason the saturation level of oxygen in seawater is not very high, and it suffers from eutrophication. The exchange of water between the Baltic Sea and the open North Sea occurs through narrow and shallow straits between Sweden and Denmark.

In the event of a gas pipeline rupture, the methane concentration in seawater with a value of 0.01 mg L^{-1} will have negative consequences caused by the poisonous influence of natural gas moving through the surface water where the early stages of fish development take place, as well as by hydrogen sulfide moving from the anaerobic zone to the surface water. Methane and other hydrocarbons have narcotic and convulsant effects on underwater organisms, and this is increased with increasing water temperature. Hypoxia is the main threat of these effects; its impact intensifies considerably with the presence of ethane, propane, butane, and other homologies of this type. A relatively harmless level of hydrogen sulfide content in water is stated in

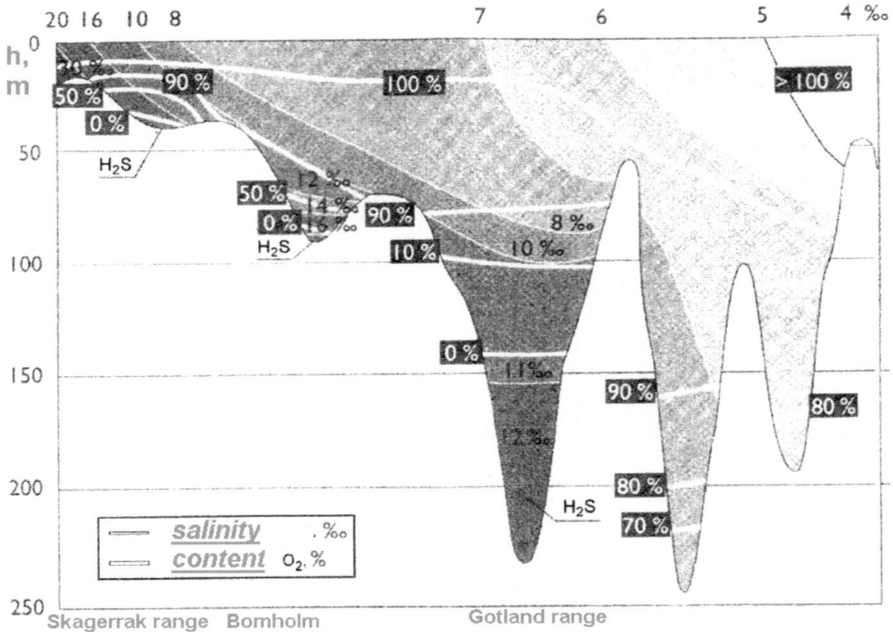

Figure 13.2. Data on vertical salinity profiles, as well as oxygen and hydrogen sulfide content in the main ports of the Baltic Sea according to Furman *et al.* (2002).

the literature (*http://www.helcom.fi*) to be 0.002 mg L^{-1}. The death of fry and fish will occur in the water with a 0.7 mg L^{-1}–1.4 mg L^{-1} concentration of methane if it is emitted for a period of about 10 hours.

In the event that a gas pipeline ruptures on the shelf, the negative impact of natural gas on the early stages of fish development will be strongly increased as a result of a massive hydrodynamic outburst which will occur in areas where the volley emission of gas is transported under high pressure.

Another negative effect of gas pipeline rupture on ichthyoplankton will be an increase in suspension concentration which forms as the result of an explosion. This effect is similar to those happening during construction, but is of shorter duration.

Another important problem as a result of NEGP construction on the bottom of the Baltic Sea is associated with the chemical weapons, conventional weapons, and explosive substances (ES) buried there as agreed by the Allies after World War II. That was not a very wise decision, and its consequences can serve as an example of ecological terrorism with respect to the Baltic Sea ecosystem and the people who live and work there (Goncharov and Pimkin, 2000). The munitions were dumped both in concentrated and loose form in the Baltic Sea in the Skagerrak and Kattegat straits, off the Swedish port of Lucechil and between the Danish island of Fjun and the mainland.

Altogether, in six areas of the water basin in Europe the Americans and British dumped the equivalent of 302,875 t of chemical weapons. A similar amount of chemical weaponry was found by Soviet troops in East Germany and these were also dumped in the Baltic Sea. They included quantities of yperite (mustard gas), chloracetophene, adamsite, diphenylchloroarsine, etc. and 7,840 tins of deadly cyclone that was used by the Nazis in 300 concentration camps for mass killing of prisoners of war and Jews in gas chambers.

Yperite is very dangerous for living beings as it tends to hydrolyze upon combining with water to form toxic substances that do not break down for a long time. Lewisite's properties are similar to those of yperite, the greater part of which lies on the sea bottom as pieces of poisonous jelly. Yperite and lewisite are arsenic-releasing substances; thus, they are ecologically hazardous as are practically all products of their transformation and their transportation in food chains.

The probability of poisonous substance penetration into the sea environment is determined by the correlation of two factors. The first is poisonous substance hydrolysis rate in seawater; the second is the intensity of poisonous substance transportation (as an impurity) by currents and during turbulent dissipation in an unconsolidated silt layer and benthos seawater. Thus, construction of special sarcophaguses for chemical weapons already dumped, and application of other measures for poisonous substance isolation and neutralization is an urgent and necessary task to ensure the ecological safety of the Baltic Sea ecosystem.

Ecological risks arising from the destruction of some chemical weapon shells containing such chemical weapons as tabun, various types of yperite, lewisite, and phosgene (some filled with up to 20 kg of chemicals) with exposure periods from 0.3 h to 11 h can result in an affected zone with a volume from 10^2 m^3 to 10^5 m^3. However, it should be noted that (at least for yperite) it is possible to neutralize chemicals by means of the bacteria *Pseudomonas duodoroffii* (Medvedeva *et al.*,1996).

As for explosive substances contained in grenades, shells, and air bombs, the affected zone resulting from their explosion can range from 5 m to 300 m depending on the amount of ES. Following the principle of least damage, the designers of the NEGP will lay it in a zone 500 m beyond the possible accessibility limits of these weapons.

All this combined with the geological features of the Baltic Sea bottom, the major navigation routes (i.e., 200,000 vessels annually), all the information from monitoring potentially hazardous areas during hydrocarbon transportation should be accumulated and stored on the basis of GIS technology to be available for ecosystem condition analysis, and in the event of an emergency to ensure decision-making to eliminate their consequences.

In the case of pipeline destruction, unless a fire starts in the initial period, the processes of emission dissipation into the environment will start and then result in the formation of gas-contaminated zones. In the case of gas concentration amounting to 5%–15% of the volume, such zones are fire-prone and may ignite if a source of fire is available, thus provoking a secondary excessive pressure wave and deflagration flame, hazardous for recipients who happen to be in such an area. In case of non-ignition, after some time the gaseous cloud tends to rise into the upper layers of the atmosphere and dissipates. Cloud dissipation is encouraged by a dramatic reduction in gas emission intensity from the open ends of the damaged pipeline, thus the contaminated area upon reaching its maximum size starts to decrease rapidly even in the first few minutes after the accident in the gas-contaminated zone.

The most dangerous accidents are when gas starts to burn in the initial period after pipeline damage. In this case the type of gas burning and the scale of the fire impact on the environment depends on a great number of factors as well as on the particular combination of some factors, the most important being working gas pressure, pipeline diameter, location of the pipeline failure, availability and location of disjunction fixtures and a means of their turning them off, method of pipeline construction, overall damage size (crack length), dimensions (length, width, depth) and shape of the new ground feature (a ditch or a pit), soil properties, and the displacement of the ends of the damaged pipeline.

This kind of project requires sub-regional cooperation between HELCOM countries aimed at ensuring expert examination, monitoring, and improvement of standards of ecological safety, concerning among other things navigation and the fishing industry in the Baltic Sea. The creation of a database of ecosystem monitoring along the NEGP route based on GIS technology is an important component of such cooperation, including both Russia's research centers and corresponding HELCOM structures.

13.5 MONITORING SYSTEM FOR HYDROCARBON TRANSPORTATION

The system of local, regional, and route monitoring of hydrocarbon transportation by ships and pipelines can be implemented on the basis of satellite remote sensing, airborne surveillance, remotely controlled unmanned airships, specialist vessels and submersibles (Kojima *et al.*,1997) including submersibles of the Mir and Remus types equipped with instrumentation for observation, monitoring, and control of the ecological safety of the sea ecosphere. The entire instrument complex and the object under investigation should be geographically positioned via GPS by means of an inertial navigation system adjustable by Doppler lag to provide exact homing and measurement of the research instrumentation carrier. Engineering research as well as geological, chemical, and ecological research should be carried out by means of an appropriate instrumental complex installed on various ships.

Table 13.2 contains a list of some tasks and corresponding instruments for the specific purpose, both for the NEGP and oilspill detection. Comprehensive ecological monitoring should be connected to a data collection center for storage and analysis concerning environmental changes in the Baltic Sea in order to make decisions about the handling of possible negative consequences during accidents occurring on hydrocarbon transportation routes. This involves cooperation with such organizations as Giprospecgas or Neftegasaerocomplex, as well as cooperation and coordination under the auspices of HELCOM.

Table 13.2. Main purposes and tasks for an instrument complex to monitor hydrocarbon transportation routes.

1.	*Purposes and tasks*	*Instrument complex*
	From underwater carriers	
2.	Identification of location of pipeline shifts and measurement	Television system, magnetic gravitation sensors, electrical and magnetic devices, acoustic profile graph, sector observation hydrolocator, GPS
3.	Identification of pipeline exposure	Hydrolocator and sector observation profile graph oriented magnetometer
4.	Inspection of the bottom terrain along the pipeline	Lateral observation hydrolocator, echolot
5.	Identification of other objects (stones, metal, chemical weapons)	Lateral observation hydrolocator
6.	Detection of transported substance leak	Acoustic profile graph, metal detector
	From ships	
7.	Investigation of shelf and sea currents, special location of main biological objects	Laboratory complex installed on the research vessel
8.	Chemical ecological investigation	Ground and water sampling with subsequent physico-chemical analysis
9.	Detection of transported substance leakage (gas, fuel, etc.)	Gas analyzer
	From air	
10.	Monitoring the blossoming dynamics of harmful micro-algae	Multiscanner MODIS
11.	Detection of transported substance leakage (gas, fuel, etc.)	Remotely controlled laser gas analyzer with a wavelength of 1.65 μm, GPS
12.	Detection of oil product spills	Radiolocator with synthesized equipment

13.6 ECOLOGICAL SAFETY OF OIL TRANSPORTATION IN THE BALTIC SEA

Cargo turnover through the ports of the Gulf of Finland and the Baltic Sea has been increasing exponentially. After commissioning the port of Primorsk, tankers with a deadweight up to 150,000 t and a loading draught over 15 m started to enter the Baltic Sea. For ships heading for Gotland, the boundary of the deepwater navigation channel runs along the 16 m–17 m depth contour, which increases the probability of their running into a shoal. Since the end of 2006 the Baltic pipeline has provided annual oil transportation of 72 Mt to Primorsk port. After the construction of new Russian oil terminals on the coast of the Gulf of Finland, including the construction of the pipeline branch from Primorsk to Vysotsk port, oil transportation will reach 78 Mt per year by 2015. Taking into consideration the fact that world oil transportation amounts to 2.2 Gt annually, the share of the Baltic Sea will be about 10% of the entire world transportation, which will result not only in the increasing intensity of navigation but also in a considerable deterioration of the ecological situation in the Baltic Sea area. Up to 10,000 t of oil products leak into the Baltic Sea annually. Such intensive development of tanker traffic in the Baltic will result in a situation by 2015 where the risk of oilspills up to 1,000 t in size will increase by 50%, while that of oilspills over 1,000 t will increase by 25%. The risk of emergency situations is especially high for oil transportation by tankers. The probability of large oilspills (over 150 t) during pipeline transportation and in the process of drilling works is reduced two to three times (Semanov, 2005).

Estimation of oilspill risks at sea implies

- identification of the potential source of oilspills in the sea;
- calculation of oilspill volumes and frequency of their occurrence;
- identification of natural resources and industrial facilities that may be contaminated as a result of oilspills;
- development of scenarios of oil behavior on the sea surface that should take into account oil spreading and weathering, depending on conditions in the spill area and the length of the affected coastal area.

Risk estimation can be the basis for designing measures to reduce emergency occurrences and their consequences, their elimination costs, and taking decisions to justify planned activities. The basic component of risk estimation is calculation of oilspill volumes and their frequency. This parameter is essential for the systemization of emergencies at sea and calculation of the resources required for oilspill elimination.

The main sources of oilspills are loading activities at oil terminals, accidents involving oil and oil product carrying tankers, illegal dumping of oil-containing wastes and accidents at oilrigs. Figure 13.1 shows cases of oilspills in the Baltic Sea occurring as a result of shipwrecks and during loading activities at oil terminals in 2005.

According to Russian legislation concerning measures pertaining to oilspills, the following classification of oilspill emergencies at sea is adopted:

- A local oilspill is an oilspill for whose elimination the resources available at the facility or its vicinity are sufficient. This spill does not exceed 500 t. It is handled by local resources or by the resources of cooperating organizations hired on a contract basis.

- A regional oilspill is one for whose elimination the resources available in the region are sufficient. Normally these are spills not exceeding 5,000 t. The Basin Administration of the Marine Rescue Service (BAMRS) is responsible for their handling and elimination. BAMRS is also involved in the elimination of local spills if they occur beyond the zone of responsibility of the organization involved in oil transportation activities or if this organization is not able to eliminate oilspills with its own resources.
- A federal oilspill is one exceeding 5,000 t and its elimination requires the involvement of resources from other basins and neighboring states.

The Federal Service of Maritime and River Transportation of Russia's Ministry of Transportation is responsible for oil collection activities in the sea.

The main sources of oilspills are loading activities at terminals where accidents, including flexible pipe rupture, loading device failure, tank overfilling, and loading tank damage, may occur during landing activities. The frequency of oilspills over 1 t per terminal can be considered equal to 5×10^{-4}, with the spill share within the 1 t–10 t range being 0.79%, that of 10 t–100 t being 0.036%, and over 1,000 t being 0.008% (i.e., 96% of all spills at terminals do not exceed 100 t; Tables 13.3 and 13.4).

Table 13.3. The probability of spilling more than 100 tons of oil during accidents involving single-hull and double-hull tankers.

Parameter	*Single-hull tankers*		*Double-hull tankers*	
Probability (*P*) of spill/accident	*P* spill under 100 t	*P* spill over 100 t	*P* spill under 100 t	*P* spill over 100 t
Shoal running	0.25	0.04	0.03	0.09
Collisions	0.25	0.04	0.03	0.09
Damage to structural elements	0.05	0.16	0.05	0.09
Fire, explosion	0.1	0.14	0.1	0.09

Table 13.4. Estimated mean volumes of oilspills.

Port	*Cargo* (10^3 t)	*Cargo* (10^3 t)	*Average oilspill* (t)
	2004	2010	
St. Petersburg	1,356	10,000	937
Primorsk	44,565	52,000	2,500
Vysotsk	1,515	14,000	1,250

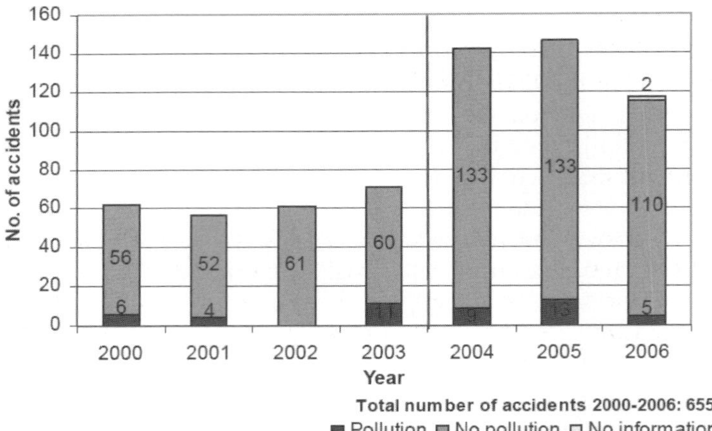

Figure 13.3. Number of reported accidents in the Baltic Sea during the period 2000–2006.

Figure 13.3 shows that the accident occurrence on oil vessels in the Baltic Sea in 2005, according to HELCOM data is most common in the Danish straits of the Baltic Sea. In 2007 while leaving Primorsk port a Greek oiltanker with a capacity of 100, 000 t was shipwrecked and only the fact that it was a double-hull tanker prevented it from causing an oilspill. As can be seen from Figure 13.3 most accidents in 2003–2005 were not accompanied by significant contamination of the environment.

Thus, according to the statistics, for every 100,000 loadings at a terminal there may be two oilspills with a mass of 100 t or more. Based on this, there is a probability that when the Primorsk terminal has achieved its planned capacity of 60 Mt per year there is expected to be one oilspill in 400 years during oil loading in tankers with a deadweight of 120,000 t. Calculation of the frequency and size of oilspills as a result of tanker accidents is based on statistics from the International Maritime Organization (IMO), according to which accident frequency (for seas with intensive navigation) includes shoal accidents 5.4 per 170 km, collisions 1.9 per 170 km, and fire or explosion 0.063 per 170 km.

To calculate the amount of damage it is necessary to estimate the volume of possible leaks (spills) resulting from potential accidents. The consequences of possible oilspills to a considerable extent will be determined by the size of oil product slick and the extent of sensitivity of the contacting components of the environment: land, water, and air.

Statistical data testify that most contaminants ending up in the water basin of the Gulf of Finland are contributions from river flows containing waste water from industrial enterprises (28%) and from ballast water (23%). This is confirmed by data from the routine practice of the emergency services. On the other hand, it is clear that oil product contribution from ship accidents does not exceed 5%–10%. However, it is these accidents that get most publicity, as in these cases thousands of tonnes of oil are spilled causing vast amounts of damage. Hydrocarbon contamination of the Baltic Sea results in its eutrophication, and according to data from the MODIS spectroradiometer encourages the concentration of blue-green algae, suspended particles in water basins experiencing the most intensive navigation, and in fish spawning areas (in particular, in the eastern part of the Gulf of Finland).

13.7 CONCLUSION

The findings of our investigation show that ecological risks involved in the construction of the NEGP on the bottom of the Baltic Sea are considerably lower than in the case of oil transportation by ships. The risk of a contamination emergency is especially high during oil transportation by tankers, and though natural gas is less hazardous than oil and its products both fuels when they get into seawater cause contamination, eutrophication, and changes in the food chains of the Baltic Sea ecosystem. Thus, ecological monitoring of hydrocarbon transportation routes should be comprehensive and regular, with permanent stations for automatic monitoring provided for the most hazardous locations of oil and gas transportation routes.

The capacities of GIS technologies (as exemplified by MapInfo) were used to provide an initial database of the Baltic Sea ecosystem with an estimation module to estimate ecological risks and potential economic damage from transported hydrocarbons, as well as for optimization of measures to eliminate the consequences of possible emergencies during the extraction, transportation, storage, and reloading of hydrocarbons.

Ensuring the ecological safety of plant and animal wildlife in the Baltic Sea area and the entire sea ecosystem should be implemented within the framework of international legislation and close cooperation between the Baltic Sea countries.

13.8 REFERENCES

Anon. (2006). Baltic Sea Day. *Seventh Int. Environmental Forum: Materials, March 22–23, 2006, St. Petersburg*. OOO Dialog, 592 pp.

Binenko V.I. (2006). Terrorism statistics in the Russian Federation: Ecological extremism and safety problems. *Problems of Safety and Emergency*, **4**, 45–56 [in Russian].

Binenko V.I. and Berkovits A.V. (2006). Ecological risks connected with transportation of hydrocarbons with estimation of the proposed construction of the North European Gas Pipeline (NEGP) and the safety of the Baltic Sea. *Problems of Safety and Emergency*, **3**, 83–96 [in Russian].

Binenko V.I., Khramov G.N., and Yakovlev V.V. (2004). *Emergency Situations in the Modern World and the Safety of Human Activity*. St. Petersburg University, St. Petersburg, 400 pp. [in Russian].

Furman E., Munsterhulm R., Salemna H., and Vjalipakka P. (eds.) (2002). *The Baltic Sea: The Environment and Ecology*. HELCOM, Digitone Oy, Helsinki, 39 pp.

Goncharov V.K. and Pimkin V.G. (2000). Forecasting the ecological consequences of PS penetration into seawater from the aged chemical weapons dumped in the Baltic Sea. *Ecological Chemistry*, **9**(3), 196–204 [in Russian].

Khristenko V.B. (2006). Russia's energy strategy: On the prospects for development and application of transportation of hydrocarbon raw materials and products. *Transportation Safety and Technology.* **4**(9), 22–29 [in Russian].

Kojima J., Kato Y., and Asakawa K. (1997). Development of autonomous underwater vehicle "Aqua Explorer-2" for inspection of underwater cables. *Proceedings of the Oceans '97 MTS/IEEE Conference, October 6–9, 1997.* World Trade and Convention Centre, Halifax, Nova Scotia, Canada, pp. 1007–1012.

Krapivin V.F. and Kondratyev K.Ya. (2002). *Global Environmental Change: Ecoinformatics.* St. Petersburg State University, St. Petersburg, 724 pp. [in Russian].

Medvedeva N.G., Sukharevich V.I., Poliak Yu.M., Zaitseva T.B., and Gridneva Yu. (1996). Russian Federation Patent No. 2103357 "Biodegradation technology for yperite-containing mixture, *Pseudomonas* bacteria yperite biodegrader, bacteria *Pseudomonas duodoroffi* 70-11-yperite biodegrader, bacteria *Corynebacterium* sp., KSB—yperite biodegrader" (AC12N1/20, C02 F 3/34). Ecological Safety Research Center of the Russian Academy of Sciences (filed 23.05.1996).

Rastoskuev V.V. and Shalina E.V. (2006). *Geoinformation Technologies for Solution of Ecological Safety Problems.* St. Petersburg University, St. Petersburg, 256 pp.

Semanov G.N. (2005). Oil spills in sea and provision of immediate response measures. Available at *http://www.secupress.ru/issue/Tb/2005-2/neft-rasliv.htm/*

Smirnova N.F. and Smirnov N.P. (2005). *Atlantic Cod and Climate.* St. Petersburg University, St. Petersburg, 222 pp. [in Russian].

Turkin V. (2004). Estimation of the ecological risk of offshore oil extraction. *Proceedings of International Conference. Modeling and Analysis of Safety and Risks: Complex Systems, MASR-2004, June 22–25, 2004, St. Petersburg*, pp. 430–433.

14

New directions in biophysical ecology

Andrey G. Degermendzhi

14.1 INTRODUCTION

It can be argued that biophysical ecology (i.e., the science concerned with studying the subject matter of ecology from the physical–mathematical point of view) is developing rather slowly. The rate of development of this science, which is highly important for developing scientifically based management of ecosystems and the biosphere, is limited by the following factors: (1) the absence of systematic experimental approaches (of the type used in physics) connected with the impossibility to make experiments with the ecological object which is unique (e.g., unique is the biosphere itself, a certain lake, river ecosystem, etc.); (2) the rare procedures for the verification of ecosystem mathematical models using field and/or experimental data; (3) the variety of interactions within ecosystems in terms of energy, matter, and control even for small-species communities; and (4) the absence of strict methods for the transfer of laboratory-scale experimental data to full scale. In this chapter we shall discuss some solutions to the situation.

We shall consider water resources as an example. The rapidly increasing consumption of water will soon make the lack of freshwater a factor that will limit the development of civilization as severely as diminishing energy resources will do. As a rule, the interests of water users are conflicting. However, almost all of them pollute water environments, seriously interfering with ecosystems and making harmful alterations to them. Aquatic ecology must be able both to predict the environmental consequences of the activities of water users and also to satisfy their needs in the best possible way.

As a fundamental science, the biophysics of aquatic ecosystems studies the physical and biochemical principles of ecological mechanisms responsible for the stability, controllability, and variability of aquatic ecosystems for short times (successions) and for long times (microevolution). The biophysics of ecosystems has three major branches with their own physical–mathematical methods: namely,

(a) monitoring the integrated parameters of ecosystems, (b) the kinetic experimental approach, and (c) mathematical modeling, which is based on the first two branches. In its methodology, the biophysics of ecosystems currently tends towards reductionism, maybe because it has been used successfully in physical sciences. Investigations address the spatio-temporal distribution and dynamics of various ecological structures of aquatic ecosystems (species, age, sex, functional structure, and trophic structure) and the hydrochemical conditions of a water body. More specifically, the biophysics of ecosystems deals with

— biochemical and population mechanisms: self-regulation of growth in aquatic communities, substrate consumption, material cycling, inter-specific relationships in the community;
— contribution of density and limiting factors to the stability of aquatic communities;
— physical principles underlying the theory of the search for limiting factors;
— laws of the stable coexistence of interacting populations;
— principles and theory of material cycling in aquatic communities;
— experiments, mechanisms, and the theory of migration behavior of aquatic organisms (plankton);
— scale-up of ecosystems;
— construction of ecosystems with tailored properties;
— ecosystems with closed material loops as models of biosphere-like systems.

The purpose of ecosystem biophysics is to reach such a level of knowledge about the elementary physical–biochemical mechanisms responsible for the functioning of aquatic ecosystems that would be sufficient to make valid prognoses of their natural and human-induced dynamics and to control their state.

A very important part of ecosystem biophysics is theoretical prediction of the development of aquatic ecosystems, including water quality. An instrument of prognosis (i.e., the theory and models of aquatic ecosystems) must be regarded as equal to the methods of biological monitoring (Kratasyuk *et al.*, 1996), including remote control, and physicochemical analysis of the state of a water body. Until recently, modeling of aquatic ecosystems has been only (and rather weakly) related to data of the classical monitoring of water bodies. The existing procedure of model identification and verification (actually fitting to field data) does not allow an extrapolation of constructed models to other water bodies, because it disguises and mixes up the errors of measurements of ecosystem inputs and the lack of knowledge of mechanisms responsible for the functioning of ecosystems. The most serious drawback of the existing method of modeling aquatic ecosystems (compared with physics) is that modeling is unrelated to experimental investigations. Thus, we cannot gain any essentially new knowledge about the mechanisms of interactions of biological components, so the heuristic significance of investigations is limited. Experimental investigations are laboratory and/or semi-field investigations of both the kinetic characteristics of aquatic organisms and the behavior of a community in special experiments. Experimental methods in biophysical ecology must, like physical ones, provide insight into the internal structure of communities and interactions between populations. The deepest insight into the structure of an ecosystem, its parts and their functioning is gained when experimental and field data are coordinated and the logical consistency of this coordination can be verified by mathematical models of various hierarchical levels. Although biologists are sometimes skeptical about the achievements of mathematical modeling, this may be the only means to strictly verify ecological hypotheses, particularly in the case of events with multi-directional processes running simultaneously, and the universal method to check the ecological efficiency of different scenarios of controlling the state of a water body (Gubanov *et al.*, 1996).

Contemporary knowledge of the structure of river, lake, and reservoir ecosystems and the practical positive control of the state of water bodies suggest more questions than answers. The answers are less profound than ecological problems. The reason is that aquatic ecology as a science encounters some objective difficulties related to the following sections (Sections 14.1.1–14.1.4)

14.1.1 Experiment in ecology

In contrast to physics, ecology is poor in experimental approaches; we do not refer to methods of field observations but rather to experimental approaches similar to physical ones (i.e., a discriminating experiment with a whole ecosystem responding to a sole experimentally calibrated impact).

14.1.2 Complexity of ecosystems

The rapid accumulation of ecological knowledge is naturally impeded by specific features of aquatic ecosystems. An ecosystem consists of numerous variously interrelated components, which are responsible for its counter-intuitive behavior (i.e., the behavior is opposite to what we can predict based on our limited knowledge, which seems to us quite complete). In ecology, this behavior has particularly grave consequences, as the human impact on aquatic ecosystems increases and there is rather limited time for thorough studies to counterbalance counter-intuitiveness. In this respect, physics has been in a better position for quite a long time. Counter-intuitive behavior can also be caused by changes in interactions between populations (due to adaptation, microevolution) that the ecology researcher is not aware of.

14.1.3 Non-trophic regulation of ecosystems

In the general case (maybe as a consequence of Section 14.1.2), we adhere to classical concepts and assume that, to make a valid prognosis, it is sufficient to know only the trophic–energy structure of an aquatic ecosystem and to have basic knowledge of the species. However, an ecosystem comprises organized fluxes of energy, matter, and control. Processes of control may be even more important for a valid prognosis than material flows. Moreover, the effective specific mechanisms of regulation that have been selected in the course of long-term evolution and that include various (e.g., chemical) special signal systems can influence all species, from bacteria to humans. Thus, when we consider the impact of pollutants, we should study not only the processes of their decomposition and biochemical transformation but also their damaging effects on regulatory interactions and their interference with regulation, including communications.

14.1.4 Hierarchy of ecosystems

Presumably, the declared hierarchical principles of the ecosystem structure must help us quickly accumulate ecological knowledge. At present, however, we cannot find an example of an actually efficient hierarchy with clearly defined rules for the formation of laws at each level. The holistic approach, as the antithesis of reductionism, must develop more rapidly and build up its own axiomatic basis. Cooperation of the holistic approach and reductionism in the research on one water body (on one problem) may essentially facilitate the establishment of workable hierarchical principles in aquatic ecology.

Investigations in the biophysics of aquatic ecosystems can be intensified along new lines as described in Section 14.2.

14.2 FUNDAMENTALS OF WATER ECOSYSTEM SIMILARITY THEORY

If we address the problem of experimenting with real aquatic ecosystems (see Section 14.1.1), leaving aside quite successful experiments with water treatment facilities, we can see that a well-developed methodology is still lacking. There is an approach based on the construction of various sizes of experimental micro-ecosystems; there are systems of continuous cultivation of microorganisms, and finally there are test-tank or aquarium-type laboratory systems. However, all these systems are deficient in principles, methodology, and methods of extrapolating the results of laboratory and semi-laboratory experiments to natural ecosystems. A mathematical theory of scaling of aquatic ecosystems could provide a scientific basis for developing the principles of such extrapolation. Scaling theory has proved to be useful in hydrodynamics and aerodynamics.

Let us recall the theory of dimensionality and scaling (Barenblatt, 1982; Sedov, 1972). The main result is contained in the "S-theorem" (short for "similarity theorem"). We suppose that physical value a depends on determining parameters and variables $a_1, \ldots, a_k, a_{k+1}, \ldots, a_n$:

$$a = f(a_1, \ldots, a_k, a_{k+1}, \ldots, a_n). \tag{14.1}$$

If $a_1, \ldots, a_k$ are independent variables then Equation (14.1) can be reduced to the relationship of dimensionless quantities:

$$S = F(1, \ldots, 1, S_{k+1}, \ldots, S_n),$$

where $S = a/a_1^h \cdots a_k^q$; $S_j = a_j/(a_1^{p_1^j} \cdots a_k^{p_k^j})$; $j = k+1, \ldots, n$, or compactly:

$$S = F(S_1, \ldots, S_{n-k}). \tag{14.2}$$

It follows from (14.2) that S really depends—not on n parameters—but rather on $n-k$ parameters.

Let us apply the S-theorem to the simplest model of an aquatic microbial ecosystem based on the principle of a chemostat. Let a population of microorganisms of biomass $x(t)$ develop in the system at specific flow rate D (the ratio of volume flux to system volume) and consume some substrate of the background concentration $S(t)$ and the input concentration S_0. An increase in biomass of 1 gram requires the consumption of y grams of substrate. The dependence of the specific growth rate (SGR) of biomass (g) is given as $g = \mu S/(K_s + S)$, where μ is the maximum SGR, and K_s is the half-saturation constant for the substrate. Then

$$\left.\begin{aligned} S &= \varphi(x(0), S(0), S_0, t, \mu, K_s, D, y) \\ x &= f(x(0), S(0), S_0, t, \mu, K_s, D, y), \end{aligned}\right\} \tag{14.3}$$

where dimensionalities are as follows:

$$[x] = [S] = [x(o)] = [S(o)] = [S_0] = [K_s] = M/L^3;$$
$$[t] = T; \quad [D] = [\mu] = T^{-1}; \quad [y] = 1.$$

As independent variables we take K_s and μ. Then, according to Equation (14.2), the dimensionless parameters are $F = x/K_s$, $W = S/K_s$, $\tau = t/\mu^{-1}$, $V = D/\mu$, etc. Equations (14.3) will be given as $W = \varphi(x(0)/K_s, S(0)/K_s, S_0/K_s, t\mu^{-1}, D/\mu, y)$ or $W = \varphi(\tau, V, y)$. Similarly, $F = f(\tau, V, y)$.

In contrast to an empirical search for Equation (14.2) type relationships, for this system there is a known mechanism, and thus dimensionless equations $W_\tau^* = (S_0/K_s - W)V - yWF/(1+W)$ and $F_\tau^* = (W/(1+W) - V)F$ can be written down. In the steady state $W = V/(1-V)$. A graph of a theoretical dimensionless relationship between the residual concentration of limiting substrate $W(= S/K_s)$ and the dimensional quantity of flow rate $V(= D/\mu; D < \mu)$ together with respective experimental values is presented in Figure 14.1. All undimensioned points are adequate (i.e., belong to) one and the same curve $W = V/(1-V)$. Even this very simple example shows that the condition of similarity between field (f) and laboratory (l) ecosystems (i.e., equality of all dimensionless similarity parameters, $\tau_f = \tau_l$; $V_f = V_l$, etc.) leads to the requirement of a certain relationship between population microbiological parameter (μ) and flow rate (D) as a hydrodynamic quantity: $D_f/\mu_f = D_l/\mu_l$. Hence, in laboratory experiments, populations growing at higher rates μ_l can be used, and thus higher flow rates D_l can be set. The dimensionless laboratory relationship between the background concentration of the limiting substrate and D/μ will be the same as the field concentration. Since $t_f\mu_f = t_l\mu_l$, laboratory time (t_l) of the identical laboratory and field dynamics of the components will be μ_l/μ_f times shorter than the field time.

Using the S-theorem, one can write down simultaneous ecological–hydrophysical equations for the dynamics of the state of an aquatic ecosystem in dimensionless

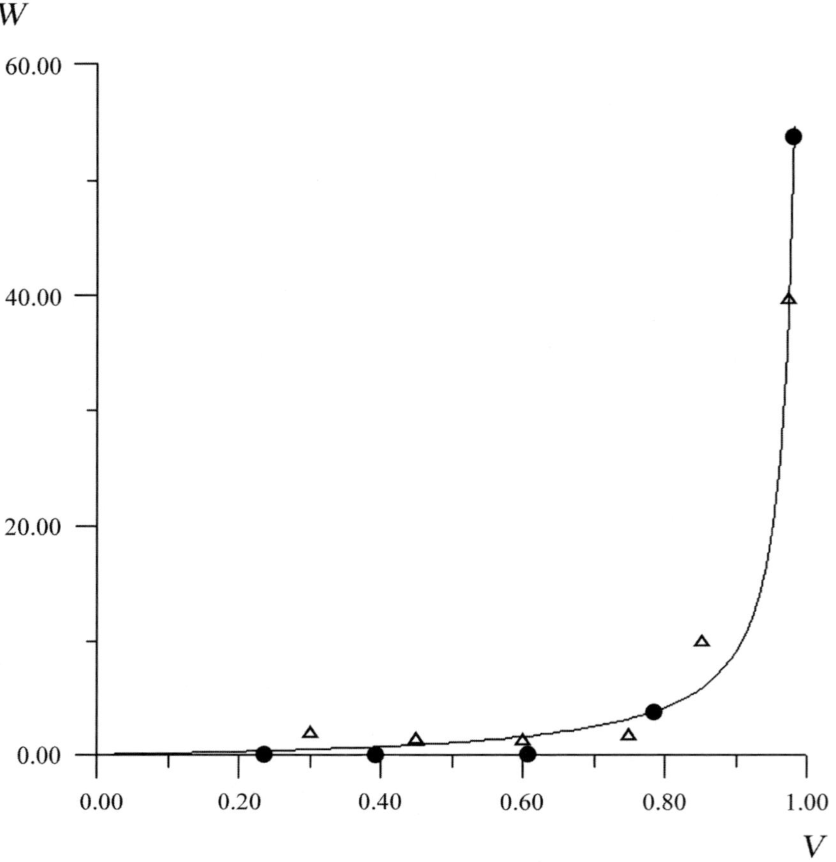

Figure 14.1. Dimensionless relationship between residual substrate concentration (W) and dimensionless flow rate (V). Experiments: • *Saccharomyces carlsbergensis*, substrate, glucose (Toda, 1976); △ mixed culture of activated sludge, substrate, glucose (Chiu *et al.*, 1972). Theory: —— $W = V/(1 - V)$.

form. Thus, new dimensionless parameters can be added to well-known ones (i.e., Reynolds', Froude's, etc.), with ecological micro-parameters used along with hydrophysical ones. The future scaling theory for aquatic ecosystems will contain a simultaneous mathematical description of the three main groups of processes: hydrodynamic, hydrochemical, and hydrobiological. The ultimate goal must be scaling of the maximally complete system of equations generally consisting of (1) a hydrodynamic unit, (2) a hydrophysical unit, and (3) an ecosystem unit. The objective of the hydrodynamic unit is to calculate the spatio-temporal dynamics of current velocity (depending on the morphometry of the water body floor, friction, slopes, water flow, and inflow). The objective of the hydrophysical unit is to calculate the dynamics of the following parameters: water temperature (depending on turbulence, heat balance with the atmosphere, and input of thermal effluents); the level of underwater irradiation (depending on the outer light flux, light absorption and reflection by microalgae and particles); sedimentation; turbidity; etc. The objective of the ecosystem unit is to calculate the dynamics of the concentrations of phytoplankton, zooplankton, bacteria, the main hydrochemical components, and pollutants in the water column, and the dynamics of bottom-water organisms (depending on biological interactions between populations, material cycling, industrial effluents, limiting factors, hydrophysical and hydrodynamic conditions, and sludge transport). The author is planning to create a computer system that will simulate these units, in dimensional and dimensionless forms, and inverse algorithms, which will reconstruct field dynamics from laboratory dynamics.

To understand the interactions between sub-systems it may be interesting to consider various correlations between characteristic relaxation times and the times of impact increase. According to the data of other natural sciences, different correlations between these times can cause various instabilities, and consequently isolated or ubiquitous occurrences of a sharp increase in the biomass of aquatic organisms or some other pronounced imbalances. These deflections from the theoretically monotonic smooth trend of the curve are crucial growth points in scaling theory.

Having undimensioned macro-parameters of the system of the abovementioned groups of equations, we can make a universal undimensioned description of the dynamics of some ecosystems. Then, varying experimental dimensional micro-parameters, we may be able to find the values of undimensioned macro-parameters equal to real ones and conduct experiments with this small ecosystem. Conversely, experimental dynamics must be converted into real dynamics for a large ecosystem, which cannot be experimented on. Accurate similarity scaling can start a new direction in the experimental modeling of very many ecologically significant phenomena (material cycling in aquatic ecosystems, self-purification, stratification of biological components, migration of plankton, microalgal blooms) together with the modeling of hydrophysical parameters (currents, light and temperature fields, etc.). It would be good to use experimental facilities that hydraulic engineers have used for similarity scaling of hydrophysical characteristics only. For the sake of similarity, it will be necessary to equip these facilities with technical systems of light radiation for microalgal photosynthesis, to prepare model effluents, etc. The great advantage of this approach is that decision-makers would clearly see the environmental consequences of a given project even before it is practically implemented. First, it would be reasonable to construct simple homogeneous ecological flow-through systems and then gradually to move up to spatially heterogeneous ones. At the same time, it would be necessary to develop an ecological–hydrophysical scaling theory, later involving the scaling of hydrochemical processes. In the course of development, theoretically grounded bans may be placed on simultaneous scaling of ecological–hydrophysical processes that produce an opposite effect on scaling parameters, as happens in hydrodynamics in the case of wave resistance to movement ($\mathrm{Fr} = v/\sqrt{lg}$) and in the case of viscose resistance ($\mathrm{Re} = \rho vl/\mu$). The main concerns of the scaling theory for aquatic ecosystems are (a) the validity of systems of equations and (b) the theoretical limits of similarity scaling.

14.3 GROWTH ACCELERATION; A NEW INTEGRAL INDEX OF THE CUMULATIVE EFFECT OF ALL THE REGULATORS IN A MONOCULTURE

As the question of the complexity of ecosystems (Section 14.1.2) is rather difficult, the question of the non-trophic regulation of ecosystems should be pursued simultaneously. To create a stock of valid models, taking into account the mechanisms of population regulation (see Section 14.1.3), it is necessary to amass experimental data on the kinetic parameters of aquatic organisms, with kinetics being defined broadly (growth rates, food spectra, types of limiting factors, death rates, nature and intensity of inter-population relationships, etc.). These kinetics must be used in models along with quantitative field observations of the dynamics of ecosystem components so as to verify and identify the structures of model ecosystems. That is why the modeler's work cannot be independent of the experimenter's and the naturalist's work. They have to design experiments together.

Experimental methods must play a special part in the development of mathematical models of natural aquatic ecosystems, and specifically of microbial aquatic communities. The most important biochemical substances are those that are responsible for the sustainability of a microbial community. First of all, these are density-dependent growth control factors (DDGCFs; i.e., substances that are released or consumed by a population and that influence the growth of this or another population; Odum, 1971). It is traditional to determine the relationship of the SGR to a specific DDGCF (e.g., a Monod-type relationship). However, the question of whether one such relationship is enough is not usually discussed (i.e., whether Liebig's bottleneck principle is valid here or the SGR depends on other DDGCFs, unknown to the researcher). In more general terms, this question can be formulated as follows. If we know the relationship of the SGR to some specific DDGCF, can we accurately quantify our knowledge of the density-dependent control of this species in a specific system? In other words, is there a way to determine the aggregate effect of all the DDGCFs on a specific population? In contrast to physics, where the types and number of forces and principles of their action are well-known, the situation in aquatic ecology is quite different. Any product of the ecosystem's metabolism (innumerable biochemical substances) can potentially be a factor controlling the stability of the community by positive or negative feedback. Even if we manage to make a complete list of all the biochemical products of metabolism, the main question remains open as to which of these substances can influence, say, the growth rate of a microbial population and how? Only these substances can be regarded as DDGCFs, which are essential for modeling.

The fundamental solution to this problem is based on an essentially physical idea. The idea is as follows. Take a separate microbial population, a monoculture, and assume that it is related to several biochemical DDGCFs by feedbacks. Microbiologists know that not only limiting substrates—but also metabolites—inhibiting or stimulating growth, can be considered to be DDGCFs. Then, what is the overall measure of the feedback level in growth control; that is, what is the estimate of the total effect produced by all the DDGCFs on population growth? As the theory developed previously states (Degermendzhy *et al.*, 1993), this is a change in the growth rate increase B (i.e., acceleration of growth). Or, in other words, it is the rate of change of SGR, g, in response to a pulse disturbance of population concentration ΔX, under an unchanged (at the moment of disturbance) chemical composition of the environment:

$$B = \left.\frac{\partial g}{\partial t}\right|^{d} - \left.\frac{\partial g}{\partial t}\right|^{u}, \tag{14.4}$$

where d is the disturbed state; u is the undisturbed state (the control state); and $[B] = T^{-2}$ (i.e. the dimensionality of growth acceleration value (B) is inverse to negative quadratic time $T(T^{-2})$; there is a similar value with similar dimensionality in Newton's mechanics called "physical object acceleration").

In the general case, for a monoculture whose SGR is determined by several (n) DDGCFs, the formula for calculation of the theoretical specific values of $B(B_{Th})$ is given as

$$B_{Th} = \sum_{i=1}^{n} (\partial g/\partial A_i)a_i, \tag{14.5}$$

where $g(A_1, A_2, \ldots, A_n)$ is the SGR of the monoculture as a function of all DDGCFs; and a_i is the coefficient of transformation of the ith DDGCF.

Thus, all the n DDGCFs make a plus or minus contribution ($a_i\, \partial g/\partial A_i$) to the total theoretical value of B_{Th}. On the other hand, the same value can be found experimentally, B_E, from a change in the growth rate increase of a disturbed population and an undisturbed one, based on the above definition of feedback and Equation (14.4). It is assumed that growth rate increase as a response to the disturbance by biomass concentration occurs without any delay, due to the density activity of microorganisms (Figure 14.2). The value of $B_E - B_{Th} \equiv \Delta B$ determines the total

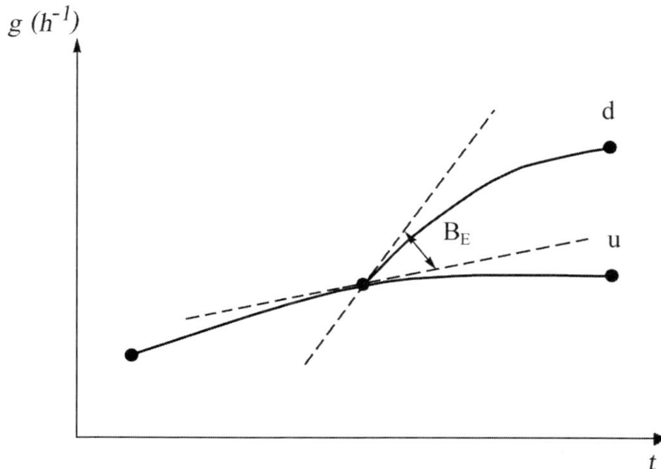

Figure 14.2. Approach to estimating the experimental level of feedback B_E.

control by unknown DDGCFs. The proportion of total unknown DDGCFs in the total control (B_E) (i.e., $\Delta B/B_E$) can be determined, too. This is a very important value, showing the magnitude of the total control efficiency contributed by the yet unknown DDGCFs. If this value amounts to several dozen percent, a search for other DDGCFs should be continued. Strict equality of the positive feedback component to the negative one, so that $\Delta B = 0$, seems unlikely. As long as this variant has not been found in reality, we will not take it into account. In principle, the proposed method can be realized experimentally (Degermendzhy *et al.*, 1993). Based on this, we can estimate the contribution of specific regulators to the integral value of feedback and the natural (seasonal) values of feedback for natural populations of aquatic microorganisms in their natural habitats. Put in simpler terms, the experimental value of the natural negative feedback shows the degree to which the growth of a population of microorganisms is limited by natural substrates in a given place at a given time. A similar value obtained as a coefficient of interactions between populations shows the degree of competition or other types of relations.

Thus, in modeling the internal structure of a microbial community, the freedom of the model should be restricted by the requirement that the calculation should be in agreement with both classical kinetic parameters (production, generation time, the first-time derivatives of biomass) and new ones, values of natural feedbacks, and interaction coefficients (the second derivatives). A necessary limitation is the requirement that the model should correspond to field monitoring data (e.g., biomass of species, or concentration of chemical substances such as zero derivatives).

14.4 BIOASSAY SYSTEM AS A NEW METHOD OF DESCRIPTION OF THE STATE AND DYNAMICS OF ECOSYSTEMS, AND THE ALTERNATIVE OF MAXIMUM PERMISSIBLE CONCENTRATION (MPC)

As we have already noted, the traditional, essentially reductionist, scheme of the mathematical model describing the state of an aquatic ecosystem and water quality (see Sections 14.1.2 and 14.1.3) is based on a rather detailed flowchart of biochemical transformations of matter and energy in the trophic links of the ecosystem, taking into account the nature and intensity of non-trophic (regulatory) interactions between populations, the relationship of kinetic characteristics to modifying factors, etc. Many of these characteristics can be obtained experimentally under laboratory or semi-field conditions. Identification and verification of these models must also involve substantial hydrochemical and hydrobiological field data, including hydrometeorological information. If the model representation is valid, calculations of numerous components are then aggregated in some characteristics or categories of water quality and compared with respective standards. This approach will be successfully applied at various water bodies for many years to come. However, for the prognosis of water quality, this approach would seem excessive, very labor-consuming, and costly. A scientific search for alternative approaches is necessary.

The increasing human impact will present a number of challenges for these approaches:

(1) the chemical range of new pollutants broadens more quickly than their instrumental control develops and the norms of their levels in the environment are established;
(2) the system of establishing the norms for individual chemical components through the Maximum Permissible Concentration (MPC) and Maximum Permissible Discharge (MPD) is also far from perfect:
    (a) the MPC does not take into account the biological consequences of interactions between chemical components;
    (b) the MPC and MPD ignore the fact that in an aquatic ecosystem, in the course of biotransformation and succession, the spectrum of substances significantly changes compared with the input (Teplyakov and Nikanorov, 1994);
(3) the broad spectrum of pollutants is a serious obstacle to obtaining experimentally a large number of necessary kinetic parameters: growth rates, rates of uptake of various substances (pollutants), coefficients of interactions between populations, etc. Existing approaches to the determination of limiting factors acting in the community and the kinetic principle of aggregation of biological components offer only a partial solution to the problem of multi-dimensionality.

These arguments encourage the development of new integrated methods of estimating the state of aquatic ecosystems, based on the holistic approach in biophysics. This is, for instance, the development of a system of so-called bioassays, based on the following:

(a) Every bioassay is a model (or rather an express model) of some target biological function (of an organism, a population) such as respiration, motility, reproduction, death, mutability (mutagenicity), etc. The number of these functions and hence of bioassays must be finite.
(b) The ultimate finite number of bioassays depends solely on the completeness of determining biologically significant target functions and should not increase with broadening the spectrum of pollutants.
(c) Bioassays must permit writing a system of equations describing their specific temporal dynamics for typical ecosystems (laboratory and natural ones). These equations, together with inputs in the form of bioassay inflows must then be used in prognoses of bioassay dynamics for aquatic ecosystems.

Below is a somewhat more detailed description of the approach. Let there be a broadening spectrum of chemical substances $(X_1, X_2, \dots, X_k) = \{X_i\}$, k is large. The set $\{X_i\}$ influences some important biological functions of a human organism or an ecosystem; for example,

F_1 the activity of the respiratory system;
F_2 the activity of the digestive system;

F_3 survival;
F_4 mutagenicity;
F_5 growth activity, etc.

Many interesting bioassays have already been prepared: luciferase-based ones (Kratasyuk *et al.*, 1996) and tests for genotoxicity (Gunderina and Aimanova, 1998; Kovaltsova and Korolev, 1996; Zakharenko *et al.*, 1997).

General considerations suggest $(F_1, F_2, \ldots, F_p) = \{F_r\}, r = 1, 2, \ldots, p$ is the set of target functions, although it can expand (p can increase in the course of investigation), but it seems that in the limit there is a basis (i.e., a set of the finite number, m, of independent functions ($p = m$) such that there cannot be $U(F_1, F_2, \ldots, F_m) = 0$); any $F_r(r > m)$ can be expressed through the basis. It is clear that target functions depend on the broadening chemical spectrum $\{X_i\} : F_r(X_1, X_2, \ldots, X_k\}$. Integral factors (let us call them bioassays) will be the factors $(T_1, T_2, \ldots, T_n) = \{T_i\}, i = 1, 2, \ldots, n$, that depend on $\{X_i\} : T_i(X_1, X_2, \ldots, X_k)$, and in turn target functions in the general case can be expressed through $T_i : F_r(T_1, T_2, \ldots, T_n), r = 1, 2, \ldots, m$. Evidently, whether the bioassays are constructed in a laboratory (e.g., luciferase-based ones; Kratasyuk *et al.*, 1996) or naturally occurring variants are used, the situation when each target function has its own bioassay (i.e., $F_i(T_i)$) is preferable.

Thus, T_j are intermediate parameters between the spectrum of substances and the target function: $\{X_i\} \rightarrow \{T_j\} \rightarrow F_r$. We can show that if $\{T_j\}$ and $\{F_r\}$ form the bases, then $m < n$ (i.e., the number of bioassays is not smaller than the number of tested functions). If there is a correlation between bioassays, which is often considered as an advantage by experimenters, then at least one of the bioassays must be excluded from the basis.

One of the obvious spheres of application for bioassays is an alarm test (i.e., an early signal of an unfavorable impact on the environment of the tested function in a given place). In this case, the place must be analyzed in detail by chemical methods to determine the chemical reason for biotoxicity.

Another, absolutely new sphere is the prediction and calculation of $\{T_i\}$ for a real ecosystem. A great potential advantage of $\{T_i\}$ is that the $\{T_i\}$ set forms a complete event and an addition of new chemical or other components does not expand the $\{T_i\}$ set. Then, if we manage to construct a closed model of the dynamics of $\{T_i\}$ for a given ecosystem, the prediction of, say, water quality can be made (calculated) directly in terms and units of bioassays $\{T_i\}$, which, through the previously determined functions $F_j(T_1, T_2, \ldots, T_n)$, will be converted into medical consequences or target functions. For F_j it is necessary to determine the tolerance range (i.e., the limits of the norm $F_{j\text{-min}} < F_j < F_{j\text{-max}}, j = 1, 2, \ldots, m$). The function of parameters $F_{j\text{-min}}$ and $F_{j\text{-max}}$ is similar to that of the MPC, but is devoid of the MPC's major drawbacks (namely, the relationship of the MPC to the simultaneous action of several substances and to the width of their chemical spectrum). Reversing this procedure, based on $F_j(T_1, T_2, \ldots, T_n)$, one can calculate the limits of the norm for bioassays $\{T_i\}$.

If we construct the equations for ecosystems that would include not only bioassays $\{T_i\}$ but also hydrochemical $\{HX_k\}$ and hydrobiological $\{HB_j\}$ components, the prediction procedure will not become significantly easier. These will be the same reductionist models. In the course of investigation, we can include all the three groups of components $\{T_i, HX_k, HB_j\}$, but the ultimate goal is a closed system of differential equations (for a homogeneous case) given as

$$dT_j/dt = R_{j\cdot v}(T_1, \ldots, T_n; T_{1\cdot o}, \ldots, T_{n\cdot o}), \quad j = 1, \ldots, n, \tag{14.6}$$

where $T_{i\cdot o}$ are inflows of bioassays into the ecosystem; and $R_{j\cdot v}$ is a certain form of equations.

These equations and their respective dynamics can be termed the ecological laws of integral bioassays. The situation is very similar to the early stages of the development of Verhulst–Pearl type models (Odum, 1971) in population ecology, when the equation of the S-shaped population growth curve was written based on experimental data, taking only numerical population dynamics $X(t)$ into account and ignoring finer mechanisms of growth limitation by substrate deficiency or inhibition by metabolites. Index v in Equation (14.6) characterizes the most important notion of ecosystem type, based on the hypothesis about a possible discreteness of the type of ecosystem functioning in the dynamics of some integral parameters such as self-purification (Gladyshev, 1997). The concept of typification of aquatic ecosystems should also involve such parameters of the trophic status as oligotrophic, mesotrophic, and eutrophic types of water bodies.

In order to write the system of Equations (14.6), we performed special laboratory experiments. The blue-green alga *Spirulina platensis* was grown in enrichment culture and a bloom was simulated. It was a model for the investigation of temporal dynamics of some model bioassays (Figure 14.3) providing a basis for the future model description of the type of Equation (14.6). Omitting a detailed interpretation, we note that (a) bioassays demonstrate good reproducibility of toxicity parameters and (b) bioassay dynamics curves have characteristic phases of toxicity decrease and increase. Strictly speaking, in such experiments, the temporal dynamics of all potentially basic bioassays F_r must be studied concurrently.

Summing up the prospects of this direction in biophysics, we can conclude that ideally, it will be possible to determine the most important parameters of an ecosystem using bioassays as integral parameters that can be measured directly. Their number will not be great, but much smaller than the number of chemical substances. Holistic prognostic models should be constructed as follows:

- all ecosystems are typified;
- for every typical ecosystem there is a closed system of equations in terms of only integral parameters, bioassays;
- ecosystem inputs are set—not in classical terms of pollutants and other substances—but rather in terms of bioassay flows.

The model is identified and verified using data from field observations of seasonal and year-to-year dynamics of bioassays in different points of the water body. Then, model prognosis is made in terms of integral parameters.

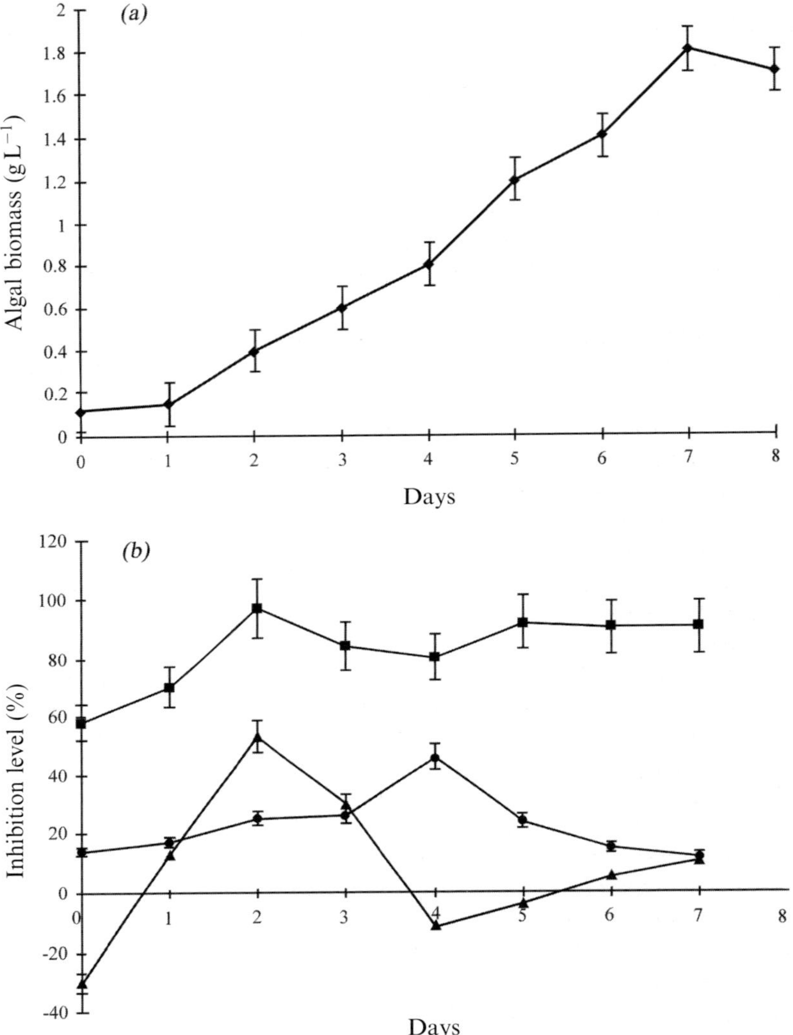

Figure 14.3. Combined dynamics of concentrations of algae (a) and bioassays (b): ◆ algal biomass; ● luciferase–reductase double-enzyme system; ■ alcohol dehydrogenase test; ▲ trypsin test

Evidently, chemical control and bioassays are not conflicting but complementary approaches. Chemical control must first of all be conducted in the zones of the water body where bioassays have given an alarm signal. Bioassays may become a new important tool of monitoring the environment in the 21st century. The strategic task is to develop a logical methodology integrated with a system of bioassays and to organically arrange a mosaic of available bioassays in it.

14.5 ARGUMENTS SUPPORTING THE STATEMENT ABOUT THE DEGREE OF DEPENDENCY OF POPULATION-SELECTIVE PARAMETERS DURING SELECTION MODELING

The approach to constructing a hierarchical scale in the cell–population chain (Section 14.1.4) can be based on statistical analysis of possible relationships between values of population micro-parameters. Problems of population microbiology (particularly those of micro-evolution) are often solved on the basis of description of population dynamics and DDGCFs by equations including various kinetic micro-parameters of populations that characterize their different relationships with the environment. These parameters are maximum specific growth rate, coefficient of substrate consumption (productivity), the Michaelis coefficients, coefficient of release for different substances, etc. Bringing together similar populations with varied micro-parameters in a model, we investigate the consequences of their encounter (expulsion, coexistence, domination). We assume that these micro-parameters are independent, and as a result the dominating population can grow more rapidly, use the substrate more efficiently, and have other advantages.

So, we should answer the basic question about the degree of the relationship of population micro-parameters in the case when one of them is varied (e.g., by mutation). If there were valid models describing interrelated variations in these micro-parameters, this question could be answered exhaustively. However, such models are unavailable. Even the most frequently used micro-parameter, the maximum SGR and the relationship between an SGR and a DDGCF, is described in several fundamentally different terms: for enzymes, populations, etc. However, even the enzymatic description suggests that the hypothesis about the independence of micro-parameters is not sound. Let us consider this in greater detail. Let the mechanism of the relationship between the SGR and the limiting substrate be described by a real enzymatic reaction of the bottleneck: $g = \mu S/(K_s + S)$, where g is SGR, μ is the maximum reaction rate ($\mu = K_3 E$, where K_3 is the rate of formation of product P (biomass) and E is total enzyme concentration), S is substrate concentration, and K_s is the Michaelis–Menten constant ($K_s = (K_2 + K_3)/K_1$, where K_1 is the rate of the formation of the enzyme–substrate complex, and K_2 is the rate of the reverse reaction). Clearly, mutations essentially change micro-parameters (K_1, K_2, K_3, and E). If, for example, K_3 is increased, μ and K_s will increase too (i.e., the SGR graph will flatten, but will go up at large S). On the other hand, a change in K_1 will affect K_s only, but μ will remain unchanged. More examples can be given. Hence, even very simple models show that micro-parameters can be independent in the case of one type of mutations (K_1, K_2, E) and interdependent in the case of another mutation (K_3).

A frontal solution for this problem depends on the type of chosen model of the cell or population level, describing a relationship among micro-parameters similar to the one described above. Since, however, we know too little of the full model of cell biosynthesis and its regulation, this way is unacceptable. There is another, phenomenological, way, currently lacking the analysis of mechanisms of relationships among parameters, but based on statistical analysis. Let us imagine that we have a collection of mutants of one strain with the measured micro-parameters. Then, using a multivariate regression parametric analysis, we can solve this problem at a phenomenological level. This can conveniently be done using electronic databases on kinetic parameters of microorganisms maintained in museum collections. Various mechanismic models of relationships between micro-parameters (or micro-processes) can be investigated concurrently.

14.6 EXPERIMENTAL MODELING OF THE PHENOMENOLOGICAL LAWS OF MIGRATION OF AQUATIC ORGANISMS

Aquatic ecology has a branch dealing with migration of aquatic organisms, in which biophysical investigations would be necessary because modeling of ecosystem dynamics lacks formalized knowledge of the mechanisms and laws of migration. The mechanisms of migrations of motile unicellular aquatic organisms exhibited as various behavioral responses, termed taxes (phototaxis, chemotaxis, gravitaxis, etc.), have been investigated for a long time though usually at qualitative and phenomenological levels for separate species. To interpret the results of the impact of different chemical substances on mobile responses of protozoa and microalgae in bioassaying and in estimating and predicting migration behavior of micro-plankton in water bodies, it is necessary to study the laws and quantitative parameters of these processes. Migrational responses of unicellular organisms are integral parameters of the effect of different environmental factors, concentration of chemical substances, light intensity, gravity, etc., on a cell and a population. They are characterized by certain parameters: direction, movement velocity, performance time, power intensity, etc. It is also interesting to investigate possible competition between different types of responses (e.g., between chemotaxis and phototaxis or gravitaxis) and to quantify them. Experimentally, migration behaviors can be investigated with special devices, partly described in the literature. Combining the biophysical laws of migration based on experiments and field observation data is a way to improve prognostic models.

14.7 CONCLUSION; THE FUTURE MONITORING OF AQUATIC ECOSYSTEMS

We have proposed approaches to the study of ecosystems under natural and human-imposed conditions. These include (1) studying the fundamentals of the ecosystem similarity theory based on the principle of scaling however complex a system of equations may be necessary and getting a new set of dimensionless macroparameters and working out experimental approaches including the method of transfer of laboratory data to field data, (2) a new experimental–theoretical approach using the growth acceleration index to estimate the integral degree of knowledge of growth regulators, (3) the idea of working out a new language to describe the state and dynamics of ecosystems with the help of specially selected bioassay indicators, and (4) based on the simple schemes of fermentation reactions, an idea is being developed about the possible dependence of a series of selective population parameters, which is important for modeling the microevolutional process. The proposed new trends of biophysical ecology will increase the physical strictness of the method.

Investigation of the functioning of an ecosystem can be evaluated as successful only if it yields a valid prognosis of the ecosystem's response to control measures (a change in freshwater flow, a change in the structure of currents, a shut-off of effluents, etc.). Therefore, there must be some test aquatic ecosystems. Test water bodies (or some of them) must be open for large-scale field experiments and for the development of experimental methods of similarity scaling of ecosystems.

Based on the fundamental knowledge obtained, the biophysics of ecosystems can solve practical problems including the control of water quality by controlling the species composition of the aquatic community and its activity; environmental impact assessment of water management projects; prognosis of the state of ecosystems and the chemical composition of the water; control of blooms; ecologic–economic elements of water use optimization; etc.

Acknowledgments. This work was supported by project N2004 0.47.011.2004.030 (the Russian Foundation for Basic Research and the Netherlands Organization for Scientific Research).

14.8 REFERENCES

Barenblatt G.I. (1982). *Similarity, Scaling, Intermediate Asymptotic*. Hydrometeoizdat, Leningrad, 255 pp. [in Russian].

Chiu S.Y., Fan L.T., Kao I.C., and Erickson L.E. (1972). Kinetic behavior of mixed populations of activated sludge. *Biotechnology and Bioengineering*, **14**(2), 179–199.

Degermendzhy A.G., Adamovich V.V., and Adamovich V.A. (1993). A new experimental approach to the search for chemical density factors in the regulation of monoculture growth. *Journal of General Microbiology*, **131**, 2027–2031.

Gladyshev M.I. (1997). On types of aquatic ecosystems and their integral kinetic characteristics. *Water Resources*, **24**(5), 526–531 [in Russian].

Gubanov V.G., Degermendzhy A.G., Bayanova Yu.N., Bolsunovsky A.Ya., Gladyshev M.I., Gribovskaya I.V., Zinenko G.K., Kalacheva G.S., Stepen A.A., Temerova T.A., Ustyugova T.T., Khromechek E.B., and Shitova L.Yu. (1996). Prognostic modeling of ecosystem dynamics and water quality based on kinetic characteristics. *Siberian Ecological Journal*, **5**, 453–472 [in Russian].

Gunderina L.I. and Aimanova K.G. (1998). Genetic consequences of γ-irradiation of *Chironomus thummi*: Aberrations of polytene chromosomes. *Genetics*, **34**, 54–62 [in Russian].

Kovaltsova S.V. and Korolev V.G. (1996). *Saccharomyces cerevisiae* strain for testing mutagens in the environment, based on interaction of rad2 and him1 mutations. *Genetics*, **32**(3), 366–372 [in Russian].

Kratasyuk V.A., Kuznetsov A.M., Rodicheva E.K., Egorova O.I., Abakumova V.V., Gribovskaya I.V., and Kalacheva G.S. (1996). Problems and prospects of bioluminescent assay in ecological monitoring. *Siberian Ecological Journal*, **5**, 397–403 [in Russian].

Odum E.P. (1971). *Fundamentals of Ecology*, Third Edition. W.B. Saunders, Philadelphia, 574 pp.

Sedov L.I. (1972). *Methods of Similarity and Dimensionality in Mechanics*. Nauka, Moscow, 440 pp. [in Russian].

Teplyakov Yu.V. and Nikanorov A.M. (1994). Simulation of heavy metal effect on fresh-water ecosystems in mesocosms and estimation of water body self-purification properties. In: N.E. Peters, R.J. Allan, and V.V. Tsirkunov (eds.), *Hydrological, Chemical and Biological Processes of Transformation and Transport of Contaminants in Aquatic Environments*. International Association of Hydrological Sciences, Wallingford, U.K., Publication No. 219, pp. 293–301.

Toda K. (1976). Invertase biosynthesis by *Saccharomyces carlsbergensis* in batch and continuous culture. *Biotechnology and Bioengineering*, **18**(8), 1103–1115.

Zakharenko L.P., Zakharov I.K., Vasyunina E.A., Karamysheva T.V., Danilenko A.M., and Nikiforov A.A. (1997). Determination of genotoxicity of fullerene C60 and fullerol by the method of somatic mosaics on cells of *Drosophila melanogaster* wing and in SOS chromotest. *Genetics*, **33**(3), 405–409 [in Russian].

15

The Earth as an open ecosystem

Lev S. Ivlev

15.1 INTRODUCTION

The most important factor that determines the existence and development of the Earth ecosystem (i.e., the climate) is the Earth–Sun interaction (Kondratyev and Ivlev, 1995; Kondratyev *et al.*, 1995; Krapivin and Kondratyev, 2002). The variations in the Earth–Sun interaction cause gradual or periodic changes of climatological conditions, commonly known as Milankovich cycles, as well as catastrophic phenomena. The cycles which correspond to the ice ages and the intervening interglacial periods were originally studied by James Croll, the son of a Scottish crofter, who had very little formal education, in the 1860s and 1870s; however, his work was largely ignored and the theory was revived and expanded by Milutin Milankovitch, a Serbian civil engineer and mathematician, in the early 20th century and they are now named after him. Catastrophic changes do not follow such a cyclical pattern. In choosing the direction for further development of our civilization the importance of the Earth–Sun interaction must be understood in emerging global changes of the environment, particularly those concerning the Earth's biosphere. In order to be able to separate anthropogenic effects on the Earth's biosphere from natural effects, it is necessary to understand the physical aspects of the formation and evolution of environmental conditions, of evolutionary processes under external influences of varied intensity and duration; this involves understanding the current status—not only of the environment—but of the Earth as a planet and of space (particularly the Earth's near space) with its physical processes.

The concept of the Earth as an open ecosystem in space has developed gradually, although it has always been clear that the most important factor of the existence of this ecosystem (the climate) is determined primarily by interactions between the Earth and the Sun. Changes in our understanding of the climate and of the role of the ecosystem in its evolution are particularly obvious when comparing today's studies with those made in the 1970s (Kondratyev *et al.*, 2006).

The essential role in the development of the concept of the Earth as an open ecosystem was emphasized by Kirill Kondratyev (1990) through his scientific and organizational work; in particular, in his support of Gorschkov's ideas of feedback effects of ecosystems influencing the climate in such a way that initiates changes which are favorable for the ecosystem (Gorschkov *et al.*, 2006). Equally important for understanding the role of the Earth–Sun interaction in this problem were annual scientific seminars on "Space ecology" held in St. Petersburg by Kondratyev and transformed in 2001 to the "Ecology and space" workshop (Ivlev, 2001a, b, 2005, 2007; Reznikov, 2007a, c), as well as "Aerosol and climate" section meetings at five international conferences on "Natural and anthropogenic aerosols" (Ivlev and Chvorostovsky, 2000a, b; Ivlev *et al.*, 2003). In this chapter we confine ourselves to a brief statement of some topics concerning the climate-changing problem and discussed at these seminars. It is natural that participants in these seminars strived to answer the most pressing issue facing human society, namely the forecast of possible changes of the environment in the near future.

15.2 EVOLUTION PROCESSES ON THE EARTH

As an open physical system the Earth is continuously exposed to solar and space radiation as well as the gravitational forces of the Sun, Moon and other space objects so that, over a long period of time under continuous external influence, a dynamic balance has developed between the Earth and space. Solar radiation creates certain physical conditions in the thin upper layer of the Earth and above it. A sharp distinction between these conditions on Mars, Earth, and Venus—all of them having relatively similar amounts of solar radiation—is in the first place due to the Earth's biosphere that has been regulating and stabilizing physical conditions on the Earth within a narrow range of values optimal for the existence of all forms of life during almost 4 billion years, and working against entropy with the aid of constant input of external energy to the Earth.

The role of the mutual adjustment of the components of the environment increases in the process of evolution, and the emergence of human civilization is a manifestation of the escape of one component—mankind—from submission to the single whole (i.e., the environment), and an attempt to bring the environment under control. All the man-made power constitutes only 10^{-4} of the solar energy flux reaching the Earth, while the information capacity of human activity is 17 orders of magnitude below the information capacity of the biosphere (Gorschkov, 1990; Gorschkov *et al.*, 2006). But the way mankind uses energy and information differs from that of the rest of the biosphere in such aspects as transfer, scale, and motivation. Each biological object gets information and acts according to its genetic code, and these actions are regulated and governed by short-range interaction and relatively simple commands; this mitigates both the destructive and constructive effects of actions of a limited society of biological objects. Information accumulated and sources of power created by human civilization can be used by a very limited part of human society, and not necessarily towards sustainable development of the Earth's biosphere and conservation of the Earth's climate at a global scale.

The simplest way of environmental forecasting (the accuracy of which depends on the timespan of the forecast and variational character of the physical parameter being investigated) is an extrapolation of the observed trends for these parameters. The analysis of paleoclimatic evidence shows that long evolutionary changes of the physical parameters often alternate with sharp fluctuations with low predictability (in some cases it is not possible to predict such a fluctuation before it begins). Relaxation processes by no means always bring the parameters back to their initial, pre-fluctuation level; and the relaxation time largely depends on the nature and scale of the process and varies by many orders of magnitude. When the influence intensity exceeds some threshold value the environment changes in an irreversible way: it either reaches a new state of stability (phase transition), or it starts to be destroyed continuously (degradation) moving towards a less ordered state (increase of entropy).

Climate is generally defined as a statistical regime, or long-term average, of short-period variations of meteorological fields (weather) that itself is subject to long-period variations (Houghton, 1984; Monin, 1986; Zuev and Titov, 1996). The statistical characteristics of the climate at any point on the Earth's surface are calculated by averaging values of the observed weather-forming factors over time. The average value may differ from one period to another, either as a result of deviations of sample averages or due to changes in the expected values. The latter can be used as reference climate elements. In this case the deviations are considered as disturbances that hinder practical climate observation.

Because these deviations (weather noise) decrease with the increase of time, the average over the longer period is more representative and is closer to the "true climate" as a limit of averages over an infinite time period. In this case the notion of weather forecast loses its meaning and problems arise due to the impossibility of studying climate changes under slowly varying external influences. This idea is also used for environmental forecast.

Authors of numerical forecast models attempt to restrict themselves to the use of empirical values of the parameters which for certain condition ranges can be considered as constant, thus limiting to a considerable degree the timespan of the correct forecast. To improve forecast accuracy some correlation between individual components of the environment, based on observation data (Zuev and Titov, 1996), is added to semi-empirical models.

Comparative surveys of climatic fluctuations show noticeable horizontal differences that indicate the dominant role of redistribution of the heat, precipitation, and atmospheric pressure as a result of changes in the system of atmospheric circulation.

The most important condition that determines the success of numerical weather and climate forecast models is the choice of model of the movement of atmospheric masses. But it is extremely difficult to solve the fundamental equations describing the movement in such detail that would make it possible to take into account effects of much smaller movements; in other words, there is the problem of the total effect of small-scale movements influencing the course of the processes of a much bigger scale.

In this way, relatively fast non-linear dynamic interactions between atmospheric currents of comparatively small scale can lead within 2 weeks to non-predictable (according to specialists in the area of numerical hydrodynamic forecasts) meteorological changes of the environment—so-called climatic noise. At the same time large-scale atmospheric movements are much more stable because of relatively slow changes in boundary conditions (Houghton, 1984).

The processes of energy transformation in the atmosphere are so diverse and closed that the effect of adding a small amount of energy to it is by no means obvious: it can either increase or decrease the system's stability. The result of long-term changes is determined by non-linear processes while for short-term weather forecasts the equations of classical physics with relevant approximations can be used at the first stage. Therefore, the dynamics of the climate and the resulting environmental conditions to a large extent have a deterministic character, and the reliability of short-term environmental forecasts depends on the accuracy and completeness of the data describing the initial status of the environment and influences affecting the system.

The observed processes are averaged over periods longer than the timespan of fast fluctuating movements but shorter than the timespan of large-scale processes—which does not fully agree with the physical essence of the processes. Temporal and spatial amplitudes of fluctuations of climatic and other characteristics of the Earth's shell caused by influences of various kinds are no less important parameters than their averaged values. The forecast of their effect on the weather and environmental conditions requires an understanding of the portion of energy hidden in the fluctuations (Gorschkov, 1990; Sakrzhewskaya and Sobolev, 2002). Single-layer models of atmospheric circulation have considerable drawbacks caused by a lack of understanding of the physical mechanisms of exchange of kinetic energy between movements of different scales. In particular, the assumption of plane motion in the mathematical treatment of fluid motion equations contradicts the mass conservation law and does not account for the role of energy influx. Hence, this assumption makes it impossible to predict the moment of transformation of the pressure field and thus decreases forecast accuracy.

Equations that best describe evolutionary problems of this type are known as 3-D Navier–Stokes equations, which were first presented 200 years ago. Solution of these fluid-dynamic equations is a major mathematical problem and the subject of a number of papers (Kropotkin, 1996).

In particular, these equations have been repeatedly used to describe climate processes, primarily movements of air masses. The atmosphere–ocean–mainland climate system is influenced by spatial inhomogeneities of different scale: global-scale inhomogeneities with characteristic dimensions 10^4 km (horizontal) and 10 km (vertical), and volume 10^9 km^3; also there are very small turbulent inhomogeneities in the ocean and atmosphere, like tree leaves and structural inhomogeneities in soils. Restricting ourselves to structures not less than 1 mm in size we get 10^{27} homogeneous mini-volumes. Each of them being characterized by ten parameters, the whole system has 10^{28} degrees of freedom, which makes individual description of its status impossible in practice. To make the description and calculations more feasible, the inhomogeneities are divided into two types: large-scale that can be described individually, and small-scale that are described statistically (Monin, 1986).

Small-scale inhomogeneities in the atmosphere and in the ocean are created by high-frequency hydrodynamic processes with periods from fractions of a second to minutes, and from minutes to hours. Their statistical regime can be parameterized and expressed analytically.

Vasilyev (2005a) suggested a mechanism of energy exchange between movements of different scale (effect of turbulent friction), provided by a cascade of dynamic rotational bifurcations. A transformed Kármán street was selected as the most suitable analogue physical model of regular vortex circulation in an unstratified (or equilibrium-stratified) liquid (Chromov and Mamontova, 1974) with excess impulsive disturbance developed behind a body immersed in the moving liquid.

The suggested effect of turbulent friction and cascade of dynamic bifurcations obtained on this basis (Vasilyev, 2005a, b) in regular vortex structures, developing in a convective stream due to heat flux, eliminates the above-mentioned drawbacks of single-layer models of atmosphere circulation. Then the atmospheric pressure systems (cyclones and anti-cyclones) would represent, according to the mass conservation law, dynamically interconnected Kármán structures. This is a major breakthrough in the understanding of the non-linearity of the physical mechanism of processes of kinetic energy transformation (generation and dissipation) in the atmosphere and exchange of kinetic energy between movements of different scales, provided by the cascade of dynamic rotational bifurcations. Understanding this mechanism makes it possible to interpret and suggest a solution of the Navier–Stokes equations for large-scale vortex structures in the atmosphere, to abandon the use of the Courant–Friedrich–Levi criterion for averaging the spacetime scales of the processes, to achieve a quantitative improvement in forecasting the development of atmosphere processes and to assess the role of non-linear processes in the atmosphere (particularly, of latent heat liberation). Further development of hydrodynamic forecast of dynamic atmospheric processes will presumably employ the model of convective cascade of dynamic bifurcations in its own phase space (Vasilyev, 2005a) .

15.3 EFFECT OF GREENHOUSE GASES AND AEROSOLS ON CLIMATE

In scientific publications special attention is given to problems such as adding radiative forcing caused by the growing concentration of greenhouse gases in the atmosphere, to numerical modeling of the climate, and taking into account the influence of radiative forcing on total ocean circulation, because the most significant effect of climate change is World ocean level rise and intensification of the global hydrological cycle (Kondratyev and Ivlev, 2001).

As regards radiative forcing (defined as a change of radiation balance of the system "underlying surface–atmosphere" caused by climate-changing factors), the portion of it determined by changes in greenhouse gas concentrations can easily be assessed. It is generally accepted that the most important roles of all greenhouse gases are played by carbon dioxide, water vapor, methane, and ozone. Global 3-D climate models have allowed validation of the concept of "global warming", caused by human actions in releasing greenhouse gases (principally CO_2) into the atmosphere. Calculations made on the assumption of continuous growth of CO_2 concentration (about 1% per year) resulted in the atmosphere overheating, so a cooling factor was introduced that accounts for the presence of sulfate aerosol in the atmosphere which acts as a scattering agent (or equivalent to its ability to decrease underlying surface albedo). But this idea of the possible effect of aerosol on climate as well as coordinating it with observations is no more than a corrective adjustment (Kondratyev, 1990).

It should be noted that some controversy exists concerning the role of carbon dioxide and water in climate change (Houghton, 1984). As regards effects on fast global changes of the Earth's climate, ozone, water vapor, and aerosols are of the most interest, as they determine to a great extent the energy balance of different atmospheric layers and, moreover, of the atmosphere with the underlying Earth's surface. At the same time their content in the atmosphere is highly variable and depends on both natural and man-made factors.

Estimates of aerosol radiative forcing, particularly of its "indirect" component that represents the effect of atmospheric aerosol on cloud cover character, are most difficult (Kondratyev and Ivlev, 2001). The value of "shortwave" radiative forcing, measured since 1850, varies within the range $0.1\,W\,m^{-2}$–$0.5\,W\,m^{-2}$ (the greenhouse effect of the same period amounts to $2.4\,W\,m^{-2}$, so the influence of greenhouse gases dominates). Estimates of the effects of various factors on climate change (seasonal, annual, decennial, and centennial), ozone content, intensity of ultraviolet radiation, chemical composition of the atmosphere, according to Earth-observing system data, testify to the significant role of aerosols in these processes. Estimates of the climate-forming effects of various types of aerosols, including indirect effects through the influence on the formation, structure, and optical properties of clouds, are of great importance.

The presence of man-made tropospheric aerosol causes growing atmosphere haziness and the formation of aerosol haze in high latitudes of the northern hemisphere. The mechanism behind gas-phase aerosol formation (the dominant mechanism for the formation of volcanic stratospheric aerosol) plays a decisive role in the interaction between the sulfur and nitrogen biogeochemical cycles and atmospheric aerosol formation processes. The problem of the aerosol effect on cloud formation and destruction deserves serious consideration, as well as the problem of smoke aerosols with strong absorption qualities generated by fires and smoke and dust aerosols generated by above-ground nuclear explosions.

The problem of anthropogenic effects on the sulfur cycle deserves particular attention. Existing estimates confirm that emissions of gaseous sulfur into the atmosphere due to fuel burning has reached the same order of magnitude as emissions caused by natural factors. Most significant are estimates of gaseous compounds of reduced sulfur of biological origin, studies of processes and anthropogenic influences that contribute to their formation, study of transformation of anthropogenic emissions of sulfur dioxide and of gaseous sulfur compounds in the atmosphere, including gas-phase reactions of sulfate aerosol formation (Ivlev, 1998a, b, 2001a, b, 2005).

Because the anthropogenic effects on biogeochemical cycles emerge slowly and are characterized by high persistency, it is very important to indicate hazardous trends in advance. Issues like the contribution of biological sources to the formation of carbon, sulfur, nitrogen, and halogen cycles, intensity of aerosol generation (dust, particularly in deserts, fuel burning, etc.) at the regional and continental scale, global distribution of major gaseous and aerosol components of chemical and hydrological cycles (water vapor, clouds, precipitation), including the photochemical processes of dry and wet sedimentation of gases and particles, play a key role in the control of the chemical composition and cycle of various components of the troposphere that determine its reaction to external influences.

Numerical modeling of the global climate taking into account not only the growing concentration of greenhouse gases due to anthropogenic impact but also the growing content of anthropogenic sulfate aerosol (the daily emission of sulfur dioxide transforming to aerosol is equal to $70–80 \times 10^6$ t of sulfur) has shown far more complex dynamics of climate formation than was suggested previously: climate cooling due to atmospheric aerosol considerably offsets the greenhouse effect, and spacetime variability of aerosol concentration determines geographical climate variability (Ivlev, 2005).

The aerosol impact on climate exhibits itself both directly and indirectly: by an increase in the Earth's albedo and through an influence on the dynamics and microstructure of clouds, increasing their albedo due to a growing number of small droplets during the generation of sulfate condensation nuclei.

Evaluation of the indirect effects of aerosols on climate is a difficult and not completely solved problem. Difficulties arise in the estimation of climate effects which are caused by varying the sensitivity of the system to radiation disturbance due to the greenhouse effect (approx. $1.7°C/(W m^{-2})$) and aerosol ($1.0°C/(W m^{-2})$). This makes it unreasonable to use simple energy balance models to assess global climate change. In this way, aircraft-flown experiments that measure fluxes and influxes of shortwave solar radiation indicated strong absorption in "dirty" clouds at $\lambda = 0.5\,\mu m$ with an optical thickness, τ_a, up to 0.15 compared with $\tau_a = 0.03$ approximately in "clean" clouds, which was not taken into account previously (Kondratyev *et al.*, 2006).

The impact of aerosols on various processes (e.g., on radiation transfer and water phase transfers) depends generally on a set of chemical and physical processes, with dependence of composition on aerosol particle size often playing a significant role. So an adequate description of actual aerosol characteristics is possible only on the basis of a complex study of its properties. One of the most extensively used types of aerosol measurement (i.e., mass concentration) is the least informative, because it does not provide any information about the sources and composition of the aerosol and its possible effects.

Aerosol cycles are closely connected with hydrological processes in the atmosphere due to the interaction between aerosol and clouds: clouds and precipitation play an important role in the generation, transformation, and removal of aerosols from the atmosphere; on the other hand, aerosols exert considerable effects on microphysical processes in the clouds and, therefore, on heat–mass exchange processes in the atmosphere. Due to this connection between clouds and aerosols it is impossible to understand completely the processes of generation and transformation of aerosols without reliable knowledge of the physical and chemical characteristics of the clouds (in particular, of nucleation mechanisms).

There are good reasons for anxiety concerning possible man-made growth of aerosol content in the atmosphere that might affect the climate by shifting the Earth radiation balance or by influencing the hydrological cycle. Strong spacetime variability of aerosol characteristics complicates the indication of the anthropogenic component; this indication requires an understanding of the reasons behind this variability of atmospheric aerosol. The role of aerosols in global Earth climate changes is diverse and not confined to the cooling effect, although the latter is the most obvious, particularly for upper-layer atmospheric aerosols. The most significant effect on radiation and global climate is exerted by volcanic stratosphere aerosol which causes strong and long-term disturbance of the radiation regime and associated climate changes.

It should be noted that spaceflight and space research has contributed to the gradual accumulation of technogenic materials in space ("space junk") and has increased the inflow of dust and gaseous particles to high atmosphere layers and considerably changed their optical and electrical properties, the changes being similar to those caused by a meteor stream entering the Earth's atmosphere.

In the papers by Ivlev and Chvorostovsky (2000a, b) and Ivlev *et al.* (2001) the effects of high-energy particle streams on cloud formation in the upper troposphere have been considered, and Ivlev *et al.* (2003) described the impact of ion charge on nucleation intensity and temperature of cloud particle crystallization. The recognition of atmospheric aerosols as an important climate-forming factor is obviously a new stage in understanding and numerical evaluation of the crucial role of aerosols in today's climate change.

15.4 THE ROLE OF WATER IN THE VARIABILITY AND EVOLUTION OF THE ENVIRONMENT

There are various human activities that are affecting the natural hydrological cycle. These include

- deforestation and the conversion of forests into agricultural land
- afforestation
- urbanization
- desertification as a result of poor land use practices
- release of ground water into the above-ground hydrological cycle
- construction of dams and artificial lakes
- diversion of rivers, irrigation, etc.

Therefore, in studying the effects of these activities on the climate, it is necessary to study the role of water in climate and the biosphere.

There is no need to linger on the well-known properties of water, like the coexistence of three phases, high values of heat capacity and latent heat, strong polarizability, and high value of its dielectric constant. Its unique physical and chemical properties determine its important role in the variability and evolution of the environment: formation of clouds and precipitation, hydrologic cycle, heat–mass transfer in the atmosphere and underlying surface, electrical phenomena. Most of these topics are discussed in detail in the work of numerous authors (Kornfeld, 1951; Derpgolz, 1971; Sazepina, 1974; Aisenberg and Kauzman, 1975). Here we shall only consider issues related to the structures of water molecules and their aggregates, and related electrical properties. The interaction between H_2O molecules in the vapor phase is governed by forces of a complex nature and is mostly determined by the electrical properties of the water molecule. Water vapor consists mainly of single H_2O molecules. But the physical and chemical properties of water imply the existence of a certain number of clusters $H^+(H_2O)_n$ (i.e., supramolecular aggregates with delocalized electron–proton states).

At temperatures below $-40°C$ tetrahedral complexes of four molecules govern the properties of ice, establishing appropriate intermolecular distances. They affect the generation of electric charge carriers in the lower atmosphere due to unbounded electrons joining to the complexes. So, generation of molecular complexes in water vapor has proved to be related to the ionization of the lower-atmosphere layer (troposphere).

Abnormally low mobility of $H^+(H_2O)_n$ clusters in the gaseous phase serves as an indication of their globular structure, caused by the ordering of molecules around impurity centers (Reznikov and Ivlev, 2005), globules with axial symmetry most probably generating around dimers. Because of the decrease of vibrational energy in the closed system a clathrate structure is possible for 20 and more interacting single molecules irrespective of their composition.

Some of the most stable supramolecular aggregates are fullerenes and fullerene-like aggregates. It is natural to suggest that such aggregates can originate from water molecular centers. A collective electronic state is typical for fullerene geometry (ES Band 4.1 eV–4.48 eV). Induction of this band is usually associated with increasing transparency (decreasing optical density D), which testifies to the change of equilibrium concentration of the atomic–molecular centers and is one of the spectral characteristics that indicates globular structure.

A short analysis of optical investigations of the supramolecular structure of water provides good grounds (Reznikov, 2007b) to consider water as a quasi-plasma capable of polarization and generation of bulk charge, domain, thread-like, and other 3-D structures. Water with its super-stochiometric concentration of surface $n(H^+–H^+)^-$ is highly adhesive and has acid characteristics. There is delocalization of H-atoms and generation of H-plasma, which dissolves in certain metals like molecular hydrogen. Accumulation of hydrogen is possible in metals with low electron affinity energy ($E_a\mathrm{M} < E_a\mathrm{H}^0$) and relatively high energy of M–H bond (Al, Cd, Ni, Ti, Fe, Zn, Mg). Concentration of the solution $(H^0 \leftrightarrow H^+)^-$ in metals is comparable with the density of collective electron states. Fast diffusion of hydrogen in metals is possible at dynamic equilibrium $(H^0 \leftrightarrow H^+)^-$ with a collective subsystem of electron states (Reznikov and Ivlev, 2005). Diffusion in dielectrics is also known for hydrogen. Virtually all electrical phenomena in the atmosphere—conductivity, silent discharge, linear and ball lightning, fireballs—involve water in one form or another.

Condensation of water on particles with negative surface charge allows for the adsorption type of interaction of (H_2O)-clusters with H-molecules connected with electronegative molecules, with stability being provided for by delocalized electron states. An H-cluster model with collective electron states implies diffusion of $n(H^+–\bar{e})$-pairs in metal and their resulting transformation to a system $n(H_2^+)^- \leftrightarrow (H^+–H^+)^-$ and coulomb adsorption on air molecules. The probability of this model, however, is low for nH_2 aggregates because $E_aH_2 < E_aO_2$, and $E_a(H_2O)_n \sim 1.3\,\text{eV}$.

Therefore, apart from the excited state of hydrogen ionization of the surface of nanodroplets there is the necessity of the precondition for luminous vapor–gas objects: the excitation of nanodroplets in an electric field reduces the surface tension and increases the relative concentration of delocalized surface $(H^+–H^+)^-$, which may provide contact or long-range H-interaction between vapor-phase particles.

Non-homogeneous geometry of the surface of the excited water droplet in the electric field can be compared with the non-homogeneous surface of a vapor–gas bubble in the area of elastic wave localization (e.g., as in the case of ultrasonic exposure). In the case of photo-excitation, on the contrary, increase of surface tension is more probable.

The equilibrium condition of ball lightning in the air and the low concentration of free hydrogen rules out the suggestion that ball lightning consists only of hydrogen, and implies involvement in its synthesis of water vapor as a source of hydrogen, which has polar properties in ball lightning. Experimentally observed density and flexible geometry of ball lightning makes it possible to model it using thin-layer water or a water-gaseous shell with varying concentration of mobile charge carriers (which increases as a result of ball lightning generation).

Volume absorption of radiation is a natural result of the increase of ball lightning mass, and dark or faintly luminous atmospheric ellipsoid objects known as bolides can be considered as ball lightning. It is coulomb forces in the atmospheric potential gradient ($\sim$130 V/m–170 V/m) rather than gravitation that determines the form of flattened ellipse. Unified model of clouds and bolides associates the fractal geometry of the former with a cluster–cluster structure (Reznikov, 2007a) of a two-phase system (adsorbate of electronegative gases on microdroplets with positive surface potential) whereas for bolides maximum fractal dimension ($D = 3$) is more probable for single-phase shells with a high concentration of dissolved gases.

The generation of atmospheric H_2O shells is conditioned upon simultaneous electro-desorption or photo-desorption of gases from the surface of subfine-dispersed water droplets and their subsequent coalescence in the course of coulomb cluster–cluster aggregation. This process is possible if the surface potential is non-homogeneous. Besides, surface tension of excited liquid-phase particles decreases to the macro-level value, the volume of the droplets decreases due to evaporation, and droplet geometry becomes more flexible. Local increase of the dispersion degree of water droplets and concentration of charged centers imply the inverse process of over-condensation that affects not only the matter within the excitation area but also the adjacent area. For the process to be massive, surface potential should have non-homogeneities of the same type that result in coulomb separation of single-component particles. The generation of ball lightning includes formation of core and periphery areas with increased density of the excited nanodroplets of water with unstable surface geometry. The initial shape of the luminous surface varies widely and depends on numerous factors, but final spherical or ellipsoid shape is characteristic of thin-layer liquid-phase films.

The capability of water, as the main carrying agent of local excited states in the lower layers of the atmosphere, to organize itself into structures with active proton–electron conductivity is evident from the fact of the existence of jellyfish (97%–99% water) as a plasma-like condition. The amount of water in a jellyfish demonstrates that water is a cementing component; this is a subject for a separate article.

At the molecular level the collective organization of ionized or polar molecules of one type is considered as the most probable clusterization mechanism. Generation of shells from amorphized micro-particles is a known phenomenon associated with collective organization of molecules of the same type. This type of reorganization involves a quasi-liquid phase and is largely influenced by intermolecular interaction. At the macro-level the collective organization of vapor-phase self-similarity elements into stable shells also suggests the presence of condensate of excited molecules of the same type or particles with flexible surface geometry. Long-range interaction between particles with the involvement of hydrogen can also follow a coulomb mechanism: the velocity of sound in gaseous phase and condensate $v = 1{,}280\,\mathrm{m/s}$–$1{,}300\,\mathrm{m/s}$ (Ivlev, 2005). A condensate of aromatic $(H_2O)_n$ clusters meets these characteristics. A clathrate model $(H_2O)_n$ implies delocalization of ten and more H atoms on the shell surface. A plasma shell model of water makes it possible to consider a delocalized H component of the water as an agent determining a number of the main properties of ball lightning (in particular, the coulomb structure of shells and their stability).

The appearance of ball lightning does not always coincide with a linear electric charge passing by and is possible without direct involvement of the latter (Stahanov, 1985), which allows for the existence of long-lived excited states $H_n(OH)_n$ and associates $H_{n-x}(OH)_n$–H_x–$H_{n-x}(OH)_n$. Dissolution of ionized gaseous molecules in a water shell increases its stability (crystalline hydrates). Generation of ozone and nitrogen oxides in an electrical field allows for a wide range of gas hydrogen cluster composition.

The stability of water shells results from the high surface tension of water. The work function is $\sim$3.3 eV for water and for clusters like H_a it is $\sim$3.3 eV ($n \geq 6$).

In this ball lightning model the density of charge carriers is higher on the inner surface of a water shell with adsorbed electronegative gas than on the outer surface, and is determined by the concentration of delocalized $(H^+–\bar{e})_n$, where the $\bar{e}$ component is shifted by coulomb interaction into the volume or is captured by traps. The outer negatively charged surface of the water–gas shell is a natural area for water vapor adsorption. Under the conditions of uniform Laplacian compression micro-droplets contain less dissolved gas than larger volumes of water. Adsorption of micro-droplets and nano-droplets (depending on relative air humidity) is equivalent to energy absorption (in terms of kilotonnes) through bulk redistribution of surface energy and increase of concentration of binding H-atoms on the outer surface. At 40%–70% relative humidity the duration of water vapor adsorbtion is proportional to the negative potential of the outer surface of the water–gas shell. When ball lightning appears in lower positively charged atmosphere layers, at the final stage of surface charge neutralization, gas adsorption and transition of ball lightning into the electronegative area are possible, which can lead to temporal and geometrical shell stabilization (bolide). Metallization of water–gas shells can take place if the temperature falls to -20°C and lower values due to the increase of water electro-conductivity by 3–5 orders of magnitude (absence or decrease of a pairwise structure of the system of H-bonds between the shells).

Reznikov (2007a) noted that water has certain temperature areas with non -monotonic temperature dependence of its physical and chemical properties, determined by changes in the geometry and structure of H_2O clusters. The concept of such a resonance nature of this non-monotonic temperature dependence is based on the suggestion of the supra-molecular size of self-similarity elements. Vibrational states of self-similarity elements or their components are observed. An aggregate of $(H^+)^-$ between self-similarity elements can be considered as a composite with variable statistical density that agrees with the $(H^+ \leftrightarrow H^0)$ model of 3-D redistribution. In terms of vibrational states this model corresponds to the resonance $\omega = \sum(160; 3; 21; 32)\,\text{cm}^{-1}$, where $\omega = 32\,\text{cm}^{-1}$ $(B_c H_2^+ \sim 30\,\text{cm}^{-1})$ is interrelated with $\Delta E = E_a H^+ \leftrightarrow E_a H^0$ (Reznikov and Ivlev, 2005).

The resonance of all the vibrational states of H-subsystems $\sum(160; 2.21; 8; 32; 44)\,\text{cm}^{-1} = 35.5\,\text{meV}$ (99.75°C) is associated with the destruction of H-bonds between self-similarity elements, which in energy terms corresponds to the resonance of vibrational states within self-similarity elements and bonding of the H-subsystem with oscillations of quasi-molecules O_2^+ ($\omega = 1{,}904\,\text{cm}^{-1}$) including H^+ precessing between them ($\omega = 8\,\text{cm}^{-1}$), and amounts to $392\,\text{cm}^{-1}$. Nominally the energy of an H-bond ($E_H = 153\,\text{meV}$) is the resonance of H-atom vibrations in an H_3 molecule (binding energy $E_B \sim 150\,\text{meV}$) that causes the interference quenching of precession between the pairs $(H^+–H^0)^-$.

Vibrational-state resonance is characteristic for plasma. Proton exchange between the binding H-subsystem and self-similarity elements is possible in the case of ion and/or donor–acceptor interaction between their components. The equilibrium of self-similarity elements in H-condensates suggests that they are hydrogen-like in terms of energy and geometry. Therefore, water can be viewed as a dynamically balanced system: $(O–H^+–O)^- \leftrightarrow (H^+ \leftrightarrow H^0)$, which corresponds to the organized plasma.

Experiments were conducted to verify the model of water as a plasma with an H-subsystem and self-similarity elements. A (glass) reservoir with water and a plate with a metallized rim that can initiate an electric discharge was put into a microwave oven. After the oven was switched on a characteristic hydrogen flame appeared. The same phenomenon was observed when a plate with metal film (silver paint) was fastened to the wall of the oven. Most convincing was the experiment with water dispersed in the oven chamber. Fine-dispersed droplets burned off completely in the flame. Repeated experiments demonstrated high reproducibility of the results. Burning efficiency markedly depends on the metal's work function, with the best results obtained when using copper. It was suggested that the experiment be performed again under strictly controlled conditions.

The facts stated above concerning the structural properties of condensed water and a suggested model of its structure will probably lead to an explanation in future of the fast relaxation of the bulk charge of a cloud (approximately 1–2 minutes) after a lightning discharge of several tens of coulombs by restructuring water supermolecular aggregates in the cloud; and also to explain such phenomena as the existence of thread-like structures and "strings" in the atmosphere at low temperatures and, in the absence of advection (Ivlev and Chvorostovsky, 2000a, b), the existence of interdependence between electric field strength, meteorological range of visibility and air humidity, charge transport in the atmosphere, and cloud burning caused by volcanic eruptions.

15.5 SUN–EARTH INTERACTION AND GLOBAL CATASTROPHES

Although occasional catastrophes appear to have been caused by the impacts of meteorites on the Earth's surface, most global catastrophes, which appear to have been associated with solar influence on Earth processes, are of special interest for mankind. The physical mechanisms of these phenomena are by no means always clear. In particular, despite problems of the effect of solar radiation on the Earth's magnetosphere, ionosphere, atmosphere, and underlying surface being thoroughly studied, problems relating to gravitational and probably electrical interaction between the Sun and the Moon with the interior of the Earth require much study. The importance of this interaction follows from its mass, and therefore the energy of the processes taking place within the Earth's core.

15.5.1 Tectonic processes

Interest aroused by powerful tectonic processes, the regularities that govern their manifestations, and possibilities of their forecast is obvious and easily understandable. Recently much progress has been made in the area of forecast of both earthquakes and volcanic eruptions, based primarily on precursors of various kinds (Pilnik, 1988; Dobrovolsky, 1991; Florsch *et al.*, 1995; Sobolev and Ponomarev, 2003). These forecasts are mainly based on changes in physical field characteristics in the period prior to the catastrophe, when a powerful tectonic process is already in progress (Linkov *et al.*, 1990; Sobolev *et al.*, 1996; Liskov and Petrova, 2002). Thus, the forecast is confined to a certain time interval.

It is also extremely important to produce well-grounded estimates of catastrophe probability for a long period of time (Kornfeld, 1951). Accumulated statistical data, mainly of a chronological character, makes it possible to draw certain conclusions about the physical mechanisms promoting these processes. Mapping of the most powerful earthquakes and volcanic eruptions and comparison of their locations with relatively young mountain ranges and the outlines of tectonic plates demonstrate that

(1) the zones of powerful earthquakes and volcanic eruptions lie at the boundaries of tectonic plates, young mountain ranges lie some distance away from the boundaries, and old ranges may lie at the plate center;
(2) the zones of these tectonic phenomena generally do not coincide.

Orogenesis, earthquakes, and volcanic eruptions result from tectonic plate shifts (Figure 15.1), usually at different stages, and are not governed by identical

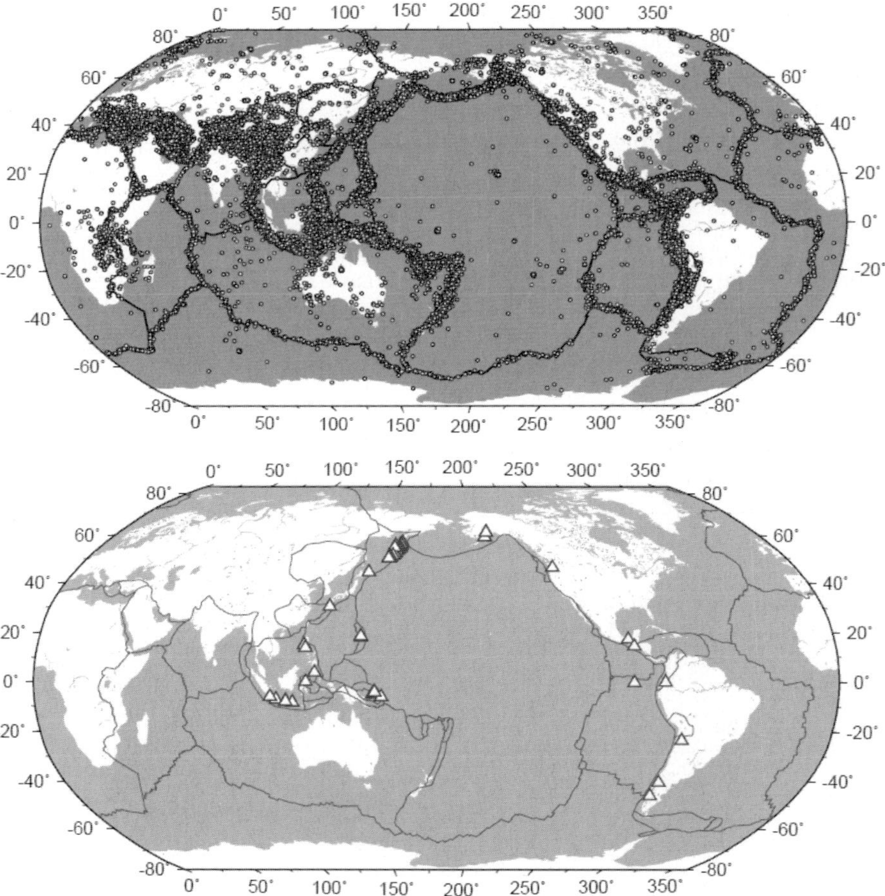

Figure 15.1. (a) Lithospheric plates and earthquake zones (black dots) and (b) intensive volcanic eruptions over a period starting from 1970.

mechanisms. All these phenomena are connected with the development and accumulation of strong elastic tensions in the Earth's crust with their subsequent relaxation and the breakthrough of magmatic mass (Kropotkin, 1996). Elastic deformation and intracrustal discontinuities take place when the pressure created by the deforming field exceeds the critical stress limits, causing shifts, faults, and ruptures of the crust. These pressures can be caused for mechanical (uneven movement of massive bodies) and thermodynamical reasons (increase of internal gas pressure).

Critical stress limits can decrease as a result of long exposure to relatively weak fields (vibrations, electrical processes, and electromagnetic radiation), heating and softening of the crust sections. The temporal and spatial characteristics of these processes both at the initial period of the catastrophic event and during its development and attenuation depend on which mechanisms play the leading role. These phenomena are accompanied by acoustic emission and the generation of elastic and electric signals which may be used for monitoring the processes mentioned above.

It has become a general practice to distinguish between three groups of stress sources in the Earth's crust: the first group is associated with endogenous (i.e., inner) processes that develop not only within the crust but also in the Earth's mantle, generating both the global Earth stress field and tectonic movements in the Earth crust; the second group of stress sources is associated with exogenous factors like inland ice, the load of water stored in reservoirs, erosion caused by rivers, pumping out of oil, gas, and water from deposits a few kilometers deep, deep mining and open-cast mining for coal and uranium and for non-fuel minerals such as diamonds and metal ores, and underground nuclear testing (these factors play a smaller role in global stress field development); the third group is associated with cosmic sources, for example, with the Earth's rotational forces or forces resulting from fast nearly stepwise change of the planet's rotational velocity, as well as with the tidal effect of the Moon and the Sun that have a regular character and are governed by gravitational forces. Because of its elasticity the Earth deforms under the influence of the tidal-rising forces that cause redistribution of masses and generate additional forces. This disturbance induces gravitational potential resulting from corresponding deformations.

It is generally accepted that the most significant contribution to the total stress field, according to initial estimates, is that of endogenous processes that generate stress fields of various grades (Artyushkov, 1979; Aplonov, 2001). The most important factor is thermo-gravitational instability of the Earth's mantle down to a depth of 2,900 km, particularly within the asthenosphere (the upper layer of the Earth's mantle, below the lithosphere) where the viscosity is 2–3 orders of magnitude lower than in the upper layers of the mantle and the Earth's crust (see Figure 15.3). The slow motion of matter within the asthenosphere layer as a result of viscous tension transfers force to the overlying mantle layer and the Earth's crust (lithosphere), causing stresses and therefore deformations in the latter. Stresses may also result from ascending and descending convective jets in the Earth's mantle, which, according to certain models, form two-layer system of convective cells (Ladyzhenskaya, 1970). The actual existence of these extremely slow jet flows in the Earth's mantle is confirmed by seismic tomography data, supported by the results of gravity force observations, the sharp anomalies of which are most clearly expressed at sites where submersion or uplift of the mantle matter is expected. These narrow positive or negative gravitational abnormal zones are confined to deep-sea trenches and young mountain ranges like the Andes, those in Indonesia, the Aleutians, Kuriles, Japan, and other islands.

Extremely strong compression at sites where oceanic—more heavy and cold—crust immerses (subducts) under more lightweight continental crust, is indicated by the presence of seismic focal zones within the Earth's crust and upper mantle, with deep earthquake foci (500 km–600 km deep). Revealed inhomogeneities of the upper mantle under mid-ocean ranges and old platforms also serve as sources of stress in the lithosphere and the crust. The present structure of the Earth's surface being determined by the motion of lithospheric plates, compression and expansion strain concentrates at sites with a corresponding geodynamic regime (Petrova *et al.*, 1996, 2007). Expansion dominates at the mid-ocean ranges, along divergent boundaries, and compression prevails in subduction zones (along convergent boundaries). The rigidity (solidity) of lithosphere plates makes it possible to transfer stress to remote areas of the plate, located several thousand kilometers away from the focus. The interaction of lithosphere plates contributes greatly to generation of the stress field in the uppermost shell of the Earth. Endogenous forces form stress fields of several hundred megapascals.

The motion of lithosperic plates is undoubtedly determined to some degree by the convective processes in magma. These processes are facilitated by the inhomogeneous structure of thermodynamic parameters within the Earth, namely temperature, pressure, and density (Figure 15.2). This is also confirmed by the two-layer structure of the Earth's crust, illustrated by Figure 15.3 from a paper by Aplonov (2001), which clearly shows a low-viscosity zone at a depth of 27 km–30 km on the vertical viscosity profile.

Numerous factors affect local stress fields. For example, one of the factors, permanent gravity force, does not perform tectonic work but affects the local stress field. Additional sources of stress in the Earth's crust are connected with areas of heating, local melting, and volcanic activity. Generated thermal tension acts within limited areas, distorting the more extensive stress field, and can become a factor triggering stress relief.

Additional stress in the Earth's crust caused by pronounced relief and growing mountain folds is one of the exogenous factors. The weight of the mountain structures affects the development of stress in neighboring lithosphere areas causing elastic reaction forces. Local stress can develop as a result of underground water flow or other fluids. This stress is far less than the one resulting from endogenous factors.

Cosmic factors, in particular rotational forces, create stresses not exceeding 0.1 Pa, and tidal forces caused by the interaction of the Moon, the Sun, and the Earth provoke stresses up to 10 Pa. Supposedly, the main external (geophysical and extraterrestrial) triggering factors that contribute to earthquakes and volcanic eruptions are: (1) solar activity; (2) Earth rotation velocity; (3) Earth tides; (4) various geomagnetic phenomena (Malinezky and Kurdyumov, 2001); (5) meteorological factors.

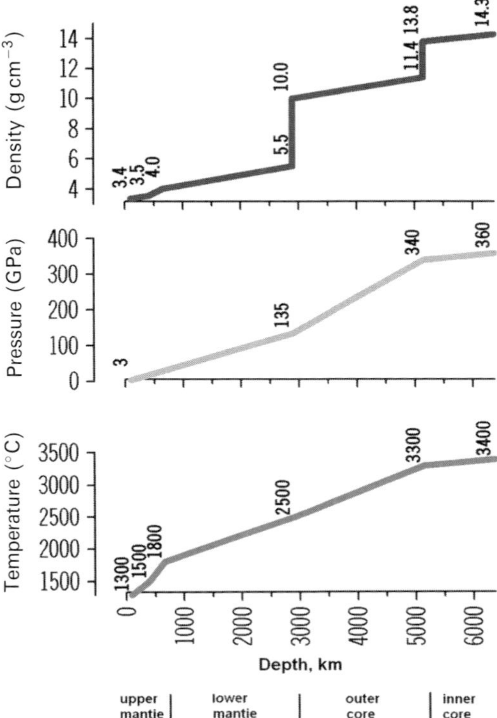

Figure 15.2. Thermodynamic characteristics of matter within the Earth.

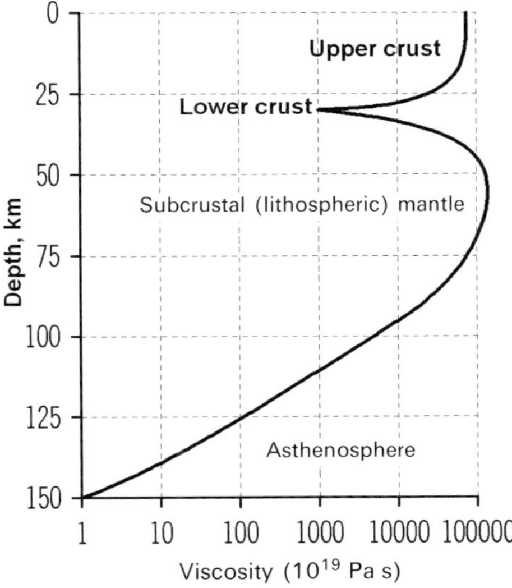

Figure 15.3. Vertical viscosity profile (from Aplonov, 2001).

In recent years the role played by processes in Earth core degassing in a number of geophysical and extraterrestrial phenomena (Sivorotkin, 2002) was suggested to be far more significant that was generally accepted before; the presence of permanent baric centers on Earth, abnormal atmospheric phenomena that correlate with geophysical processes (Petrova and Lyubimzev, 2006). Most important for the development of mathematical models of tectonic processes are hypotheses stating that the mobility of tectonic plates is higher than was generally accepted (i.e., tens of centimeters per year instead of 0.1 mm–1 mm per year, Trubizin and Rykov, 1998), and pulsation theory that suggests the Earth's volume oscillations (increasing and decreasing) are slowly expanding (Larin, 1980).

15.5.2 Earthquakes

In recent years important results have been obtained in the area of laboratory modeling of the processes of deformation stress accumulation, their localization, development of preconditions for macroscopic rupture, initiation of dynamic shifts by weak physical interactions, rupture development; the evolution of physical field characteristics during these processes (Linkov *et al.*, 1990; Sobolev *et al.*, 1996; Liskov and Petrova, 2002) has also been studied.

The study of the local deformation field using various rock samples has revealed an asymmetric pattern of compression and expansion areas, though abnormal areas of dilatancy have not been clearly identified. But it is possible to state that even initially homogeneous materials in the course of deformation develop stress-related inhomogeneities, reflecting both the hidden structure of the material and the initiation of rupture. This inhomogeneity increases with deformation as a result of material property change in the vicinity of developing cracks.

Even in brittle material, the localization of any unstable deformation develops gradually. The place of rupture (fracture nucleus) is indicated based on subsequent migration of the deformation from the nucleolus to the periphery and back. The polarity of local deformation discontinuities depends not only on the local inhomogeneities of the environment structure, but also on the more fundamental process of division of the material into areas of unstable and elastic deformation (Sobolev *et al.*, 1996).

Study of the velocities of lengthwise and transverse elastic waves generated by rock deformation and their dependence upon the rock type, rate of deformation, and sample humidity, has demonstrated that it is possible to study in natural conditions the type of rock (velocity ranges from 2,000 m s^{-1} to 4,000 m s^{-1}), degree of deformation (the velocity of elastic waves decreases if deformation exceeds 1.5%), and water saturation (the velocity of elastic waves is less for dry samples), duration, and intensity of rock exposure to mechanical factors.

Changes of electrical resistance in the area of a developing rupture indicate that its variations often have a sign opposite to the sign of variations in outer space, that the variation amplitudes at short distances are larger than at long ones, and that the dispersion of changes increases with the development of a macro-fracture. Regularities in acoustic and electromagnetic emission in the course of deformation were also revealed. An extensive study of variations in the electric field of the ground preceding earthquakes has been carried out by Varotsos and Alexopoulos (1984a, b, 1987) and Varotsos *et al.* (2002, 2003a, b) as the basis of an earthquake prediction system.

Laboratory study of the initiation of dynamic shifts demonstrated that an elastic impulse reduces the threshold value of the external stress necessary for shift initiation. There is a time delay between the initial impulse and the shift that exceeds the impulse duration and this is tens of times more than the travel time of elastic waves. It decreases when the impulse amplitude increases. Probably, the delay is caused by the fact that the generation of an unstable shift in one of the contacts due to its destruction causes creeping faults. Its speed depends on irregularities at the contact and is far less than the elastic wave velocity (Sobolev *et al.*, 1996). The energy and frequency composition of the acoustic radiation associated with the initiated shift is much more than without the initiation. The maximum intensity of radiation is observed at the initial stage of an unstable shift (Petrova *et al.*, 1996). The tendency for the acoustic activity of the deformed model to increase under exposure to electric impulses by about 1% of the initial level has also been observed.

The initial stage of an earthquake is associated with quasi-periodic changes in stress within the Earth's crust caused, for example, by tides and atmospheric pressure variation. Numerical estimates of the trigger effect of sinusoidal oscillations on shifts are given in the book by Sobolev and Ponomarev (2003).

The most developed model of earthquakes, based on the results of natural and laboratory experiments, is obviously a model of avalanche-unstable crack formation (AUCF), developed and presented by the Institute of Earth Physics (Russian Academy of Science) (Sobolev, 1993). The model is based on the interaction between the stress fields of the fractures (defects) and localization of crack formation processes. The number and size of the fractures grow steadily under permanent subcritical stresses; after a certain critical density (number of fractures per unit area) the rock enters the phase of fast macro-destruction. The hierarchy of fractures (coalescence of smaller fractures into bigger ones) is a significant aspect, and ensuring the similarity of processes in natural and laboratory situations is essential.

On the basis of this model, criteria concerning the seismic regime in certain regions have been developed involving

- a concentration criterion of seismogenic ruptures based on the kinetic concept of rigidity (accumulation of fractures);
- a criterion of localization of numerous fractures during gradual migration of weak earthquakes to the nodal plane of the future main rupture (seismic foci occur in those areas most weakened by preceding lesser earthquakes); and
- a criterion of decrease in seismic activity as a result of a creeping fault (development of a macro-fracture in the altered stress field requires accumulation and integration of lesser fractures).

MacDonald (1975) suggested a hypothesis explaining the observed super long period (seismic and gravitational) oscillations of the Earth's surface by natural oscillations of lithosphere plates that experience stretching of various magnitudes caused by deformation and stress changes to the planet as a whole. Daily variations in Earth's rotational velocity, its slowing down and acceleration, generate additional stresses at plate boundaries, stimulating the natural oscillations of the plates with nonzero amplitudes at the point of force application. If boundary stress causes accumulation of energy, natural oscillation frequencies increase according to the fundamental variational principle for elastic body eigenvalues. Elastic plates are not isolated: zones of contact of the rigid edge of a plate with the next plate (or plates) are small compared with plate size. Plate shift resulting from convection in the mantle and changing the Earth's rotational velocity (Pilnik, 1988; Zharov *et al.*, 1991; Sidorenkov, 2002) generates local stress that can result in elastic energy accumulation. Then two scenarios are possible. In the first one (the evolutionary scenario) this accumulated energy dissipates through, for instance, a crack formation mechanism, affecting the composition and structure of aerosols in the near-field zone and acoustic emission characteristics. In the second scenario (the catastrophic one) energy does not discharge gradually, in parallel with energy accumulation, but in an explosive way after a certain stress level is achieved, or as a result of some triggering event. This relatively long-term accumulation of a significant amount of elastic energy will lead, according to the fundamental variational principle for eigenvalues, to an increase in a plate's natural oscillation frequencies. Considering this scenario of elastic energy accumulation we suggest that the increase of certain eigenvalues of a biharmonic operator, caused by local stress within a relatively small zone at the plate edge, can be interpreted as an indication of a catastrophic scenario.

15.5.3 Explosive volcanic eruptions

First of all it should be noted that there are two different types of volcanic eruptions: explosive and extrusive. The extrusive type is characterized by relatively quiet and prolonged extrusion of magma through crust fractures, accompanied by small eruptions, which are probably caused by water getting into the conduit. Explosive eruptions are characterized by violent ejection of volcanic material that rises 30 km and more vertically or laterally into the atmosphere, as well as having large amounts of water in the ejected material.

Powerful explosive eruptions contribute greatly to variation in the optical characteristics of the atmosphere and therefore in climate change (MacDonald, 1975; Marchinin, 1980; Ivlev, 1998a). An interconnection between solar activity, the number of sunspots, and eruptions has been revealed. Figure 15.4 shows data over the period 1970–2003 relating to the number of sunspots and aerosol backscattering in the atmosphere caused by the eruption of El Chichón and Pinatubo volcanoes. Volcanic material ejected during the eruption, according to solar radiometry data, can stay in the atmosphere for a year and longer.

Dust particles themselves drift down to the troposphere; this process can take up to several months depending on particle size and the height of initial ejection. But the stratosphere actually remains dust-laden for a longer time due to the lengthy processes of generation of extremely small-sized aerosol particles from the gaseous

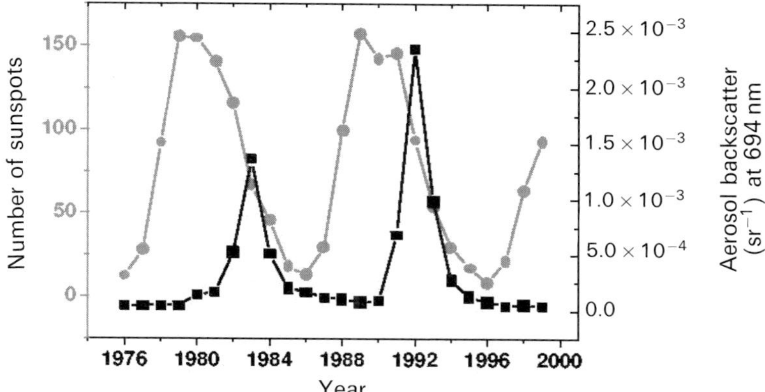

Figure 15.4. Connection between the number of sunspots and the amount of dust in the atmosphere caused by volcanic eruptions (Ivlev, 1998b).

phase of volcanic material (sulfur dioxide, water vapor, chlorides, oxides of nitrogen, etc.). Another reason for aerosol material to stay for a long time in the upper atmosphere is successive eruptions of one of several volcanoes after the first powerful eruption.

Mass-spectrometry analysis shows that water vapor constitutes 95% of all the volcanic gases (Marchinin, 1980). The percentage of water affects magma viscosity, significantly decreasing it. Increase of alkalinity of the volcanic material, decrease of the SiO_2 and Al_2O_3 concentration and increase of the temperature affect it in the same way; magma from depths is less viscous.

Explosive volcanoes are generally situated on islands or near the ocean shore, often close to underwater crust fractures and accompanied by hydrothermal springs. The latter are usually confined to zones of slow (from 1 cm–2 cm to 18 cm–20 cm per year) sliding of huge blocks of Earth's crust (lithospheric plates) that move in the upper layer of the semi-liquid Earth shell (i.e., in the mantle). It is now generally accepted that ocean water penetrates into the Earth's interior through fractures in young crust, mixes with magma, becomes saturated with chemical elements and very hot (several hundred degrees), and eventually effuses from conduits of "black smokers" at the ocean floor. This suggests that ocean water penetrates at least 10 km into the crust, overcoming a pressure of more than 1,000 atmospheres. However, this is an unrealistic suggestion, along with the idea of a capillary mechanism of water penetration. More probably it is another mechanism: water is forced to the area of low pressure that forms during fast generation of a crust fracture. Also there should be some canal through which water can pass. If the canal is not filled with water, and if air gets into the canal, the pressure within it slowly increases with depth from 1 atm to approximately 8 atm. In this case ocean water from depths greater than 100 meters easily penetrates the Earth's interior through fractures in the young crust, becomes saturated with minerals, heats up, and returns to the ocean through hydrothermal springs. These volcanic formations lie in zones where there is relatively slow separation between blocks of the Earth's crust. Probably "catastrophic" separation of the Earth's crust can also take place.

The uppermost layers of the Earth's crust and ocean floor (i.e., soil and the ground surface) act as very deep and intense filters, functioning as a blow-off valve that blocks the water flow and at the same time as a drain. In particular, clays, being products of secondary transfer, under the pressure of excess water become saturated and turn into thixotropic suspension. The fluidity of this system depends on water content in a non-linear way, and starting from a certain threshold value the suspension flows nearly as freely as water, the transition to fluid state being a "water hammer"-type phenomenon. Further growth in water pressure results in efflux of material from a filtering zone, with avalanche-like development of erosion in the moraine mass as well as formation of drainage canals and cavities. Water flows wash out debris from the drainage canal walls and makes the conduits wider until water pressure at the top layer is balanced by the strength of materials at the drain level; after this balance is reached the water content in the mud stream decreases and the fluidity of the latter also falls to the threshold value. Canals and ducts begin to clog up spontaneously one after another, preventing water from flowing back. In this way a drainage canal forms spontaneously and after a certain time clogs up spontaneously and disappears.

It is more difficult to imagine the formation of a giant cavity, several cubic kilometers in volume, with relatively low pressure. But if a continental or ocean plate tens or hundreds of square kilometers in area rises by just one meter, a cavity of this volume could be formed. The Moon's tide causes a rise in the Earth's surface of 55 cm, thus explaining the formation of a cavity of the required volume. A tsunami can form a hump 10 m high, or more, on the ocean surface when it approaches the land. If an underwater plate at that place bends inward at the same time, the formation of a cavity is also possible. Filling of the cavity with water and not with far more viscous magma is possible if the plate splits. To fill a cavity of 1 cubic kilometer with water through an opening about 10^3 m^2 with a flow rate of about 10 m s^{-1} requires 10^5 seconds. Water flowing down a gradient plane would travel in this time as far as a thousand kilometers. Subsequent heating and the high pressure of the environment is known to change the physical and chemical properties of water, which acquires certain acidic properties. This water dissolves the lowest layer of the crust above the magma, making it thinner. Possibly water lines are formed under the plates.

The interconnection between tsunamis and explosive eruptions has recently been studied. The following connection can be suggested for the most significant tsunami and eruptions, with a 3-year to 5-year delay between them: tsunami in the Aleutian Basin in 1946 and the Kelut volcano in 1951; tsunami near Kamchatka in 1952 and the Bezymianny volcano in 1956; tsunami in the Pacific Ocean in 1960 and the Agung volcano in 1963; tsunami in the Pacific Ocean in 1965 and the Avu volcano in 1966 and possibly the Ferdinando volcano in 1969. If such a dependence exists in reality, the tsunami of 2004 in the Indian Ocean would initiate an eruption in the Indonesian archipelago in 2007–2009. Because the distances between the tsunami center and volcano are very significant in these cases, it is too early to reach a conclusion about the mechanism involved in the evolution and motion of a water bubble under the Earth's crust.

Analysis of powerful explosive eruptions (rated at VEI 4) that took place during the last 200 years has not revealed any cyclicity of the eruptions or any connection between them. However, with a certain amount of caution it can be suggested that some cyclicity with a period of 6–38 years is possible, and some regularities can be found in the distribution of the duration of periods without powerful eruptions and periods with numerous successive eruptions. It can be taken as a hypothesis that eruptions are interdependent if the time delay between them is less than or equal to 30 months. It is natural to consider such phenomena (i.e., powerful eruptions) as non-random events. Under these suggestions all eruptions can be classified according to their number in the succession of interdependent eruptions. There were five single eruptions during this period, four double eruptions, three triple, two quadruple, two quintuple, two sextuple, three septuple, two octuple eruptions, and 1 nonuple eruption.

In two cases the duration of periods without eruptions constituted approximately 13 years, in five cases 5 to 7 years, and in 12 cases from 2.5 to 4 years, so the ratio of periods is 4 : 2 : 1. The ratio for the duration of periods of successive eruptions is more fuzzy but also is close to 2 : 1 which is characteristic of a cascade process.

In almost all successions of eruptions one or two eruptions were rated VEI 5 or higher. In four successions with $n > 6$ the first eruption was rated VEI 5 or higher. In two cases the last eruption in the succession was the strongest one. In eight cases the most powerful eruption rated VEI 5 or higher took place in the middle of the succession. (Six successions did not include eruptions rated VEI 5 or higher.)

An important role in the mechanism of disturbance transfer (provoking an eruption) is played by the primary disturbance, its duration, and relative spatial position of the disturbance origin and receiver. Spatial classification of powerful active volcanoes was performed in order to make analyzing these factors easier. At the present time the main eruptive zones are: (1) *A* America; (2) *B* North Pacific Ocean (Kamchatka, Kurile Islands, Japan, and probably the Aleutian Range); (3) *C* Pacific Ocean islands, Indonesia, and the Philippines. Other regions make only a small contribution to the statistics of powerful volcanic eruptions: in Europe *E1* Iceland and *E2* Mediterranean Sea (seven strong eruptions over 170 years); Africa *F* one; and New Zealand and surrounding islands *C5* one.

The data analyzed clearly demonstrate the weak dependence of the time period between initial disturbance and its effect on distance from the disturbance origin to the receiver: in some cases the effect manifested itself within several months several thousand kilometers away from the disturbance origin, and in other cases a distance of just a thousand kilometers was covered in more than a year. The maximum speed of signal passage is about 2.0 m/s–2.5 m/s. The longest time delay between the initial impulse and the eruption constitutes 1–2 years.

The efficiency of the impact (trigger-off) depends on the direction of disturbance propagation. Eastward has a significant advantage, and southward a slight one; this regularity is more pronounced in short successions. Long successions of eruptions often have a loop route of impact transfer.

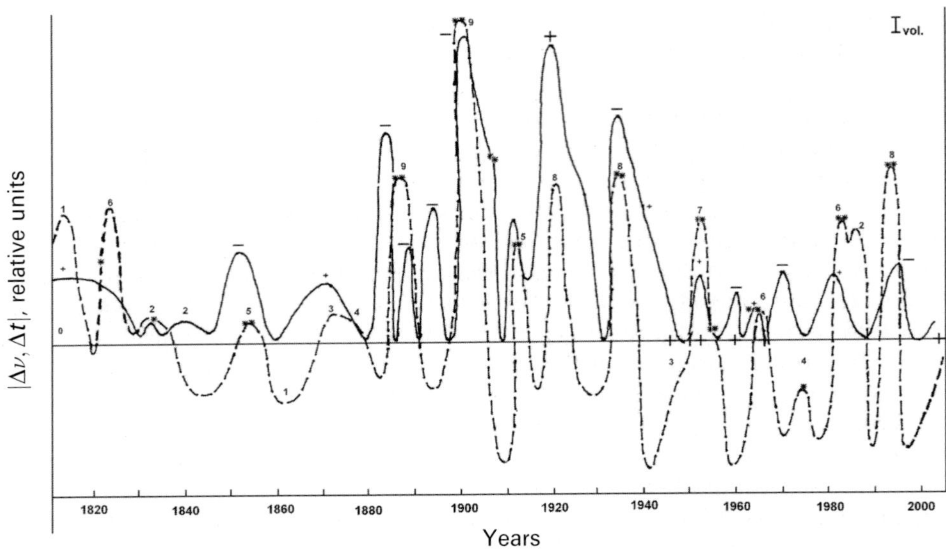

Figure 15.5. Time characteristic of intensity of volcanic activity and $\Delta\nu, \Delta t$, the change in the Earth's rotation rate (numeral = number of volcanic eruptions; + = acceleration of the Earth's rotation rate; − = slowing down).

The interaction between eruptions in three main zones is not exactly symmetric. The most intensive interaction of volcanic eruptions is observed in the American continent where the southward direction of propagation is most clearly expressed. In the Pacific zone mutual triggering of eruptions is relatively symmetric, with the certain dominance of the Indonesian eruptive zone affecting other regions (westward propagation).

To establish connections between volcanic activity and changes in the Earth's rotational velocity, data from the monograph by MacDonald (1975) were used. We studies the relation between the amplitude of volcanic activity (taking into account the number of powerful eruptions and their intensity) and the rate of the Earth's rotation (Figure 15.5). Data from the period 1860–1866 show a clear correlation between changes in the Earth's rotation rate and intensity of explosive eruptions. No correlation was revealed between seasonal changes in the Earth's rotation rate, which is lowest in April and November and highest in January and July, and the intensity of explosive eruptions.

The role of tides (interplanetary interaction between the Moon, the Earth, and the Sun) and geomagnetic phenomena is not understood because of lack of observational data. Man-made factors do not play any part in powerful volcanic eruptions. In this way, analysis of the spacetime characteristics of powerful volcanic eruptions provides the basis for the large-scale study of the mechanism and regularities of propagation of disturbances leading to eruptions.

15.6 DETERMINED CHAOS OF THE TEMPORAL–SPATIAL STRUCTURE OF GEOPHYSICAL FIELDS

The intensive development of computer technologies in recent decades has made it possible to process huge arrays of various geophysical data to give a fresh outlook on the structure of geophysical fields, reflecting the hierarchical structure of the geological environment, its ability to accumulate and redistribute incoming external energy, and the non-linearity of processes developing in this environment. The temporal–spatial variability of environmental parameters combines both regular determined and irregular stochastic characters. The stochastic nature is conditioned by a large number of mutually independent factors (which are unexplored for the most part) and the openness of the environment to external influences, lithosphere activity, and the non-linearity of processes within it.

Processes of this type are dealt with by dynamic systems theory, where the main parameters are the realization dimension D and Kolmogorov's entropy K (Sobolev, 1993). The realization dimension serves as a measure of the number of degrees of freedom in the studied process (i.e., the number of equations necessary for description of the dynamic system). It can be evaluated as a number of observable environment parameters, $X_1, X_2, \dots, X_m$, that can be correlated to the system of equations of the following type:

$$\frac{\partial X_i}{\partial t} = F_i(X_1, X_2, X_3, \dots, X_m, t), \quad i = 1, 2, \dots, m,$$

where the F_i are, generally speaking, non-linear functions; and m is the dimension of the attractor (stable manifold of points formed by the solution trajectories). Solutions can be represented by both smooth and chaotic functions. If $m \leq 2$ the system has only smooth solutions describing monotonic or strictly periodic parameter variation.

In other cases chaotic solutions determined by the instability of the equations (i.e., by exponential growth with time of an arbitrary small disturbance of some variable, other variables being fixed) are observed. This chaos is generated by a small number of factors and is known as determined or dynamic chaos, and the system dimension is known as the realization dimension (number of degrees of freedom). For instance, there are systems that consists of only three non-linear differential equations of the first order. In the general case D and m are related by the inequality $D \leq m \leq 2D + 1$.

The parameter K, Kolmogorov's entropy, reflects the rate of growth of the uncertainty, with relative error $\delta(t)$, of the forecast of system behavior, with system state at a given instant of time t_0 determined with a certain accuracy (δ_0); that is, the rate of distance growth between the attractor points $\delta(\Delta t)\, e^{K\Delta t}$. The characteristic time $1/K$ is called the predictability time or time of deterministic behavior of the dynamic system. For a purely deterministic process (solution of a stable system) $1/K \to \infty$, and for a purely chaotic process $1/K = 0$; that is, these processes are extreme, special cases of deterministic chaos.

The work of Sobolev *et al.* (1996) shows that the chaotic character of data variations can be explained by the interplay of a small number of non-linearly related parameters, if components depending on certain external periodic factors and trends (comparable or exceeding the length of the array row used) are excluded from time data arrays. This implies that the mechanisms underlying determined chaos play an important role in variations of the parameters of geophysical fields.

All studied geophysical fields can be divided into two groups: fields that directly reflect deformation processes, whose modeling requires systems of equations (with dimension 3–4), and fields that reflect chaotic physical processes (electroconductivity, radon emanation, alignment variations) with dimension 2.7–3.5, dynamic description of which requires systems of relatively high dimension (4–7), which agrees with the more complex nature of the latter processes.

So the analysis of tectonic processes that lead to catastrophic phenomena shows that, regardless of the great number, complexity, and ambiguity of the physical factors determining the scale, duration, and external manifestation of these phenomena, it is possible to identify those factors that are most important for a particular process, and to forecast its characteristics with a given probability.

15.7 CONCLUSION

To summarize, climate as the most important factor in the existence and development of an ecosystem is determined in the first place by the Sun–Earth interaction, which can cause both gradual or periodic climate changes and catastrophic phenomena. The immediate cause of earthquakes and volcanic eruptions is resonance phenomena initiated by gravity forces, by exposure to hard solar and cosmic radiation, by variations in the magnetic and electric fields of the Earth and near-Earth space, and by meteorological processes. Earthquakes and eruptions are accompanied by accumulated stress relief. Some mechanisms of these phenomena are considered in this chapter. Besides the immediate damage to the environment, earthquakes and eruptions also release a great amount of materials affecting natural processes within the ecosystem. Relaxation processes can be rather lengthy. This can be explained, in particular, by gases and aerosol particles lingering in the atmosphere and significantly affecting the physical and chemical properties of the environment (mainly in the atmosphere), and therefore heat–mass exchange and dynamic processes. The physical mechanisms involved are the same as those that cause gradual or periodic climate change: variation in environmental radiation characteristics (albedo, radiation absorption), and in water-phase transition processes (condensation and crystallization, evaporation, cloud system generation, precipitation).

Space research has contributed to the gradual accumulation of technogenic material in near space, increased the inflow of dust and gaseous particles to high atmosphere layers and considerably changed its optical and electrical properties—the changes being similar to those caused by meteor streams breaking through the Earth's atmosphere. In our previously published papers the effects of high-energy particle streams on cloud formation in the upper troposphere have also been considered.

15.8 REFERENCES

Aisenberg D. and Kauzman B. (1975). *Structure and Properties of Water*. Hydrometeoizdat, St. Petersburg, 280 pp. [in Russian].

Aplonov S.V. (2001). *Geodynamics*. St. Petersburg University, 360 pp. [in Russian].

Artyushkov E.V. (1979). *Geodynamics*. Science, Moscow, 216 pp. [in Russian].

Chromov S.P. and Mamontova L.I. (1974). *Meteorological Dictionary*. Hydrometeoizdat, Leningrad, 568 pp. [in Russian].

Derpgolz V.F. (1971). *Water in the Universe*. Nedra, Leningrad, 223 pp. [in Russian].

Dobrovolsky I.P. (1991). *The Theory of Tectonic Earthquake Preparation*. Academy of Science of U.S.S.R., Moscow, 217 pp. [in Russian].

Florsch N., Legros H., and Hinderer J. (1995). The search for weak harmonic signals in a spectrum with applications to gravity data. *Phys. Earth Planet. Int.*, **90**, 197–210 [in Russian].

Gorschkov V.G. (1990). *Energy of the Biosphere and the Stability of Environment*, Theoretical and General Questions of Geography Series. ARISTI, Moscow, 237 pp. [in Russian].

Gorschkov V.G., Dovgalyuk Yu. A., and Ivlev L.S. (2006). *Principles of Physical Ecology*. St. Petersburg University, 251 pp. [in Russian].

Houghton J.T. (ed.) (1984). *The Global Climate*. Cambridge University Press, Cambridge, U.K., 581 pp.

Ivlev L.S. (1998a). Aerosols produced by the Earth's deformations. *Proc. Int. Conf. "Natural and Anthropogenic Aerosols", September 29–October 4, 1997, St. Petersburg*. St. Petersburg University, pp. 73–82 [in Russian].

Ivlev L.S. (1998b). About connection between volcanic activity and climatic characteristics. *Proc. Int. Conf. "Natural and Anthropogenic Aerosols", September 29–October 4, 1997, St. Petersburg*. St. Petersburg University, pp. 64–72 [in Russian].

Ivlev L.S. (2001a).The causes and mechanisms behind space debris emergence. In L.S. Ivlev (ed.), *Ecology and Space*. St. Petersburg University, pp. 20–21 [in Russian].

Ivlev L.S. (2001b). Increase of intensity of dust sedimentation in high atmospheric layers. In L.S. Ivlev (ed.), *Ecology and Space*. St. Petersburg University, pp. 38–39 [in Russian].

Ivlev L.S. (2005). Physical principles of biosphere ecology of the Earth as a planet. In L.S. Ivlev (ed.), *Ecology and Space*. St. Petersburg University, pp. 7–13 [in Russian].

Ivlev L.S. (2007). About disturbance propagation velocity from one powerful volcanic eruption to another. *Proc. Fifth Int. Conf. "Natural and Anthropogenic Aerosols", October 6–9, 2006, St. Petersburg*. St. Petersburg University, pp. 71–182 [in Russian].

Ivlev L.S. and Chvorostovsky S.N. (2000a). Study of space radiation influence on microstructure parameters and optical properties of lower atmosphere in middle and high latitudes, 1: Influence of charges on atmospheric heterogeneous processes. *Optics of Atmosphere and Ocean*, **13**(12), 1423–1431 [in Russian].

Ivlev L.S. and Chvorostovsky S.N. (2000b). Study of space radiation effect on microstructure parameters and optical properties of lower atmosphere in middle and high latitudes, 2: Heterogeneous processes during influence of high-energy particles flows. *Optics of Atmosphere and Ocean*, **13**(12), 1432–1437 [in Russian].

Ivlev L.S., Kondratyev K.Ya., and Chvorostovsky S.N. (2001). Influence of space debris on composition, optical properties and physical processes in high and middle atmosphere. *J. Optics*, **68**(4), 3–12 [in Russian].

Ivlev L.S., Klingo V.V., and Chvorostovsky S.N. (2003). Emergence of crystal clouds in high troposphere as a result of aerosols charging with molecular ions. *Proc. Third Int. Conf.*

"Natural and Anthropogenic Aerosols", September 24–27, 2001. St. Petersburg University, pp. 279–287 [in Russian].

Kondratyev K.Ya. (1990). *Key Problems of Global Ecology*. ARISTI, Moscow, 454 pp. [in Russian].

Kondratyev K.Ya. and Ivlev L.S. (1995). About influence of anthropogenic aerosols on climate. *Proc. Russian Academy of Sciences*, **340**(1), 98–100 [in Russian].

Kondratyev K.Ya. and Ivlev L.S. (2001). Space ecology problems and aerosols monitoring. *J. Optics*, **68**(7), 3–8 [in Russian].

Kondratyev K.Ya., Ivlev L.S., and Galindo I. (1995). Application of enrichment factor concept for study of volcanic eruption products. *Proc. Russian Academy of Sciences*, **394**(6), 581–583 [in Russian].

Kondratyev K.Ya., Ivlev L.S., Krapivin V.F., and Varotsos C.A. (2006). *Atmospheric Aerosol Properties: Formation, Processes and Impacts*. Springer/Praxis, Chichester, U.K., 572 pp.

Kornfeld M. (1951). *Elasticity and Stability of Liquids*. Technical-Theoretical Literature, Moscow, 107 pp. [in Russian].

Krapivin V.F. and Kondratyev K.Ya. (2002). *Global Changes of Environment: Ecoinformatics*. St. Petersburg University, 723 pp. [in Russian].

Kropotkin P.N. (1996). Tectonics: Tensions in the Earth's crust. *Geotectonics*, **2**, 3–5 [in Russian].

Ladyzhenskaya O.A. (1970). *Mathematical Problems of Viscous Incompressible Liquid Dynamics*. Nauka, Moscow, 287 pp. [in Russian].

Larin V.N. (1980). *Hypothesis of Originally Hydride Earth*. Nedra, Moscow, 123 pp. [in Russian].

Linkov E.M., Petrova L.N., and Osipov K.S. (1990). Seismic and gravitational Earth pulsations and atmospheric perturbations as possible precursors of powerful earthquakes. *Proc. Russian Academy of Sciences*, **313**(5), 1095–1098 [in Russian].

Liskov A.I. and Petrova L.N. (2002). About the Earth's crust oscillations. *Herald of St. Petersburg State University, Ser. 4*, **2**(12), 99–102 [in Russian].

MacDonald G. (1975). Volcanoes. Nedra Publ., Moscow, 435pp. [in Russian].

Malinezky G.G. and Kurdyumov S.P. (2001). Nonlinear dynamics and forecasting problems. *News of Russian Academy of Sciences*, **71**(3), 210–232 [in Russian].

Marchinin E.K.(1980). *Volcanoes and Life: Biovolcanology Problems*. Mysl, Moscow, 196 pp. [in Russian].

Monin A.S. (1986). *Introduction to Climate Theory*. Hydrometeoizdat, Leningrad, 218 pp. [in Russian].

Petrova L.N. and Lyubimzev D.V. (2006) Planetary character of seismogravitational Earth pulsations. *Earth Physics*, **2**, 26–36 [in Russian].

Petrova L.N., Belyakov A.S., and Nikolaev A.V. (1996). Acoustic emission and seismogravitational oscillations of the Earth as related components in studies of internal plate dynamics. *J. Earthquake Prediction Res.*, **5**(2), 211–223.

Petrova L.N., Pavlov B.S., and Ivlev L.S. (2007). Interpretation of seismogravitational Earth pulsations: Distubance of laminar eigenfrequencies under the influence of point tension at boundaries. *Earth Physics*, in press [in Russian].

Pilnik G.P. (1988). Nonregularities in Earth's diurnal rotation. *Astron. J.*, **65**, 184–189 [in Russian].

Reznikov V.A. (2007a). Water: Bolides and ball lightning. In L.S. Ivlev (ed.), *Ecology and Space*. St. Petersburg University, pp. 242–248 [in Russian].

Reznikov V.A. (2007b). Water as oxygen shells in hydrogen plasma. In L.S. Ivlev (ed.), *Ecology and Space*. St. Petersburg University, pp. 248–254 [in Russian].

Reznikov V.A. (2007c). Discrete change of hydrogen connections in water. In L.S. Ivlev (ed.), *Ecology and Space*. St. Petersburg University, pp. 255–257 [in Russian].

Reznikov V.A. and Ivlev L.S. (2005). Collectivization of electronic states in the course of intermolecular interactions. In L.S. Ivlev (ed.), *Ecology and Space*. St. Petersburg University, pp. 109–113 [in Russian].

Sakrzhewskaya N.A. and Sobolev G.A. (2002). Probable effect of magnetic storms on seismicity. *Physics of the Earth*, **4**, 103–109 [in Russian].

Sazepina G.N. (1974). *Properties and Structure of Water*. Moscow State University, 167 pp. [in Russian].

Sidorenkov N.S. (2002) *Atmospheric Processes and Earth's Rotation*. Hydrometeoizdat, St. Petersburg, 366 pp. [in Russian].

Sivorotkin V.L. (2002). *Deep Earth's Degassing and Global Catastrophes*. Geoinformcenter, Moscow, 250 pp. [in Russian].

Sobolev G.A. (1993). *Basic Principles Underlying Earthquake Forecast*. Science, Moscow, 313 pp. [in Russian].

Sobolev G.A. and Ponomarev A.V. (2003). *Physics of Earthquakes and Precursors*. Nauka, Moscow, 270 pp. [in Russian].

Sobolev G.A., Ponomarev A.V., Koltsov A.V., and Smirnov V.B. (1996). Simulation of trigger earthquakes in the laboratory. *Pageoph*, **147**(2), 345–355.

Stahanov I.P. (1985). *About Physical Nature of Ball Lightning*. Energoatomizdat, Moscow, 209 pp. [in Russian].

Trubizin V.P. and Rykov V.V. (1998). Global tectonics of floating continents and oceanic lithospheric plates. *Proc. Russian Academy of Sciences*, **359**(1), 109–111 [in Russian].

Varotsos C. and Vasilyev S.L. (2005a). Anisotropy of unstably stratified atmosphere. *Proc. Int. Conf. "Natural and Anthropogenic Aerosols", October 6–9, 2003, St. Petersburg*. St. Petersburg University, pp. 344–357 [in Russian].

Varotsos P. and Alexopoulos K. (1984a). Physical properties of the variations of the electric field of the earth preceding earthquakes, I. *Tectonophysics*, **110**, 73–98.

Varotsos P. and Alexopoulos K. (1984b). Physical properties of the variations of the electric field of the earth preceding earthquakes, II. Determination of epicenter and magnitude. *Tectonophysics*, **110**, 99–125.

Varotsos P. and Alexopoulos K. (1987). Physical properties of the variations in the electric field of the earth preceding earthquakes, III. *Tectonophysics*, **136**, 335–339.

Varotsos P., Sarlis N., and Skordas E. (2002). Long-range correlations in the electric signals that precede rupture. *Phys. Rev. E*, **66**, 011902 (7).

Varotsos P., Sarlis N., and Skordas E. (2003a). Long-range correlations in the electric signals that precede rupture: Further investigations. *Phys. Rev. E*, **67**, 021109 (13).

Varotsos P., Sarlis N., and Skordas E. (2003b). Electric fields that arrive before the time-derivative of the magnetic field prior to major earthquakes. *Phys. Rev. Lett.*, **91**, 148501 (4).

Vasilyev S.L. (2005b). Analysis of regular motion in unstratified flow using Karman street as an example. *Proc. Int. Conf. "Natural and Anthropogenic Aerosols", October 6–9, 2003, St. Petersburg*. St. Petersburg University, pp. 358–363 [in Russian].

Zharov V.E., Konov A.S., and Smirnov V.B. (1991). The variations of Earth rotation parameters and their connection with the most powerful earthquakes in the world. *Astron. J.*, **68**(1), 187–196 [in Russian].

Zuev V.E. and Titov G.A. (1996). *Atmospheric Optics and Climate*. Spectr, Tomsk, 271 pp. [in Russian].

16

Problems of the sustainable development of ecological–economic systems

Gennadiy A. Ougolnitsky and Anatoliy B. Usov

16.1 INTRODUCTION

One of the most important global problems of modern society is the continuing growth of environmental pollution and its effect in aggravating the ecological situation. The development of industrial and agricultural production intensifies many problems of the ecological safety of exploitation of the environment. At present it is very important to estimate the consequences of the economic activity of industrial enterprises and whole regions.

It is necessary to determine the most efficient economic actions which will improve the ecological situation in various regions and to know how to forecast changes in the ecological systems. The interaction between human beings and nature is formalized by means of the "ecological–economic system". The ecological–economic system is a complex of the interconnected economic, technical, social, and natural factors in the world surrounding human beings. Elaboration of the control mechanism in the ecological–economic system is a very important problem.

The general theory of decision-making in an ecological–economic system is not reduced to global optimization. This theory must provide more complex procedures and take into account the interests of all participating sides. Many papers are devoted to this problem (e.g., Fathutdinov and Sivkova, 1999; Ougolnitsky, 1999; and Ryumina, 2000). These authors point out the necessity for co-ordination of the interests of all participating sides in the management process.

The importance and the role of a well-organized control system for the ecological–economic system are now increasing. Such systems create the conditions for decision-makers to respond adequately to processes that influence the social and economic situation in the world.

Determination of the optimum structure of the control system is one of the most important problems of the sustainable development of ecological–economic systems.

What is the optimum number of hierarchically subordinated subjects in the management system of the ecological–economic object? What is the nature of their relations? Answers to these and many other similar questions are of interest to leaders at various management levels.

The resolution of these and other multiple problems is impossible without a complex approach to the situation. This approach implies setting up a general concept for management of the ecological–economic system. The main management problems, the structure of management systems, and the principles governing their organization must be stated in this concept, which is based on the notion of sustainable development of ecological-economic systems and on the use of hierarchical management methods.

16.2 THE NOTION OF SUSTAINABLE DEVELOPMENT

For the first time the necessity for sustainable development of ecological–economic systems was announced at the meeting of UNESCO in Paris in 2002. This is the key notion behind formalization of the relations between society and nature. Practically, the notion of sustainable development implies a transition from the problem of preserving the environment at the expense of economic growth to the problem of ensuring simultaneous economic development and preservation of the environment.

The term "sustainable development" is better expressed as "ecologically sustainable economic development". Complex studies of the problems of sustainable development of the ecological–economic system have been conducted by Danilov-Daniliyan and Losev (2000), Dreyer and Losi (1997), Kondratyev *et al.* (1996, 2002), Koptyug *et al.* (1999), Ryumina (2000), among others.

The concept of sustainable development has made it possible to take a new look at the notion of "cost–performance". If long-term economic projects take into account natural regularities, then finally they will be economically efficient and profitable. But, if economic projects are realized without regard to the permanent ecological consequences, then they will be unprofitable.

The notion of the sustainable development of ecological–economic systems includes the following obligatory requirements:

(1) satisfaction of the requirements of both economic development and ecological balance;
(2) the observance of these requirements for ever, or at least for a very long time;
(3) the necessity for hierarchical management of sustainable development. Only this form of management is able to satisfy all the different interests in the essential performance of key requirements.

16.3 THE HIERARCHICAL APPROACH TO MANAGEMENT

A good way of interpreting the ecological–economic system is as a hierarchically controlled dynamic system when its sustainable development is scrutinized (see Ougolnitsky, 1999).

Single-level models are used to analyze the conditions of ecological–economic systems. But these models do not completely take into account the structure of modern management systems, all the various relationships between economic and the ecological sub-systems, and all types of influence on ecological systems. So, the notion of hierarchically controlled dynamic systems is used in the analysis.

Use of the concept of a hierarchically controlled dynamic system takes into account the details of a management mechanism to describe the operation process of the real management system. It leads to better decisions about the practical problems facing preservation and rational use of the environment.

The necessity for hierarchical management regarding sustainable development of the various ecological–economic systems is conditioned by the following main factors:

(1) the mismatch between the objective strategic interests of the whole system and the subjective short-term private interests of different participants;
(2) the complex structure of modern management systems. This structure defines the necessity for additional co-ordination of the system, the group, and the individual interests of the various participants; and
(3) the management system must offer up a decision and find solutions to the various and even inconsistent purposes and problems of the participants.

Therefore, modern management systems are made on the hierarchical principle. There are several hierarchical subordinated controlling subjects. Their interests are different, occasionally even the opposite of one another.

The simplest hierarchical management system is a two-level system, comprising (Ougolnitsky, 1999):

- a source of influence at the top level (i.e., FC or federal center);
- a source of influence of the bottom level (i.e., IE or industrial enterprise);
- a controlled dynamic system (DS).

The main reasons behind introducing the notion of a hierarchically controlled dynamic system in the present context are the following. Every industrial enterprise (IE) acts upon the ecological–economic system for its own purposes. These purposes, in general, do not meet the demands of sustainable development. As a rule, each IE strives to maximize its own current income, received as a result of the production activity.

The ecological–economic system is an inanimate object. So, this system cannot act "to defend its own interests". Its return reactions are spontaneous and are capable of leading to disastrous consequences. So, a federal center (FC) is needed. The FC holds sway over the IE obliging the IE to ensure sustainable development. The FC keeps the interests in mind of a more broad system than the IE (e.g., society as a whole).

Most often, state authorities use an organ such as the FC to oversee the ecological–economic system. The whole ecological system of a certain region or its separate components (e.g., a river) is called the controlled dynamic system (DS).

It is assumed that relations between the elements of this system are organized as follows. The FC holds sway over the IE and the DS, while the IE only acts upon the DS. The FC and the IE are the total source of influence on the DS. This source has a hierarchical structure. The IE pursues private purposes.

The management system is adaptive and has feedback. Information about the DS reaches management subjects. The role of the FC consists in providing sustainable development of the DS.

This two-level scheme of the organization of the management system is often adequate. For the broad class of ecological–economic systems this scheme allows the simplest organization of the management system. But often this scheme of management organization does not take into account all the possible relationships and the details of the interaction between all the management objects and subjects. In these cases it is necessary to take into account additional hierarchical subordinated management levels. They allow the structure of the relations between various governing subjects of the real systems to be pictured more exactly.

We shall consider three-level management systems, involving the influence source of the upper level (federal center or FC), the influence source of the average level (local management body or MB) and the influence source of the lower level (industrial enterprise or IE). All these levels affect the control of the dynamical system (DS; e.g., the river). It is assumed that relations between the elements of this system are organized as follows: the FC acts upon the MB and the DS, the MB acts upon the IE and the DS, and last the IE acts only upon the DS.

Moreover, the real management system can become complicated. The FC, MB, and IE can lose their monolithic structures. In this case, management subjects are divided by the set of purposeful management subjects. All these subjects have private purposes and possibilities for their achievement. So, more complex tree-type, multi-function hierarchical structures appear instead of a linear hierarchical chain.

16.4 MANAGEMENT METHODS

The purpose of management consists in the provision of sustainable development of ecological–economic systems. The purpose of sustainable development can be reached in different ways. So, the question of the choice of the best way for the FC to operate arises. In other words, the FC follows one or several optimum criteria when the sustainable development conditions of the ecological–economic system are executed. These criteria reflect the additional preferences of the FC on the set of sustainable development strategies.

The achievement of sustainable development can be realized by different methods of hierarchical management. It is possible to select the following methods of management depending on the degrees of freedom of the individual as the object of management:

(1) *Compulsion*. The subject forces the object to promote the achievement of the subject.
(2) *Impulsion*. The subject creates conditions for the object to make it economically advantageous for the object to promote achievement of the subject's purposes.
(3) *Conviction*. Subject–object interaction is organized so that the object strives to achieve strategic objectives together with the subject of management. The object acts of its own accord. The relations between the subject and the object of management acquire a subject–subject character.

The compulsion method expects that the FC influences the set of the IE's possible management strategies. As a result, this set is narrowing, with the result that the IE has to choose only those strategies that provide for sustainable development of the whole system. The FC's influence is administrative–legislative in character. The FC chooses a strategy from the set of possible management strategies according to its own criteria. For instance, the FC can strive to minimize its own costs. Of course, compulsion is realized only if the FC has the significant possibility of exercising administrative influence on the IE. With compulsion the economic mechanisms of influence are not used.

In modern conditions the compulsion method, as a rule, is inefficient and not economically beneficial for the sustainable development of the whole system.

The incentives method is more flexible than the compulsion method and is a more progressive method of management. It gives more freedom to all management subjects. The FC acts upon the IE by economic measures only. It stimulates the IE by influencing the IE payoff function. The thinking underlying the incentives method is the following. As a result of incentives the IE's optimum strategy guarantees sustainable development of the ecological–economic system. For this purpose an economic mechanism with feedback is used. This mechanism provides the IE with encouragement (privileges, subsidies, etc.) in the case of system sustainable development, and it provides the IE with punishment (fines, increased taxes, etc.) otherwise. The FC knows the IE's payoff function. So, the FC can foresee the IE's optimum reaction. The FC can choose the optimum strategy, thus guaranteeing sustainable development of the DS. Unlike the compulsion method, in this case the IE can choose the strategies, breaking the conditions for system sustainable development. So, if the possibility offered by the FC is not enough to stimulate the IE, then the incentives strategy cannot provide sustainable development of the ecological–economic system.

In the case of incentives the FC does not influence the set of the IE's possible management options.

There are multi-function mechanisms of management: the incentives–compulsion and the compulsion–incentives. In these cases the FC simultaneously acts upon the set of the IE possible managements and its payoff function. The FC uses a combination of administrative and economic methods. These management methods are most profitable for the FC. They put serious constraints on the activity of the IE. The effect is to bring maximum income to the FC and to bring about sustainable development of the system.

The conviction method implies voluntary cooperation between all management subjects. Together they provide sustainable development of the ecological–economic system, and together maximize the total payoff function. In this case their total income is divided according to some cooperative principle. The conviction method is psychological in character and is the best approach to the problem of sustainable development of the ecological–economic system. The essence of this method is to transform the hierarchical relationship to a cooperative relationship and at the same time convert the IE from being a subordinate to being an ally of the FC. It is in exactly this way that the conviction method presents the most interest. It allows organization of the operation of the ecological–economic system in the most efficient way and leads to the best economic effect. Detailed research on the different methods of management was provided by Fathutdinov and Sivkova (1999), Ougolnitsky (2002), Ougolnitsky and Usov (2004), among others.

The triad of hierarchical management methods (i.e., compulsion–incentives–conviction) is ranked according to

(1) decrease in the acerbity of the FC's influence on the IE;
(2) decrease in the degree of IE dependence on the FC required for realization of the method; and
(3) the growth of progressions of management patterns.

At the same time, aspiration for the use of conviction does not exclude the use of incentives and even compulsion in concrete practical situations.

We shall illustrate the application of these methods of hierarchical management in the following model of quality management of the water in a river.

16.5 MATHEMATICAL MODEL OF A MANAGEMENT SYSTEM OF WATER QUALITY

We shall consider three-level management systems, involving the influence source of the upper level (federal center or FC), the influence source of the average level (local management body or MB) and the influence source of the lower level (industrial enterprises or IE). All these levels affect the dynamical system (DS) under consideration (e.g., our river).

The aim of an IE is to maximize its income obtained from its industrial work after deduction of all expenses. The IE discharges a polluting substance (PS) into the river. The MB defines the allowed amount of the discharge, the payment for allowing this, and the minimum amount of the PS which must be removed by the IE in the process of sewage treatment. The FC is concerned to achieve sustainable development of the DS, but it cannot act upon the DS directly. Indirect influence of the FC on the DS is achieved in the following way. The FC determines the amount of cash that flows from the IE to the MB. The aim of the MB is to maximize these cash flows. The interests of the FC and the MB are different, sometimes even opposite.

Under different methods of management the FC's aim is to create conditions that are advantageous to the MB and consequently to the IE, to keep the DS in a sustainable state. If the quality standards of river water and sewage are achieved then the DS is in a sustainable state.

The FC payoff function is

$$J_a = \int_0^{\Delta} \{ - C_F(y_c, y_n) + \sum_{i=1}^{N} (H_i^c F_i^c(T_i^c)(1 - P_i^c) W_i^c + H_i^n F_i^n(T_i^n)(1 - P_i^n) W_i^n)\} \, dt \to \max(\{H_i^{c,n}\}_{i=1}^n)$$

$$y_k = \sum_{i=1}^{N} (1 - P_i^k) W_i^k; \quad k = c, n; \tag{16.1}$$

where t is the time; $T_i^k(F_i^k(T_i^k))$ are the fees levied (resp., the charge function) for discharging of carbon-containing ($k = c$) and nitrogen-containing ($k = n$) PSs by the IE at time t for IE number i; W_i^k and $(1 - P_i^k)W_i^k$ are the amounts of carbon-containing and nitrogen-containing ($k = c$ and $k = n$, respectively) PSs discharged into the river by the IE before and after, respectively, drainage water treatment in the unit of time; $P_i^k(t)$ is the share of carbon-containing ($k = c$) and nitrogen-containing ($k = n$) PSs removed by the IE in the process of sewage treatment; Δ is a terminal moment of time; $H_i^{c,n}(t)$ is the FC share of the IE's payment for discharging PSs in the river; C_F is the part of the FC's payoff function, reflecting the expenses incurred for river water quality improvement.

The aim of the MB is to maximize cash flows from the IE minus the expenses for river water quality improvement.

The MB's payoff function is of the form

$$J_0 = \int_0^{\Delta} \{ - C_o(y_c, y_n) + \sum_{i=1}^{N} ((1 - H_i^c) F_i^c(T_i^c)(1 - P_i^c) W_i^c + (1 - H_i^n) F_i^n(T_i^n)(1 - P_i^n) W_i^n)\} \, dt \to \max(\{T_i^{c,n}; q_i^{c,n}\}_{i=1}^n). \tag{16.2}$$

In (16.2) the term $C_0(y_c, y_n)$ reflects the expenses of the MB for the improvement of the quality of the river water; $q_i^{c,n}$ are the minimum possible share of carbon-containing and nitrogen-containing PSs, removed by the IE in the process of sewage treatment. The values of $q_i^{c,n}$ enter into restrictions on the IE's management.

The IE's payoff function is ($i = 1, 2, \ldots, N$)

$$J_i = \int_0^{\Delta} \{z_i R_i(\Phi_i) - C_p^c(P_i^c) W_i^c - F_i^c(T_i^c)(1 - P_i^c) W_i^c - C_p^n(P_i^n) W_i^n - F_i^n(T_i^n)(1 - P_i^n) W_i^n\} \, dt \to \max(\{P_i^{c,n}\}_{i=1}^n). \tag{16.3}$$

In (16.3) the term $C_p^k(P_i)$ reflects the expenses of IE number i for drainage of a unit for water treatment at moment of time t ($k = c, n$); Φ_i is the IE's production income; $R_i(\Phi_i)$ is the IE number i production function; $z_i(t)$ is the IE profit from realization of the unit-made product at moment of time t.

The dynamics of the IE's production income is described by a common differential equation:

$$\frac{d\Phi_i}{dt} = -k_i\Phi_i + Y_i, \quad \Phi(0) = \Phi_0 = \text{const}, \tag{16.4}$$

where k_i is the coefficient of amortization of the IE production income; Y_i are the investments; and the value of Φ_0 is given.

The dependence of the amount of PS discharged by the IE (before treatment) on the amount of the IE's production is linear. IE production functions are $(i = 1, \ldots, N)$

$$W_i = \beta_i R_i(\Phi_i), \qquad R_i(\Phi_i) = \gamma_i \Phi_i^{0.5}; \qquad \gamma_i, \beta_i = \text{const}. \tag{16.5}$$

The main characteristics of water quality are the concentrations of carbon and nitrogen biochemical consumption of oxygen (B^c, B^n) and the concentration of oxygen dissolved in the water (B^0). In the case of spatial spottiness along the riverbed only these quantities are described by non-linear differential equations:

$$\frac{\partial B^k}{\partial t} + v_x \frac{\partial B^k}{\partial x} = \frac{1}{A}\frac{\partial}{\partial x}\left[EA\frac{\partial B^k}{\partial x}\right] - k_x B^k + \frac{(W^k)^0(1-(P^k)^0)}{A}; \quad k = c, n \tag{16.6}$$

$$\frac{\partial B^0}{\partial t} + v_x \frac{\partial B^0}{\partial x} = \frac{1}{A}\frac{\partial}{\partial t}\left[EA\frac{\partial B^0}{\partial x}\right] - k_c B^c - k_n B^n + k_0[B^o_{\text{sat}} - B^o(x,t)]$$
$$+ F_0 - F_1 - F_2, \tag{16.7}$$

where x is the space coordinate along the riverbed, $0 \leq x \leq L$; L is the river's length; E is the coefficient of dispersion; A is the area of cross-section of the river; v_x is river water velocity; $k_c B^c, k_n B^n$ are the functions of carbon and nitrogen biochemical consumption of oxygen because of decomposition; $k_0[B^o_{\text{sat}} - B^o]$ is the addition of oxygen dissolved in consequence of re-aeration; B^o_{sat} is the oxygen concentration of saturation; F_0 is the addition as a result of photosynthesis; F_1 is the consumption of the dissolved oxygen by breathing; F_2 is bottom oxygen consumption.

The functions $(W^k)^0, (P^k)^0$ reflect the presence of PS sources. These are given by:

$$(P^k)^0(x,t) = \begin{cases} 0 & \text{if } x \neq x_i; i = 1, 2, \ldots, N \\ P_i^k(t) & \text{in other cases} \end{cases}; \quad k = n, c;$$

$$(W^k)^0(x,t) = \begin{cases} 0 & \text{if } x \neq x_i; i = 1, 2, \ldots, N \\ W_i^k(t) & \text{in other cases.} \end{cases}$$

The IE is disposed in the points of x_i $(i = 1, 2, \ldots, N)$.

Equations (16.6) and (16.7) are considered using initial and boundary conditions.

Optimization problems (16.1)–(16.3) are solved with the following limitations on IE controls

$$q_i^k \le P_i^k < 1 - \varepsilon; \quad i = 1, 2, \ldots, N; \quad 0 \le t \le \Delta; \quad k = c, n \tag{16.8}$$

on MB controls

$$0 \le q_i^k < 1 - \varepsilon; \quad 0 \le T_i^k \le T_{\max}; \quad i = 1, 2, \ldots, N; \quad 0 \le t \le \Delta; \quad k = c, n \tag{16.9}$$

and on FC controls

$$0 \le H_i^c \le 1; \quad 0 \le H_i^n \le 1; \quad i = 1, 2, \ldots, N; \quad 0 \le t \le \Delta; \quad k = c, n, \tag{16.10}$$

where the constant $T_{\max}$ is given; and the constant ε is defined by the technological possibility of IE's sewage treatment.

The State's standards for river water quality are in the form

$$0 \le B^k \le B_{\max}^k, \qquad B_{\min}^0 \le B^0; \qquad k = c, n; \qquad 0 \le t \le \Delta. \tag{16.11}$$

The state standards for IE sewage quality discharged into the river are in the form

$$\sum_{i=1}^{N} \frac{W_i^c(t)[1 - P_i^c(t)] + W_i^n(t)[1 - P_i^n(t)]}{Q_i^0(t)} \le Q_{\max}; \qquad 0 \le t \le \Delta, \tag{16.12}$$

where $Q_i^0(t)$ is the IE water expense at moment of time t; and constants $B_{\max}, B_{\min}^0, Q_{\max}$ are given.

We shall solve the problem (16.1)–(16.12) by the different methods of hierarchical managements. The ecological–economic system (16.1)–(16.12) is found to be sustainable if conditions (16.11) and (16.12) are achieved. We shall consider the different methods of hierarchical management that enable us to obtain sustainable development of the ecological–economic system.

16.6 FORMALIZATION OF THE METHODS OF HIERARCHICAL MANAGEMENT

Any problem can be corrected if the FC has sufficient economic instruments to influence the MB. If the values $H_i^{c,n} = 1$ then execution of conditions (16.11) and (16.12) must be economically profitable for the MB.

16.6.1 Compulsion

The MB defines the minimum possible amount of carbon-containing and nitrogen-containing PS to be removed by the IE in the process of sewage treatment. The values T_i^k $((k = n, c); i = 1, 2, \ldots, N)$ are constants.

The algorithm for reaching compulsion equilibrium is described as follows:

(1) The IE's optimum strategies are found as a result of minimizing the payoff function (16.3) under constraints (16.8). They depend on the MB's management

$$(P_i^k)^*(T_i^k, q_i^k), \qquad i = 1, 2, \ldots, N; \qquad k = c, n.$$

(2) The IE optimum strategy (see Step 1) is substituted in (16.2). Furthermore, maximization of the payoff function (16.2) on $\{q_i^k(H_i^k)\}_{i=1}^N$; $k = n, c$ using condition (16.9) is accomplished. The values T_i^k $(k = n, c)$; $i = 1, 2, \ldots, N$ are constants.

As a result, optimum MB controls depend on the FC's strategy. They are defined as

$$\{(q_i^k)^*(H_i^k)\}_{i=1}^N; \qquad k = c, n.$$

(3) The payoff function (16.1) under condition (16.10) is maximized. It is accepted that

$$P_i^k = (P_i^k)^*(q_i^k, (T_i^k)^*); \qquad T_i^k = (T_i^k)^*; \qquad k = c, n$$

in (16.1) and (16.10). The values $\{(H_i^k)^*\}_{i=1}^N$ $(k = c, n)$ are the optimum for the FC. They give the FC maximum income under conditions (16.11) and (16.12).

(4) The incentives equilibrium is defined as the set of values

$$\{(H_i^k)^*, (q_i^k)_*, (P_i^k)_*\}_{i=1}^N; \qquad k = n, c,$$

where

$$(q_i^k)_* = (q_i^k)^*((H_i^k)^*); \qquad (P_i^k)_* = (P_i^k)^*((q_i^k)^*, T_i^k).$$

In the general case the compulsion balance is defined as a result of limitations. We shall consider several private cases.

Suppose that

$$\left.\begin{aligned} C_p^k(Y) &= D^k \frac{Y}{1 - Y}; \\ C_F(x, y) &= C_1^c x + C_1^n y; \\ C_o(x, y) &= C_2^c x + C_2^n y; \\ F_i^k(T) &= A^k T; \\ D^k, C_1^k, C_2^k &= \text{const}; \\ k &= c, n. \end{aligned}\right\} \qquad (16.13)$$

If

$$D^k < F_i^k(T_i^k); \qquad i = 1, 2, \ldots, N; \qquad k = n, c,$$

then the critical strategies of the FC, MB, and IE are defined by the equalities

$(k = n, c)$:

$$(P_i^k)_1^0 = q_i^k;$$

$$(P_i^k)_2^0 = 1 - \sqrt{\frac{D^k}{F_i^k(T_i^k)}} \quad \text{in the case of} \quad q_i^k < 1 - \sqrt{\frac{D^k}{F_i^k(T_i^k)}};$$

$$(H_i^k)_1^0 = 1;$$

$$(H_i^k)_2^0 = 1 - \frac{C_2^k}{1 - F_i^k(T_i^k) - \varepsilon_1}; \qquad \varepsilon_1 \ll 1$$

$$(q_i^k)_1^0 = 1 - \varepsilon;$$

$$(q_i^k)_2^0 = 1 - \sqrt{\frac{D^k}{F_i^k(T_i^k)}}.$$

If

$$\varepsilon < \sqrt{\frac{D^k}{F_i^k(T_i^k)}} \frac{F_i^k(T_i^k) - C_1^k - C_2^k}{|F_i^k(T_i^k) - C_1^k|}; \qquad i = 1, 2, \ldots, N$$

and the State's standards for river water quality and sewage quality are achieved in the case of

$$P_i^k = 1 - \sqrt{\frac{D^k}{F_i^k(T_i^k)}}; \qquad k = n, c; \qquad i = 1, 2, \ldots N; \qquad t \in [0, \Delta],$$

then the compulsion balances are defined as

$$(H_i^k)^* = 1 - C_2^k/F_i^k(T_i^k) - \varepsilon_1; \qquad (q_i^k)_* = (P_i^k)_* = 1 - \sqrt{\frac{D^k}{F_i^k(T_i^k)}}. \tag{16.14}$$

In the opposite case, equalities are executed at least once at some moment of time

$$(H_i^k)^* = 1; \qquad (q_i^k)_* = (P_i^k)_* = 1 - \varepsilon; \qquad k = n, c. \tag{16.15}$$

If

$$D^k > F_i^k(T_i^k); \qquad i = 1, 2, \ldots N; \quad k = n, c,$$

then the critical strategies of the FC, MB, and IE are defined by the equalities

$$(P_i^k)^0 = q_i^k; \qquad (q_i^k)_1^0 = 1 - \varepsilon; \qquad (q_i^k)_2^0 = 0; \qquad (H_i^k)_1^0 = 1; \qquad k = n, c$$

$$(H_i^k)_2^0 = 1 - \frac{C_2^k}{1 - F_i^k(T_i^k) - \varepsilon_1}.$$

Then, if

$$\varepsilon < \frac{-F_i^k(T_i^k) + C_1^k + C_2^k}{|F_i^k(T_i^k) - C_1|}; \qquad i = 1, 2, \ldots, N; \quad k = n, c$$

and the State's standards for river water quality and sewage quality are achieved in the case of

$$P_i^k = 0; \qquad k = n, c; \qquad i = 1, 2, \ldots, N; \qquad t \in [0, \Delta]$$

then the compulsion balances are defined as

$$(H_i^k)^* = 1 - C_2^k / F_i^k(T_i^k) - \varepsilon_1; \qquad (q_i^k)_* = (P_i^k)_* = 0. \tag{16.16}$$

In the opposite case, equalities (16.15) are executed at least once at some moment of time.

Example 1. Let us research models (16.1)–(16.12) in the case of (16.13) for the following input data (euro, the cost; da, a day; m, a meter; mg, a milligram; L, a liter):

$$\gamma_i = 0.2 \text{ euro}; \qquad k_i = 0.001 \frac{\text{L}}{\text{da}}; \qquad Q_i^o = 10^6 \frac{\text{m}^3}{\text{da}}; \qquad F_i^c(T) = 0.5T;$$

$$T_i^n = 20 \text{ euro}; \qquad T_i^c = 50 \text{ euro}; \qquad F_i^n(T) = 2T; \qquad i = 1, 2; \qquad A = 100 \text{ m}^2;$$

$$T_{\max} = 1{,}000 \text{ euro}; \qquad \beta_i^n = 0.003 \frac{\text{mg}}{\text{da euro}}; \qquad \beta_i^c = 0.003 \frac{\text{mg}}{\text{da euro}};$$

$$B^o(x, 0) = B^o(L, t) = B^o(0, t) = 10 \frac{\text{mg}}{\text{L}}; \qquad N = 2; \qquad L = 100 \text{ m};$$

$$B(x, 0) = B(L, t) = B(0, t) = 5 \frac{\text{m}}{\text{L}}; \qquad \Phi_0 = 10^{15}; \qquad E = 24{,}000 \frac{\text{m}^2}{\text{da}};$$

$$C_1^n = 60 \frac{\text{da euro}}{\text{mg}}; \qquad C_1^c = 40 \frac{\text{da euro}}{\text{mg}}; \qquad C_2^c = 30 \frac{\text{da euro}}{\text{mg}}; \qquad Y_i = 0 \text{ euro};$$

$$B_{\text{sat}}^o = 18 \frac{\text{mg}}{\text{L}}; \qquad B_{\max} = 14 \frac{\text{mg}}{\text{L}}; \qquad C_2^n = 70 \frac{\text{da euro}}{\text{mg}}; \qquad B_{\min}^o = 4 \frac{\text{mg}}{\text{L}};$$

$$F_0 = F_1 = F_2 = 0 \frac{\text{mg}}{\text{L da}}; \qquad x_1 = 20 \text{ m}; \qquad \varepsilon = 0.001; \qquad Q_{\max} = 0.4;$$

$$v_x = 3{,}500 \frac{\text{m}}{\text{da}}; \qquad \Delta = 365 \text{ da}; \qquad k_c = 0.03 \frac{\text{L}}{\text{da}}; \qquad x_2 = 60 \text{ m}; \qquad z_1 = 25 \frac{\text{L}}{\text{da}}$$

$$z_2 = 20 \frac{\text{L}}{\text{da}}; \qquad D^n = 5; \qquad D^c = 30; \qquad k_n = 0.01 \frac{\text{L}}{\text{da}}; \qquad k_o = 0.2 \frac{\text{L}}{\text{da}}.$$

In this case the compulsion balance is defined by the formulas of (16.13) and (16.15). What is more,

$$R_F = 473 \text{ euro}; \qquad R_y = -1.6 \times 10^4 \text{ euro};$$

$$R_1 = -2.3 \times 10^{11} \text{ euro}; \qquad R_2 = -2.5 \times 10^{11} \text{ euro}$$

(R_F, R_y, R_1, R_2 are the profits of the FC, MB, two IEs, respectively).

Example 2. In the case of input data from Example 1 and

$$T_i^n = 333 \text{ euro}; \qquad T_i^c = 100 \text{ euro}; \qquad i = 1, 2,$$

the compulsion balance is defined by the equalities of (16.15) for carbon-containing PS, and by the equalities of (16.14) for nitrogen-containing PS:

$$(H_i^c)^* = 1; \qquad (H_i^n)^* = 0.8948; \qquad (q_i^c)_* = (P_i^c)_* = 0.999; \qquad i = 1, 2;$$
$$(q_i^n)_* = (P_i^n)_* = 0.9134.$$

Moreover,

$$R_F = 6.476 \times 10^4 \text{ euro}; \qquad R_y = 1.16 \times 10^5 \text{ euro};$$
$$R_1 = -2.034 \times 10^{11} \text{ euro}; \qquad R_2 = -2.14 \times 10^{11} \text{ euro}.$$

Example 3. In the case of input data from Example 2 and $T_i^c = 390$ euro ($i = 1, 2$) the compulsion balance is defined by formulas (16.14) ($i = 1, 2$):

$$(H_i^c)^* = 0.8461; \qquad (H_i^n)^* = 0.8948; \qquad (q_i^c)_* = (P_i^c)_* = 0.6078;$$
$$(q_i^n)_* = (P_i^n)_* = 0.9134.$$

Moreover,

$$R_F = 1.65 \times 10^7 \text{ euro}; \qquad R_y = -6.17 \times 10^5 \text{ euro};$$
$$R_1 = 5.16 \times 10^{10} \text{ euro}; \qquad R_2 = 4.1 \times 10^{10} \text{ euro}.$$

Example 4. In the case of input data from Example 3 and

$$\beta_i^c = 0.08 \frac{\text{mg}}{\text{da euro}}; \qquad i = 1, 2,$$

we get

$$(H_i^c)^* = 1; \qquad (H_i^n)^* = 0.8948; \qquad (q_i^c)_* = (P_i^c)_* = 0.999;$$
$$(q_i^n)_* = (P_i^n)_* = 0.9134; \qquad R_F = 7.09 \times 10^6 \text{ euro};$$
$$R_y = -2.76 \times 10^6 \text{ euro}; \qquad R_1 = -5.06 \times 10^{12} \text{ euro}; \qquad R_2 = -5.1 \times 10^{12} \text{ euro}.$$

Example 5. In the case of input data from Example 1 and

$$T_i^n = 5 \text{ euro}; \qquad T_i^c = 10 \text{ euro}; \qquad i = 1, 2; \qquad C_2^n = 3 \frac{\text{da euro}}{\text{mg}};$$
$$C_1^n = 2 \frac{\text{da euro}}{\text{mg}}; \qquad C_1^c = 1 \frac{\text{da euro}}{\text{mg}}; \qquad C_2^c = 2 \frac{\text{da euro}}{\text{mg}}$$

we get

$$(H_i^c)^* = 0.6; \qquad (H_i^n)^* = 0.7; \qquad (q_i^c)_* = (P_i^c)_* = 0;$$
$$(q_i^n)_* = (P_i^n)_* = 0.2929; \qquad R_F = 5.909 \times 10^5 \text{ euro};$$
$$R_y = 2.9 \times 10^5 \text{ euro}; \qquad R_1 = 5.33 \times 10^{10} \text{ euro}; \qquad R_2 = 4.26 \times 10^{10} \text{ euro}.$$

Example 6. In the case of input data from Example 1 and

$$T_i^n = 150\text{ euro}; \qquad T_i^c = 100\text{ euro}; \qquad i = 1,2; \qquad F_2^n(T) = 0.5T;$$
$$F_1^c(T) = 2T; \qquad C_2^n = 20\frac{\text{da euro}}{\text{mg}}; \qquad C_1^m = 10\frac{\text{da euro}}{\text{mg}}; \qquad F_2^c(T) = T;$$
$$\beta_i^n = 0.06\frac{\text{mg}}{\text{da euro}}; \qquad \beta_i^c = 0.025\frac{\text{mg}}{\text{da euro}}; \qquad i = 1,2$$

the compulsion balance is defined by formulas (16.13) and (16.14) with the exception of carbon-containing PS on a second IE at a certain moment in time (t_k):

$$(H_1^c)^* = 0.8499; \qquad (H_1^n)^* = 0.9332; \qquad (H_2^c)^* = 0.7; \qquad (H_2^n)^* = 0.733;$$
$$(q_1^c)_* = 0.6127; \qquad (P_1^c)_* = 0.6127; \qquad (q_2^c)_* = (P_2^c)_* = 0.4522;$$
$$(q_1^n)_* = (P_1^n)_* = 0.8709; \qquad (q_2^n)_* = 0.7418; \qquad (P_2^n)_* = 0.7418;$$
$$(H_2^c)^*(t_1) = 1; \qquad (q_1^c)_*(t_1) = (P_1^c)_*(t_1) = 0.999; \qquad t_1 = \text{const}; \qquad 0 \le t_1 \le \Delta;$$

Moreover,

$$R_F = 1.025 \times 10^8\text{ euro}; \qquad R_y = -1.24 \times 10^7\text{ euro};$$
$$R_1 = 3.75 \times 10^{10}\text{ euro}; \qquad R_2 = -3.38 \times 10^{11}\text{ euro}.$$

16.6.2 Incentives

The MB setsout the allowed amount and the cost of discharging PS into the river. Building up an incentives balance makes sense if conditions (16.11) and (16.12) are not satisfied in the case of $P_i^k = q_i^k$ $(k = n, c)$; $i = 1, 2, \ldots N$. Otherwise, the discussion leads—not to incentives—but to compulsion.

The algorithm for detecting incentives equilibrium is described as follows:

(1) The IE's optimum strategies are found similarly to the method of compulsion, as a result of minimizing the payoff function (16.3) under constraints (16.8). They depend on the values T_i^k $(k = n, c)$.
(2) The IE's optimum strategy (see Step 1), is substituted in (16.2). Furthermore, maximization of the payoff function (16.2) on $\{T_i^k(H_i^k)\}_{i=1}^N$; $k = n, c$ using condition (16.9) is accomplished. The values q_i^k $(k = n, c)$; $i = 1, 2, \ldots N$ are constants. As a result, the optimum MB strategy depends on the FC's strategy. They are defined as $\{(T_i^k)^*(H_i^k)\}_{i=1}^N$; $k = c, n$.
(3) The payoff function (16.1) under condition (16.10) is maximized. It is accepted that $P_i^k = (P_i^k)^*(q_i^k, (T_i^k)^*)$; $T_i^k = (T_i^k)^*$; $k = c, n$ in Equations (16.1) and (16.10). The values $\{(H_i^k)^*\}_{i=1}^N$; $k = c, n$ are the optimum for the FC. They give to the FC the maximum income under conditions (16.11) and (16.12).
(4) The incentives equilibrium is defined as the set of values

$$\{(H_i^k)^*, (T_i^k)_*, (P_i^k)_*\}_{i=1}^N; \qquad k = n, c,$$

where

$$(T_i^k)_* = (T_i^k)^*((H_i^k)^*); \qquad (P_i^k)_* = (P_i^k)^*((T_i^k)^*, q_i^k).$$

In the case of (16.13) the optimum strategies of the IE, MB, and FC are defined as follows.

If

$$D^k < F_i^k(T_{\max}^k); \qquad k = n, c; \qquad i = 1, 2, \dots, N$$

and the State's standards for river water quality and sewage quality are achieved in the case of

$$P_i^k = 1 - \sqrt{\frac{D^k}{F_i^k(T_{\max}^k)}},$$

then

$$(H_i^k)^* = 1; \qquad (T_i^k)_* = T_{\max}^k; \qquad (P_i^k)_* = 1 - \sqrt{\frac{D^k}{F_i^k(T_{\max}^k)}}.$$

If

$$D^k < F_i^k(T_{\max}^k); \qquad k = n, c; \qquad i = 1, 2, \dots, N$$

and the State's standards for river water quality and sewage quality are not achieved in the case of

$$P_i^k = 1 - \sqrt{\frac{D^k}{F_i^k(T_{\max}^k)}},$$

then the incentives method is not realized.

If

$$D^k > F_i^k(T_{\max}^k); \qquad k = n, c; \qquad i = 1, 2, \dots, N,$$

then

$$(H_i^k)^* = 1; \qquad (T_i^k)_* = T_{\max}^k; \qquad (P_i^k)_* = 1 - \varepsilon.$$

In the general case the incentives equilibrium is defined by means of computer simulations.

Example 7. In the case of input data from Examples 1–3 we get

$$R_F = 3.5 \times 10^7 \text{ euro}; \qquad R_y = -1.6 \times 10^4 \text{ euro};$$
$$R_1 = 5 \times 10^{10} \text{ euro}; \qquad R_2 = 4 \times 10^{10} \text{ euro}.$$

Example 8. In the case of input data from Example 4 the method of incentives does not work.

Example 9. In the case of input data from Example 5 we get

$$R_F = 3.7 \times 10^7 \text{ euro}; \qquad R_y = -0.9 \times 10^4 \text{ euro};$$
$$R_1 = 5 \times 10^{10} \text{ euro}; \qquad R_2 = 4 \times 10^{10} \text{ euro}.$$

Example 10. In the case of input data from Example 6 we get

$$R_F = 4.7 \times 10^8 \text{ euro}; \qquad R_y = -1.1 \times 10^7 \text{ euro};$$
$$R_1 = 1.7 \times 10^{10} \text{ euro}; \qquad R_2 = 6.2 \times 10^9 \text{ euro}.$$

16.6.3 Conviction

If the MB and IE understand the importance of improving the ecological situation in the region and strive to achieve conditions (16.11) and (16.12), then all the management subjects have one general criterion instead of criteria (16.1)–(16.3):

$$J = \int_0^\Delta [-C_F(y_c(t), y_n(t)) - C_y(y_c(t), y_n(t)) + \sum_{i=1}^{N} (z_i R_i(\Phi_i) - C_p^c(P_i^c)W_i^c - C_p^n(P_i^n)W_i^n)]\, dt \to \max(\{P_i^c, P_i^n\}_{i=1}^N).$$

In this case $q_i^k = 0$; $k = n, c$; $i = 1, 2, \ldots, N$.

This criterion is considered using conditions and correlations of (16.4)–(16.12).

The sharing of the joint profit between the IE, MB, and FC is realized according to some cooperative distribution. The balances as a result of conviction are found by way of imitations.

In the case of (16.13) and input data from Examples 1–3 the joint profit of the IE, MB, and FC (R_c) from conviction balances is defined by the formula $R_c = 9.5 \times 10^{10}$ euro.

In the case of input data from Examples 4–6 we get $R_c = 4.6 \times 10^8$ euro, $R_c = 3.1 \times 10^{11}$ euro, and $R_c = 1.1 \times 10^{11}$ euro, respectively.

All the examples considered are researched by means of scenarios through simulation modeling. Equations (16.6) and (16.7) are solved by the method of finite differences with the first approximations on the spatial variable and on time, in analogy with Ougolnitsky and Usov (2004).

Therefore, in the case of compulsion for the IE it is advantageous that the fee for discharging PSs is more than a certain value $C_{\min}$ and less than a certain value $C_{\max}$. For such a fee the maximal possible degree of discharge water treatment by the IE is disadvantageous to the FC, and consequently to the MB. But the fee for discharging PS must not be too much from the IE's standpoint.

In Examples 1–3 the increase in the fee led to increasing the IE's profit. A further increase of the fee reduces the IE's profit. In Example 3, by contrast with Examples 1 and 2, the increase in values T_i^c ($i = 1, 2$) leads to changing the FC's optimum strategies of management (i.e., transition from strategies (16.15) to strategies (16.14)). The increase in values of T_i^m ($i = 1, 2$; $m = n, c$) can increase the FC's profit and simultaneously reduce the MB's profit. This fact is particularly interesting. It illustrates the direct opposition of FC and MB interests in certain cases.

The main purpose of the FC consists in the provision of sustainable development of the ecological–economic system. The FC recalls that its main objective is achieved if the ecological situation in the river occurs. In this case the FC abandons its optimum strategy and returns to strategy (16.15), though in this case its profit is less than in the case of strategy (16.14).

The method of compulsion can be brought about when the method of incentives cannot be applied (Example 4). The method of incentives gives greater economic freedom to the IE by contrast with the compulsion method.

In most cases, but not always, the incentives method is more economically profitable for the IE and society as a whole (but not for the MB), by contrast with the compulsion method (see the examples). The third level of management is spare in the case of incentives. In this case the FC does not take into account all the MB's interests and reserves all the IE's fees for itself. In most cases the compulsion method is more profitable for the FC than the incentives method.

The management process in a three-level hierarchical system is clumsier, less flexible, and consequently less efficient than in two-level management systems, which were researched, for instance, by Ougolnitsky and Usov (2004). The total profit of all management subjects in a three-level management system is sharply reduced by contrast with a two-level management system.

In a real management system, use of the intermediate level of management must be motivated. Its use must follow from the nature of the system and from the impossibility of making decisions on all management questions at just two levels. In the opposite cases the management system must not contain the additional intermediate level of management. The efficiency of the hierarchical management system falls with an increase in the number of intermediate levels in the system.

16.7 CONCLUSIONS

Relations between the subjects of an ecological-economic management system are hierarchical in character. So, the provision of sustainable development of the ecological–economic systems requires the use of a hierarchical management mechanism. This mechanism has administrative, economic, and psychological components.

If the FC knows exactly the values of the parameters for ecological–economic system sustainable development then it can use further developed mechanisms of management as a result of feedback. Such mechanisms reduce the probability of the FC's losses when sustainable development is provided by the methods of compulsion or incentives.

The compulsion method allows support of the DS in a sustainable condition regardless of the strategies of other management subjects. Under compulsion the FC imposes restrictions on the possible management sets of the various management subjects. But such an approach to management can be unprofitable, by contrast with the incentives method. The incentives method gives greater economic freedom to all management subjects.

The best method of providing sustainable development for ecological–economic systems is the conviction method. This method of hierarchical management implies voluntary cooperation between the subjects of management. In this case, achievement of the purpose of sustainable development for ecological–economic systems is realized in the most efficient way for all management subjects.

The choice of hierarchical management method in real management systems depends on objective and subjective conditions. The objective conditions are defined by the FC's purposes and the FC's ability to influence the IEs. Subjective conditions are defined by the level of culture of management subjects and the scale of their thinking.

16.8 REFERENCES

Danilov-Danilyan V.I. and Losev K.S. (2000). *The Ecological Call and Sustainable Development*. Progress-Tradition, Moscow, 415 pp. [in Russian].

Dreyer O.K. and Losi V.A. (1997). *Ecology and Sustainable Development*. Science, Moscow, 222 pp. [in Russian].

Fathutdinov R. and Sivkova L. (1999). Personnel management. *Interperiodika*, **2**, 32–40 [in Russian].

Kondratyev K.Ya., Donchenko V.K., Losev K.S., and Frolov A.K. (1996). *Ecology: An Economic Policy*. Scientific Center of Russian Academy of Science, St. Petersburg, 827 pp. [in Russian]. Available at *http://www.nwicpc.ru/gec.htm*

Kondratyev K.Ya., Krapivin V.F., and Phillips G.W. (2002). *Global Environmental Change: Modelling and Monitoring*. Springer-Verlag, Heidelberg, 317 pp. Available at *http://www.nwicpc.ru/gec.htm*

Koptyug V.A., Matrosova V.M., and Levashova V.K. (eds.) (1999). *The New Paradigm of Russian Development: Complex Studies of the Problems of Sustainable Development*. Academia, Moscow, 459 pp. [in Russian].

Ougolnitsky G.A. (1999). *Management of the Ecology–economic System*. High School Book Publications, Moscow, 132 pp. [in Russian].

Ougolnitsky G.A. (2002). The game-theoretical modeling of hierarchical management methods for sustainable development. *Game Theory and Applications*, **1**, 92–97 [in Russian].

Ougolnitsky G.A. and Usov A.B. (2004). Hierarchical management methods of the quality of river water. *Journal of Water Resourses*, **31**(3), 375–382 [in Russian].

Ryumina E.V. (2000). *Analysis of the Interaction between Ecology and Economy*. Science, Moscow, 158 pp. [in Russian].

17

Sustainable development problems in the context of global ecoinformatics

Arthur P. Cracknell, Vladimir F. Krapivin, and Costas A. Varotsos

17.1 INTRODUCTION

Nowadays, the problems of globalization and sustainable development have been discussed in a vast scientific literature. Even more numerous are the concerns raised in the mass media. However, discussions of such problems contain many contradictions and disagreements. First of all, globalization and sustainable development problems are treated, as a rule, as independent and separate. Nevertheless, the priority of the globalization problem and subordinate importance of sustainable development problems as one of the most important aspects of globalization processes raises few doubts.

A constructive view on the globalization and sustainable development problem was first broached by Kondratyev (Kondratyev *et al.*, 1992, 1994, 1997). The development of globalization theory took a number of different directions that were often contradictory and colored by socio-political considerations (Friedman, 2005; Mander and Goldsmith, 2006; Wijen *et al.*, 2005). Global socio-political issues have a long history beginning with the era of colonialism, then independence. After 1917 there existed two periods: the pre-Cold War and post-Cold War eras. The post-Cold War era led to the increasing influence of what some people these days call quasi-governments (such as the International Monetary Fund or the World Bank). This delimitation served for a certain time as the cause of the principal differences between the existing concepts of the globalization process. At its most basic, there is nothing mysterious about globalization in spite of these differences. Really there exist two main aspects of globalization:

(i) Globalization that brings prosperity to certain countries while impoverishing other countries.

(ii) Globalization as the objective process behind development of the nature/society system, taking into account existing global population dynamics.

In the first case, many experts around the world identify globalization with "Americanization" of economics and culture (Friedman, 2005). The rest of the world seems to be following the U.S.A. and leaving behind their own ways of life. "Americanization" is the contemporary term used for the influence that the U.S.A. has on the culture and economics of other countries, replacing their way of life by the American way. This process is managed by the many international structures under the control of the U.S.A. Significant social networks and international forums have been organized with the purpose of proliferating capitalist interests. Some of these forums date from earlier in the 20th century, such as the International Chamber of Commerce, the International Organization of Employers, the Center for Environmental Diplomacy, and the Bretton Woods Committee. They were created as a specific response to questions about globalization (Devkota, 2005; Furth, 1965). These and other organizations and forums continue to support the interests of developed countries in the 21st century as well.

Reformists of global social democracy put forward the theory of neoliberal globalization which proposes the transformation of social ambience in the world. The countries of the former Soviet Union have felt the negative consequences of this theory fully. Therefore, the question "what is globalization?" is becoming an important one in recent times. Many experts see it as a primary economic phenomenon involving the increasing integration of national economic systems through the growth in international trade, investment, and capital flows. Other experts define globalization as the social, cultural, political, and technological exchanges between countries. As the result, two alternatives were formed: anti-globalization and pro-globalization.

The alternatives to globalization promoted by anti-globalists include different approaches to the dynamics of world structure:

- recognition of the importance of free trade and investment for economic growth and development, but there is also some evidence that trade and investment have in fact increased poverty and inequality;
- sustainable development is best achieved through a single framework, integrating environmental protection and the promotion of economic growth and social equity;
- innovation is necessary for sustainable living;
- support for the accumulation of power at the local level;
- implementation of trade barriers to protect local production and renunciation of international trade unless goods or services cannot be produced locally;
- free trade usually benefits wealthy countries at the expense of poor countries; and
- environmental protection is the key to sustainable development and is imperative for economic development.

The argument of pro-globalization groups is based on the promotion of free trade as the key to eliminating poverty and ensuring effective development and that market instruments such as intellectual property rights are necessary to protect the environment and promote development. Supporters of globalization argue that it can be rolled back and point to the period between the First and Second World Wars as evidence. The increase in world trade as a proportion of world GDP was proportionately greater between 1870 and 1914 that it has been since 1975. That expansion was stopped, not just by the First World War, but by the loss of support for free trade that followed. This led directly to the 1930s depression and indirectly to the Second World War. All these changes took place when the world population did not exceed 2 billion. At the present time the situation is rapidly changing with the exponential growth of world population that now exceeds 6.6 billion. Therefore, the globalization process is becoming an inevitable phenomenon of world life. It is for this reason that the list of key questions on globalization can be extended:

- Is there any alternative to globalization in the coming decades?
- What are the environmental impacts of globalization?
- How does globalization affect culture and religion?
- What is globalization and when did it start and how does it depend on population size?
- Who are the players in the interactions between nature and society?
- Why is there global inequality, and is it getting worse?
- What are the costs and benefits of free trade?
- What is the role of the internet and communications technology in globalization?
- Is globalization shifting power from nation states to undemocratic international organizations?
- Is globalization resulting in industries in developed countries being undermined by industries in developing countries with inferior labor standards?
- What does globalization mean to separate countries like Australia?

Nevertheless, when considering the prospects for life on Earth, it is necessary to proceed from the human criteria of assessing the level of environmental degradation, since over time local and regional changes in the environment become global ones. The amplitude of these changes is determined by mechanisms that manage nature/society system dynamics and structure (Ehlers and Kraft, 2005; Gorshkov *et al.*, 2000, 2004; Guista and Kambhampati, 2006). Humankind deviates more and more from this optimality in the way it interacts with the surrounding inert environment. At the same time, human society as an element of the nature/society system tries to understand the character of large-scale relationships with nature, directing the efforts of many sciences to study cause-and-effect feedbacks from this system. The basic item of a global human society is the country (i.e., considered as a nature/society system component) (Kondratyev and Krapivin, 2005; Olenyev and Fedotov, 2003).

National safety under present-day conditions is estimated on the basis of many criteria, mainly military, economic, ecological, and social in character. Development of an efficient method of objective analysis of the problem of national security requires use of the latest methods to collect and process data on various aspects of the functioning of the world system. Such methods are provided by GIMS technology (Kondratyev *et al.*, 2002).

The development and realization of an efficient technology to assess ecological safety on a global scale may be possible through the International Center on Global Geoinformation Monitoring (ICGGM). This may make it possible to understand the mechanisms behind nature/society system co-evolution. The basic mechanism will be geared towards new technologies of data processing based on progress in evolutionary informatics and global modeling. Here the point is about realizing an approach developed by some authors to modeling the processes in conditions of inadequate *a priori* information about their parameters and the presence of principally unavoidable information gaps.

This chapter expounds ideas, methods, and information technologies that can help to solve the many problems of global ecodynamics.

17.2 GLOBAL ECOINFORMATICS AS THE SCIENCE OF THE NATURE/SOCIETY SYSTEM

17.2.1 A new approach to the study of the nature/society system

On a planetary scale, all living beings in the biosphere are closely interconnected by the ways in which the mechanisms that regulate energy fluxes and cycles of substances are organized: a single biocybernetic system of the highest rank. Within the continents and oceans (the structural units making up the biosphere) processes of energy and substance transformation take place automatically. Land biogeocenoses are characterized by distributed productivity, a function in many territories that is under the control of humans and therefore depends on the development of scientific–technical progress. The World's oceans currently provide about 1% of the resources consumed by humankind and remain one of the few elements of the biosphere not under human control. This low level is connected with insufficiently studied production processes in the oceans.

Anyhow, the interconnection of human living standards and natural processes has recently become of fundamental importance. Study of the biosphere as a complicated hierarchically organized unique system has become an urgent problem of humankind because human life depends completely on the state of the biosphere. In this study, a central role is played by system ecology, the science of using numerical modeling methods and computers to study how biospheric ecological systems function. Development of this direction has led to new ideas in the sphere of global change studies as a result of realization of numerous anthropogenic projects. As a result various scientific disciplines have appeared, such as Geographic Information Systems (GIS), global modeling, Geoinformation Monitoring Systems (GIMS), survival theory, and systemology (Fleishman, 1982; Nitu *et al.*, 2004). However, each of these directions has limited possibilities to study the dynamics of natural sub-systems with different spatio-temporal limitations. Therefore, there is a need to develop an interdisciplinary direction of science that combines GIS, global modeling, GIMS, expert systems, and takes into account the socio-economic aspects of nature protection activity at the same time. The term ecoinformatics has been introduced to describe such a development (Kondratyev *et al.*, 2004; Krapivin and Kondratyev, 2002).

Many international conferences were organized towards the end of the 20th century whose subject matter was closely connected with the problems of ecoinformatics and with certain features of Earth sciences that need to be considered in order to solve problems about environmental protection and to give recommendations about how to organize global ecological monitoring. Analysis of this subject matter shows that it is only in Kondratyev's publications (see list of references) that a brand new approach has been proposed, which is aimed at producing a complex information system that both describes biospheric and climatic processes and takes into account anthropogenic activity trends at the same time. The main idea of this approach consists in creating information technologies that would permit several models to be developed despite the fragmentary and distorted information that is available about processes taking place both in nature and in society.

The basic idea of ecoinformatics consists in creating a model, universal in its thematic content, which describes the interaction of natural and anthropogenic processes. Modeling social relationships and how they change depending on the environmental state is most complicated and insufficiently studied. The most complicated task for global modeling is the problem of prediction of these processes on a global scale in order to work out optimal behavioral policies at a governmental level. Nevertheless, ecoinformatics can suggest an approach that may well solve this problem too.

17.2.2 Indicator of the nature/society system state

The problem of interaction between various elements and processes in the global nature/society system has recently attracted the attention of many scientists (Kondratyev *et al.*, 2002, 2003, 2004; Nitu *et al.*, 2004). Attempts to estimate and predict the dynamics of this interaction have been made in different scientific spheres. One of these attempts is the program "Biocomplexity" produced in the U.S.A. by the National Science Foundation, within which plans were made for the period 2001–2005 to study and understand relationships between the dynamics of complexity of biological, physical, and social systems and trends in changes of the present environment. Within the framework of this program, the complexity of the system somehow interacting with the environment is connected with phenomena appearing as a result of global-scale contact of a living system with the environment.

Biocomplexity is a derivative of biological, physical, chemical, social, and behavioral interactions of environmental sub-systems, including living organisms and the global population. As a matter of fact, the notion of biocomplexity in the environment is closely connected with the rules that govern biosphere functioning as the sum of its constituent ecosystems and natural–economic systems of different scales, from local to global. Therefore, to determine biocomplexity and to assess it, a combined formalized description is needed of biological, geochemical, geophysical, and anthropogenic factors and processes taking place at a given level of the spatio-temporal hierarchy of units and scales.

Biocomplexity is a characteristic feature of all systems of the environment connected with life. Elements of this manifestation are studied within the framework of the theory of stability and vitality of ecosystems. Note should be taken here that biocomplexity includes indicators of the degree of mutual modification of interacting systems; this means that biocomplexity should be studied by taking both the spatial and biological levels of organization into account. The difficulty of this problem is explained by the complicated behavior of the object under study, especially if the human factor is considered, due to which the amount of stress situations in the environment is constantly growing.

Humankind has accumulated a considerable amount of knowledge about environmental systems. Use of this knowledge to study biocomplexity is possible within the framework of the synthesis of a global model that reflects the laws of interactions between environmental components and makes it possible to assess the "efficiency" of realizing scenarios of the development of human society, based on the actual data of ground and satellite measurements. It it this problem that serves the basis of all questions set forth by the program "Biocomplexity".

Studies of the process of the interaction between humankind and nature are aimed, as a rule, at understanding and assessing the consequences of this interaction. The reliability and accuracy of these assessments depend on criteria that serve as the basis for conclusions, expert examination, and recommendations. At present, there is no agreed method to select such criteria because of the lack of a single scientifically substantiated approach to ecological normalization of economic forcings on the environment. The choice of such criteria determines the accuracy of the ecological expertise of existing and planned productivity as well as the representativeness of global geoinformation monitoring data.

Processes taking place in the environment can be represented as the totality of interactions between its sub-systems. Since a human being is one of its elements, it is impossible to definitely divide the environment, for instance, into the biosphere and society; everything on the Earth is correlated and interconnected. The point is to find mechanisms to describe such correlations and interdependences that would reliably reflect environmental dynamics and answer the questions formulated in the program "Biocomplexity":

1. How does the complexity of biological, physical, and social systems in the environment appear and change?
2. What are the mechanisms behind spontaneous development of numerous phenomena in the environment?
3. How do the systems of the environment react with living components, including those created by humans, and adjust themselves to stress situations?
4. In what ways do information, energy, and matter move within the systems of the environment and through their levels of organization?

5. Is it possible to predict system adaptability and to give prognostic estimates of changes in it?
6. How does humankind affect and respond to biocomplexity in natural systems?

One can add many other, no less important, questions. For instance, up to what level of complexity should spaceborne observation systems be improved in order that their information is enough to estimate reliably the state of the environment, at least at the moment of receiving this information? No less important is the question about optimal allocation of the means of the geoinformation monitoring at different levels of its organization.

17.2.3 Biocomplexity indicator determination

Environmental biocomplexity is to some extent an indicator of the interconnection of its systems. In this connection, one can introduce the unit Ξ of biocomplexity varying with conditions when all interactions in the environment are reduced to the level where they correspond to the natural process of evolution. Thus, we obtain an integral indicator of the state of the environment as a whole, taking into account biocapability, biodiversity, and biosurvivability (Kondratyev *et al.*, 2002, 2004; Krapivin *et al.*, 2005). Such an indicator characterizes all manner of interactions between environmental components. For instance, at the biological interaction connected with relationships of the type "prey–predator" or "competition for energy resource", there is a minimum level of food availability, when it becomes practically inaccessible, and the consumer–producer interaction ceases. The chemical and physical processes of the interaction between environmental elements depend also on a set of certain critical parameters.

All this demonstrates that biocomplexity refers to categories difficult to measure and estimate quantitatively. Let us use a formalized estimation. To transfer to gradations of the unit Ξ with the digital scale, we state that there are relationships of the type $\Xi_1 < \Xi_2$, $\Xi_1 > \Xi_2$, or $\Xi_1 \equiv \Xi_2$ between two values of the unit indicator. In other words, there is always a value of this unit ρ that determines the level of biocomplexity $\Xi \to \rho = f(\Xi)$, where f is some transformation of the notion of biocomplexity into a number.

Let us try to find a satisfactory model that would put the descriptive portrait of biocomplexity into notions and indicators, following the formalized description and transformation. With this aim in view, we select in the nature/society system m elements–sub-systems of the lower level, the interaction between which we determine from a binary matrix function: $A = \|a_{ij}\|$, where $a_{ij} = 0$, if elements i and j do not interact; and $a_{ij} = 1$, if elements i and j interact. Then any point $\xi \in \Xi$ is determined as the sum

$$\xi = \sum_{i=1}^{m} \sum_{j>i}^{m} a_{ij}. \tag{17.1}$$

Here an ambiguity appears, to overcome which it is necessary to complicate unit Ξ, for example, by introducing the weight coefficients for each element of the nature/

society system. The character of these coefficients depends on the nature of elements. Therefore, we select in the nature/society system two main types of elements: living (plants included) and non-living elements. The living elements are characterized by density estimated in species number per unit area (volume) or in biomass concentration. Vegetation is characterized by the type and share of the covered area. Non-living elements are divided by the level of their concentrations related to the area or volume of space. Generally, to each element i some characteristic k_i is ascribed that corresponds to its significance. As a result, we obtain a specified formula at transition from the notion of biocomplexity to unit Ξ of its indicator:

$$\xi = \sum_{i=1}^{m} \sum_{j>i}^{m} k_j a_{ij}. \tag{17.2}$$

Clearly, $\xi = \xi(\varphi, \lambda, t)$, where φ and λ are geographical latitude and longitude, respectively, and t is the current time. For some territory Ω the indicator of biocomplexity is determined as an average:

$$\xi_\Omega(t) = (1/\sigma) \int_{(\varphi,\lambda)\in\Omega} \xi(\varphi, \lambda, t)\, d\varphi\, d\lambda, \tag{17.3}$$

where σ is the area of territory Ω.

Thus, indicator $\xi_\Omega(t)$ becomes an integral indicator of nature/society system complexity, reflecting the individual character of its structure and behavior at each time moment t in a space of Ω. Following the laws of natural evolution, a decrease (increase) of the ξ_Ω value will follow an increase (decrease) of biodiversity and survivability of natural–anthropogenic systems. Since a decrease of biodiversity breaks the closure of biogeochemical cycles and leads to an increase in the load on non-renewable resources, the binary structure of the matrix A moves to a state in which the level of its survivability lowers.

17.2.4 Nature/society system biocomplexity

The nature/society system consists of elements–sub-systems B_i $(i = 1, \dots, m)$, whose interaction forms in time depending on many factors. The biocomplexity of the nature/society system consists in the structural and dynamic complexity of its constituents. In other words, nature/society system biocomplexity is formed in the process of interaction of its constituents $\{B_i\}$. With time, sub-systems B_i can change their state, and hence the topology of the connections between them will change, too. The evolutionary mechanism of the adjustment of sub-systems B_i to their environment makes it possible to develop a hypothesis that each sub-system B_i, independent of its type, has structure $B_{i,S}$, behavior $B_{i,B}$, and goal $B_{i,G}$. Thus, $B_i = \{B_{i,S}, B_{i,B}, B_{i,G}\}$. Goal $B_{i,G}$ of sub-system B_i is to reach certain states that are preferable for itself. The expediency of structure $B_{i,S}$ and purposeful behavior $B_{i,B}$ of sub-system B_i is estimated from the efficiency of achieving goal $B_{i,G}$.

As an example, we take the process of migration of the elements of nekton. Fish migrate towards a maximum of the gradient of nutrients taking into account possible limits to water basin parameters (temperature, salinity, concentration of oxygen, contamination, etc.). Hence, the elements of nekton have goal $B_{i,G}$ to increase their rations, and their behavior $B_{i,B}$ is aimed at determining the trajectory of movement which would help to achieve the goal. There are also structural changes in the process of formation of shoals, which for each kind of nekton element can be presented in terms of $B_{i,S}$.

Since the interaction of sub-systems $\{B_i\}$ is connected with chemical and energy cycles, it is natural to suppose that each sub-system B_i organizes the geochemical and geographic transformations of matter and energy in order to remain stable. The formalized approach to this process consists in the supposition that in nature/society system structure, exchanges of some amounts of spent resources V for some amounts of consumed resources W (i.e., (V, W) exchange) take place between sub-systems B_i. The goal of sub-system B_i is to reach the most profitable (V, W) exchange (i.e., for a minimum of V to obtain maximum possible amounts of W), which is a function of the structures and behaviors of interacting sub-systems $W = W(V, B_i, \{B_k, k \in K\})$, where K is the multitude of sub-systems in contact with sub-system B_i. Let $B_K = \{B_k, k \in K\}$. Then the interaction of sub-system B_i with its environment B_K results in the following (V, W) exchanges:

$$W_{i,0} = \max_{B_i} \min_{B_K} W_i(V_i, B_i, B_K) = W_i(V_i, B_{i,\text{opt}}, B_{K,\text{opt}}) \tag{17.4}$$

$$W_{K,0} = \max_{B_K} \min_{B_i} W_K(V_K, B_i, B_K) = W_K(V_K, B_{i,\text{opt}}, B_{K,\text{opt}}). \tag{17.5}$$

Hence, there is some smearing of the goal of sub-system B_i when determining levels V_i and V_K. Since in nature limiting factors are active, in this case it is natural to suppose the presence of some threshold $V_{i,\min}$, and when reaching this threshold, the energy resource of the sub-system stops being spent on the extraction of external resources (i.e., at $V_i \le V_{i,\min}$ sub-system B_i operates in the regime of generation of the internal resource). In other words, at $V_i \le V_{i,\min}$ the indicator of biocomplexity $\xi_\Omega(t)$ decreases due to breaking the connections between sub-system B_i and other sub-systems. In general, $V_{\min}$ is a structural function of the step-function (i.e., the transition of a_{ij} from state $a_{ij} = 1$ into state $a_{ij} = 0$ does not for all j take place simultaneously). In any trophic pyramid, the "predator–prey" relationship ceases with decreasing concentrations of victims below some critical level. In other cases the interaction of sub-systems $\{B_i\}$ can cease depending on various combinations of other parameters. A formalized description of possible situations of interaction of the sub-systems $\{B_i\}$ can be realized within the framework of a simulation model of the functioning of the nature/society system.

17.2.5 Global model of nature/society system biocomplexity

The nature/society system is a self-organizing and self-structuring system, the correlation of whose elements in time and space are ensured by the process of natural evolution. The anthropogenic constituent in this process is aimed at breaking this integrity. Attempts to formally parameterize the process of co-evolution of nature and humans as elements of the biosphere are connected with a search for a single description of all the processes in the nature/society system that would unite the efforts of various branches of knowledge about the environment. Such synergism serves as the basis of many studies on global modeling.

Let us cover the Earth's surface Ω with the geographical grid $\{\varphi_i, \lambda_j\}$ with digitization steps $\Delta\varphi_i$ and $\Delta\lambda_j$ by latitude and longitude, respectively, so that within a cell of land surface $\Omega_{ij} = \{(\varphi, \lambda) : \varphi_i \leq \varphi \leq \varphi_i + \Delta\varphi_i; \lambda_j \leq \lambda \leq \lambda_j + \Delta\lambda_j\}$ all processes and elements are considered as homogeneous and are parameterized by point models. In the case of the water surface in the territory of cell Ω_{ij} water masses are stratified into layers Δz_k thick (i.e., 3-D volumes are selected $\Omega_{ijk} = \{(\varphi, \lambda, z) : (\varphi, \lambda) \in \Omega_{ij}, z_k \leq z \leq z_k + \Delta z_k\}$), inside which all elements of the ecosystem are distributed uniformly. Finally, the atmosphere over site Ω_{ij} at height h is digitized either by levels of atmospheric pressure or by layers Δh_s thick.

Interactions in the nature/society system are considered as interactions between natural and anthropogenic components within these spatial structures and between them. The complex model of the nature/society system realizes the spatial hierarchy of hydrodynamic, atmospheric, ecological, and socio-economic processes with the division of the whole volume of the environment into structures Ω_{ij} and Ω_{ijk}. The cells of this division are the supporting grid in numerical schemes for solutions based on dynamic equations or in the synthesis of data series in learning procedures of the evolutionary type.

Pixels Ω_{ij} and Ω_{ijk} are heterogeneous in parameters and functional characteristics. Through this heterogeneity the global model is referenced to databases. Moreover, to avoid an excess structure of the global model, it is supposed *a priori* that all its elements taken into account in the model and in nature/society system processes have a characteristic spatial digitization. Ambiguity of spatial digitizations in various units of the global model is removed at an algorithmic level of agreement of data fluxes from the system of monitoring. As a result, the model's structure is independent of the structure of the database, and hence does not change when the latter changes. A similar independence between the model's units is also provided. This is realized by data exchange between them only through inputs and outputs under the control of the basic data. When turning off one or several units their inputs are identified with corresponding inputs in the database. Then the model operating in the regime of a simulation experiment can be schematically represented by the process, where by means of the user's choice a spatial image is formed of the modeled medium and of the regime of control of the simulation experiment. Of course, in this case the user should have a certain knowledge base and know how it is structured (e.g., by using a list of key problems of global ecology or lists of nature/society system elements recommended for studies).

The character of the spatial structure of the global model is determined by the database. The simplest version of the point model is realized using initial information in the form of averaging over the land surface and all the World's ocean basins. Spatial heterogeneity is considered through various forms of space digitization. The base form of the spatial division of land and oceans is a heterogeneous grid $\Delta\varphi \times \Delta\lambda$. A real version of the use of the model is provided by integrating cells Ω_{ij} so that various forms of the spatial structure of the elements and biospheric processes considered can be present in each unit. Such a flexible setting of the spatial structure of the biosphere makes it possible to easily adapt the model to heterogeneities in databases and to perform simulation experiments with realistic values of the various parameters in each grid cell.

Depending on special features of the natural process under consideration, the structure of regional division can be identified by climatic zones, continents, latitudinal belts, socio-administrative structure, and natural zones. For climatic processes, many scientists work with regions with dimensions $\Delta\varphi = 4^\circ$ and $\Delta\lambda = 5^\circ$, biogeocenotic processes are studied at $\Delta\varphi = \Delta\lambda = 0.5^\circ$, the socio-economic structure is represented by nine regions, atmospheric processes in the biogeochemical cycles of long-lived elements are approximated by point models, the functioning of ocean ecosystems is described by heterogeneous digitization of the shelf zone into pixels Ω_{ij} by selecting four parts of the World Ocean.

The structure of division of the Earth's surface into regions Ω_{ij} covers all enumerated versions. This means that the general scheme of digitization of the processes in the nature/society system foresees a hierarchy of levels including global, continental, regional, landscape, local, etc. The scheme of independent inclusion of units at all these levels with their combination through parametric interfaces does not prevent an increase in the number of the model's units due to the introduction of new components that specify models of the processes under consideration. The model of the upper level can serve as an information base for the model of the lower level and *vice versa*. The results of modeling at the lower level can be used to form the information base for models at higher levels. This mechanism of information exchange between the models of various levels reduces the requirements of the global database and broadens the capabilities of the nature/society system model.

The structure of the global model includes some auxiliary units that provide for interaction of the user with the model and operate with the database. In particular, these are units that realize the algorithms of spatio-temporal interpolation or coordinate the user's actions with the bank of scenarios. Note that some scenarios can be transformed by the user's wish into the rank of model units. Such a duality (excessiveness) is characteristic of scenarios of climate, demography, anthropogenic activity, scientific–technical progress, and agriculture. The user's interface makes it possible to select the structure (Ω_{ij}) in default mode or the required spatial structure can be formed from base elements by averaging and interpolation.

Thus, synthesis of the global model version requires preliminary analysis of the present situation with global databases and knowledge bases. Here specialists face serious difficulties; mainly, the absence of adequate knowledge about climatic and biospheric processes as well as an uncoordinated database on global processes on the land, in the atmosphere, and in the oceans.

Another serious difficulty is connected with the inability of modern science to formulate the requirements for global databases needed to reliably assess the state of the environment and to give a reliable forecast of its development for a sufficiently long period. Moreover, there is no technology to form databases aimed at creating the global model.

Many scientists have made attempts to answer these questions (Krapivin, 1993; Tianhong *et al.*, 2003). One of the efficient ways to solve these problems is a single planetary adaptive geoinformation monitoring system (GIMS), which has a hierarchic structure of data collection and forms a multi-level global database. An adaptive character of this system is provided by correcting the regime of data collection and by changing the parameters and structure of the global model.

The global GIMS can be created with an account of the existing structure of databases whose formation continues within the framework of the IGBP and numerous national ecological and nature protection programs. The developed system of World data centers favors the rapid use of accumulated information about global processes and simplifies GIMS synthesis. However, significant progress in this direction connected with large economic expenses cannot lead to a successful solution of the problem of global environmental control, trough this phase cannot be avoided.

According to Kondratyev (1998), to control the global geobiosystem of the Earth, regular observations of specified key variables are needed. With an increasing probability of drastic global changes, the spectrum of these variables will vary, and the global prediction system should be constantly modernized. The choice of variables to provide adequate information for the monitoring system can be objective only in the case of GIMS functioning. Many of the chosen variables can be calculated using the respective models, and there is no need to measure them. However, so far, measurements are planned in parallel with model development, and, for the present at least, continue to be necessary. As follows from some studies (Kondratyev and Galindo, 2001; Kondratyev *et al.*, 2002), bases of global knowledge and data will make it possible to synthesize and develop GIMS series (Kondratyev *et al.*, 2004).

Inclusion of the global model into the GIMS structure enables one to consider it as an expert system. This means that there is a possibility of complex analysis of numerous elements of the nature/society system in conditions suitable for realizing hypothetic situations, which can appear for natural or anthropogenic reasons. Figure 17.1 reflects the basic elements taken into account in the global model of the nature/society system (GMNSS). Concrete realization of each unit of the GMNSS is determined by the level of knowledge of the processes reflected in the unit. The units responsible for modeling biogeochemical and biogeocenotic processes are described using balance equations.

Let $\psi_s(t)$ be the information content of element ψ in medium S at moment t. Then, following the law of conservation of matter and energy, we write the following balance equation

$$\frac{d\psi_s}{dt} = \sum_j H_{js} - \sum_i H_{si}, \tag{17.6}$$

where fluxes H_{js} and H_{si} are, respectively, incoming and outgoing fluxes with respect to medium S. Summation is made by external media i and j interacting with S. In fact, medium S implies elements of digitization of the environment by latitude φ, longitude

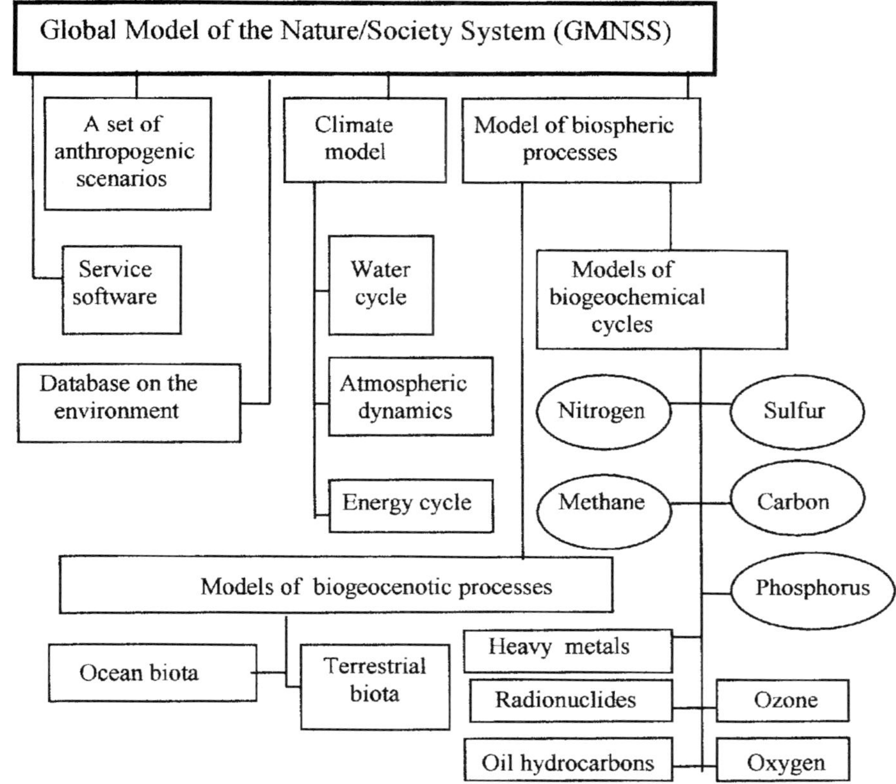

Figure 17.1. The concept of global modeling as a means to study cause-and-effect feedbacks in the environment. Detailed description of the GMNSS is given by Kondratyev *et al.* (2004).

λ, depth z, and height h. A variety of functional parameterizations of fluxes H_{pq} is determined by the level of knowledge of the physical, chemical, and biological features of element ψ.

Parameterization of the processes of photosynthesis, dying off, and respiration of plants in land ecosystems is based on knowledge of phytocenology, which includes information about the external and internal system connections of the vegetation community. These are the temperature dependences of photosynthesis and evapotranspiration of plants, gas exchange processes between plants and the atmosphere, impacts of solar radiation energy on the processes of growth and exchange, relationships between plants and processes in the soil, and the interaction of vegetation covers with the hydrological cycle.

The GMNSS units responsible for parameterizing climatic and anthropogenic processes are complex in character (i.e., partially described by the equations of motion and balance, and partially an evolutionary model is constructed for them that is based only on observational data).

17.2.6 Simulation results and discussion

The present administrative division of the world numbers 271 nations with total populations of about 6.6 billion and annual population growth rate of 1.14%. The real annual growth rate of gross domestic product is evaluated as 3.8% with distribution by sectors: agriculture 4%, industry 32%, and services 64%. Regional data about natural resources, population age structure, economics, and environmental parameters are obtainable from *http://www.cia.gov/cia/publications/factbook/index.html*

The introduced indicator ξ_Ω is used to describe the world dynamics in the framework of the realization of some scenarios of what is possible in the nearest future. Figures 17.2 and 17.3 show the results of such calculations. Some of the global hazardous tendencies are deforestation, over-fishing threatening marine-living populations, rapid urbanization, and over-exploitation of soils. These and others negative processes in the nature/society system change the biogeochemical interactions between biospheric and geospheric elements as well as the influence on the level of (V, W) exchanges. This is reflected in biocomplexity indicator dynamics.

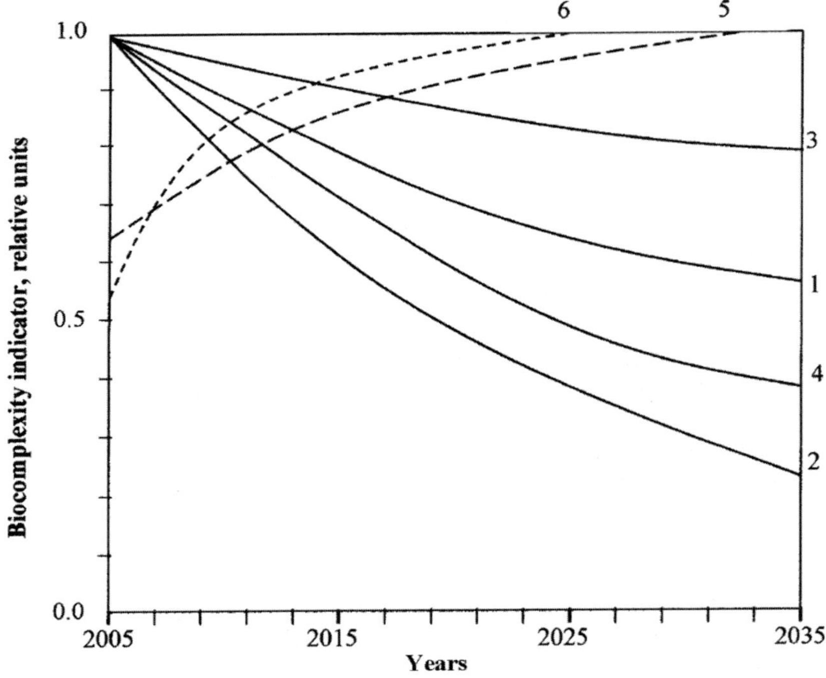

Figure 17.2. Dependence of the biocompexity indicator on the human strategy toward forests: 1, rate of change in forest areas remains the same as that in 1970–2000 (mean value); 2, by 2050 forests have totally disappeared; 3, by 2050 the area of forest is reduced by 10%; 4, by 50%; 5, by 2050 forest areas increase by 10%; 6, by 30%.

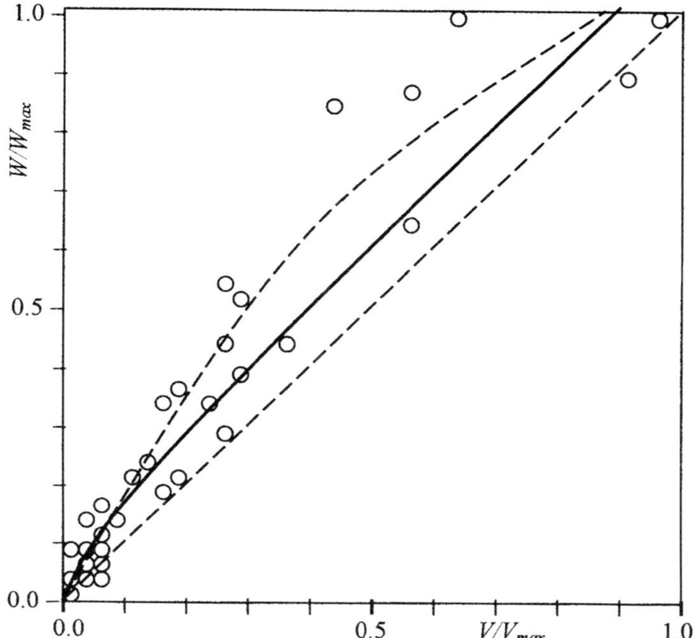

Figure 17.3. Correlations between exports and imports depending on environmental strategy. Circles correspond to real correlations (a fragment is given in Table 17.2). Solid curve is an approximation of existing (V, W) exchanges. Broken curves correspond to the range of variations in (V, W) exchanges depending on the changes in regional GDP by $\pm 5\%$.

Figure 17.2 shows the role of forest vegetation in global biocomplexity dynamics. This correlation is displayed through the dependence of biogeochemical cycles of greenhouse gases on the planetary forest cover. At the present time the forested area is approximately 40.3 million–41.8 million km^2 (Watson *et al.*, 2000) with 1% constituting national parks and forest reserves. As seen from Figure 17.2, the increasing rate of deforestation raises considerably the concentration of CO_2 in the atmosphere (by 31%) and destabilizes the biogeochemical cycles of greenhouse gases. It causes disturbance in the global structure of (V, W) exchanges. Even with a 10% reduction of forest areas by 2050 compared with the present time, atmospheric CO_2 could increase by 44% by the end of the 21st century. In contrast, a 10% increase in forested areas decreases the concentration of atmospheric CO_2 only by 15%. With a 30% increase of forested areas by 2050, the decrease of atmospheric CO_2 by 2100 will constitute 53% relative to its possible value, with the scale of impacts on forest ecosystems observed at the end of the 20th century preserved. Hence, variations in forested areas in the biosphere even within $\pm 10\%$ can substantially change the dynamics of numerous components of the global ecosystem. The biocomplexity indicator reflects the integral effects of scenario realization. The GMNSS really does permit evaluating the spatial dependencies between different processes taking place in the nature/society system. A more detailed discussion of these dependencies is given by Kondratyev *et al.* (2002, 2003, 2004).

The curves on Figure 17.3 are a good representation of the complex processes within the nature/society system when different interactions between its components of a biological, physical, chemical, geophysical, economic, and social character reflect their joint effect in export/import processes. We see that the real data about these processes are situated along the solid curve. Variations in regional GDP really do not change this dependence.

Table 17.1 gives a comparative analysis of the dynamics of biocomplexity distribution by different region. We see that the main causes of negative development of regional environments are deforestation, basin pollution, soil contamination, and urbanization. During the next 45 years only Australia, Canada, Central Africa, China, Japan, and South-East Asia have a chance to have favorable environmental conditions for sustainable development. Other regions are characterized by a decrease in their complexity related to different interactions with the environment, and hence their survivability is reduced. This follows from the decrease in intensity of the (V, W) exchange (Kondratyev *et al.*, 2002). Figure 17.3 demonstrates how regional (V, W) exchanges could change under GDP variations. We see that exports and imports are drawing closer to one other when GDP is rising and are moving away from the stable state when the GDP growth rate is decreasing. This shows that there exists a dependence between economic parameters and the biocomplexity indicator of a given region, and this is a function of global environmental strategy. Study of this

Table 17.1. Comparative analysis of biocomplexity indicators for different regions within existing regional anthropogenic strategies. Time-dependent dynamics of normalized biocomplexity indicators $\xi^*_\omega = \xi_\omega/\xi_{\omega,\max}$ and $\xi_{\omega,\max} = \max_{\omega\in\Omega} \xi_\omega$.

Region	ξ^*_ω				*Commentaries concerning key reason for biocomplexity change*
	2005	*2010*	*2020*	*2050*	
Australia	0.56	0.57	0.58	0.59	Urbanization and desertification limit environment diversity
Belgium	0.41	0.39	0.38	0.34	Repercussions on neighboring countries reduce biodiversity
Brazil	0.58	0.63	0.59	0.53	Deforestation in Amazon Basin, land and wetland degradation destroys the habitat
Bulgaria	0.24	0.22	0.21	0.18	Acid rain and soil contamination from heavy metals intensify deforestation
Canada	0.57	0.59	0.61	0.60	Acid rain affects lakes and lowers forest productivity

(*continued*)

Table 17.1 (*cont.*)

Region	ξ_{ω}^{*}				*Commentaries concerning key reason for biocomplexity change*
	2005	*2010*	*2020*	*2050*	
Central Africa	0.58	0.58	0.59	0.61	Anthropogenic deforestation and flash floods prevent regional development
Central Asia	0.39	0.38	0.35	0.33	Desertification predominates
China	0.78	0.81	0.84	0.87	Population growth causes negative consequences for natural systems
France	0.29	0.28	0.26	0.23	Agricultural runoff and acid rain limit the country's progress
Germany	0.31	0.30	0.28	0.27	Flora and fauna are badly affected by sulfur emissions
Japan	0.22	0.23	0.24	0.24	Threat to aquatic life and degradation of water quality create obstacles to progress
Mexico	0.49	0.47	0.43	0.39	Deforestation, widespread erosion, and deteriorating agricultural lands condemn the country's development to delay
North Africa	0.19	0.17	0.16	0.14	Soil degradation process leads to the simplification of natural ecosystems
Russia	0.95	0.91	0.87	0.81	Uncontrolled deforestation and urbanization and air pollution accelerate environmental degradation
South Africa	0.39	0.37	0.34	0.31	Soil erosion and river pollution by agricultural runoff lead to losses in biodiversity
South-East Asia	0.88	0.89	0.93	0.96	Deforestation, soil erosion, and overgrazing reduce the rate of regional progress
Spain	0.43	0.41	0.39	0.36	Effluents from the offshore production of oil and gas and deforestation are the cause of biocomplexity decrease
Ukraine	0.51	0.48	0.45	0.41	Poor use of arable land brings large deviations from optimal (V, W) exchange
U.S.A.	0.63	0.62	0.60	0.59	Growth in fossil fuels and non-perfect management of natural resources cause degradation of the environment

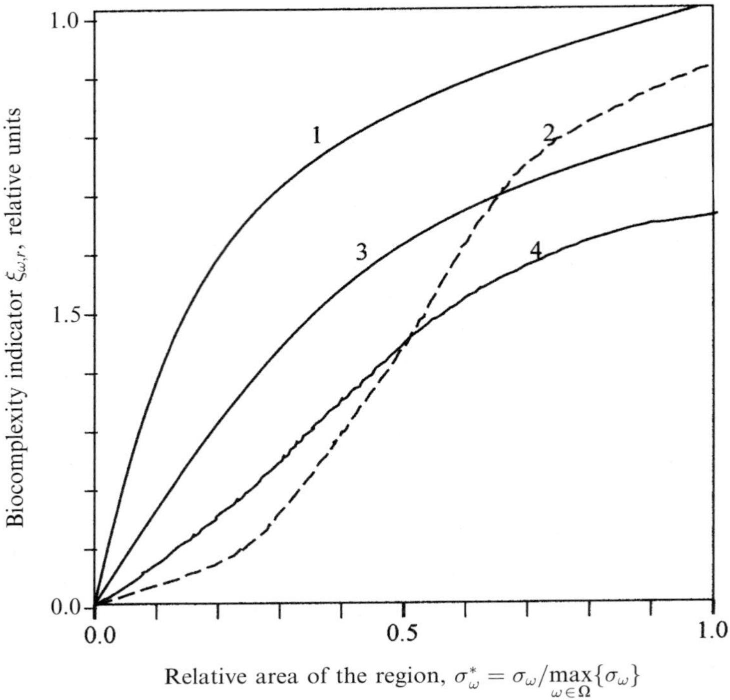

Figure 17.4. Correlations between regional biocomplexity and different state of a region. Numbers on the curves correspond to the following scenarios: 1, development regions with the area of arable land less than 20%; 2, regions with investment less than 20% of GDP; 3, regions oriented toward agricultural production (area of arable lands more than 20%); and 4, regions with investment more than 20% of GDP.

dependence really does demand consolidating World data, which at the present time still remains to be done.

Finally, the results shown in Figure 17.4 indicate the existence of the non-simple dependence of biological complexity of regional parts of the nature/society system on the correlation between arable lands and investments. It can be seen that the influence of investment strategy on regional development is non-linear. This depends weakly on investments in small areas of developed regions where investments are less than 20% of GDP, and this dependence increases sharply in other areas.

17.2.7 Conclusion

Biocomplexity is clearly an important characteristic of nature/society system dynamics. It has importance for complex study of the interactions between living and non-living elements of the environment, and more significantly it can make valuable contributions to the understanding and solution of key socio-economic and environmental problems. It is reasonable to expect that biocompexity will soon be used as an informative indicator analogous to such indicators as the normalized difference vegetation index (NDVI), leaf area index (LAI), etc. It appears that the only satisfactory way to develop an appropriate definition of the biocomplexity indicator is to summarize the many structural ideas in a series of global biospheric models. The synthesis of these models requires not only their coexistence with global databases, but also interconnections between different sources of data. This chapter proposes a global model and biocomplexity indicator in only one category in which biospheric processes are considered to be predominating. Further study should be directed to the expansion of the information already taken into account in the global model, and it is necessary to make correlation dependences between socio-economic and biospheric components more precise.

17.3 BASIC MODEL OF SUSTAINABLE DEVELOPMENT

17.3.1 Principle determination

Nature, N, and human society, H, constitute a single planetary system. Therefore, separating them when developing global or regional models should be considered a conditional step. Systems N and H have hierarchical structures $|N|$ and $|H|$, goals $\underline{N}$ and $\underline{H}$, behaviors $\overline{N}$ and $\overline{H}$, respectively. From the mathematical point of view, interactions between systems N and H can be considered a random process $\eta(t)$ with an unknown distribution law, representing the level of tension in the interaction of these systems or assessing the state of one of them. The goals and behaviors of the systems are functions of the indicator $\eta(t)$. There are ranges of $\eta(t)$ in which the system's behavior can be antagonistic, indifferent, and cooperative. The main goal of system H is to reach a high living standard with guaranteed long-term survival. Similarly, the goal of system N can be defined in terms of survival. The behavior of system N is determined by the objective laws of co-evolution. In that sense, the selection of H and N is conditional and can be interpreted as the separation of a multitude of natural processes into controllable and non-controllable. Without dwelling upon the philosophic aspects of this separation, we shall consider systems H and N as being symmetrical, in the sense of their description given above, and open. System H disposes of technologies, science, economic potential, industrial and agricultural production, sociological structure, population size, etc. The interaction between systems H and N leads to a change in $\eta(t)$, the level of which affects the structure of vectors $\underline{H}$ and $\overline{H}$. There exists a threshold for η_{max} beyond which humankind ceases to exist but nature survives. The asymmetry of systems H and N causes a change in the goal and strategy of system H. Apparently, under present conditions of the interaction between these systems $\eta(t) \to \eta_{max}$ at a rapid rate, and therefore some components of vector $\underline{H}$ can be attributed to the class of "cooperative". Since the present socio-economic structure of the world is represented by the totality of countries, we shall consider the country as a functional element of system H. Function $\eta(t)$ reflects the result of the interaction of countries between themselves and nature. The totality of the results of these interactions we shall describe by matrix $\mathbf{B} = \|b_{ij}\|$, each element of which has a concept of its own:

$$b_{ij} = \begin{cases} + & \text{for cooperative behavior,} \\ - & \text{for antagonistic relationships,} \\ 0 & \text{for indifferent behavior.} \end{cases} \tag{17.7}$$

A country H_i has m_i possible ways to achieve goal $\underline{H}_i$. In other words, it uses a row of strategies $\{\overline{H}_i^1, \ldots, \overline{H}_i^{m_i}\}$. The weight of each strategy $\overline{H}_i^j$ is determined by p_{ij} ($\sum_{j=1}^{m_i} p_{ij} = 1$).

The resulting quantity of parameter $\eta(t)$ is a function of the indicated characteristics, and overall the situation at any moment is described by the game theory model.

17.3.2 Common view of global model

Objective assessment of the environmental dynamics $N = \{N_1, N_2\}$ is possible by making certain assumptions using models of the biosphere N_1 and climate N_2. Such models have been developed by many authors and accumulated experience covers examples of point, regional, box, combined, and spatial models. This experience makes it possible to move toward synthesizing a global model of a new type covering the key relationships between the levels of hierarchy of natural processes on land and in the World Ocean. Relationships between global model components are provided by parameters that are shared between the components. The units of climate and the hydrosphere are basic, since the main circulation of substance and energy is realized through their components. The multiplicity of spatial digitizations in these units provides adaptive flexibility of the global model by coordinating the data fluxes between them. As a result, the model's structure is independent of the structure of a global database, and hence does not change when the latter changes. Versions of space digitization include:

- random setting of regions and water basins;
- division of the planetary surface into sites with constant steps in latitude and longitude;
- superposition of this grid only on land and the shelf zone of the World Ocean with point presentation of its pelagic zones; and
- taking the administrative structure as the basis of nature/society system digitization.

This universality of space digitization is possible due to presentation of the global model in the form of a set of units connected only through inputs and outputs by the principle of open systems. In other words, let $A_s(t)$ be the content of chemical matter or energy of element A in medium S at time moment t. Then, following the law of conservation of mass and energy, we write the balance equation:

$$dA_s/dt = \sum_j Q_{js} - \sum_i Q_{si}, \tag{17.8}$$

where fluxes Q_{js} and Q_{si} are, respectively, input and output. Summing up is made by external media i and j interacting with medium S.

Biogeocenotic processes in the global model consist of photosynthesis, dying off of plants, their respiration, and growth. A change in biomass P is approximated by the dynamic equation:

$$dP/dt^{-} = \min\{\rho P, R\} - M - U - V, \tag{17.9}$$

where ρ is the maximum production/biomass ratio for the plant type considered; R is the real productivity of plants; M and U are the amounts of dying off and expenditure on energy exchange with the medium; and V is the loss of biomass for anthropogenic reasons.

The distribution of the types of soil–plant formations over land has been well studied. Data on assessment of the productivity, biomass supplies, and dead organic matter on land and in the ocean make it possible to parameterize the components of Equations (17.8) and (17.9), by taking the whole structure of energy exchange in the environment into account. The equation of balance between energy and matter in water ecosystems is written as:

$$\partial P_i/\partial t^{-} + Y = R_i - H_i - M_i - \sum \gamma_{ji} R_j, \tag{17.10}$$

where $\|\gamma_{ji}\|$ is the matrix of the efficiency of trophic bonds in the water ecosystem; and Y is the hydrodynamic component.

The monographs by Kondratyev *et al.* (2002, 2003, 2004, 2006a-c) describe in detail the global model of the nature–society system (GMNSS) which takes into account the interaction of the nature/society system components shown in Figure 17.5.

17.3.3 Sustainability criterion

In a general case, the state of systems H and N can be described by vectors $x_H(t) = \{x_H^1, \ldots, x_H^n\}$ and $x_N(t) = \{x_N^1, \ldots, x_N^m\}$, respectively. The combined trajectory of these systems in $n+m$-dimensional space is described by the function $\eta(t) = F(x_H, x_N)$, which is determined by solving Equations (17.1)–(17.3) and other ratios from the global model. The form of $F(x_H, x_N)$ is determined from the laws of co-evolution, and therefore there is a wide field here for studies in different spheres of knowledge. Available estimates of $F(x_H, x_N)$ indicate a relationship between the notions "survivability" and "sustainability". According to Ashby (1956), a dynamic system is in a "living state" in the time interval (t_a, t_b) if its determining phase coordinates are within "permissible limits": $x^i_{H,\min} \leq x^i_H \leq x^i_{H,\max}$; $x^j_{N,\min} \leq x^j_N \leq x^j_{N,\max}$. Since systems H and N have a biological basis and limited energy resources, one of these boundary conditions is unnecessary: that is, for the components of vector $x = \{x_H, x_N\} = \{x_1, \ldots, x_k\}$ $(k = n+m)$, conditions $x_{\min} \leq \eta = \sum_{i=1}^{k} x_i$ should be satisfied. This simple scheme includes requirements for both consideration of total energy in the system and the diversity of its components.

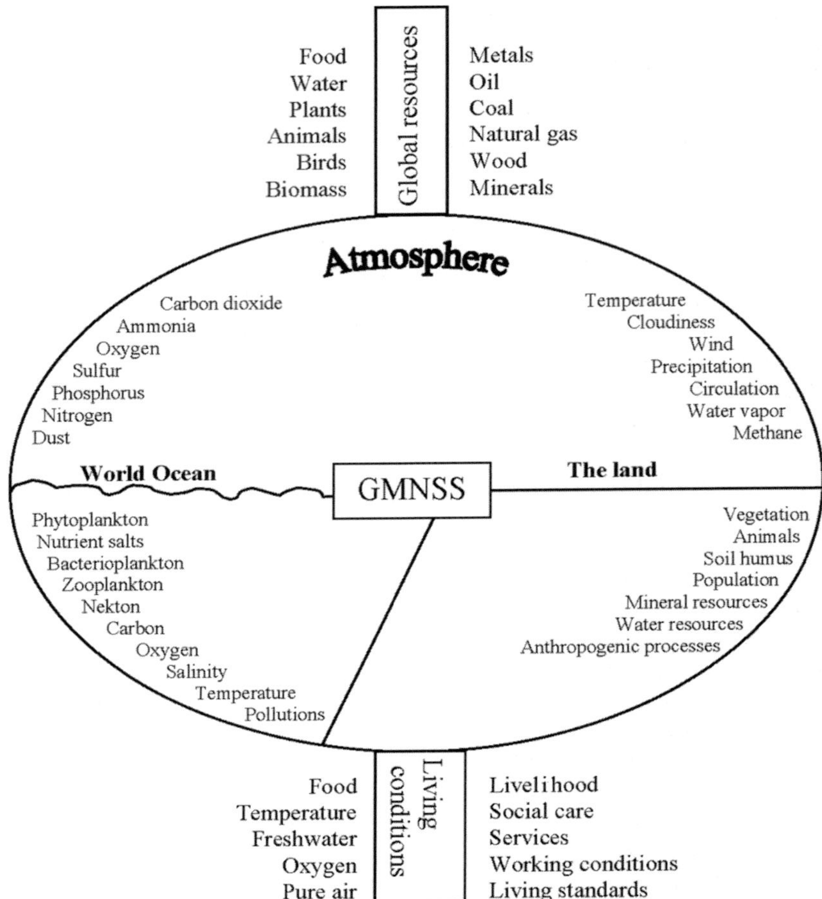

Figure 17.5. Key elements of the nature/society system and the energy components that are taken into account when formulating a global model of ecodynamic forecast.

Of course, the notion of system sustainability is more capacious and informative. In system ecology, many authors use this term bearing in mind the stability and integrity of the system, meaning the system's ability to resist external forcings. In other words, sustainability is measured by the trend of the system to suppress large oscillations in its structure and components, returning the system to its equilibrium state. Thus, system sustainability is supposed to mean its ability to actively withstand external forcings, to preserve its characteristics (i.e., probability of survival of the states in which it remains able to live), and to function with certain applied methods and under certain conditions when it is being exploited.

The conceptual scheme of a new version of the global model covering the vital components of the environment has now been developed by many scientists; in particular, at the Potsdam Institute for Climate Impact Research (Boysen, 2000), at the Institute of Biophysics of the Siberian Branch of the Russian Academy of Sciences (Degermendzhi *et al.*, 2008). A new version of the global model based on the use of the open systems theory has also been worked out in Russia (Kondratyev *et al.*, 2004; Krapivin and Kondratyev, 2002). The structure of connections between the model's units foresees an information exchange between them via their inputs and outputs which ensures not only the use of the evolutionary selection among these units, but also accomplishment of structural synthesis of the whole global model. Its elements are either represented by a set of earlier models or described by an observational data series of their characteristics. This structure of the global model ensures its independence of the procedure of changing some units and their interconnection, such that the relationship between global model units is ensured by the parametric compatibility of all its units and nothing else.

The spatial structure of the environment is described by a geographical grid with steps $\Delta\varphi$ in latitude and $\Delta\lambda$ in longitude. With a homogeneous surface the atmosphere is divided into two layers: a mixed layer h_1 in height and an upper layer that is h_2 thick; the oceans are represented by a multi-layer structure with selection in each cell $\Delta\varphi \times \Delta\lambda$ of upper layers Δz_1 thick down to the thermocline, then Δz_2 thick to a photically effective depth $z(\varphi, \lambda)$, and the deep ocean. Inside a cell pixel $\Delta\varphi \times \Delta\lambda$, all components are assumed to be homogeneous. Biotic, biogeochemical, and demographic processes are described by balanced finite-difference equations, while anthropogenic, climatic, and socio-economic processes are represented either by sets of scenarios or by models of evolutionary type. A computer version of the global model gives the possibility to realize numerical experiments that consider numerous problems in the study of biospheric processes. The user can choose the domain of studies, adjusting the whole model to the chosen scenario. There is a possibility here to choose between a standard set of scenarios or to synthesize a new scenario both in the sphere of anthropogenic activity and in the analysis of other processes in the biosphere and climate system. The choice of territory for analysis of the environmental processes taking place over this territory is realized by indicating in the menu a concrete object, or with the help of a special procedure by formation of the inner structure of subject-thematic identifiers $\{Y_k\}$ with the use of which a model is synthesized either for all the environmental elements of the territory or for the whole biosphere. Figure 17.6 explains this procedure.

17.4 BIOSPHERIC EVOLUTION, RISKS, AND NATURAL DISASTERS

The notion of risk is closely connected with the notion of uncertainty of prediction of the development of events in the nature/society system that are undesirable for humans. By the definition of Burgman (2005), risk is the chance of occurrence of an unfavorable event within some time interval: for the nature/society system such intervals are measured in centuries. Society primarily wants to know the prospects for improving living standards. However, the state of current environmental science does not guarantee solution of this problem, and therefore it is necessary to find

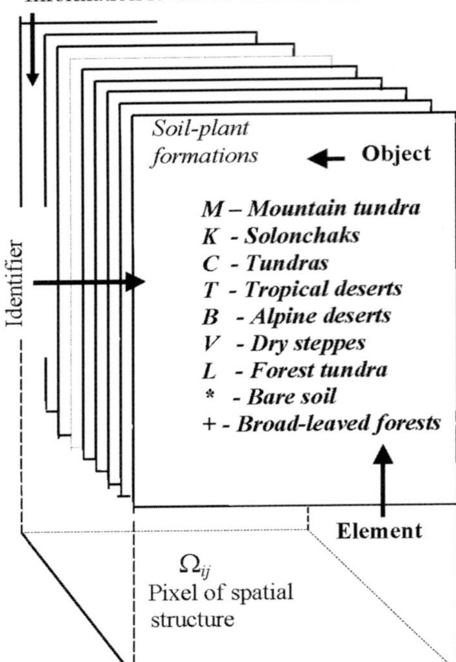

Figure 17.6. Information levels of the global model database and their cartographic identification in the GMNSS.

constructive technologies of risk assessment. A reliable and constructive approach to solution of this problem could be brought about by expanding GMNSS functions. In particular, the number of uncertainties hindering reliable prediction of global ecodynamics is so large that without their purpose-oriented analysis and selection their interactions cannot all be formally taken into account in the global model. For instance, of the millions of chemical compounds and species of living organisms, only thousands have been adequately studied.

One of the many factors involved in evolutionary process development in the nature/society system is sudden change in environmental characteristics, which induces stress in living organisms or even death. In various periods of evolution, the scale and significance of individual factors are known to change. After the tragedy of the tsunami in Asia as a result of the magnitude-9 earthquake on December 26, 2004 near the northwestern coast of Sumatra, the problem of extreme natural phenomena prediction has become urgent. It is clear that modern geophysical science can only comment on the causes of earthquakes, putting forward various hypotheses that explain them as shifts in the Earth's crust. The most complicated problem facing present-day science is arguably earthquake prediction (see also Section 15.5.1). Despite the existence of specialized centers that can identify minute oscillations in the Earth's crust, the progress of the scientific community in studying the laws governing planetary development is still negligible. Nevertheless, some progress in

Table 17.2. Key characteristics of (V, W) exchange to be taken into account in the GMNSS for selected regions.

Significant index of regional development	*Region*						
	World	*Russia*	*U.S.A*	*China*	*Japan*	*Mozambique*	*Brazil*
Global Domestic Product growth rate (%)	3.8	7.3	3.1	9.1	2.7	7.0	−0.2
Industrial production growth rate (%)	3.0	7.0	0.3	30.4	3.3	3.4	0.4
Investment (% of GDP)	—	18.2	15.2	43.4	23.9	47.8	18.0
Exports/imports (U.S.$ billion)	6,421/6,531	134.4/74.8	714.5/1,260	436.1/397.4	447.1/346.6	0.795/1.142	73.28/48.25
Electricity (billion kWh/yr):							
production	14,930	915	3,719	1,420	1,037	7.19	321.2
consumption	13,940	773	3,602	1,312	964.2	1.39	335.9
exports		21.16	18.17	10.30	0	5.8	0
imports		7.00	38.48	1.80	0	0.5	37.2
Oil production/consumption (Mbbl/da)	75.57/75.57	7.3/2.6	8.1/19.7	3.3/4.6	0.02/5.3	0/0.09	1.6/2.2
Natural gas production/consumption (billion m^3/da)	2,578/2,555	581/408	548/641	30.3/27.4	2.52/80.42	0.06/0.06	5.95/9.59
Arable lands (million km^2)	15.981	1.246	1.753	1.436	0.046	0.040	0.589

predicting other types of natural disasters has been achieved due to development of the theory of climate and global ecodynamics. However, assessment and prediction are only possible with certain scenarios of climate and strategies of human development. Therefore, it is important that these scenarios are based on nature/society system historical considerations (Jolliffe and Stephenson, 2003; Lawrence, 2003; Vaitheeswaran, 2005; Yue *et al.*, 2005).

One approach to the prediction of earthquakes and volcanic eruptions involves using the statistics from natural disasters as input information to the GMNSS. The prediction of random successions with the help of the method of evolutionary modeling enables one to determine with some probability the time of occurrence of the next event. Prediction of other types of natural disasters using the GMNSS is possible because the model considers all direct connections and feedbacks in the biosphere–climate system. To make such predictions it is necessary to prescribe scenarios of the potential development of the interaction between society and the environment. The diversity of such scenarios complicates the problem, though using evolutionary technology here makes it possible to reveal the most probable trends in this interaction. Let us consider some outcomes of SRES scenarios (Edmonds *et al.*, 2004). The most pessimistic scenarios are A1G MiniCAM and A2ASF which lead to an increase in the concentration of CO_2 in the atmosphere by 2020 up to 390 ppm–410 ppm and by 2100 up to 520 ppm–550 ppm. As a result, the pH of the upper layer of the oceans, especially its coastal basins, decreases, which leads to changes in trophic relationships between ecosystem elements. A characteristic example is the ecosystem of the Peruvian upwelling (Krapivin, 1996) whose trophic pyramid under standard conditions is characterized by its spatial binary nature. The curves in Figure 17.7 demonstrate the state of survivability of this ecosystem evaluated by the criterion:

$$v(t) = \sum_{i=1}^{m} B_i(t) \Big/ \sum_{i=1}^{m} B_i(t_0), \tag{17.11}$$

where m is the number of trophic levels; $t_0 = 1999$; and $B_i(t)$ is the total biomass of the ith trophic level over the water basin. It is assumed that minimum concentrations of nutrients not assimilated at other levels constitute 10% of their initial values.

Figure 17.7 demonstrates the response of the system to an increase in upper-layer temperature. Calculations show that an increase in temperature of 0.4°C is harmless, and that a greater increase would result in the system changing its phase state. In the latter case, the effect of spatial binary relations in the trophic pyramid disappears and the system enters the phase of unstable functioning. In general, the GMNSS makes it possible to study the behavior of land and marine ecosystems in the various scenarios indicated in Table 17.3. In the case of scenarios A1T Message and B1 Message the climatic situation in the Peruvian upwelling basin does not change substantially and the ecosystem changes trophic structure only in the coastal zone between El Niño periods. In the case of scenarios B2 Message and A1 AIM, periods occur when there is a prolonged increase in upper-layer temperature, which causes some imbalance between energy fluxes in the ecosystem, but on the whole its stability is preserved. In the third case, when scenarios A1G MiniCAM or A2 ASF are included, the

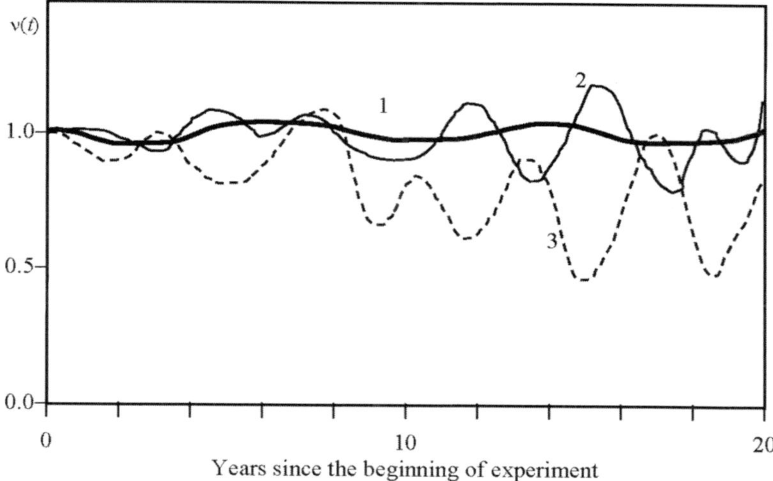

Figure 17.7. Assessment of survivability of the Peruvian upwelling ecosystem with different scenarios of global ecodynamics. Notation: 1, A1T Message; 2, A1 AIM; 3, A2 ASF (Edmonds *et al.*, 2004).

ecosystem starts moving into another state characterized by long-term reduction in total biomass. Additional experiments show that considerable water temperature oscillations principally change the state of the ecosystem. Phase trajectories of the ecosystem form quasi-periodic structures of the type of standing waves with a shift of the center of masses toward a decrease of $v(t)$. The system can withstand an increase in temperature of more than 5°C for no longer than 190 days. Oscillations in the

Table 17.3. General characteristics of scenarios of the SRES series by the rate of development of technologies of extraction, re-equipment, and distribution of energy resources (Arnell, 2004; Fenhann, 2000; Nakicenovic and Swart, 2000; Nicholls, 2004).

Class of scenario	*Coal*	*Oil*	*Gas*	*Non-fossil fuels*
A1B	Average	High	High	High
A2	Average	Low	Low	Low
B1	Average	Average	Average	Medium high
B2	Low	Below average	Medium high	Average
A1G	Low	Very high	Very high	Average
A1C	High	Low	Low	Low
A1T	Low	High	High	Very high

concentration of dissolved oxygen, which decreases with increasing temperature, should not be beyond $0.2\,\mathrm{mL\,L^{-1}}$ for longer than 100 days, and the rate of vertical advection should not be below $0.5 \times 10^{-4}\,\mathrm{cm\,s^{-1}}$. On the whole, assessment of the vitality of the Peruvian upwelling ecosystem shows that with long-term slow changes in environmental conditions the community re-arranges the structure and intensity of energy fluxes between trophic levels. One of the factors of the ecosystem's high stability is the vertical shift of the biomasses of ecosystem components, which makes it possible to preserve the phase pattern of the community for a long time even with substantial changes in environmental parameters as, for instance, in the case of the A2 ASF scenario.

Studies accomplished by many authors (Kondratyev *et al.*, 2003, 2004; Krapivin and Kondratyev, 2002) show that the study of global ecodynamics requires the development of a mathematical tool that can fulfill the interdisciplinary needs of biology, geophysics, economics, sociology, climatology, and biocenology. The GMNSS only partially meets these requirements. One feature of the GMNSS is the possibility to study the processes of interaction between natural and anthropogenic factors that take into account the broad spectrum of direct connections and feedbacks between nature/society system components. The principal non-linearity in parametric presentation of these connections complicates analysis of the laws of global ecodynamics and poses additional problems for evaluation of the numerous parameters that depend on time and space coordinates. Therefore, the reliability of any values and predictions depends on the accuracy of assumptions and scenarios.

Another prediction that can be made using the GMNSS is assessment of the variability in global water balance components. Taking the IPCC IS92a scenario that foresees the growth of World population to 11 billion by 2100 as our basis, we can predict that increased rain rates will be observed in northwestern Europe by 2020, which will cause a decrease in atmospheric moisture flow from the European continent to America of about $400\,\mathrm{km^3\,da^{-1}}$. In other regions, the water cycle will vary within $\pm 7\%$ with a gradual increase in amplitude by 2100. As a result, by the end of the century the rain rate will increase near the Pacific coastline of the U.S.A., northeastern India, southwestern China, and the zone of heavy rains in Europe will extend northward. Hence, floods in these regions will be more frequent. At the same time, the rain rate will decrease along the eastern coastline of North America, in the countries of Middle Asia and the Near East, and the regime of the contrasting alternation of wet and dry seasons will change in southeastern Asia. For the European continent, a negative fact will be a marked decrease of rainfall in Greece, Italy, and the Caucasus. In Central Europe the regime of precipitation will change by no more than 3%.

The GMNSS can also assess the potential risks of any greenhouse effect. These assessments are exemplified in Figure 17.8. Comparison of the results of predicted temperature changes obtained with the Hadley Centre model and the GMNSS demonstrates the efficiency of GIMS technology (Krapivin *et al.*, 2006) and points to the need for further GMNSS modernization by extending its units, especially those connected with parametrrizing the interactive mechanisms involved in climate regulation.

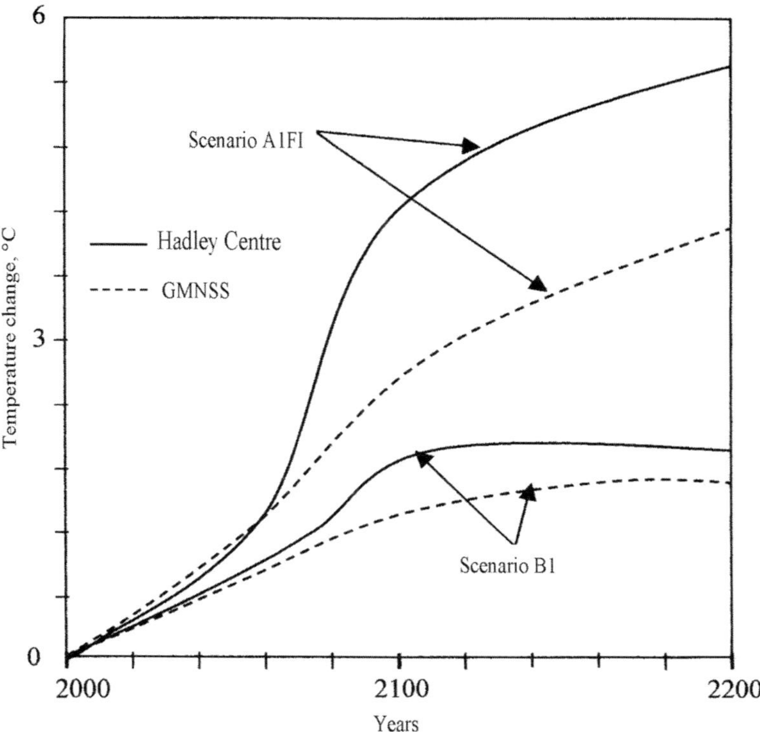

Figure 17.8. Forecasts of global mean temperature change using the Hadley Centre climate model and the GMNSS with two scenarios of energy use.

17.5 CONCLUDING REMARKS

Lomborg (2001, 2004) is of course right to reject the apocalyptic predictions of global ecodynamics based on an exaggerated fear of limited natural resources and the environmental state. Lomborg's opinions and assessments are confirmed by the data in Table 17.4 compiled by Holdren (2003), which characterize both real and potential global energy resources. Energy units are expressed here (in case of non-renewable energy sources) in terawatts per year, which is equivalent to 31.5 exaJ (1 TW = 1 TW-yr yr^{-1} = 31.5 exaJ yr^{-1}). It should be added that global energy consumption in 2000 constituted about 15 TW or 15 TW-yr yr^{-1} which is expected to increase up to 60 TW-yr yr^{-1} by 2100.

Despite the optimistic data in Table 17.4, present global ecodynamics shows that existing consumption levels in society have no future (Kondratyev *et al.*, 2003, 2004). Therefore, at the World Summit on Sustainable Development held in Johannesburg in 2002, the need to carry out 10-year programs to bring about stable production and consumption was emphasized and the following recommendations were made (Starke, 2004):

Table 17.4. Global energy resources (IEA, 2005a–c; WEO, 2006).

	Terawatts per year
Non-renewable resources	
Standard oil and natural gas	1,000
Non-standard oil and gas, except clusters of methane	2,000
Clusters of methane	20,000
Shale	30,000
Geothermal sources:	
Vapor and hot water	4,000
Hot dry rocks	1,000,000
Uranium:	
In reactors with light water	3,000
In breeder reactors	3,000,000
Thermonuclear energy:	
Heavy hydrogen limited with lithium	140,000,000
Heavy hydrogen–heavy hydrogen	250,000,000,000
Renewable resources	
Hydroenergy	15
Use of biomass	100
Wind energy	2,000
Solar energy	
On land surface	26,000
Over the globe	88,000

- developed countries should take the leading role to bring about stability between production and consumption;
- these goals should be achieved on the basis of common responsibility;
- stability between production and consumption should play the key role;
- the young must take part in solution of the problem of sustainable development;
- the "polluter pays" principle should be practiced;
- control over the complete cycle in a product's evolution from production, consumption, right through to disposal in order to raise the production efficiency;
- support should be given to political parties favoring the output of ecologically acceptable products and rendering of ecologically adequate services;
- to develop more ecological and effective methods of energy provision and eliminate energy subsidies;
- to support the free-will initiatives of industry aimed at raising its social and ecological responsibility; and
- to study and introduce means of ecologically pure production, especially in developing countries and in small-sized and medium-sized businesses.

Though these recommendations are rather declarative, they still clearly point to the necessity to change the paradigm of socio-economic development (primarily in developed countries) from a consumption society to priorities of public and spiritual values. Concrete analysis of the ways of such development requires the participation of specialists in the field of social sciences. Some related opinions were expressed by Corcoran (2005). Therefore, the question as to whether humans can change climate still needs further studies (Borisov, 2005).

Finally, one can draw the following conclusions:

- existing climate models cannot be used to make decisions and assess the risk of future anthropogenic scenarios becoming a reality;
- the level of uncertainty in climate forecasts can be reduced by giving broader consideration in global models to interactive bonds in the nature/society system and to the mechanisms of biotic regulation of the environment, in addition to improving global monitoring system; and
- the use of hydrocarbon energy sources in the 21st century will not lead to a catastrophic climate change if the Earth's land covers are preserved and the World Ocean is protected from pollution.

17.6 REFERENCES

Arnell N.W. (2004). Climate change and global water resources: SRES emissions and socio-economic scenarios. *Global Environmental Change*, **14**(1), 31–52.

Ashby W.R. (1956). *An Introduction to Cybernetics*. Chapman & Hall, London, 334 pp.

Borisov P. M. (2005). *Can Man Change Climate?* Science, Moscow, 270 pp. [in Russian].

Boysen M. (ed.) (2000). *Biennial Report 1998 & 1999*. Potsdam Institute for Climate Impact Research, Potsdam, Germany, 130 pp.

Burgman M. (2005). *Risks and Decisions for Conservation and Environmental Management*. Cambridge University Press., Cambridge, U.K., 488 pp.

Corcoran P.P. (ed.) (2005). *The Earth Charter in Action: Toward a Sustainable World*. KIT, Amsterdam, 192 pp.

Degermendzhi A.G., Bartsev S.I., Gubanov V.G., Erokhin D.V., and Shevirnogov A.P. (2008). Forecast of biosphere dynamics using small-scale models [Chapter 10 of this book].

Devkota S.R. (2005). Is strong sustainability operational? An example from Nepal. *Sustainable Development*, **13**(5), 297–310.

Edmonds J., Joos F., Nakicenovic N., Richels R.G., and Sarmiento J.L. (2004). Scenarios, targets, gaps, and costs. In: C.B. Field and M.R. Raupach (eds.), *Global Carbon Cycle: Integrating Humans, Climate, and the Natural World*. Island Press, Washington, D.C., pp. 77–102.

Ehlers E. and Kraft T. (eds.) (2005). *Earth System Science in the Anthropocene: Emerging Issues and Problems*. Springer-Verlag, Heidelberg, Germany, 300 pp.

Fenhann J. (2000). Industrial non-energy, non-CO_2 greenhouse gas emissions. *Technological Forecasting and Social Change*, **63**(2–3), 313–334.

Fleishman B.S. (1982). *The Principles of Systemology*. Radio & Communication, Moscow, 250 pp. [in Russian].

Friedman T.L. (2005). *The World Is Flat: A Brief History of the Twenty-first Century*. Farrar, Straus & Giroux, New York, 496 pp.

Furth J.H. (1965). Professor James on the theory of monetary policy. *Journal of Economics*, **25**(1/2), 199–203.

Gorshkov V.G., Gorshkov V.V., and Makarieva A.M. (2000). *Biotic Regulation of the Environment: Key Issues of Global Change*. Springer-Praxis, Chichester, U.K., 367 pp.

Gorshkov V.G., Esenin B.K., Karibayeva K.N., Kurochkina L.Ya., Losev K.S., Makarieva A.M., and Shukurov E.D. (2004). Scientific fundamentals of strategic directions for nature protection policy. *Ecology and Education*, **1**, 29–31 [in Russian].

Guista M.L. and Kambhampati U.S. (eds.) (2006). *Critical Pespectives on Globalization*. Edward Elgar, London, 656 pp.

Holdren J.P. (2003). Environmental change and human condition. *Bull. Amer. Acad. Arts. Sci., New York*, **57**(1), 25–31.

IEA (2005a). *Energy Policies of IEA Countries*. IEA Books, Paris, France, 588 pp.

IEA (2005b). *Key World Energy Statistics*. IEA Books, Paris, France, 82 pp.

IEA (2005c). *World Energy Outlook: Middle East and North Africa Insights*. IEA Books, Paris, France, 600 pp.

Jolliffe I.T. and Stephenson D.B. (2003). *Forecast 2002 Verification*, Wiley, London, 254 pp.

Kondratyev K.Ya. (1998). *Multidimensional Global Change*. Wiley/Praxis, Chichester, U.K., 771 pp.

Kondratyev K.Ya. and Galindo I. (2001). *Global Change Situations: Today and Tomorrow*. Universidad de Colima, Colima, Mexico, 164 pp.

Kondratyev K.Ya. and Krapivin V.F. (2005). Civilization development and its ecological limitations: Numerical modelling and monitoring. *Research of the Earth from Space*, **4**, 3–23 [in Russian].

Kondratyev K.Ya., Ortner J., and Preining O. (1992). Priorities of global ecology now and in the next century. *Space Policy*, **8**(1), 39–48.

Kondratyev K.Ya., Moreno-Pena F., and Galindo I. (1994). *Global Change: Environment and Society*. Universidad de Colima, Colima, Mexico, 147 pp.

Kondratyev K.Ya., Moreno-Pena F., and Galindo I. (1997). *Sustainable Development and Population Dynamics*. Universidad de Colima, Colima, Mexico, 128 pp.

Kondratyev K.Ya., Krapivin V.F., and Phillips G.W. (2002). *Global Environmental Change: Modelling and Monitoring*. Springer-Verlag, Heidelberg, Germany, 316 pp.

Kondratyev K.Ya., Krapivin V.F., and Savinykh V.P. (2003). *Perspectives of Civilization Development: Multi-dimensional Analysis*. Logos, Moscow, 574 pp. [in Russian].

Kondratyev K.Ya., Krapivin V.F., Savinykh V.P., and Varotsos C.A. (2004). *Global Ecodynamics: A Multidimensional Analysis*. Springer/Praxis, Chichester, U.K., 658 pp.

Kondratyev K.Ya., Ivlev L.S., Krapivin V.F., and Varotsos C.A. (2006a). *Atmospheric Aerosol Properties: Formation, Processes and Impacts*. Springer/Praxis, Chichester, U.K., 572 pp.

Kondratyev K.Ya., Krapivin V.F., Lacasa H., and Savinykh V.P. (2006b). *Globalization and Sustainable Development: Ecologicasl Aspects*. Science, St. Petersburg, 241 pp. [in Russian].

Kondratyev K.Ya., Krapivin V.F., and Varotsos C.A. (2006c). *Natural Disasters as Components of Ecodynamics*. Springer/Praxis, Chichester, U.K., 625 pp.

Krapivin V.F. (1993). Mathematical model for global ecological investigations. *Ecological Modelling*, **67**(2/4), 103–127.

Krapivin V.F. (1996). The estimation of the Peruvian current ecosystem by a mathematical model of biosphere. *Ecological Modelling*, **91**(1), 1–14.

Krapivin V.F. and Kondratyev K.Ya. (2002). *Global Environmental Changes: Ecoinformatics.* St. Petersburg University, St. Petersburg, 724 pp [in Russian].

Krapivin V.F., Mkrtchan F.A., and Trong B.D. (2005). Microwave radiometry technology for the nature–society system biocomplexity assessment. *Proceedings of the 26th Asian Conference on Remote Sensing (ACRS-2005), November 7–11, 2005, Hanoi, Vietnam*, pp. 43–47.

Krapivin V.F., Shutko A.M., Chukhlantsev A.A., Golovachev S.P., and Phillips G.W. (2006). GIMS-based method for vegetation microwave monitoring. *Environmental Modelling and Software*, **21**(3), 330–345.

Lawrence D.P. (2003). *Environmental Impact Assessment: Practical Solutions to Recurrent Problems.* Wiley, New York, 562 pp.

Lomborg B. (2001). *The Sceptical Environmentalist: Measuring the Real State of the World.* Cambridge University Press, Cambridge, U.K., 496 pp.

Lomborg B. (ed.) (2004). *Global Crisis, Global Solutions.* Cambridge University Press, Cambridge, U.K., 670 pp.

Mander J. and Goldsmith E. (eds) (2006). *The Case against the Global Economy.* IFG, San Francisco, CA, 560 pp.

Nakicenovic N. and Swart R. (eds.) (2000). *Special Report on Emissions Scenarios.* Cambridge University Press, Cambridge, U.K., 570 pp.

Nicholls R.J. (2004). Coastal flooding and wetland loss in the 21st century: Changes under the SRES climate and socio-economic scenarios. *Global Environmental Change*, **14**(1), 69–86.

Nitu C., Krapivin V.F., and Pruteanu E. (2004). *Ecoinformatics: Intelligent Systems in Ecology.* Magic Print, Onesti, Bucharest, Romania, 411 pp.

Olenyev V.V. and Fedotov A.P. (2003). Globalistics on the verge of the 21st century. *Problems of Philosophy*, **4**, 18–30 [in Russian].

Starke L. (ed.) (2004). *State of the World—2004: Progress towards a Sustainable Society.* Earthscan, London, 246 pp.

Tianhong L., Yanxin S., and An X. (2003). Integration of large scale fertilizing models with GIS using minimum unit. *Environmental Modelling*, **18**(3), 221–229.

Vaitheeswaran V. V. (2005). *Power to the People: How the Coming Energy Revolution Will Transform Industry, Change Our Lives, and Maybe Even Save the Planet.* Earthscan, London. 368 pp.

Watson R.T., Noble I.R., Bolin B., Ravindranath N.H., Verardo D.J, and Dokken D.J. (eds.) (2000). *Land Use, Land-use Change, and Forestry.* Cambridge University Press, Cambridge, U.K., 377 pp.

WEO (2006). *Energy Market Reform; Energy Policy; Energy Projections.* IEA, London, U.K., 600 pp.

Wijen F., Zoeteman K., and Pieters J. (eds) (2005). *A Handbook of Globalization and Environmental Policy: National Government Interventions in a Global Arena.* Edward Elgar, London, 768 pp.

Yue T.X., Fan Z.M., and Liu J.Y. (2005). Changes of major terrestrial ecosystems in China since 1960. *Global and Planetary Change*, **48**(4), 287–302.

18

"Sustainability—no hope!" or "Sustainability—no hope?"

Arthur P. Cracknell

18.1 INTRODUCTION, DEFINING SUSTAINABILITY

Unlike most discussions of sustainability (including Chapter 17 of this book) this chapter is concerned with the long-term future of humanity, let us say in 500 or 1,000 years time when, almost certainly, the main fossil fuels are likely to be exhausted. Uranium, which of course is not a fossil fuel but which is nevertheless a non-renewable fuel, may or may not be exhausted by that time, but eventually it too will be exhausted. The title of this chapter is actually two alternative titles and it proved impossible to decide between them because they represent totally different views, namely those of the pessimist and those of the optimist. The significance of the difference lies in the exclamation mark, which indicates the view of the school of the pessimists, who hold that our present way of life is definitely unsustainable and will come to some sort of "sticky end", and in the question mark which indicates the view of the school of the optimists, who hold that maybe our present way or life, or some scaled-down version of it, might be possible. In terms of the title of this chapter the pessimist sees that that in, say, 500 years time our present lifestyle will have vanished, while the optimist, choosing the second alternative title of this chapter "Sustainability—no hope?", believes that some sort of tolerable, if not luxurious, lifestyle could be possible for our descendants in 500 years time.

Let us begin by considering the definition of sustainability. We begin, not by surveying the very extensive literature on "sustainability" but by going back to the simple definition to be found in the *Oxford Dictionary* from which we extract the following:

> "Sustainable (adjective): 1: able to be sustained. 2: (of industry, development or agriculture) avoiding depletion of natural resources."

and

> "Sustain (verb) ... keep (something) going over time or continuously."

Let us consider the words "over time" of the definition quoted above. How long is the time period we should consider? Harold Wilson, a mid-20th-century British prime minister, once said that a week was a long time in politics. Most elected politicians cannot see any farther than their next election after, say, four or five years. The rest of us think in terms of the remainder of our lifetimes, our children's lifetimes, and possibly our grandchildren's lifetimes; say a few decades and probably less than a century. On these timescales the fuel minerals are not (all) going to run out nor are the important non-fuel minerals going to run out either. But in this chapter we propose to consider a longer timescale and one on which these fuel and non-fuel minerals will be exhausted. We cannot estimate precisely when that will be; it could be 500 years from now or 1,000 years or even 2,000 years from now. We can use 500 years as a working figure.

There are three useful principles of sustainability as it relates to resources and pollutants due to Daly (1990):

1. For a *renewable resource* the sustainable rate of consumption/use can be no greater than the rate of regeneration of the resource.
2. For a *non-renewable resource* the sustainable rate of use can be no greater than the rate at which a renewable resource, used sustainably, can be substituted for it.
3. For a *pollutant* the sustainable rate of emission can be no greater than the rate at which the pollutant can be recycled, absorbed, or rendered harmless in its sink.

Another useful definition is that of the ecological footprint of humanity. As defined by Wackernagel *et al.* (2002) this is

> "the area of biologically productive land and water requried to produce the resources consumed and to assimilate the wastes generated by humanity, under the predominant management and production practices in any given year."

According to Wackernagel *et al.* this corresponded to 70% of the capacity of the global biosphere in 1961 but had grown to 120% (i.e., overshoot) in 1999.

If we look at our way of life, we are consuming fuel resources at an annual rate of 4.8 billion tonnes of coal, 3.4 billion tonnes of oil, and 110 million tonnes of gas (these figures are for 2002, see Table 18.1). There is no stretch of the imagination by which we can regard as sustainable this use of energy which was stored up from the energy of the Sun in these minerals over hundreds of millions of years and is being consumed over a mere few centuries. It will "over time" (see the above definition) be completely used up. In discussing fuel resources we should perhaps include uranium, as being both a mineral and a (non-replaceable) nuclear fuel (see Table 18.1). Table 18.1 also includes estimates of the world's proven reserves of coal, oil, gas, and uranium. In terms of Daly's second principle we are (slowly) attempting to achieve sustainability

Table 18.1. Annual consumption and proven reserves of fuel minerals.

	Consumption in 2002 (10^6 tonnes)	*Proven reserves (2002)* (10^6 tonnes)
Coal	4,800	909,000
Oil	3,400	148,000
Gas	110	
Uranium	0.036	3.2[1] 9.8[2]

Notes:
[1] At a price of up to US\$130 kg^{-1}.
[2] Estimated additional resources at US\$130 kg^{-1}.
Source: World Energy Council's website *http://www.worldenergy.org/wec-geis/publications/reports/foreword.asp* (accessed May 21, 2007).

but we are a very long way from achieving that target. We should perhaps not include as fuel all the oil that we consume, since some of it is used as feedstock for the petrohemicals industry; indeed, there are those who would argue that oil is too valuable to burn as a fuel for electrical power generation and that we should keep it for special uses, primarily as feedstock for the petrochemicals industry and for transportation, for which it is not easy to find a substitute.

To the question of fossil fuel energy resources we should also add some consideration of our consumption of non-fuel minerals, mainly metals or their ores, see Table 18.2 where figures are given for our current consumption rates and the estimated world reserves of aluminum, cobalt, gold, iron, nickel, silver, and tin. These minerals were formed long ago and once they are gone then, to state the obvious, they are gone for ever. Of course recycling can help but there comes a point beyond which a material is so scattered that recycling becomes unfeasible. In terms of Daly's second principle, we are a very long way from achieving sustainability by replacing metals by renewable resources and it is very difficult to envisage how we could ever completely satisfy this principle.

There is a vast literature on what is described as sustainability, but it is mostly what it would be fairer to describe as "reducing unsustainability"; it does not take as its starting point the position that we have just described (i.e., that our present lifestyle is unsustainable in the long term, and therefore how much of it can be salvaged once the mineral resources, both fuel and non-fuel, have run out?). We should not denigrate all the work that has been done on reducing unsustainablity, increasing efficiency, and avoiding waste; this is important and postpones the "evil day".

We can put this in the context of the late Academician Kirill Kondratyev's later writings on global warming. Over the last 15 or so years of his life, Kondratyev was concerned with climate change as well as the many various aspects of ecology, all in relation to the sustainability of life, especially human life, on Earth. Before that Kondratyev had pioneered research in a number of fields related to these problems, namely atmospheric physics, satellite meteorology, climatology, and global change. It is now widely accepted that global warming, due to anthropogenic greenhouse gas emissions, represents a threat to the sustainability of human life on Earth. However, as Kondratyev was at pains to point out (see Kondratyev *et al.*, 2004 and many other references too), there are many other threats that are, potentially, just as serious; these include atmospheric pollution, ozone depletion, water pollution, the degradation and pollution of agricultural land, deforestation, depletion of the world's mineral resources, and population growth.

Over the last nearly 20 years of his life Kondratyev had stood out against the conventional wisdom adopted by many climatologists and politicians, as embodied in the IPCC (the Intergovernmental Panel on Climate Change). The IPCC concentrated the resources of hundreds of climatologists on the question of anthropogenically produced greenhouse gases and their consequences in terms of global warming. Undoubtedly the achievement of the IPCC is that now most sensible people accept that human activities do lead to global warming and that it is occurring at an increasing rate (see, for instance, Stern, 2007). But the downside is that other threats to the existence of life and our standard of living have been largely ignored. Kondratyev stood out against that and argued that the various forms of pollution, degradation and consumption of the world's mineral resources, and population growth are all part of global change and pose a very serious combined threat to the future of (human) life on Earth.

In terms of widespread influence, a milestone in making people aware that development or economic growth cannot continue for ever came in 1972 with the publication of *Limits to Growth* (Meadows *et al.*, 1972). This was followed 20 years later by further developments in *Beyond the Limits* (Meadows *et al.*, 1992) and after a further 10 years or so by a 30-year update (Meadows *et al.* 2005). Meadows *et al.* consider physical resources (i.e., energy and raw materials), population, economic theory, and social policy. They argue that unrestrained growth is leading the Earth towards ecological overshoot and impending disaster. The success of the work of Meadows *et al.* stems from the fact that not only do they consider the factors and mechanisms of the threats associated with "development", but they also put the various factors into computer models that enable them to vary the parameters and examine the consequences of changing the values of these parameters. Meadows *et al.* consider the driving force and especially the difference between exponential growth and linear growth, which to a considerable extent underlies the classic work of Malthus on population (see Section 18.9). Meadows *et al.* then consider the limits to growth and study the twin problems of (a) reaching a sustainable way of life and (b) reaching a fair distribution of resources to provide a common standard of living for everyone on the planet. This common level in the standard of living is almost certainly considerably below that of the present "advanced" industrial societies. According to the publicity material for the 30-year update (i.e., Meadows *et al.*, 2005), the first

> "book went on to sell millions of copies and ignited a firestorm of controversy that burns hotter than ever in these days of soaring oil prices, wars for resources and human-induced climatic change. This substantially revised, expanded and updated edition ... marshalling a vast array of new, hard data and more powerful computer modelling, and incorporating the latest thinking on sustainability, ecological footprinting and limits, presents future overshoot scenarios and makes an even more urgent case for a rapid readjustment of the global economy towards a sustainable path."

People who study these problems, including the several authors of Meadows *et al.* (1972, 1992, 2005) are divided into two camps: the optimists and the pessimists. The optimists come in two general classes. There are what we might call the "woolly" optimists, who are usually economists and the like, who suppose that someone or something will make it turn out alright for us in the end: *Deus providebit* (God will provide), Gaia, or Technology will come to our rescue, etc. They believe that human beings can make the transition to a fair and just society at a common standard of living in a peaceful and ordered manner. The others are the cautious optimists, usually scientists rather than economists, who acknowledge that there is a problem and realize that it needs political and social willpower to implement the technology necessary to solve it (e.g., Monbiot, 2006; Stern, 2007; Walker and King, 2008). The pessimists, who are usually physical or environmental scientists, fear that as resources become more scarce we shall drift into conflict, mega-deaths, and a spectacular degradation of our lifestyle. On a geological timescale (i.e., in terms of millions of years), humanity is doomed to extinction. But on the timescale we are considering just now, say 500–2,000 years, complete extinction seems unlikely. However, it seems highly likely that many parts of our present civilization will pass away; and we can assume that those living sophisticated urban lives will be more likely to perish and that those living in simple conditions much closer to nature will be more likely to survive. For a highly pessimistic view of just the next 100 years one could read chapter 10 of the book *Peak Everything: Waking up to the Century of Decline in Earth's Resources* by Richard Heinberg (2007). This chapter is a fictitious letter written in 2107 by someone who was born in 2007 and who has seen the complete collapse of industrial civilization as we know it; it is well worth reading. The fictitious writer says that while "attempting to pursue the career of a historian" circumstances forced him to "learn and practice the skills of farmer, forager, guerilla fighter, engineer ..." He describes the energy crisis:

> "Folks then thought it would be brief, that it was just a political or technical problem, that soon everything would get back to normal. They didn't stop to think that 'normal', in the longer-term historical sense, meant living on the energy budget of incoming sunlight and the vegetative growth of the biosphere. Perversely, they thought 'normal' meant using fossil energy like there was no tomorrow ..."

He describes how energy shortages led to economic recession and endless depression, the collapse of currencies, inflation, deflation, the return of barter. "We went from global casino to village flea market." Manufacturing collapsed, transportation collapsed. Supermarkets were empty. People scavenged through all our landfill sites "looking for anything that could be useful". He castigates us for taking "billions of tons of invaluable, ancient, basic resources and turn(ing) them into mountains of stinking garbage, with almost no measurable period of practical use in between!" There were purges, wars ("the generals managed to kill a few million popple ... it could have been tens or hundreds of millions, even billions ..."), epidemics and famine. And so on.

Another serious pessimist is Lord (Martin) Rees, whose main arguments are (a) that in many fields of scientific and technological research, notably biology and particle physics, we are now tinkering with deep fundamentals that are not completely understood and (b) in some of the new technologies there is the opportunity for small disaffected groups or individuals to cause enormous damage (Rees, 2004). An example of the first was provided by the fear in 1945 that a nuclear explosion might trigger an enormously destructive chain reaction beyond the initial explosion (actually, as we now know, it did not do that) and an example of the second is provided by in the destruction of the World Trade Center in New York on 11 September 2001.

18.2 GLOBAL WARMING

It is worthwhile giving some consideration to global warming since this is, for many people, the starting point of worries about the future of life on Earth. A few years ago there was a political lobby, especially backed by the big oil companies in the U.S.A., that sought to discredit the science of the idea of human-induced global warming. By now things seem to have moved on and now almost everyone accepts that the climate is changing as a result of human activities, principally the release of greenhouse gases (CO_2, etc.) by the burning of fossil fuels (coal, oil, and gas). So we shall consider briefly the scientific evidence for global warming and discuss the consequences in terms of the melting of glaciers and of the ice in the polar regions, the rise of sea level, and so on. Over millions of years of the Earth's existence the climate has varied, due of course to natural events. Recently, however, (on a geological timescale) what is new is that the human population has expanded and human activities have expanded in such a manner that the climate is now influenced by various human activities as well.

Until around 10,000 years ago, or maybe slightly longer ago, human beings across the globe lived in small groups, subsisting as "hunter-gatherers" over a huge range of environments, from tropical Africa to the polar regions. What happened 10,000 years or so ago was the development of settled communities and settled food production by the development of agriculture in terms of (deforestation and) planting crops and keeping livestock animals. Human beings have been making more and more changes to the surface of the Earth and coming more and more to affect the climate ever since. Human activities have escalated in recent years, due to population expansion and industrialization (which are not un-connected to one another). The problem now is that we have both natural events still affecting the climate and we have human activities that are also affecting the climate, and it is very difficult to separate the two of them.

18.2.1 Climate change

Let us consider the causes of climate change (see, for example, Cracknell, 1994). It is convenient to consider three categories of events that affect the climate: (a) events that occur outside the Earth, (b) natural events on the surface of the Earth, and (c) human activities. It is then convenient to distinguish three separate components to the events that occur outside the Earth: these are (a) variations in the intensity of the radiation emitted by the Sun, (b) changes in the transmission properties of space between the Sun and the Earth, and (c) changes in the Sun–Earth distance. Changes in the Sun-Earth distance arise, in turn, from three causes: namely, variations in the eccentricity of the Earth's orbit (with a period of around 100,000 years), oscillation of the tilt of the Earth's axis (with a period of about 40,000 years) and precession of the equinoxes (with a period of about 22,000 years). These changes cause a (slight) variation in the intensity of the solar energy arriving at the Earth, and it has long been known that there is a good (anti-)correlation between the intensity of sunlight reaching the Earth and the volume of ice in the polar regions (i.e., with the occurrence of ice ages at intervals of around 100,000 years). The first suggestion that ice ages were related to the Earth's orbit around the Sun appears to have been made by Joseph Adhémar, a mathematics teacher in Paris, in 1842; he concentrated on the 22,000-year period. The theory was extended to include changes in the eccentricity of the Earth's orbit by James Croll, the son of a Scottish crofter, who had very little formal education. He stumbled on this idea and spent his spare time in the 1860s and 1870s working on the idea; he estimated that the last ice age ended about 80,000 years ago. There was some interest in Croll's theory at the time; however, because he was of low birth and not part of the fashionable circles of the day and because it became apparent that the last ice age ended only about 10,000 years ago rather than 80,000 years ago, his ideas were largely forgotten by the end of the 19th century. The cycles due to orbital changes are now known as Milankovitch cycles or wobbles, after Milutin Milankovitch, a Serbian mathematician who revived and extended Croll's ideas in the early 20th century (Cox, 2005; Pearce, 2006a).

We now turn to natural events on the surface of the Earth. These include (a) plate tectonics, (b) expansion and contraction of the polar ice caps, (c) volcanic eruptions, and (d) ocean circulation. The first three of these are important because they affect the balance between, on the one hand, the energy reaching the Earth's surface (plus the very small amount of energy rising up from within the Earth's interior) and, on the other hand, the energy leaving the Earth and passing into outer space. In the case of ocean circulation the major ocean currents transport heat around the Earth's surface.

As to human activities that affect the climate, there are several. The one of which we are most generally aware in terms of its effect on the global environment is the generation of carbon dioxide by the burning of fossil fuels, coal, oil, and gas. The concentration of CO_2 in the atmosphere has risen from about 280 ppm at the time of the Industrial Revolution to about 385 ppm at present. The growth in CO_2 concentration is accelerated by other factors too, principally by deforestation, which removes trees which convert CO_2 into oxygen, and also by the manufacture of cement. The main effect of increasing the concentration of CO_2 is to change the balance between incoming radiation from the Sun and outgoing radiation from the Earth. CO_2, along with various other gases (principally water vapor, but also a number of other gases, see Section 2.4), acts like the glass in a greenhouse allowing in short-wavelength radiation (visible and near-infrared) but blocking outgoing longer wavelength radiation (emitted thermal infrared). Basically, the greenhouse effect is good, it means the average temperature of the Earth's surface is about 288 K (15°C) instead of about 255 K (−18°C) which it would be without the Earth's atmosphere. But adding more CO_2 will increase the temperature.

The second most well-known effect of human activity after CO_2 production is depletion of the ozone layer. Basically ozone is created by the action of sunlight (ultraviolet radiation) on oxygen molecules, O_2, and it is destroyed by oxides of nitrogen, referred to generally as NO_x and by the decomposition products of CFCs (chlorofluorocarbons, principally CCl_3F and CCl_2F_2). The importance of stratospheric ozone is that it absorbs incoming UV which would be harmful to human beings (and other lifeforms). There is a general depletion of ozone all over the Earth and a spectacular decrease in Antarctica in the spring, known as the ozone hole. The ozone hole was discovered in the early 1980s by scientists at the British Antarctic Survey's base in Halley Bay in Antarctica; its existence has subsequently been confirmed by data from Earth-observing (remote-sensing) satellites and it is now being monitored continuously by satellites.

The replacement of tropical rainforest, or other forest, by agricultural land and the degradation and erosion of good land that becomes semi-desert or desert areas affects the weather and the climate. So also does irrigation, the damming of rivers to form reservoirs for hydroelectric schemes, and the development of urban land with huge surface areas of concrete and tarmac. This is because all these things change the balance of the exchange of heat and moisture between the land and the atmosphere.

As already mentioned, the greenhouse effect in itself is good; what is bad is the enhanced greenhouse effect (i.e., the *extra* contribution to the greenhouse effect arising from the production of greenhouse gases by human activities. A lot of effort has gone into estimation of the effect of these human activities, and this is done by climate modeling. The difficulty is to separate natural variations in the climate and the influences of human activities on the climate. This is done by using what are known as general circulation models (GCMs) of the atmosphere, or strictly speaking GCMs of the atmosphere and the oceans. Different versions of general circulation models are used for numerical weather forecasting, on the one hand, and for climate modeling, on the other hand (see also Section 2.3). What is done is to construct a grid of points, perhaps with spacing of 1° of latitude and 1° of longitude and, say, about 20 levels in the vertical and to specify the appropriate atmospheric parameters at each of these points. The parameters involved include pressure, temperature, motion (wind speed), composition (water vapor, cloud, CO_2, trace gases, etc.). A more pictorial

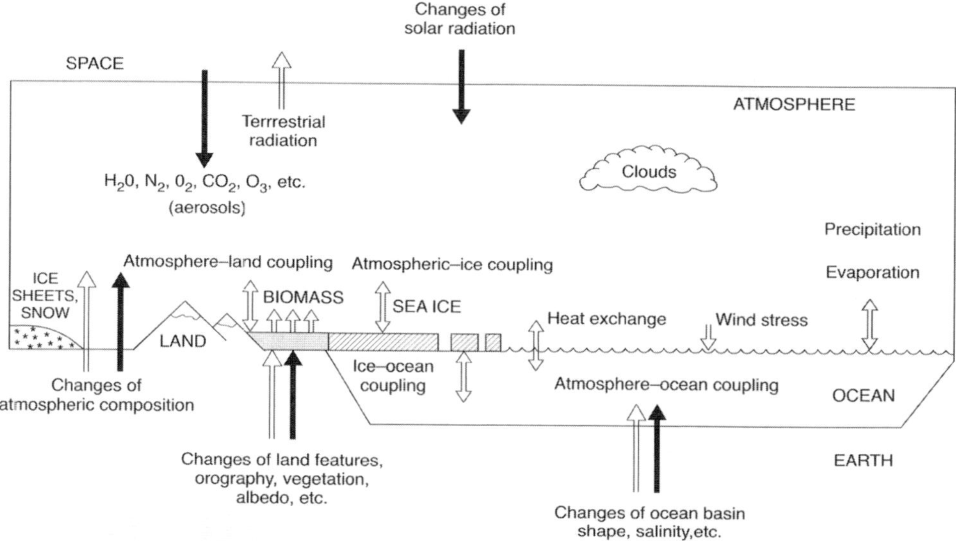

Figure 18.1. Diagrammatic representation of the climate system (from Houghton, 1984).

representation of what is involved is shown in Figure 18.1. One then writes down the equations that describe or follow from the physics of these parameters. This produces a (complicated) set of simultaneous equations, which are in fact non-linear integro-differential equations. We also suppose that meteorological measurements will provide us with the starting values of all of these parameters, and then these equations have to be solved to give the values of these parameters at subsequent times. This, of course, is a massive computing problem and only organizations with very powerful computing facilities can contemplate attempting to do this. Since we cannot separate human-induced effects from natural effects, what is commonly done is to take the present climatic conditions, run the model for, say, the equivalent of 100 years assuming the present concentration of greenhouse gases in the atmosphere. Then the model is run a second time assuming the concentration of the greenhouse gases to be doubled (this is actually using the carbon dioxide equivalent of all the greenhouse gases). Doubling on that timescale is not an unreasonable estimate of what is likely to happen unless we (i.e., all of us collectively) do something rather drastic in terms of reducing our production of greenhouse gases. Then, by subtracting the results of one model from those of the other one can determine the effect of the extra greenhouse gases. There are differences between the results from different models used in different laboratories around the world, but the general conclusion is that if we go on allowing the concentration of greenhouse gases to increase at much the present rate then after a hundred years the mean temperature of the surface of the Earth will be somewhere in the region of 2°C or 3°C or 4°C higher than it is now. This may not sound very serious. However, one should recall that at the height of the last ice age the average temperature was only about 4.5°C lower than it is now. And we are talking now about something of the same sort of magnitude, admittedly a rise rather than a fall. The consequences could be quite dramatic.

18.3 THE IPCC

In Section 2.3 we mentioned briefly the award of the 2007 Nobel Peace Prize to the IPCC and Al Gore. There is now widespread acceptance of the idea that the climate is changing as a result of human actions. A lot of the credit for this education of the public must go to the IPCC for its relentless pursuance of the scientific evidence, digesting it, and presenting it, without overstating its case, and to Al Gore for his writings (Gore, 1992, 2006) and other presentations on the subject. But there is a downside to this undoubted success. This is that it leads people to think that the emission of greenhouse gases and the consequent global warming is the major threat, or even the only threat, to the continuation of human life, or at least of our highly sophisticated society, on the planet. But this is not the case (as we pointed out in Section 2.3); one person responsible for laboring the point that this is not the case was the late and very great Russian scientist Academician Kirill Kondratyev. Apart from releasing greenhouse gases into the atmosphere we are polluting the atmosphere in other ways, we are polluting the water supplies, we are degrading the soil, we are destroying the forests and other natural habitats, and we are causing many species of animals, birds, and plants to become extinct. We are creating ecological disasters in many ways. We are consuming non-renewable resources: fossil fuels (oil, gas, and coal) and various non-fuel minerals, mostly metals and metal ores. Kondratyev wrote quite a lot about all this, but we shall also refer later on to some other more popular writings of Jared Diamond about his studies of ancient civilizations, some of which collapsed and some of which survived. We face problems similar to the problems that our ancestors faced and we face some additional new problems as well. Maybe we can learn from history.

18.4 THE CONSEQUENCES OF CLIMATE CHANGE

18.4.1 The consequences of climate change based on IPCC predictions

What comes out of climate model calculations, on which the IPCC has placed such importance, is that the likely climate change will vary a great deal from place to place. Nevertheless, a number of general conclusions can be drawn. Quite a good source for the discussion of the consequences of global warming is the book *Global Warming: The Complete Briefing* by Sir John Houghton (1997), while a particularly detailed study of the economic consequences of global warming is given in the *Stern Review* (Stern, 2007). The main questions are (a) how much will sea level rise and what effect will it have? (b) How will water resources be affected? (c) What will be the impact on agriculture and the food supply? (d) Will natural ecosystems suffer change? (e) How will human health be affected? We shall consider these briefly in turn, taking the rise in sea level first.

During the warm period before the last ice age, about 120,000 years ago, the global mean temperature was a little higher (perhaps 2°C) than at present. Average sea level was about 5 m or 6 m higher than it is now. At the height of the last ice age, about 18,000 years ago, mean sea level was over 100 m lower than now. The rise in sea level over the last century has been about 10 cm. The sea level rises because ice melts and also because if temperature rises the water in the oceans expands. Estimates of the anticipated rise in mean sea level over the next 100 years, based on several different models, range from about 30 cm to about 120 cm (1.2 m). The main contributions to sea level rise come from (a) the melting of the Antarctic ice sheet, (b) the melting of the Greenland ice sheet, (c) the melting of mountain ice caps and glaciers, and (d) the thermal expansion of seawater due to the rise in temperature. Thermal expansion is the largest single contributing factor. We could be looking at rises of about 12 cm by 2030 and 50 cm by 2100. These may not sound very much but to areas that are close to sea level (e.g., the Netherlands or Bangladesh), they can be very significant. However, it is not just a simple figure like 30 cm or 50 cm that is important. As well as sea level rise, there is subsidence of the land, often made worse by the extraction of ground water. And this gives predicted sea level rises of around 1 m (30 cm due to global warming and 70 cm due to subsidence) by 2050 and nearly 2 m (70 cm due to global warming and 1.2 m due subsidence) by 2100. In the Netherlands, over the centuries, people have built huge dykes (walls) all along the coastline and the water is pumped out, which is one of the reasons the Netherlands was a country of windmills. But such a system of sea defences is impractical for Bangladesh. The most obvious effect will be the loss of good agricultural land; in a country where 85% of the people depend on agriculture for their livelihood, many are at the very edge of subsistence. There are other serious problems too such as storm surges and saline intrusion. The storm surge of November 1970 in Bangladesh caused the loss of around a quarter of a million lives and that of April 1991 caused the loss of about another 100,000 lives. Finally, because the oceans take centuries to adjust to a change in surface temperature, then even if the concentrations of greenhouse gases were stablilized now, so that, later on, anthropogenic-induced climate change was halted, sea level rise would continue for many centuries as the oceans adjust to the new climate.

The global water cycle will also be affected, though in different ways in different places. Studies of regional climate models, rather than global models, therefore are important in relation to future water resources. Desertification is an ongoing process; there is already a progressive loss of good agricultural land and this is likely to continue. Drylands (defined as those areas where precipitation is low and where rainfall typically consists of small, erratic, short, high-intensity storms) cover about 40% of the total land area and support one-fifth of the World's population. Desertification in these drylands is due to the degradation of land because of decreased vegetation, reduction of available water, reduction of crop yields, and erosion of soil. It results from excessive land use generally because of increased population, increased human needs, or political or economic pressures (e.g., the need to grow cash crops to raise foreign currency). It is often triggered or intensified by a naturally occurring drought. The current rate of desertification is estimated to be about 60,000 km^2 per year or 0.1% of the total area of drylands. It is a potential threat to 70% of these drylands (i.e., to 25% of the world's land area; Houghton, 1997).

We turn to the consideration of agriculture, food supply, and forest resources. Temperature and rainfall are key factors in making decisions about what crops to grow. Thus, agriculture will need to adapt to changes in the climate. This may not be a serious problem for crops that mature over one or two years (or less). But it may be a problem for trees with periods of maturing of decades or even of centuries. Houghton cites the case of farmers in Peru adjusting their choice of crops to plant based on forecasts of the presence or absence of the El Niño phenomenon that year. There is talk of carbon dioxide fertilization (i.e., increased carbon dioxide concentration) leading to increased photosynthesis; but the evidence seems to suggest that the effect is very small indeed. There is no simple answer to "what is the effect of climate change on agriculture?" because the answer is that agricultural success or failure depends to a considerable extent on human decisions, human management operating within the natural environment. Changing climate is one factor that will affect these decisions. With regard to human health the direct effect of global warming can be handled relatively easily. On the other hand, deaths due to stress in times of extremely high temperature or to hypothermia in times of extremely low temperature (since the amplitude of fluctuations about the mean seems set to increase) are likely to increase.

18.4.2 Paleoclimatic information; catastrophic changes

There are quite a few laboratories around the world which are involved in running their own climate models on large and powerful computers, and inevitably they do not all produce exactly the same results. Working Group 1 of the IPCC, as we indicated in Section 2.3, attempted to take into account the results of all these models and to make the best estimates possible of the anticipated rise in temperature, and the changes in precipitation and in soil moisture for various different scenarios of future greenhouse gas emissions. Kondratyev was critical of the way these models took into account atmospheric aerosols and cloud screening. Paleoclimatic information is an important source of data for comparative analysis of the present climate and the paleo-climate. Analysis of the data of paleoclimatic observations reveals large-scale abrupt climate changes taking place in the past in conditions when the climate system had exceeded certain threshold levels.

What the IPCC does and what we have just described in Section 18.4.1 is based on the idea of slow gradual responses in the climate. However, one of the problems with the assumption of gradual change and the use of computer models to predict the future climate is the inability to predict sudden changes. Take, for example, the melting of ice (Gore, 2006; Pearce, 2006a). It is commonly assumed that a glacier or an ice shelf just melts from absorbing radiation at its surface (i.e., rather slowly). However, in reality, cracks develop in the ice, meltwater pours into the cracks, and the whole melting process accelerates and the water pressure in the cracks acts like wedges and forces the ice to break up. Spectacular situations occur like the break-up of the Larsen B ice shelf in Antarctica in early 2002. The Larsen B ice shelf was about 150 miles (270 km) long and 30 miles (54 km) wide, and it suddenly broke up and floated away in fragments over a period of about one month and released around 500 billion tonnes of ice into the ocean. Following the melting of the ice, the water surface uncovered has a much lower albedo (reflectivity) than the ice; it therefore absorbs more heat and provides a positive feedback mechanism that enhances global warming. A second example of positive feedback is associated with drought. The withering or death of plants causes a decrease in evapotranspiration and hence attenuates precipitation which further increases drought.

In the conclusion to his book *The Last Generation: How Nature Will Take Her Revenge for Climate Change*, Pearce (2006a) says that he "called this book 'The Last Generation', not because I believe we humans are about to become extinct, but because we are in all probability the last generation that can rely on anything close to a stable climate in which to conduct our affairs." What Pearce is saying is that people have been overlooking positive feedback mechanisms, of which we have just mentioned two examples, and this positive feedback can lead to sudden precipitous swings in the climate. While the models, and the IPCC's general approach, can handle catastrophic change *after* the event by making adjustments to the parameters in the models, they cannot *predict* such sudden events. With so much emphasis having been placed on the climatic implications of the growth of greenhouse gas concentrations in the atmosphere, less effort has been made to study possible sudden climate change of natural origin and intensified by anthropogenic forcings. Pearce claims that there are many instances of sudden swings in the past, and therefore presumably this can happen in the future too. He argues that whereas the IPCC is talking about a rise in global mean surface temperature of 3°C–4°C in 100 years it could be far worse than that: it could be 10°C which would take us way beyond the changes that we have outlined in Section 18.4.1. The most important aspect of these problems are the potential effects of abrupt climatic change on ecology and economy since past estimates were generally based, as a rule, on the assumption of slow and gradual change. Pearce's general thesis is that nature often flips suddenly from one state to another and that therefore the consequences of climate change may be quite different from and much more serious than the rather simple kind of gradual changes outlined, for example, in Section 18.4.1.

Apart from possible human-induced rises that may very much exceed the 3°C–4°C rise in the next 100 years predicted by the climate models, there is the possibility of a substantial decrease in temperature that would correspond, as a result of natural causes, to a return to a new ice age. Until recently, and in the absence of any evidence to the contrary, geologists and paleoclimatologists had assumed that climate changes in the past had always been slow, and therefore in relation to human-induced global warming the opposite effect, a natural cooling leading to a new ice age, has been assumed to be on a much longer timescale than a century or so. However, it has recently emerged that rather than slow and gradual transitions between ice ages and interglacial periods there have been many abrupt changes in the climate. General discussions will be found, for instance, in the books by Cox (2005) and Pearce

(2006a). What Cox is concerned with is the recent evidence that has been found of sudden—rather than gradual—changes in the climate in the past. The terms sudden and gradual need to be clarified a little; in the framework of geological time a sudden transition from a warm interglacial period such as the present to a full-blown ice age in a period of 100 years would be regarded as sudden, whereas gradual would imply a period of 1,000 years or several thousand years. In *Climate Crash: Abrupt Climate Change and What It Means for Our Future*, Cox (2005) examines the records of past climate changes. Evidence of past climate variations have been obtained from ice cores drilled from the Greenland ice sheet, supported by some other ice cores from elsewhere, and from ocean sediment cores from various parts of the world. The study of the ice cores and marine sediment cores involved the development of special drilling techniques and the use of highly sensitive techniques for analysis of the cores. The ratio of the ^{18}O isotope to ^{16}O, the common isotope, in the ice was found to be very sensitive to the temperature of the snow, and therefore of the atmosphere, when the snow was precipitated thousands or tens of thousands of years ago. In marine sediments the ratio of ^{18}O to ^{16}O in the $CaCO_3$ derived from the shells of foraminifera was found to be sensitive to the temperature of the seawater when the material was formed. Ice cores from Greenland have provided records for more than the last 100,000 years, and marine sediments give even longer records of several hundred thousand years and including several past ice ages and intervening interglacial periods.

Considerable abrupt changes in the regional climate in the last ice age have been detected from paleoclimatic reconstructions that manifested themselves as changes in the frequency of occurrence of hurricanes, floods, and especially droughts. Evidence of more than 20 oscillations, known as Dansgaard–Oeschger oscillations, has been observed in the Greenland ice core record of the last ice age (between 110,000 and 23,000 years before the present). In each of these Dansgaard–Oeschger oscillations there was a sudden sharp rise in temperature of between 2°C and 10°C over a period of a decade or so and this was followed by a slow cooling over several centuries, on average about 1,500 years. These changes in climate appear to have occurred suddenly in the past, over a few years or perhaps a decade or two and certainly on a different timescale from the 3° or 4° per century predicted by climate models. While the evidence for abrupt changes is quite clear, the mechanisms driving these changes is less clear and is still the subject of very active research. Even if the causes of these changes were known it seems unlikely that computer models would ever predict sudden changes.

18.5 THE COST OF GLOBAL WARMING

Attempts have been made to estimate the cost of global warming and the first impression is that we might be able to buy our way out of the problem. However, there are two factors that are important. "We have only tended to look at the next 50 years or so and with the hope that greenhouse gases emissions will not go too far out of control." The effects are likely to be very serious for many people living already at subsistence level in developing countries. If as a result of desertification or sea level rise their land becomes uninhabitable they will wish to migrate and will become environmental refugees. Houghton (1997) quotes numbers of 3 million a year, or 150 million between now and 2050. There are two other sources that are worth mentioning briefly here. One is the 2007 BBC Reith Lectures by Prof. Jeffrey Sachs and the other is the so-called *Stern Review* (Stern, 2007).

In the 2007 Reith Lectures, according to Sue Lawley's introduction to the first one, Jeffrey Sachs explains

> "how he believes that with global co-operation our resources can be harnessed to create a more equal and harmonious world. If we cannot achieve this . . . we will face catastrophe; we'll simply be overwhelmed by disease, hunger, pollution, and the clash of civilisations."

The title of the whole series of five lectures was "Bursting at the Seams" and the individual lectures were "Lecture 1: Bursting at the Seams; Lecture 2: Survival in the Anthropocene; Lecture 3: The Great Convergence; Lecture 4: Economic Solidarity for a Crowded Planet; Lecture 5: Global Politics in a Complex Age". The term "Anthropocene" is a term which was coined in 2000 by Paul Crutzen to describe the last two centuries of the history of the Earth; that is, the (very brief) geological era in which a single species, *Homo sapiens*, "is in charge of the planet, altering its features almost at will" (Pearce, 2006a). The discussions in these Reith Lectures were not limited to the question of sustainability but covered a wide range of social and political questions; indeed, scientific aspects were not considered in depth at all. The lectures seem to convey an incredible optimism that many people would find difficult to share; this is on two fronts. First, Sachs assumes that human beings will cooperate for the general good, although there is plenty of evidence from history that does not support such optimism. Second, he has probably not looked quantitatively enough at the details of what is involved in replacing fossil fuels as our main source of energy or of what is involved in sustainability in relation to non-fuel mineral resources, principally metal ores and oil as the feedstock for the petrochemicals industry.

The *Stern Review* was commissioned by Gordon Brown, then the Chancellor of the Exchequer in the U.K., in July 2005 to report to the Prime Minister and the Chancellor of the Exchequer by Autumn 2006; the report was now been published (Stern, 2007). To quote from the summary of the *Review* itself:

> "There is now clear scientific evidence that emissions from economic activity, particularly the burning of fossil fuels for energy, are causing changes to the Earth's climate. A sound understanding of the economics of climate change is neeeded to order to underpin an effective global response to this challenge. *The Stern Review* is an accessible, independent, and comprehensive analysis of the economic aspects of this crucial issue. ... (it) considers all aspects of the issue, including the nature of the economics and the science; the impact of climate change on growth and development in both rich and poor countries; the economics of cutting emissions and stabilising greenhouse gas emissions in the

atmosphere; the components of policy on both mitigation and adaptation; and the challenges of achieving sustained international collective action. *The Review* will help to promote a greater understanding of the impact and effectiveness of national and international policies and arrangements in reducing emissions in a cost-effective way, and promoting a dynamic, equitable and sustainable global economy."

Although it is primarily a document for the UK, it does consider, to some extent, the global problem. The *Stern Review* runs to over 650 pages. It is impossible to summarize it here. It takes it as given that human activities are now causing global warming and argues that the sooner we take remedial action the better; the longer we leave it before tackling the problem the more drastic will be the steps required to correct the situation. The first half of the *Review* considers the evidence on the economic impacts of climate change itself and the economics of stabilizing greenhouse gas emissions. The second half looks at the policy response, which is outside the scope of the present book. To some extent the *Stern Review* is probably open to the same criticism that we have just made of Jeffrey Sachs' Reith Lectures, namely it underestimates how drastic a change we need in human activities to obtain a truly sustainable lifestyle after the fuel and non-fuel resources are (effectively) exhausted. Pearce (2006a) takes the view that the only people who are optimistic about the future are economists and that physical and environmental scientists tend to be pessimistic. The *Stern Review* is also open to the criticism that since, at present at least, it is not possible to predict abrupt changes in the climate, whether as a result of human activities or due to natural causes, the *Review* has been unable to take these into account; they could be far more serious than the steady changes predicted on the basis of climate models and the IPCC approach.

This brings us to the social and political problems. Science can only go so far. It can try to make sure that the best advice is available to people and governments. Beyond that it is out of the hands of the scientists.

18.6 "OUR" WAY OF LIFE

It is commonly held that global warming presents a threat to "our way of life". We should examine this question a little more closely. We often think of the world as being divided into three groups of countries, (a) developed or industrialized countries, (b) newly emerging or advanced developing countries, and (c) Third World countries at various stages of distance from advanced development. We know roughly what these categories mean: (a) means the U.S.A., Japan, Western Europe, Australia, New Zealand, Singapore if one reads Lee Kuan Yew's book *From Third World to First* (Lee, 2000) etc.; (b) includes a whole host of countries, Malaysia (with its aim to be developed by 2020), China, India, some South American countries like Brazil, Eastern European countries; and (c) includes most African countries. The divisions are blurred and it does not do to define things too rigidly. Those countries in (b) and (c) aim to achieve "developed" status as quickly as they can. A good indication of the level of "development" is the use of fossil fuels and the resulting production of carbon dioxide per head of population per annum:

U.S.A	20.0 tonnes
U.K.	9.5 tonnes
China	2.7 tonnes.

However, it can be argued that "our" present lifestyle is unsustainable. Everyone wants to achieve the condition of what one might describe as luxury or extravagant luxury of the countries in the developed world. One can regard the American lifestyle as not just luxury but as extravagant luxury which is unsustainable; it relies heavily on extremely cheap oil. If everyone in the world, all 6 billion or so of us, lived in the lifestyle of people in the U.S.A. then the planet would be wrecked very rapidly. A discussion of the economics and politics related to climate change written from the point of view of one very large deleloping country (India) is given by Toman *et al.* (2003).

As Diamond (2005) points out, it is not just the number of people on the planet but their impact on the environment which is important. Our numbers only cause problems insofar as each of us consumes resources and generates waste. On average, each citizen of the U.S., Western Europe, and Japan consumes 32 times more resources and puts out 32 times more waste, than do the inhabitants of the Third World. If we only maintained world population at its present level the average environmental footprint would increase because of economic development in various countries. People in other countries see films or watch TV about life in the developed countries, they see advertisements for First World consumer products sold in their countries and they observe First World visitors to their countries. Not unnaturally, they want to achieve the same lifestyle. One of the problems is how to achieve this without wrecking the planet:

> "... low-impact people are becoming high-impact people for two reasons: rises in living standards in Third World countries ...
>
> ... and immigration, both legal and illegal, of individual Third World inhabitants in the First World, driven by political, economic, and social problems at home. Immigration from low-impact countries is now the main contributor to the increasing populations of the U.S. and Europe. ...
>
> ... the biggest problem is the increase in total human impact, as the result of rising Third World living standards, and of Third World individuals moving to the First World and adopting First World living standards ..."

Diamond (2005) continues by pointing out that, in addition to their own aspirations, Third World countries are encouraged to follow the path of development

> "by First World and United Nations development agencies, which hold out to them the prospect of achieving their dream if they will only adopt the right policies, like balancing their national budgets, investing in education and

infrastructure, and so on.

But no one at the U.N. or in First World governments is willing to acknowledge the dream's impossibility: the unsustainability of a world in which the Third World's population were to reach and maintain current First World living standards. It is impossible for the First World to resolve that dilemma by blocking the Third World's efforts to catch up: South Korea, Malaysia, Singapore, Hong Kong, Taiwan, and Mauritius have already succeeded or are close to success; China and India are progressing rapidly by their own efforts; and the 15 rich Western European countries making up the European Union have just extended Union membership to 10 poorer countries of Eastern Europe, in effect thereby pledging to help those 10 countries to catch up. Even if the human populations of the Third World did not exist, it would be impossible for the First World alone to maintain its present course, because it is not in a steady state but is depleting its own resources as well as those imported from the Third World. At present, it is untenable politically for First World leaders to propose to their own citizens that they lower their living standards, as measured by lower resource consumption and waste production rates. What will happen when it finally dawns on all those people in the Third World that current First World standards are unreachable for them and that the First World refuses to abandon those standards for itself? Life is full of agonizing choices based on trade-offs, but that's the cruelest trade-off that we shall have to resolve: encouraging and helping all people to achieve a higher standard of living, without thereby undermining that standard through overstressing global resources."

18.7 THE END OF FOSSIL FUELS AND OTHER MINERALS

In the 1920s many geologists were warning that world oil supplies would be exhausted within a few years. But then the huge new discoveries which were made in various places, east Texas, the (Persian) Gulf, etc., made such predictions laughable. Each year more oil was being discovered than was being extracted. Many people assumed that this could go on for ever. But during the 1950s, 1960s, and 1970s there came a great geologist, Marion King Hubbert, and his predictions that the fossil fuel era would prove to be very brief. How did he make his predictions? The theory is simple, one starts with the known quantity of oil in the ground, takes the current extraction rate, predicts future extraction rates, and from then on it is just simple mathematics. But it is not that simple. First, we do not know how much oil there is in the ground with any reliable accuracy. Second, as time goes on it becomes progressively more difficult to extract the oil from any given well. Initially it may just gush to the surface, but later on it has to be pumped and then, later still, water has to be pumped in to push the oil out. So the extent to which a field is exploited becomes an economic/financial decision and not just a technical matter. So the future extraction rate is not easy to determine.

In 1956 Hubbert made the best estimates that he could for the U.S.A. and predicted that crude oil production in the U.S.A. would peak between 1966 and

1972; in the event it actually peaked in 1971. As an aside, we can note that the decline after 1971 led the U.S.A. to seek to assure supplies from overseas, something which lies behind a great deal of U.S. foreign policy—not to mention its wars—in recent decades. Hubbert then turned his attention to world oil supplies and predicted that the peak would come between 1990 and 2000. Current preditions are a little later, but it is very difficult to make accurate predictions because estimates of reserves of oil in the ground are notoriously "flexible"; oil companies and some governments increase or decrease their estimates for financial or political reasons without much, if any, geological evidence.

Hubbert appears to have believed that society, if it is to avoid chaos during the energy decline, must give up its antiquated, debt-and-interest monetary system and adopt a system of accounts based on matter–energy, an inherently ecological system that would acknowledge the finite nature of essential resources. Hubbert is quoted as saying that we are in a

> "crisis in the evolution of human society. It's unique to both human and geologic history. It has never happened before and it can't possibly happen again. You can only use oil once. You can only use metals once. Soon all the oil is going to be burned and all the metals mined and scattered."

We have, he believed, the necessary know-how. If society were to develop solar-energy technologies, reduce its population and its demands on resources, and develop a steady-state economy to replace the present one based on unending growth, our species' future could be rosy indeed. "We are not starting from zero," he emphasized, "we have an enormous amount of existing technical knowledge. It's just a matter of putting it all together. We still have great flexibility but our manoeuvrablility will diminish with time". His optimism has not been shared by everyone.

When it comes to calculating the lifetime of non-fuel mineral resources the principles involved in making the calculations are similar to those for fuel minerals. The difficulties are similar too, namely there are the problems of estimating the resources in the ground and in estimating future rates of consumption. Meadows *et al.* (2005) have made calculations for some important metals using what they call identified reserves and the resource base; the figures they used for the reserves are different from those we gave in Table 18.2 and they assume a growth rate of 2% per annum in consumption. Their results are shown in Table 18.3. While there are quite a few popular books on the question of the decline in oil resources (e.g., Heinberg, 2006; Leggett, 2005; Roberts, 2004) there is much less written about the decline in the sources of non-fuel minerals (Tanzer, 1980, which although a bit dated now, the general ideas are still sound).

We can consider application of the Daly principles (see section 18.1) to both fuel and non-fuel minerals. If we apply the second of these principles to the case of fuel resources then we see that the question is to what extent we can meet continued demands for energy by (a) economizing on the use of energy and (b) using renewable energy (very nearly all of it ultimately derived from the Sun) in place of non-renewable fuels. However, when it comes to non-fuel minerals (i.e., metal ores), it

Table 18.2. Annual production, reserves, and reserve base of some important non-fuel minerals.

	Production in 2005 (10^6 tonnes)	*Reserves* (10^6 tonnes)	*Reserve base* (10^6 tonnes)
Iron ore	1,500	160,000	370,000
Bauxite (Al ore)	169	25,000	32,000
Nickel	1.5	64	140
Cobalt	0.058	7	13
Tin	0.290	6.1	11
Silver	0.0193	0.270	0.570
Gold	0.0025	0.042	0.090

Note: As defined by the U.S. Geological Survey, the *reserves* are that part of the *reserve base* which could be economically extracted or produced at the time of determination; it does not signify that extraction facilities are actually in place and operative. The term *reserve base* refers to that part of an identified resource that meets specified minimum physical and chemical criteria related to current mining and production practices, including those for grade, quality, thickness, and depth. It includes reserves that are currently economic, reserves that are marginally economic, and reserves that are currently sub-economic.
Source: U.S. Geological Survey, Mineral Commodity Summaries, *http://minerals.usgs.gov/minerals/pubs/mcs/* (accessed May 21, 2007).

Table 18.3. Life expectancy of some non-fuel minerals.

Mineral	*Life expectancy of identified reserves* (years)	*Life expectancy of resource base* (years)
Bauxite	81	1,070
Copper	22	740
Iron	65	890
Lead	17	610
Nickel	30	530
Silver	15	730
Tin	28	760
Zinc	20	780

Data from Meadows *et al.* (2005).

is much more difficult to envisage finding renewable substitutes for metals. Recycling can help, but of course recycling requires energy which needs to be taken into account. In some situations metals can be replaced by plastics, and plastics could presumably be made from renewable oil sources (in competition with food and biofuels). But metals have some very unique properties and there are some situations in which it is very difficult to imagine metals as ever being able to be replaced by renewable resources (e.g., as conductors of electricity).

18.8 CAN THE PARTY CONTINUE?

Given that the end of oil will come sooner, or not much later, can we replace it so as to maintain our present lifestyle? There are now many general books that address this subject, but particularly worthy of mention are *The Party's Over: Oil, War and the Fate of Industrial Societies* (Heinberg, 2003), *Heat: How to Stop the Planet Burning* (Monbiot, 2006), *The Hot Topic: How to Tackle Global Warming and Still Keep the Lights On* (Walker and King, 2008), and several others. Heinberg's book is focused largely on the U.S.A. and Monbiot's book is focused on the British situation, while Walker and King are particularly emphatic on the need for international agreement on reducing CO_2 emissions. There is a wealth of good quantitative material in these books, but the underlying message is that if we modify our lifestyle and reduce our population then we could *perhaps* manage an acceptable lifestyle after the oil (and the gas and the coal) run out. But some things would have to be given up: air travel for a start! Planes have to be run on kerosene or something very similar; they cannot be run on coal, electricity, gas (the cylinders would be too heavy), wood, or even cow dung. What about biofuels one might ask? Basically there is a production problem there, of serious competition with food production, and we shall discuss biofuels shortly.

The reasons that oil is so useful are simple; namely, it is because oil is

- easily transported (much more so than solids such as coal or gases such as methane);
- energy-dense (gasoline contains approximately 40 kWh per gallon);
- capable of being refined into several fuels, gasoline, kerosene, and diesel suitable for a variety of applications;
- suitable for a variety of uses including transportation, heating, and as feedstock for the petrochemicals industry (fertilizers, plastics, etc., etc.)

One very important concept, and something which is often neglected in the discussion of renewable energy resources, is the energy return on energy invested (EROEI). In the early days of oil, prior to 1950, it is estimated that the EROEI was in the region of 100 : 1; one just drilled a hole in the ground and the oil gushed out. By the 1970s it is estimated that the EROEI for oil production had dropped to around 30 : 1. Energy had to be supplied for exploration, drilling, building of rigs,

Table 18.4. Energy return on energy invested (EROEI).

Energy source	*EROEI*
Oil, pre-1950	~100
Oil 1970	~30
Oil present	~8–11
Coal, U.S. average	9.0
Coal, western surface coal	6.0
Coal, ditto with scrubbers	2.5
Natural gas, onshore	10.3
Natural gas, offshore	6.8
Ethanol, sugar cane	0.8–1.7
Ethanol, corn	1.3
Palm oil	1.06
Wind, aerogenerators	~2–...
Nuclear	4.5
Hydropower	10.0
Geothermal	13.0
Solar, photovoltaics	1.7–10.0

Data from Heinberg (2003).

transportation, housing of production workers, etc., etc. Heinberg gives an extensive table of values of EROEI for various non-renewable and renewable sources of energy and a few of them are extracted in Table 18.4.

One can argue that most of the present work on sustainability is not concerned with real sustainability but is only tinkering at the edges and is more like "(slightly) reducing the unsustainability of our present way of life". For true sustainability there will have to be economies no doubt but some substitution should be possible. Many renewable sources have (hidden) costs in terms of the energy used in setting them up. So these values of EROEI are therefore very important. In this section we shall attempt to consider what happens when the fossil fuels have run out, maybe in 500 years or (in the case of coal, longer perhaps) 1,000 years.

So in, say, 500 years time when the oil is all gone what can we use in substitution?

Gas. One can dismiss gas immediately. The gas will probably run out before the oil. The present substitution in U.K. electricity generation, of gas for oil or coal, is driven by cost and helps to meet the country's Kyoto target on CO_2 emissions. It is not a long-term solution.

Coal. One can dismiss this too, though perhaps not quite so quickly. The world's coal reserves are enormous. But not only will they run out too, though on a much longer timescale, but also coal is very polluting (less so with scrubbers on power stations, but many power stations are not fitted with scrubbers). At present we use oil to mine coal. Opencast mines use relatively few "miners" and they use giant earth-moving machines that can consume up to 400 L–500 L of diesel per hour. Hence the EROEI is not particularly good. As the oil becomes less available the energy used to mine coal will have to come from coal or from some other source. And as near-surface coal runs out we shall have to turn to deep mining again and there are considerable energy requirements there and so the EROEI would reduce quite severely in the future.

Nuclear. The future for nuclear energy is very unclear, but some features can be noted:

(a) it depends on uranium and this is a finite resource;
(b) naively one can say it is clean as far as CO_2 emissions go, which is true;
(c) it has a poor image in terms of safety; and
(d) the EROEI is not very good.

However, a considerable amount of energy (coming at present from oil) is used in mining the uranium and that produces CO_2. Also the processes involved in constructing power stations involve energy coming from oil and producing CO_2 emissions and they involve concrete, which we shall come to shortly.

One may say "there is fusion (even cold fusion!)"; maybe in future this will be possible but we are a long way from being there in terms of getting nuclear fusion to work as a source of energy.

Renewables. As an example, the position of renewables in the U.S.A. at present is as follows: gas 23%, coal 23%, oil 40%, nuclear 8%, renewables 6%, and of the renewables these break down into solar 1%, biomass burning 47%, geothermal 5%, hydroelectricity 45%, and wind 2% (Heinberg, 2003). The renewable contribution is tiny.

Hydropower and geothermal. From Table 18.4 we see that from the point of view of the EROEI these energy sources look particularly good. But they can only be exploited in certain situations. Norway is more than self-sufficieint in hydropower for generating electricity and exports some of it to Sweden, but that is unusual. In Scotland, the big hydropower developments shortly after World War II provide about 10% of that country's electricity. But there is not much scope for any more. In some countries, such as England, there is very little scope for the development of hydroelectric schemes. Even such a huge project as the Three Gorges Dam in China makes only a small contribution to that country's total requirements for electricity. As for geothermal energy, there are a few places where it can be exploited, most notably in Iceland, but also to a modest extent elsewhere, New Zealand and Geyserville in northern California. But elsewhere in general, there is probably not much scope for the development of geothermal energy.

Wind. This is perhaps the most promising of the renewable resources currently being worked on. The EROEI in Table 18.4 is probably unduly pessimistic; Danish studies quoted by Heinberg suggest an EROEI of 50 or more. Windpower has possibilities, but (a) it needs a lot of investment, (b) there are limits imposed by environmental considerations, etc., and (c) the wind does not blow all the time and it is necessary to have backup generation facilities that can be brought onstream quickly and automatically.

Solar. Of course, wind and hydropower ultimately derive their energy from the Sun, but we have just dealt with them; we confine ourselves just now to direct heating and photovoltaics. Photovoltaics have their uses and the EROEI in Table 18.4 looks reasonable enough; it is used for small supplies of electricity or in isolated locations: on parking ticket machines in Edinburgh, in isolated Third World villages, in the tourist hostel near the top of Mt. Kilimanjaro, etc. There are many solar panel installations to produce hot water, rather than electricity, from the Sun. There are also a few installations where solar energy is concentrated to produce steam for the generation of electricity, but there are not many of them.

Biofuels. Extensive accounts of the technical, economic, and political details of biofuels are given by Pahl (2005) and Worldwatch (2007). We consider ethanol and biodiesel separately.

Brazil provides the best known example of the production and use of ethanol for cars starting in the 1980s, though the U.S.A. is stepping up its ethanol program now. Following the steep rise in oil prices in the 1970s, Brazil turned to ethanol produced from sugar cane, and in 1985 91% of cars produced in Brazil ran on ethanol. But as world prices fell and sugar cane prices rose, the demand for alcohol-fueled cars subsided. Brazil could afford its ethanol program because of its very favorable ratio of the area of cropland to the number of cars, even if topsoil was being lost and energy was being used in the process. There are also rumors of exploitation of child labor in the Brazilian sugar cane fields too. There are disputes about the EROEI of ethanol production but it is not particularly good. Heinberg does a little calculation of what would be involved if the U.S.A. tried to repeat the Brazilian experiment, but using corn oil rather than sugar cane because that is what they can grow (working in American units not SI units!):

— The U.S.A. has 400 million acres of cropland and about 200 million cars.
— American farmers produce about 7,110 pounds of corn per acre per year and an acre of corn yields about 341 gallons of ethanol.

- A typical American driver would burn 852 gallons of ethanol per year requiring the production from 2.5 acres.
- Thus, ethanol production from corn would need 500 million acres of cropland, or 25% more than the total area available in the U.S.A.

Clearly there is a problem of competition with agricultural resoures for food production.

Then there is biodiesel. Biodiesel is a substiute for what we might call "petroleum diesel" or "mineral diesel". It is made from vegetable oil and methanol. Various oils can be used but the one with the best yield per hectare is palm oil. Malaysia and Indonesia are the world's leading producers of palm oil. Originally the oil was exported as such but now the emphasis is on value-adding within the country and the first biodiesel plants in Malaysia are now in operation. Biodiesel is used in two ways. By modifying a diesel engine it can run on biodiesel and there are some vehicles that are modified in this way; Prince Charles is said to keep some on the Balmoral Estate in Scotland, for example. Alternatively, a small percentage of biodiesel can be added to "ordinary" diesel and engines need no modification to handle this. President George W. Bush's 2005 Energy Policy Act obliges fuel companies to sell 7.5 billion gallons of biodiesel and ethanol a year. There is an EU directive that 5.75% of the EU's transport fuel should come from renewable resources by 2010. The British govenrment has reduced the tax on biofuels by 20p a liter, and the EU is paying farmers an extra €45 a hectare to grow the crops to make biofuels. To quote Monbiot "At last, it seems a bold environmental vision is being pursued in the world's richest nations." But, then he goes on to do for the U.K. and road fuel what Heinberg did for the U.S.A.:

- Road transport in the U.K. consumes 37.8 Mtonnes of petroleum products per year.
- For oilseed rape, the most productive oil crop for the U.K. climate, the yield is between 3 tonnes and 3.5 tonnes per hectare.
- One tonne of rapeseed produces 415 kg of biodiesel.
- One hectare yields 1.45 tonnes of road fuel.
- Therefore, to provide 37.8 Mt would require 25.9 Mha.
- But there are only about 5.7 Mha of arable land in the U.K.
- Switching entirely to "green" fuel requires about $4\frac{1}{2}$ times that.
- The EU target of 20% by 2020 would consume almost all of the U.K.'s cropland, leaving virtually none for food production.

So in practice what is likely to happen is that the countries of the EU will import palm oil, or biodiesel, from Malaysia, Indonesia, etc. or ethanol from Brazil. This will be at the price of major deforestation. According to Friends of the Earth, quoted by Monbiot (2006),

> "Between 1985 and 2000 the development of oil palm plantations was responsible for an estimated 87% of deforestation in Malaysia."

In Sumatra and Borneo some 4 million ha of forest have been converted to oil palm estates. And a further 6 Mha in Malaysia and 16.5 Mha in Indonesia are scheduled for clearance. Apart from the usual problems arising from deforestation there are the frequent out-of-control fires in Indonesia that spread haze over several surrounding countries (Singapore, Malaysia, Thailand, etc.). Monbiot is scathing:

> "The decision by governments in Europe and North America to pursue the development of biofuels is, in environmental terms, the most damaging they have ever taken. Knowing that the creation of this market will lead to a massive surge in imports of both palm oil from Malaysia and Indonesia and ethanol from rainforest land in Brazil; knowing that there is nothing meaningful they can do to prevent them [i.e., these imports], and knowing that these imports will accelerate rather than ameliorate climate change; our governments have decided to go ahead anyway."

If all the developed world does the same thing as the EU the competition between land for biofuels and land for food production could become very serious. Quoting from Monbiot again:

> "If the same thing is to happen throughout the rich world, the impact could be great enough to push hundreds of millions of people into starvation, as the price of food rises beyond their means. If, as some environmentalists demand, it is to happen worldwide, then much of the arable surface of the planet will be deployed to produce food for cars, not people. The market reponds to money, not need. People who own cars—by definition—have more money than people at risk of starvation; their demand is 'effective', while the groans of the starving are not. In a contest between cars and people, the cars would win. Something like this is happening already. Though 800 million people are permanently malnourished, the global increase in crop production is being used mostly to feed animals: the number of livestock on earth has quintupled since 1950. The reason is that those who buy meat and dairy products have more purchasing power than those who buy only subsistence crops."

After all that, there is also the point that the EROEI for biodiesel (or for ethanol) is not particularly impressive (see Table 18.4), though there are arguments about the actual value. Some people even argue that the EROEI is less than 1! Instead of the term EROEI, Worldwatch (2007) in its discussion of biofuels calls it the energy balance; in this case it is the ratio of the energy in the biofuel (in joules) to the energy (in joules) used by people to plant the seeds, produce and spread agricultural chemicals, and to harvest, transport, and process the feedstock. Worldwatch (2007) also defines another quantity, the energy efficiency of biofuels; this includes the energy contained in the feedstock itself in the denominator. The energy balance quoted by Worldwatch for ethanol from sugar cane is $\sim$8 and for biodiesel from palm oil it is $\sim$9, both of which are considerably higher than the values quoted in Table 18.4 from Heinberg (2003). In addition to the question of the energy balance, or energy efficiency, another problem is that there would be a need for large supplies of methanol and a large mountain or lake of the main byproduct, glycerol. On a small scale, for disposing of used cooking oil from (fast food) restaurants making biodiesel is fine and on a small scale there is a good market for glycerol, but on a large scale biofuels have limited possibilities and the market for glycerol is not unlimited.

Waves, tidal systems, etc. Wave technology is still at the research and development stage. One or two tidal systems exist, but the number of suitable sites is very small and there are costs in terms of disturbing fisheries and ecosystems.

Hydrogen and fuel cells. This is more a secondary issue, connected with how to replace oil for transportation (i.e, in cars, etc.). One cannot think of hydrogen as a primary source of energy; it would have to be made by electrolysis of water, and the energy to generate the electricity would have to come from some sustainable source.

Cement/concrete. There is a special problem of cement (and therefore of concrete) that most writers on CO_2 and global warming ignore; they concentrate on the burning of fossil fuels and ignore other aspects of cement production. Apart from the energy consumed in extracting and transporting the raw materials, in heating the kilns to about 1,450°C, and transporting the raw materials and the product, there is a special feature of cement production. This is that we are reversing the CO_2 sequestration that occurred millions of years ago when the marine micro-organisms that became chalk or limestone were formed. As far as CO_2 is concerned, it is as if we were turning limestone or chalk into quicklime:

$$CaCO_3 \longrightarrow CaO + CO_2.$$

In fact, of course, cement is not quicklime; it consists of silicates but the effect, as far as limestone and CO_2 are concerned, is the same:

$$5CaCO_3 + 2SiO_2 \longrightarrow (3CaO, SiO_2)(2CaO, SiO_2) + 5CO_2.$$

(The SiO_2 comes from sand.) The consequence of this reaction is that CO_2 from the energy used in the production process for 1 tonne of cement, plus CO_2 emitted from this reaction comes to around 814 kg. Add in the quarrying and the transport and the result is something like 1 tonne of CO_2 being produced for each tonne of cement manufactured!

18.9 POPULATION

Malthus (an Englishman who was born in 1766 and died in 1834) is very widely referred to in relation to population expansion, and it is worthwhile giving some consideration to his work. In *An Essay on the Principle of Population*, first published in 1798 and subsequently republished in various editions right up to the present time, Malthus made the famous prediction that population would outrun food supply, leading to a decrease in food per person. His Principle of Population was based on the idea that population if unchecked increases according to geometric progression (i.e., 2, 4, 8, 16, etc.) whereas food supply only increases according to arithmetic progression (i.e., 1, 2, 3, 4, etc.). Therefore, since food is an essential component of human life, population growth in any area or on the planet, if unchecked, would lead to starvation. He even went so far as to specifically predict that this must occur by the middle of the 19th century, a prediction which failed for several reasons. This failure has led many people to be complacent about the problems of population growth that Malthus had highlighted. However, Malthus also argued that there are preventative checks and positive checks on population that slow its growth and keep the population from rising exponentially for too long, but still poverty is inescapable and will continue. Positive checks are those, according to Malthus, that increase the death rate. These include disease, war, disaster, and finally, when other checks don't reduce population, famine. Malthus felt that the fear of famine or the development of famine was also a major impetus to reduce the birth rate. He indicates that potential parents are less likely to have children when they know that their children are likely to starve:

> "The power of population is so superior to the power of the earth to produce subsistence for man, that premature death must in some shape or other visit the human race. The vices of mankind are active and able ministers of depopulation. They are the precursors in the great army of destruction, and often finish the dreadful work themselves. But should they fail in this war of extermination, sickly seasons, epidemics, pestilence, and plague advance in terrific array, and sweep off their thousands and tens of thousands. Should success be still incomplete, gigantic inevitable famine stalks in the rear, and with one mighty blow levels the population with the food of the world."

The ideas that Malthus developed came before the industrial revolution, and they are focused on plants, animals, and grain as the key components of diet. Therefore, for Malthus, available productive farmland was a limiting factor in population growth. With the industrial revolution and increase in agricultural production, land has become a less important factor than it was during the 18th century.

There was more to Malthus' work than a simple consequence of exponential growth vs. linear growth.

His example of population growth doubling was based on the preceding 25 years of the brand-new United States of America. Malthus felt that a young country with fertile soil like the U.S. would have one of the highest birth rates around. He liberally estimated an arithmetic increase in agricultural production of one acre at a time, acknowledging that he was overestimating but he gave agricultural development the benefit of the doubt.

18.10 THE COLLAPSE OF FORMER CIVILIZATIONS

There have been a number of former civilizations which flourished and then collapsed. It would be foolish to suggest that our present civilization is going to collapse in the same way that any of them did; if it collapses it will be in a different way. But, nevertheless, there may be lessons to be learned. On this topic it is worth mentioning the book *Collapse: How Societies Choose to Fail or Survive* (Diamond, 2005) and in particular to concentrate briefly on his discussion of the demise of the Easter Island civilization.

Diamond's definition of collapse is

> "... a drastic decrease in human population size and/or political/economic/social complexity, over a considerable area, for an extended time. The phenomenon of collapses is thus an extreme form of several milder types of decline, and it becomes arbitrary to decide how drastic the decline of a society must be before it qualifies to be labeled as a collapse."

He cites examples of societies which, in his view, most people would regard as having collapsed rather than just suffering minor declines:

- The Anasazi and Cahoka in modern U.S.A.
- The Maya cities of Central America
- The Moche and Tiwanaku societies in South America
- Mycenean Greece in Europe
- Minoan Crete in Europe
- The Greenland Norse settlement in Europe
- Great Zimbabwe in Africa
- Angkor Wat and the Harappan Indus Valley cities in Asia
- Easter Island in the Pacific Ocean.

The ruins left by many of these civilizations are very impresive, and they testify to the existence in the past of highly populous and organized civilizations that have now simply vanished.

In studying the collapse of various former civilizations Diamond considers a framework of five contributing factors:

- Environmental damage
- Climate change
- Hostile neighbors
- (Disappearance of) friendly trade partners
- Society's response to environmental problems.

He argues that many of these collapses were triggered, partly at least, by ecological problems, unintended ecological suicide, or "ecocide" as he calls it (i.e., destroying the environental resources on which their societies depended). This is not the only possible reason and the factors involved vary from case to case.

Among the past societies that collapsed, that of Easter Island is the one which Diamond describes as being

> "as close as we can get to a 'pure' ecological collapse, in this case due to total deforestation that led to war, overthrow of the elite and of the famous stone statues, and a massive population die-off."

The other factors seem either not to have been relevant or not to have been particularly important; there were no hostile neighbors or trade partners (Easter Island is just so far away from anywhere else), and there is no particular evidence of climate change over the period of the rise and fall of their civilization.

Other civilizations which collapsed and which Diamond discusses include Pitcairn and Henderson Islands (also in the Pacific), the Anasazi in the U.S. southwest, the Maya in central America and the Norse settlement in Greenland, this last one being "the one for which we have most information (because it was a well-understood literate European society)". He also discusses a number of past societies which faced many (similar) problems but which did not collapse: Norse colonies in Orkney and Iceland, which unlike the Norse Greenland colony did survive, and three other survivors: Tikopia, the New Guinea highlands, and Japan of the Tokugawa Era. Several chapters in his book describe modern societies that are collapsing or face high risks of collapse. His choice is interesting; some of the examples would come as no surprise but some of them are, at first sight, surprising (Australia, for instance!); to find out about all this one needs to read his book.

18.11 EASTER ISLAND

Easter Island is located in the Pacific Ocean at 109°20′W and 27°8′S and it is roughly the size and shape of Singapore or the Isle of Wight, but upside down.

> "It is the most remote habitable scrap of land in the world. The nearest lands are the coast of Chile 2,300 miles to the east and Polynesia's Pitcairn Islands 1,300 miles to the west" (quotes in this section are all from Diamond, 2005).

It appears that Easter Island was settled by Polynesian peoples coming from the west and arriving somewhat before AD 900. Recent research has shown that, for hundreds of thousands of years before human arrival and still during the early days of human settlement, Easter Island was not at all a barren wasteland, as it appeared to the early European explorers. The island was "discovered" by the Dutch explorer, Jakob Roggeveen, on Easter Day (hence the modern name of the island), April 5th, 1722. It was instead a diverse sub-tropical forest of tall trees and woody bushes. These included palm trees very similar to, but slightly larger than, the world's largest existing palm tree, the Chilean wine palm which grows to over 20 m in height and one meter in diameter.

> "Thus Easter (Island) used to support a diverse forest ... The overall picture for Easter (Island) is the most extreme example of forest destruction in the Pacific and among the most extreme in the world; the whole forest has gone and all of its tree

species are extinct. The deforestation must have begun some time after human arrival by AD 900 and reached its peak around 1400, and been virtually complete by dates that varied locally between the early 1400s and the 1600s."

Easter Island is famous for its statues which are mostly 5 m to 7 m high and the largest of which weigh about 270 tonnes. These statues must have been carved, transported, and erected by a large, well-organized, and prosperous population who had no power tools and no modern construction, lifting, or transportation machinery. Estimates of the population of Easter Island in its heyday range from 6,000 to 30,000; Diamond prefers the higher end of this range! The statues' sheer number and size suggest a population much larger than the estimated one of just a few thousand people encountered by European visitors in the 18th and early 19th centuries. What happened to the former large population? What went wrong?

In the first few centuries after the original settlers arrived, the forests were all cut down and the immediate consequences of deforestation were losses of raw materials and losses of wild-caught foods. Crop yields also decreased because deforestation led locally to soil erosion by rain and wind, while other damage to soil that resulted from deforestation and reduced crop yields included desiccation and nutrient leaching. Farmers found themselves without most of the wild plant leaves, fruit, and twigs that they had been using as compost. These were the immediate consequences of deforestation and other human environmental impacts. Various species of fish were also fished out. Further consequences start with starvation, leading to civil war, a population crash, and a descent into cannibalism.

A small number of people survived, eking out a very meager existence; the survivors adapted as best they could. When the European explorers arrived they could not understand how these people could have erected all those statues.

18.12 CURRENT ENVIRONMENTAL THREATS

As Diamond shows, the collapse of the Easter Island society followed swiftly upon the society's reaching its peak of population, monument construction, and environmental impact; he claims that the collapse was almost entirely due to environmental problems and the society's response to environmental problems. When he comes to the end of the book and seeks to draw conclusions, or messages for us and our society, and just considering the environmental aspects he classifies the most serious environmental problems facing past and present societies into 12 groups (see Table 18.5).

> "Eight of the twelve were significant already in the past, while four (numbers 5, 7, 8 and 10: energy, the photosynthetic ceiling, toxic chemicals and atmospheric changes) became serious only recently. The first four of the 12 consist of destruction or losses of natural resoures; the next three involve ceilings on natural resources; the three after that consist of harmful things that we produce or move around and the last two are population issues" (Diamond, 2005).

Table 18.5. Twelve environmental threats.

1.	Destruction of natural habitats: forests, wetlands, coral reefs
2.	Loss of wild food stocks: fish, shellfish
3.	Loss of biodiversity
4.	Soil erosion and degradation, salinization, loss of nutrients
5.	Ceiling on energy: oil, gas, coal
6.	Water: shortage, pollution
7.	Ceiling on photosynthesis
8.	Toxic chemicals
9.	Introduction of alien species
10.	Generation of greenhouse gases and ozone-destroying chemicals
11.	Growth of population
12.	Increasing environmental impact of people

Diamond (2005).

Let's look at these briefly.

#1 Destruction of natural habitats: forests, wetlands, coral reefs, and the ocean bottom. This is obvious and indisputable.

#2 Wild foods: fish, shellfish. In theory wild fish stocks could be managed sustainably—but by and large this doesn't happen. According to Diamond the great majority of valuable fisheries already either have collapsed or are in steep decline. Past societies that overfished included Easter Island and some other Pacific islands.

#3 Loss of biodiversity. Species are becoming extinct on a daily basis. One can take a moral view and say that other species have a right to existence. Or one can take a pragmatic view and cite the benefits to humankind of the diversity of species available, for agriculture, horticulture, silviculture (forestry), medicines, etc.

#4 Soil erosion and degradation (salinization), loss of nutrients, acidification, or alkalinization. We touched on this in discussing the consequences of global warming. It is a serious problem in many places and it has been going on for a long time. For instance, lands in Iraq and north Africa which were the bread basket of the Roman Empire are now semi-arid or full-blown desert.

#5 Ceiling on energy (oil, gas, coal . . .). Oil, gas, and coal will run out sometime, sooner or later.

#6 Water. There are well-known and serious problems involved in supplying clean unpolluted water to people in many parts of the world.

#7 Photosynthetic ceiling. We talk about biofuels as an alternative to fossil fuels (e.g., using Malaysian palm oil to produce biodiesel or Brazilian sugar cane or U.S. peanuts to produce ethanol for cars). But there are limits to production. And, of course, there is competition for land between biofuels and food production.

#8 Toxic chemicals. Insecticides, pesticides, and herbicides; mercury and other metals, fire-retardant chemicals, refrigerator coolants, detergents, and components of plastics are all being discarded into the environment where many of them survive for a long time (e.g., DDT and PCBs) and the metals mostly for ever.

#9 Introduction of alien species (animals and plants). Some alien species introduced by humans are obviously valuable to us as crops, domestic animals, and landscaping. But others have devastated populations of native species because the native species had no previous evolutionary experience of them and were unable to resist them.

#10 Generation of greenhouse gases that cause global warming and of gases that damage the ozone layer.

#11 The world's population is growing. More people require more food, water, shelter, space, energy, and other resources.

#12 However, what really counts is not the number of people but their impact on the environment (see Section 18.6).

These 12 sets of problems are not really separate from each other. They are linked and one problem exacerbates another or makes its solution more difficult. But any 1 of these 12 problems of non-sustainability would suffice to limit our lifestyle within the next several decades. As Diamond says:

> "They are like time bombs with fuses of less than 50 years ... People often ask, 'What is the single most important environental/population problem facing the world today?' A flip answer would be, 'The single most important problem is our misguided focus on identifying the single most important problem!' That flip answer is essentially correct, because any of the dozen problems if unsolved would do us grave harm, and because they all interact with each other. If we solved 11 of the problems, but not the 12th, we would still be in trouble, whichever was the problem that remained unsolved. We have to solve them all."

It will be noticed that human-induced global warming as a result of burning fossil fuels is only 1 of these 12 threats to our way of life. This supports Kondratyev's objection to the widespread concentration on CO_2 emissions to the exclusion of other environmental threats.

18.13 REFERENCES

Cox J.D. (2005) *Climate Crash: Abrupt Climate Change and What It Means for Our Future*. Joseph Henry Press, Washington, D.C., 215 pp.

Cracknell A.P. (1994) Climate change: The background. In: R.A. Vaughan and A.P. Cracknell (eds.), *Remote Sensing and Global Climate Change*. Springer-Verlag, Berlin, pp. 1–33.

Daly H. (1990). Towards some operational principles of sustainable development. *Ecological Economics*, **2**(1), 1–6.

Diamond J. (2005) *Collapse: How Societies Choose to Fail or Survive*. Allen Lane, London, 575 pp.

Gore A. (1992). *Earth in the Balance: Forging a New Common Purpose*. Earthscan, London, 407 pp.

Gore A. (2006). *An Inconvenient Truth*. Bloomsbury, London, 325 pp.

Heinberg R. (2003). *The Party's Over: Oil, War and the Fate of Industrial Societies*. Clairview Books, Forest Row, U.K., 306 pp.

Heinberg R. (2006). *The Oil Depletion Protocol: A Plan to Avert Oil Wars, Terrorism and Economic Collapse*. Clairview Books, Forest Row, U.K., 194 pp.

Heinberg R. (2007). *Peak Everything: Waking up to the Century of Decline in Earth's Resources*. Clairview Books, Forest Row, U.K., 212 pp.

Houghton J.T. (1984) *The Global Climate*. Cambridge University Press, Cambridge, U.K., 256 pp.

Houghton J.T. (1997) *Global Warming: The Complete Briefing*. Cambridge University Press, Cambridge, U.K., 251 pp.

Kondratyev K.Ya., Losev K.S., Ananicheva M.D., and Chesnokova I.V. (2004) *Stability of Life on Earth: Principal Subject of Scientific Research in the 21st Century*. Springer/Praxis, Chichester, U.K., 165 pp.

Krapivin V.F. and Varotsos C.A. (2007) *Globalization and Sustainable Development: Environmental Agendas*. Springer/Praxis, Chichester, U.K., 304 pp.

Lee K.Y. (2000) *From Third World to First—The Singapore Story: 1965-2000*. Times Media, Singapore, 778 pp.

Leggett J. (2005) *Half Gone: Oil, Gas, Hot Air and the Global Energy Crisis*. Portobello, London.

Meadows D.H., Meadows D.L., Randers J., and Behrens W.W. (1972) *The Limits to Growth: A Report or the Club of Rome's Project on the Predicament of Mankind*. Universe Books, New York, 205 pp.

Meadows D.H., Meadows D.L., and Randers J. (1992) *Beyond the Limits*. Chelsea Green, Post Mill, VT, 300 pp.

Meadows D.H., Randers J., and Meadows D.L. (2005) *Limits to Growth: The 30-year Update*. Earthscan, London, 338 pp.

Monbiot G. (2006) *Heat: How to Stop the Planet Burning*. Allen Lane, London, 304 pp.

Pahl G. (2005). *Biodiesel: Growing a New Energy Economy*. Chelsea Green, White River Junction, VT, 281 pp.

Pearce F. (2006a) *The Last Generation: How Nature Will Take Her Revenge for Climate Change* (Eden Project Book). Transworld Publishers, London, 324 pp.

Pearce, F. (2006b) *When the Rivers Run Dry: What Happens When Our Water Runs Out?* Eden Project Book, Transworld Publishers, London, 368 pp.

Rees M. (2004). *Our Final Century: Will the Human Race Survive the Twenty-first Century? Heinemann, Oxford, 266 pp.*

Roberts P. (2004). *The End of Oil: The Decline of the Petroleum Economy and the Rise of a New Energy Order*. Bloomsbury, London, 399 pp.

Stern N. (2007) *The Economics of Climate Change: The Stern Review*. Cambridge University Press, Cambridge, U.K., 692 pp.

Tanzer M (1980). *The Race for Resources: Continuing Struggles over Minerals and Fuels*. Heinemann, London, 285 pp.

Toman M.A., Chakravorty U., and Gupta S. (2003). *India and Global Climate Change: Perspectives on Economics and Policy from a Developing Country*. Resources for the Future, Washington, D.C., 366 pp.

Wackernagel M., Schulz N.B., Deumling D., Callejas Linares A., Jenkins M., Kapos V., Monfreda C., Loh J., Myers N, Norgaard R., and Randers J. (2002) Tracking the ecological overshoot of the human economy. *Proceedings of the Academy of Science*, **99**(14), 9266–9271.

Walker G. and King D. (2008). *The Hot Topic: How to Tackle Global Warming and Still Keep the Lights On*. Bloomsbury, London.

Worldwatch (2007). *Biofuels for Transport: Global Potential and Implications for Sustainable Energy and Agriculture*. Earthscan, London.

Index

Printing: Mercedes-Druck, Berlin
Binding: Stein + Lehmann, Berlin